The First Galaxies

For further volumes:
http://www.springer.com/series/5664

Tommy Wiklind • Bahram Mobasher
Volker Bromm
Editors

The First Galaxies

Theoretical Predictions and Observational Clues

Springer

Editors
Tommy Wiklind
Joint ALMA Observatory
Vitacura, Santiago
Chile

Bahram Mobasher
Department of Physics and Astronomy
University of California
Riverside, CA, USA

Volker Bromm
Department of Astronomy
College of Natural Sciences
University of Texas
Austin, TX, USA

ISSN 0067-0057
ISBN 978-3-642-32361-4 ISBN 978-3-642-32362-1 (eBook)
DOI 10.1007/978-3-642-32362-1
Springer Heidelberg New York Dordrecht London

Library of Congress Control Number: 2012953094

Cover illustration: The cover image is a montage of a simulated JWST image of the Hubble Ultra Deep Field, showing distant galaxies as JWST may see them in the near-infrared, some of the ALMA antennas present on the Chajnantor in 2012 (image credit: Babak Tafreshi), and an image of the James Webb Space Telescope. The simulated Hubble Utra Deep Field was made by deconvolving the original HST near-infrared image with the corresponding point-spread function, and then convolving with the expected JWST point-spread function. This process brings out finer details in the image compared with what can be seen in the original, but does not add new features that were too faint to be seen with the HST but that will be visible with JWST.

Printed on acid-free paper

Springer is part of Springer Science+Business Media (www.springer.com)

Preface

About 300,000 years after the big bang, at a redshift of $z \sim 1{,}100$, the Universe entered the Dark Ages. At this time the primordial gas had cooled enough to become neutral, but possessed a very smooth density distribution. This is evident in images of the cosmic microwave background, showing density fluctuations only of the order 1/100,000. In this smooth and rapidly expanding primordial gas, no structures capable of producing ionizing photons were present or even close to being formed. Fast-forward 700 million years, to a redshift of $z \sim 7$, and the Universe is close to fully ionized, with stars, galaxies, and even supermassive black holes present. Through dedicated studies, using the latest space- and ground-based facilities, we can now investigate the physics of the primordial gas at the very start of the Dark Ages. We can also study the properties of galaxies and quasars as they emerge towards the end of the re-ionization epoch. Very little, however, is known about the processes taking place during the 700 million years in between. This is the epoch where the first stars form, the first metals are produced, and the first generation of galaxies and supermassive black holes are assembled. Detailed understanding of the history of the Universe during this time is amongst the most outstanding questions in modern astrophysics and cosmology.

There have been significant efforts from both theoretical and observational sides to learn more about the physical processes taking place during the Dark Ages. Nevertheless, as we push the observational boundaries to higher redshifts and further back in cosmic history, we have yet to find a primordial star or galaxy, as predicted by theoretical models and simulations. Observing objects this close to the Dark Ages has proved to be very difficult, requiring new instruments and larger telescopes. This is a very active field of observational research, on which the design of future generations of ground- and space-based observatories is focused. A related and equally challenging problem that needs explaining from both an observational and theoretical perspective is the presence of supermassive black holes at the end of the re-ionization epoch. However, the emergence of primordial galaxies and the first supermassive black holes may very well be intrinsically linked, as revealed by the detection of dust and metals in quasars at the very highest redshifts. In the end, the properties of the first galaxies as well as the first supermassive black holes are likely to be shaped in large part by the very first generation of stars.

While observers often strive to find the most distant galaxies in order to learn about the early Universe, we now know that there exist stars in our Milky Way halo and in some of our satellite dwarf galaxies that are old enough to have formed during the first few hundred million years of the age of the Universe, possibly in the epoch of Dark Ages. These stars offer a unique probe of the physical conditions that prevailed when they were formed. With the aid of new sensitive detectors and large telescopes, high-resolution spectroscopy of these systems has just become possible. This is the emerging science of astro-archaeology, done in our own backyard.

Despite significant recent progress on both the theoretical and observational fronts, there are many questions that remain unanswered. In fact, how do we even define the first galaxies? Can we simulate the conditions in the early Universe and learn about star and galaxy formation through theory? What came first, the supermassive black holes that we know existed when the Universe was only a few hundred million years old, or the galaxies that host them? We know that some of the early quasars contain large amounts of dust and metals produced in stellar processes, suggesting a close connection between stars and black holes.

Of equal importance to the observations of distant galaxies is the interpretation of the data. Due to large distances and highly redshifted light, as well as the weakness of the absorption lines in ultra-metal-poor stars in our own backyard, observations are difficult to obtain and are often not as complete as we would wish. A high redshift means that the ultraviolet and optical light, which is the part of the electromagnetic spectrum where stars emit most of their energy, is shifted to infrared wavelengths. For the most part, it is impossible to obtain spectra of large samples of high redshift galaxies and we have to rely on broadband photometry and comparison with synthetic spectra. By combining the light output from a set of stars with different ages, different metallicities and with a specified star formation history, possibly adding some dust extinction, it is possible to model how the spectral energy is distributed as a function of wavelength. This can then be compared with observations and, depending on how well the synthetic template fits the broadband observations, we can deduce various properties of the distant galaxies. This process, on which a large part of our current knowledge about the most distant objects rely, has pitfalls and contains assumptions that may or may not be justified. It is therefore important to critically discuss how we extract information from the observational data in order to understand how much credence should be given to the results.

These and many similar questions are dissected in this book: *The First Galaxies: Theoretical Predictions and Observational Clues*. Here we bring together observers, theorists as well as experts on stellar population modeling and astro-archaeology for a discussion of where we stand today in terms of observations and theory, and what can be learned in the near future as new and evermore sensitive observational facilities become available. The aim of this book is to use observations to challenge current theories and use theories to direct future observations. The book grew out of a Joint Discussion at the XXVIIth General Assembly of the International Astronomical Union in Rio de Janeiro in 2009, where leading experts assembled to give the current status of this research field. This book is not a summary of this meeting, but rather a comprehensive exposition of the underlying concepts, principles, and techniques.

The research concerning the first galaxies and the early Universe is progressing rapidly and new advances can be found in the literature every week, with new instruments planned to tackle these problems rapidly becoming available. Because of this rapid advance of the field, the authors were tasked with not just summarizing where we stand today but also projecting which direction the research in the various areas is likely to take in the future. In addition, the authors describe the tools and techniques that are being used (or will be used) to pursue this research. The aim is to make the book a resource for graduate students wanting to acquire the technical background as well as being exposed to the forefront research. We believe this book will be a valuable resource for the next generation of astrophysicists and for anyone interested to learn about this research area and quickly become acquainted with the state-of-the-art research.

The book is divided into three main themes:

1. **The First Sources of Light:** What are the sources of the very first light in the Universe? How was the Universe re-ionized and when did this happen? This part also gives an overview of the theoretical framework for understanding this epoch of cosmic history as well as explores the potential for observing the primordial hydrogen gas as it is becoming ionized.
2. **The First Galaxies and Normal Stellar Populations:** Here a critical review is presented of the observational and theoretical modeling that are currently providing us with data on the first normal galaxies and supermassive black holes as they emerge from the Dark Ages. The galaxies and their content as we see them 700 million years after the big bang appear to be quite mature in terms of development, hence the term "normal," but what are the properties of these galaxies, their stellar and gaseous content? What clues do they hold regarding processes taking place in the Dark Ages?
3. **Tools & Techniques:** This theme describes how astronomers use the techniques of stellar archaeology and fitting synthetic spectral energy distributions to broadband photometric data to extract information on the early Universe. This part also contains a chapter on the potential of future facilities to observe the first galaxies.

The coming decade promises to be a very exciting one, most likely allowing us to answer one of the fundamental questions: *What are our cosmic origins?* Most major discoveries leading to paradigm shifts have not been predicted beforehand. It is therefore very likely that serendipity will play a crucial role in our quest for understanding how the first stars, galaxies and supermassive black holes formed and what role they played in re-ionizing the Universe. It is thus important to equip oneself with a comprehensive set of tools, such as the material covered in this book. After all, there is wisdom to the old adage that fortune favors the prepared mind.

Santiago, Chile — Tommy Wiklind
Riverside, CA, USA — Bahram Mobasher
Austin, TX, USA — Volker Bromm
July 2012

Contents

Part I
The First Sources of Light

The Basic Theoretical Framework

Abraham Loeb

Abstract Cosmology is by now a mature experimental science. We are privileged to live at a time when the story of genesis (how the Universe started and developed) can be critically explored by direct observations. Looking deep into the Universe through powerful telescopes, we can see images of the Universe when it was younger because of the finite time it takes light to travel to us from distant sources. Existing data sets include an image of the Universe when it was 0.4 million years old (in the form of the cosmic microwave background), as well as images of individual galaxies when the Universe was older than a billion years. But there is a serious challenge: in between these two epochs was a period when the Universe was dark, stars had not yet formed, and the cosmic microwave background no longer traced the distribution of matter. And this is precisely the most interesting period, when the primordial soup evolved into the rich zoo of objects we now see. The observers are moving ahead along several fronts. The first involves the construction of large infrared telescopes on the ground and in space, that will provide us with new photos of the first galaxies. Current plans include ground-based telescopes which are 24–42 m in diameter, and NASA's successor to the Hubble Space Telescope, called the James Webb Space Telescope. In addition, several observational groups around the globe are constructing radio arrays that will be capable of mapping the three-dimensional distribution of cosmic hydrogen in the infant Universe. These arrays are aiming to detect the long-wavelength (redshifted 21-cm) radio emission from hydrogen atoms. The images from these antenna arrays will reveal how the non-uniform distribution of neutral hydrogen evolved with cosmic time and eventually was extinguished by the ultra-violet radiation from the first galaxies. Theoretical research has focused in recent years on predicting the expected signals for the above instruments and motivating these ambitious observational projects.

A. Loeb (✉)
Department of Astronomy, Harvard University, 60 Garden St., Cambridge, MA, 02138, USA
e-mail: aloeb@cfa.harvard.edu

T. Wiklind et al. (eds.), *The First Galaxies*, Astrophysics and Space Science Library 396,
DOI 10.1007/978-3-642-32362-1_1,

1 Introduction

1.1 Observing Our Past

When we look at our image reflected off a mirror at a distance of 1 meter, we see the way we looked 6.7 nanoseconds ago, the light travel time to the mirror and back. If the mirror is spaced 10^{19}cm $\simeq$ 3pc away, we will see the way we looked 21 years ago. Light propagates at a finite speed, and so by observing distant regions, we are able to see what the Universe looked like in the past, a light travel time ago (Fig. 1). The statistical homogeneity of the Universe on large scales guarantees that what we see far away is a fair statistical representation of the conditions that were present in in our region of the Universe a long time ago.

This fortunate situation makes cosmology an empirical science. We do not need to guess how the Universe evolved. Using telescopes we can simply see how it appeared at earlier cosmic times. In principle, this allows the entire 13.7 billion year cosmic history of our universe to be reconstructed by surveying the galaxies and other sources of light to large distances (Fig. 2). Since a greater distance means a fainter flux from a source of a fixed luminosity, the observation of the earliest sources of light requires the development of sensitive instruments and poses challenges to observers.

As the universe expands, photon wavelengths get stretched as well. The factor by which the observed wavelength is increased (i.e. shifted towards the red) relative to the emitted one is denoted by $(1+z)$, where z is the cosmological redshift. Astronomers use the known emission patterns of hydrogen and other chemical elements in the spectrum of each galaxy to measure z. This then implies that the universe has expanded by a factor of $(1+z)$ in linear dimension since the galaxy emitted the observed light, and cosmologists can calculate the corresponding distance and cosmic age for the source galaxy. Large telescopes have allowed astronomers to observe faint galaxies that are so far away that we see them more than twelve billion years back in time. Thus, we know directly that galaxies were in existence as early as 500 million years after the Big Bang, at a redshift of $z \sim 10$ or higher.

We can in principle image the Universe only if it is transparent. Earlier than $400,000$ years after the big bang, the cosmic hydrogen was broken into its constituent electrons and protons (i.e. "ionized") and the Universe was opaque to scattering by the free electrons in the dense plasma. Thus, telescopes cannot be used to electromagnetically image the infant Universe at earlier times (or redshifts $>10^3$). The earliest possible image of the Universe was recorded by the COBE and WMAP satellites, which measured the temperature distribution of the cosmic microwave background (CMB) on the sky.

The CMB, the relic radiation from the hot, dense beginning of the universe, is indeed another major probe of observational cosmology. The universe cools as it expands, so it was initially far denser and hotter than it is today. For hundreds of thousands of years the cosmic gas consisted of a plasma of free protons and electrons, and a slight mix of light nuclei, sustained by the intense thermal motion

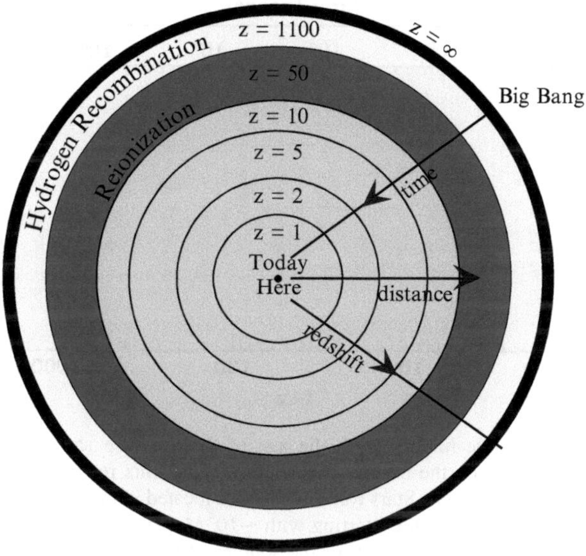

Fig. 1 Cosmic archaeology of the observable volume of the Universe, in comoving coordinates (which factor out the cosmic expansion). The outermost observable boundary ($z = \infty$) marks the comoving distance that light has travelled since the Big Bang. Future observatories aim to map most of the observable volume of our Universe, and improve dramatically the statistical information we have about the density fluctuations within it. Existing data on the CMB probes mainly a very thin shell at the hydrogen recombination epoch ($z \sim 10^3$, beyond which the Universe is opaque), and current large-scale galaxy surveys map only a small region near us at the center of the diagram. The formation epoch of the first galaxies that culminated with hydrogen reionization at a redshift $z \sim 10$ is shaded grey. Note that the comoving volume out to any of these redshifts scales as the distance cubed. **Figure credit:** Loeb (2010)

of these particles. Just like the plasma in our own Sun, the ancient cosmic plasma emitted and scattered a strong field of visible and ultraviolet photons. As mentioned above, about $400,000$ years after the Big Bang the temperature of the universe dipped for the first time below a few thousand degrees Kelvin. The protons and electrons were now moving slowly enough that they could attract each other and form hydrogen atoms, in a process known as cosmic recombination. With the scattering of the energetic photons now much reduced, the photons continued traveling in straight lines, mostly undisturbed except that cosmic expansion has redshifted their wavelength into the microwave regime today. The emission temperature of the observed spectrum of these CMB photons is the same in all directions to one part in $100,000$, which reveals that conditions were nearly uniform in the early universe.

It was just before the moment of cosmic recombination (when matter started to dominate in energy density over radiation) that gravity started to amplify the tiny fluctuations in temperature and density observed in the CMB data. Regions that started out slightly denser than average began to develop a greater density contrast with time because the gravitational forces were also slightly stronger

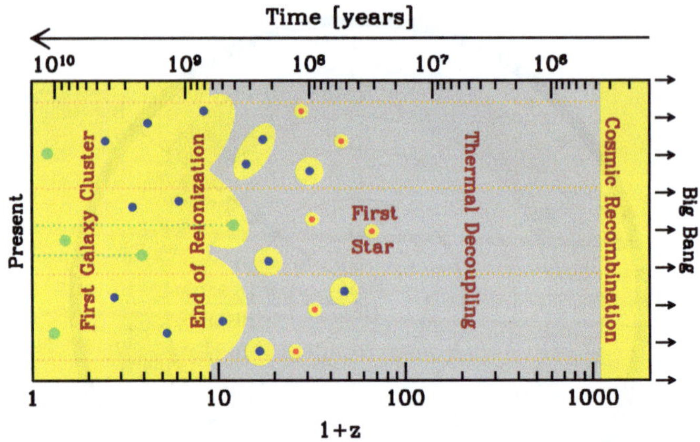

Fig. 2 Overview of cosmic history, with the age of the universe shown on the top axis and the corresponding redshift on the bottom axis. Yellow represents regions where the hydrogen is ionized, and gray, neutral regions. Stars form in galaxies located within dark matter concentrations whose typical mass grows with time, starting with $\sim 10^5 M_\odot$ (*red circles*) for the host of the first star, rising to 10^7–$10^9 M_\odot$ (*blue circles*) for the sources of reionization, and reaching $\sim 10^{12} M_\odot$ (*green circles*) for present-day galaxies like our own Milky Way. Astronomers probe the evolution of the cosmic gas using the absorption of background light (*dotted lines*) by atomic hydrogen along the line of sight. The classical technique uses absorption by the Lyman-α resonance of hydrogen of the light from bright quasars located within massive galaxies, while a new type of astronomical observation will use the 21-cm line of hydrogen with the cosmic microwave background as the background source. **Figure credit:** Barkana and Loeb (2007)

than average in these regions. Eventually, after hundreds of millions of years, the overdense regions stopped expanding, turned around, and eventually collapsed to make bound objects such as galaxies. The gas within these collapsed objects cooled and fragmented into stars. This process, however, would have taken too long to explain the abundance of galaxies today, if it involved only the observed cosmic gas. Instead, gravity is strongly enhanced by the presence of dark matter—an unknown substance that makes up the vast majority (83 %) of the cosmic density of matter. The motion of stars and gas around the centers of nearby galaxies indicates that each is surrounded by an extended mass of dark matter, and so dynamically-relaxed dark matter concentrations are generally referred to as "halos".

According to the standard cosmological model, the dark matter is cold (abbreviated as CDM), i.e., it behaves as a collection of collisionless particles that started out at matter domination with negligible thermal velocities and have evolved exclusively under gravitational forces. The model explains how both individual galaxies and the large-scale patterns in their distribution originated from the small initial density fluctuations. On the largest scales, observations of the present galaxy distribution have indeed found the same statistical patterns as seen in the CMB, enhanced as expected by billions of years of gravitational evolution. On smaller scales, the model describes how regions that were denser than average collapsed due to their enhanced gravity and eventually formed gravitationally-bound halos, first on small spatial scales and later on larger ones. In this hierarchical model of galaxy formation, the small galaxies formed first and then merged or accreted gas to form larger galaxies. At each snapshot of this cosmic evolution, the abundance of collapsed halos, whose masses are dominated by dark matter, can be computed from the initial conditions using numerical simulations. The common understanding of galaxy formation is based on the notion that stars formed out of the gas that cooled and subsequently condensed to high densities in the cores of some of these halos.

Gravity thus explains how some gas is pulled into the deep potential wells within dark matter halos and forms the galaxies. One might naively expect that the gas outside halos would remain mostly undisturbed. However, observations show that it has not remained neutral (i.e., in atomic form) but was largely ionized by the UV radiation emitted by the galaxies. The diffuse gas pervading the space outside and between galaxies is referred to as the intergalactic medium (IGM). For the first hundreds of millions of years after cosmological recombination, the so-called cosmic "dark ages", the universe was filled with diffuse atomic hydrogen. As soon as galaxies formed, they started to ionize diffuse hydrogen in their vicinity. Within less than a billion years, most of the IGM was re-ionized. We have not yet imaged the cosmic dark ages before the first galaxies had formed. One of the frontiers in current cosmological studies aims to study the cosmic epoch of reionization and the first generation of galaxies that triggered it.

1.2 The Expanding Universe

The modern physical description of the Universe as a whole can be traced back to Einstein, who assumed for simplicity the so-called "cosmological principle": that the distribution of matter and energy is homogeneous and isotropic on the largest scales. Today isotropy is well established for the distribution of faint radio sources, optically-selected galaxies, the X-ray background, and most importantly the cosmic microwave background (hereafter, CMB). The constraints on homogeneity are less strict, but a cosmological model in which the Universe is isotropic but significantly inhomogeneous in spherical shells around our special location, is also excluded.

In General Relativity, the metric for a space-time which is spatially homogeneous and isotropic is the Friedman-Robertson-Walker metric, which can be written in the form

$$ds^2 = c^2 dt^2 - a^2(t)\left[\frac{dr^2}{1-k\,r^2} + r^2\left(d\theta^2 + \sin^2\theta\, d\phi^2\right)\right] \tag{1}$$

where c is the speed of light, $a(t)$ is the cosmic scale factor which describes expansion in time t, and (r,θ,ϕ) are spherical comoving coordinates. The constant k determines the geometry of space; it is positive in a closed Universe, zero in a flat Universe (Euclidean space), and negative in an open Universe. Observers at rest remain at rest, at fixed (r,θ,ϕ), with their physical separation increasing with time in proportion to $a(t)$. A given observer sees a nearby observer at physical distance D receding at the Hubble velocity $H(t)D$, where the Hubble constant at time t is $H(t) = d\,a(t)/dt$. Light emitted by a source at time t is observed at $t = 0$ with a redshift $z = 1/a(t) - 1$, where we set $a(t = 0) \equiv 1$ for convenience.

The Einstein field equations of General Relativity yield the Friedmann equation

$$H^2(t) = \frac{8\pi G}{3}\rho - \frac{k}{a^2}, \tag{2}$$

which relates the expansion of the Universe to its matter-energy content. For each component of the energy density ρ, with an equation of state $p = p(\rho)$, the density ρ varies with $a(t)$ according to the thermodynamic relation

$$d(\rho c^2 r^3) = -p d(r^3). \tag{3}$$

With the critical density

$$\rho_c(t) \equiv \frac{3H^2(t)}{8\pi G} \tag{4}$$

defined as the density needed for $k = 0$, we define the ratio of the total density to the critical density as

$$\Omega \equiv \frac{\rho}{\rho_c}. \tag{5}$$

With Ω_m, Ω_Λ, and Ω_r denoting the present contributions to Ω from matter (including cold dark matter as well as a contribution Ω_b from ordinary matter ["baryons"] made of protons and neutrons), vacuum energy (cosmological constant), and radiation, respectively, the Friedmann equation becomes

$$\frac{H(t)}{H_0} = \left[\frac{\Omega_m}{a^3} + \Omega_\Lambda + \frac{\Omega_r}{a^4} + \frac{\Omega_k}{a^2}\right], \tag{6}$$

where we define H_0 and $\Omega_0 = \Omega_m + \Omega_\Lambda + \Omega_r$ to be the present values of H and Ω, respectively, and we let

$$\Omega_k \equiv -\frac{k}{H_0^2} = 1 - \Omega_m. \tag{7}$$

In the particularly simple Einstein–de Sitter model ($\Omega_m = 1$, $\Omega_\Lambda = \Omega_r = \Omega_k = 0$), the scale factor varies as $a(t) \propto t^{2/3}$. Even models with non-zero Ω_Λ or Ω_k approach the Einstein–de Sitter scaling-law at high redshift, i.e. when $(1 + z) \gg |\Omega_m{}^{-1} - 1|$ (as long as Ω_r can be neglected). In this high-z regime the age of the Universe is

$$t \approx \frac{2}{3H_0\sqrt{\Omega_m}}(1+z)^{-3/2} \approx 10^9\text{year}\left(\frac{1+z}{7}\right)^{-3/2}. \tag{8}$$

Recent observations confine the standard set of cosmological parameters to a relatively narrow range. In particular, we seem to live in a universe dominated by a cosmological constant (Λ) and cold dark matter, or in short a ΛCDM cosmology (with Ω_k so small that it is usually assumed to equal zero) with an approximately scale-invariant primordial power spectrum of density fluctuations, i.e., $n \approx 1$ where the initial power spectrum is $P(k) = |\delta_{\mathbf{k}}|^2 \propto k^n$ in terms of the wavenumber k of the Fourier modes $\delta_{\mathbf{k}}$ (see Sect. 2.1 below). Also, the Hubble constant today is written as

$$H_0 = 100h \text{ km s}^{-1}\text{Mpc}^{-1}, \tag{9}$$

in terms of h, and the overall normalization of the power spectrum is specified in terms of σ_8, the root-mean-square amplitude of mass fluctuations in spheres of radius 8 h^{-1} Mpc. For example, the best-fit cosmological parameters matching the WMAP data together with large-scale surveys of galaxies and supernovae are (Komatsu et al. 2011): $\sigma_8 = 0.81$, $n = 0.96$, $h = 0.72$, $\Omega_m = 0.28$, $\Omega_\Lambda = 0.72$ and $\Omega_b = 0.046$.

2 Galaxy Formation

2.1 Growth of Linear Perturbations

As noted in the Introduction, observations of the CMB show that the universe at cosmic recombination (redshift $z \sim 10^3$) was remarkably uniform apart from spatial fluctuations in the energy density and in the gravitational potential of roughly one part in $\sim 10^5$. The primordial inhomogeneities in the density distribution grew over time and eventually led to the formation of galaxies as well as galaxy clusters and large-scale structure. In the early stages of this growth, as long as the density fluctuations on the relevant scales were much smaller than unity, their evolution can be understood with a linear perturbation analysis (see Mo et al. 2010, and references therein).

As before, we distinguish between fixed and comoving coordinates. Using vector notation, the fixed coordinate $\mathbf{r}$ corresponds to a comoving position $\mathbf{x} = \mathbf{r}/a$. In a homogeneous Universe with density ρ, we describe the cosmological expansion in terms of an ideal pressureless fluid of particles each of which is at fixed $\mathbf{x}$, expanding with the Hubble flow $\mathbf{v} = H(t)\mathbf{r}$ where $\mathbf{v} = d\mathbf{r}/dt$. Onto this uniform expansion we impose small perturbations, given by a relative density perturbation

$$\delta(\mathbf{x}) = \frac{\rho(\mathbf{r})}{\bar{\rho}} - 1, \tag{10}$$

where the mean fluid density is $\bar{\rho}$, with a corresponding peculiar velocity $\mathbf{u} \equiv \mathbf{v} - H\mathbf{r}$. Then the fluid is described by the continuity and Euler equations in comoving coordinates:

$$\frac{\partial \delta}{\partial t} + \frac{1}{a}\nabla \cdot [(1+\delta)\mathbf{u}] = 0 \tag{11}$$

$$\frac{\partial \mathbf{u}}{\partial t} + H\mathbf{u} + \frac{1}{a}(\mathbf{u}\cdot\nabla)\mathbf{u} = -\frac{1}{a}\nabla\phi. \tag{12}$$

The potential ϕ is given by the Poisson equation, in terms of the density perturbation:

$$\nabla^2\phi = 4\pi G\bar{\rho}a^2\delta. \tag{13}$$

This fluid description is valid for describing the evolution of collisionless cold dark matter particles until different particle streams cross. This "shell-crossing" typically occurs only after perturbations have grown to become non-linear, and at that point the individual particle trajectories must in general be followed. Similarly, baryons can be described as a pressureless fluid as long as their temperature is negligibly small, but non-linear collapse leads to the formation of shocks in the gas.

For small perturbations $\delta \ll 1$, the fluid equations can be linearized and combined to yield

$$\frac{\partial^2\delta}{\partial t^2} + 2H\frac{\partial\delta}{\partial t} = 4\pi G\bar{\rho}\delta. \tag{14}$$

This linear equation has in general two independent solutions, only one of which grows with time. Starting with random initial conditions, this "growing mode" comes to dominate the density evolution. Thus, until it becomes non-linear, the density perturbation maintains its shape in comoving coordinates and grows in proportion to a growth factor $D(t)$. The growth factor in the matter-dominated era is given by

$$D(t) \propto \frac{\left(\Omega_\Lambda a^3 + \Omega_k a + \Omega_m\right)^{1/2}}{a^{3/2}} \int_0^a \frac{a'^{3/2}\,da'}{\left(\Omega_\Lambda a'^3 + \Omega_k a' + \Omega_m\right)^{3/2}}, \tag{15}$$

where we neglect Ω_r when considering halos forming in the matter-dominated regime at $z \ll 10^4$. In the Einstein–de Sitter model (or, at high redshift, in other models as well) the growth factor is simply proportional to $a(t)$.

The spatial form of the initial density fluctuations can be described in Fourier space, in terms of Fourier components

$$\delta_{\mathbf{k}} = \int d^3x\,\delta(x)e^{-i\mathbf{k}\cdot\mathbf{x}}. \tag{16}$$

Here we use the comoving wave-vector **k**, whose magnitude k is the comoving wavenumber which is equal to 2π divided by the wavelength. The Fourier description is particularly simple for fluctuations generated by inflation. Inflation generates perturbations given by a Gaussian random field, in which different **k**-modes are statistically independent, each with a random phase. The statistical properties of the fluctuations are determined by the variance of the different **k**-modes, and the variance is described in terms of the power spectrum $P(k)$ as follows:

$$\left\langle \delta_{\mathbf{k}} \delta^{*}_{\mathbf{k}'} \right\rangle = (2\pi)^3 \, P(k) \, \delta^{(3)} \left(\mathbf{k} - \mathbf{k}' \right), \tag{17}$$

where $\delta^{(3)}$ is the three-dimensional Dirac delta function. The gravitational potential fluctuations are sourced by the density fluctuations through Poisson's equation.

In standard models, inflation produces a primordial power-law spectrum $P(k) \propto k^n$ with $n \sim 1$. Perturbation growth in the radiation-dominated and then matter-dominated Universe results in a modified final power spectrum, characterized by a turnover at a scale of order the horizon cH^{-1} at matter-radiation equality, and a small-scale asymptotic shape of $P(k) \propto k^{n-4}$. The overall amplitude of the power spectrum is not specified by current models of inflation, and it is usually set by comparing to the observed CMB temperature fluctuations or to local measures of large-scale structure.

Since density fluctuations may exist on all scales, in order to determine the formation of objects of a given size or mass it is useful to consider the statistical distribution of the smoothed density field. Using a window function $W(\mathbf{r})$ normalized so that $\int d^3 r \, W(\mathbf{r}) = 1$, the smoothed density perturbation field, $\int d^3 r \delta(\mathbf{x}) W(\mathbf{r})$, itself follows a Gaussian distribution with zero mean. For the particular choice of a spherical top-hat, in which $W = 1$ in a sphere of radius R and is zero outside, the smoothed perturbation field measures the fluctuations in the mass in spheres of radius R. The normalization of the present power spectrum is often specified by the value of $\sigma_8 \equiv \sigma(R = 8h^{-1}\mathrm{Mpc})$. For the top-hat, the smoothed perturbation field is denoted δ_R or δ_M, where the mass M is related to the comoving radius R by $M = 4\pi \rho_m R^3/3$, in terms of the current mean density of matter ρ_m. The variance $\langle \delta_M \rangle^2$ is

$$\sigma^2(M) = \sigma^2(R) = \int_0^\infty \frac{dk}{2\pi^2} \, k^2 P(k) \left[\frac{3 j_1(kR)}{kR} \right]^2, \tag{18}$$

where $j_1(x) = (\sin x - x \cos x)/x^2$. The function $\sigma(M)$, plotted in Fig. 3, plays a crucial role in estimates of the abundance of collapsed objects.

Different physical processes contributed to the perturbation growth. In the absence of other influences, gravitational forces due to density perturbations imprinted by inflation would have driven parallel perturbation growth in the dark matter, baryons and photons. However, since the photon sound speed is of order the speed of light, the radiation pressure produced sound waves on a scale of order the cosmic horizon and suppressed sub-horizon perturbations in the photon density. The baryonic pressure similarly suppressed perturbations in the gas below the (much smaller) so-called baryonic *Jeans* scale. Since the formation of hydrogen at recombination had decoupled the cosmic gas from its mechanical drag on the CMB, the baryons subsequently began to fall into the pre-existing gravitational potential wells of the dark matter.

Spatial fluctuations developed in the gas temperature as well as in the gas density. Both the baryons and the dark matter were affected on small scales by

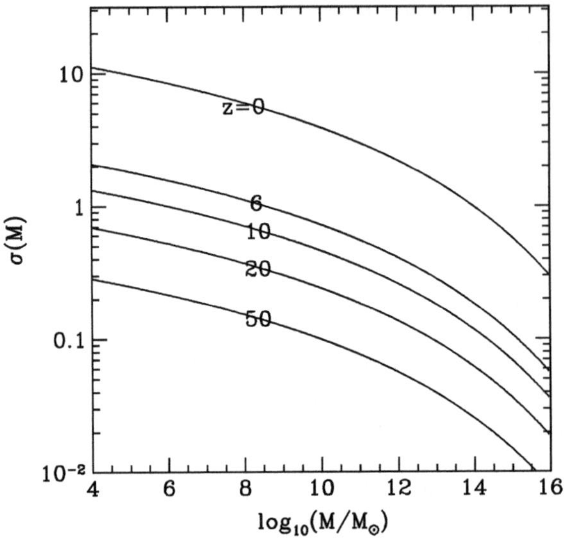

Fig. 3 The root-mean-square amplitude of linearly-extrapolated density fluctuations σ as a function of mass M (in solar masses $M_\odot$, within a spherical top-hat filter) at different redshifts z. Halos form in regions that exceed the background density by a factor of order unity. This threshold is only surpassed by rare (many-σ) peaks for high masses at high redshifts. **Figure credit:** Loeb (2010)

the temperature fluctuations through the gas pressure. Compton heating due to scattering of the residual free electrons (constituting a fraction $\sim 10^{-4}$) with the CMB photons remained effective, keeping the gas temperature fluctuations tied to the photon temperature fluctuations, even for a time after recombination. The growth of linear perturbations can be calculated with the standard CMBFAST code (http://www.cmbfast.org), after a modification to account for the fact that the speed of sound of the gas also fluctuates spatially.

After recombination, two main drivers affect the baryon density and temperature fluctuations, namely, the thermalization with the CMB and the gravitational force that attracts the baryons to the dark matter potential wells. The density perturbations in all species grow together on scales where gravity is unopposed, outside the horizon (i.e., at $k < 0.01$ Mpc^{-1} at $z \sim 1,000$). At $z = 1,200$ the perturbations in the baryon-photon fluid oscillate as acoustic waves on scales of order the sound horizon ($k \sim 0.01$ Mpc^{-1}), while smaller-scale perturbations in both the photons and baryons are damped by photon diffusion and the drag of the diffusing photons on the baryons. On sufficiently small scales the power spectra of baryon density and temperature roughly assume the shape of the dark matter fluctuations (except for the gas-pressure cutoff at the very smallest scales), due to the effect of gravitational attraction on the baryon density and of the resulting adiabatic expansion on the gas temperature. After the mechanical coupling of the baryons to the photons ends at $z \sim 1,000$, the baryon density perturbations gradually grow towards the dark matter perturbations because of gravity. Similarly, after the thermal coupling ends at $z \sim 200$, the baryon temperature fluctuations are driven by adiabatic expansion towards a value of 2/3 of the density fluctuations. By $z = 200$ the baryon infall into the dark matter potentials is well advanced and adiabatic expansion is becoming increasingly important in setting the baryon temperature.

2.2 Halo Properties

The small density fluctuations evidenced in the CMB grow over time as described in the previous subsection, until the perturbation δ becomes of order unity, and the full non-linear gravitational problem must be considered. A dark matter halo forms in a region where matter relaxes dynamically to a make a stable object that is much denser than the mean cosmic density and is held together by its own self-gravity. The dynamical collapse of a dark matter halo can be solved analytically only in cases of particular symmetry. If we consider a region which is much smaller than the horizon cH^{-1}, then the formation of a halo can be formulated as a problem in Newtonian gravity, in some cases with minor corrections coming from General Relativity. The simplest case is that of spherical symmetry, with an initial ($t = t_i \ll t_0$) top-hat of uniform overdensity δ_i inside a sphere of radius R. Although this model is restricted in its direct applicability, the results of spherical collapse have turned out to be surprisingly useful in understanding the properties and distribution of halos in models based on cold dark matter.

The collapse of a spherical top-hat perturbation is described by the Newtonian equation (with a correction for the cosmological constant)

$$\frac{d^2r}{dt^2} = H_0^2 \Omega_\Lambda r - \frac{GM}{r^2}, \qquad (19)$$

where r is the radius in a fixed (not comoving) coordinate frame, H_0 is the present-day Hubble constant, M is the total mass enclosed within radius r, and the initial velocity field is given by the Hubble flow $dr/dt = H(t)r$. The enclosed δ grows initially as $\delta_L = \delta_i D(t)/D(t_i)$, in accordance with linear theory, but eventually δ grows above δ_L. If the mass shell at radius r is bound (i.e., if its total Newtonian energy is negative) then it reaches a radius of maximum expansion and subsequently collapses. As demonstrated in the previous section, at the moment when the top-hat collapses to a point, the overdensity predicted by linear theory is $\delta_L = 1.686$ in the Einstein–de Sitter model, with only a weak dependence on Ω_m and Ω_Λ. Thus a tophat collapses at redshift z if its linear overdensity extrapolated to the present day (also termed the critical density of collapse) is

$$\delta_{\rm crit}(z) = \frac{1.686}{D(z)}, \qquad (20)$$

where we set $D(z = 0) = 1$.

Even a slight violation of the exact symmetry of the initial perturbation can prevent the tophat from collapsing to a point. Instead, the halo reaches a state of virial equilibrium by violent relaxation (phase mixing). Using the virial theorem $U = -2K$ to relate the potential energy U to the kinetic energy K in the final state (implying that the virial radius is half the turnaround radius—where the kinetic energy vanishes), the final overdensity relative to the critical density at the collapse redshift is $\Delta_c = 18\pi^2 \simeq 178$ in the Einstein–de Sitter model, modified in a Universe with $\Omega_m + \Omega_\Lambda = 1$ to the fitting formula

$$\Delta_c = 18\pi^2 + 82d - 39d^2, \tag{21}$$

where $d \equiv \Omega_m^z - 1$ is evaluated at the collapse redshift, so that

$$\Omega_m^z = \frac{\Omega_m(1+z)^3}{\Omega_m(1+z)^3 + \Omega_\Lambda + \Omega_k(1+z)^2}. \tag{22}$$

A halo of mass M collapsing at redshift z thus has a virial radius

$$r_{\rm vir} = 0.784 \left(\frac{M}{10^8\, h^{-1}\, M_\odot}\right)^{1/3} \left[\frac{\Omega_m}{\Omega_m^z}\,\frac{\Delta_c}{18\pi^2}\right]^{-1/3} \left(\frac{1+z}{10}\right)^{-1} h^{-1}\ {\rm kpc}, \tag{23}$$

and a corresponding circular velocity,

$$V_c = \left(\frac{GM}{r_{\rm vir}}\right)^{1/2} = 23.4 \left(\frac{M}{10^8\, h^{-1}\, M_\odot}\right)^{1/3} \left[\frac{\Omega_m}{\Omega_m^z}\,\frac{\Delta_c}{18\pi^2}\right]^{1/6} \left(\frac{1+z}{10}\right)^{1/2} {\rm km\ s^{-1}}. \tag{24}$$

In these expressions we have assumed a present Hubble constant written in the form $H_0 = 100\,h$ km s^{-1}Mpc^{-1}. We may also define a virial temperature

$$T_{\rm vir} = \frac{\mu m_p V_c^2}{2k} = 1.98 \times 10^4 \left(\frac{\mu}{0.6}\right) \left(\frac{M}{10^8\, h^{-1}\, M_\odot}\right)^{2/3} \left[\frac{\Omega_m}{\Omega_m^z}\,\frac{\Delta_c}{18\pi^2}\right]^{1/3} \left(\frac{1+z}{10}\right) {\rm K}, \tag{25}$$

where μ is the mean molecular weight and m_p is the proton mass. Note that the value of μ depends on the ionization fraction of the gas; for a fully ionized primordial gas $\mu = 0.59$, while a gas with ionized hydrogen but only singly-ionized helium has $\mu = 0.61$. The binding energy of the halo is approximately[1]

$$E_b = \frac{1}{2}\frac{GM^2}{r_{\rm vir}} = 5.45 \times 10^{53} \left(\frac{M}{10^8\, h^{-1}\, M_\odot}\right)^{5/3} \left[\frac{\Omega_m}{\Omega_m^z}\,\frac{\Delta_c}{18\pi^2}\right]^{1/3} \left(\frac{1+z}{10}\right) h^{-1}\ {\rm erg}. \tag{26}$$

Note that the binding energy of the baryons is smaller by a factor equal to the baryon fraction Ω_b/Ω_m.

Although spherical collapse captures some of the physics governing the formation of halos, structure formation in cold dark matter models proceeds hierarchically. At early times, most of the dark matter is in low-mass halos, and these halos continuously accrete and merge to form high-mass halos (see Fig. 4). Numerical

[1]The coefficient of $1/2$ in Eq. (26) would be exact for a singular isothermal sphere with $\rho(r) \propto 1/r^2$.

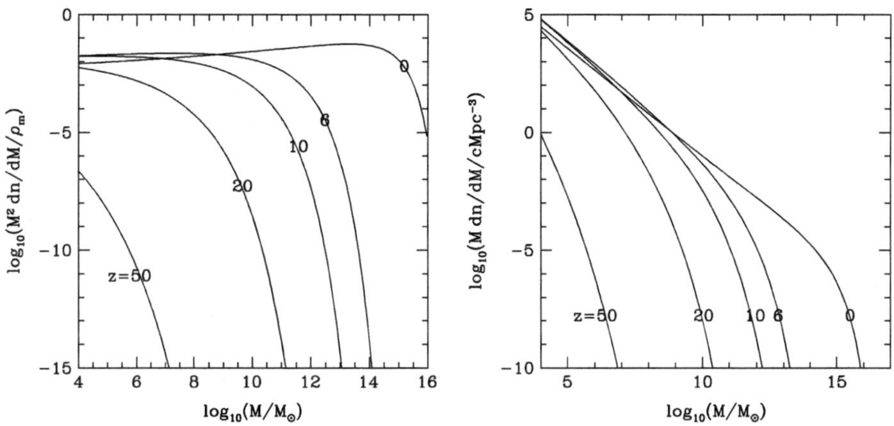

Fig. 4 *Left panel:*The mass fraction incorporated into dark matter halos per logarithmic bin of halo mass $(M^2 dn/dM)/\rho_m$, as a function of M at different redshifts z. Here $\rho_m = \Omega_m \rho_c$ is the present-day matter density, and $n(M)dM$ is the comoving density of halos with masses between M and $M + dM$. The halo mass distribution was calculated based on an improved version of the Press-Schechter formalism for ellipsoidal collapse that better fits numerical simulations (Sheth and Tormen 2002). *Right panel:* Number density of halos per logarithmic bin of halo mass, $M dn/dM$ (in units of comoving Mpc^{-3}), at various redshifts. **Figure credit:** Loeb (2010)

simulations of hierarchical halo formation indicate a roughly universal spherically-averaged density profile for the resulting halos, though with considerable scatter among different halos. The typical profile has the form

$$\rho(r) = \frac{3H_0^2}{8\pi G}(1+z)^3 \frac{\Omega_m}{\Omega_m^z} \frac{\delta_c}{c_N x(1+c_N x)^2}, \tag{27}$$

where $x = r/r_{\mathrm{vir}}$, and the characteristic density δ_c is related to the concentration parameter c_N by

$$\delta_c = \frac{\Delta_c}{3} \frac{c_N^3}{\ln(1+c_N) - c_N/(1+c_N)}. \tag{28}$$

The concentration parameter itself depends on the halo mass M, at a given redshift z.

2.3 *Formation of the First Stars*

Theoretical expectations for the properties of the first galaxies are based on the standard cosmological model outlined in the Introduction. The formation of the first bound objects marked the central milestone in the transition from the initial simplicity (discussed in the previous subsection) to the present-day complexity (e.g., Barkana and Loeb 2001; Bromm et al. 2009). Stars and accreting black holes output copious radiation and also produced explosions and outflows that brought into the IGM chemical products from stellar nucleosynthesis and enhanced magnetic fields. However, the formation of the very first stars, in a universe that had not yet suffered such feedback, remains a well-specified problem for theorists (for reviews, see Bromm and Larson 2004; Glover 2005).

Stars form when large amounts of matter collapse to high densities. However, the process can be stopped if the pressure exerted by the hot intergalactic gas prevents outlying gas from falling into dark matter concentrations. As the gas falls into a dark matter halo, it forms shocks due to converging supersonic flows and in the process heats up and can only collapse further by first radiating its energy away. This restricts this process of collapse to very large clumps of dark matter that are around $100,000$ times the mass of the Sun. Inside these clumps, the shocked gas loses energy by emitting radiation from excited molecular hydrogen that formed naturally within the primordial gas mixture of hydrogen and helium.

The first stars are expected to have been quite different from the stars that form today in the Milky Way. The higher pressure within the primordial gas due to the presence of fewer cooling agents suggests that fragmentation only occurred into relatively large units, in which gravity could overcome the pressure. Due to the lack of carbon, nitrogen, and oxygen—elements that would normally dominate the nuclear energy production in modern massive stars—the first stars must have condensed to extremely high densities and temperatures before nuclear reactions were able to heat the gas and balance gravity. These unusually massive stars produced high luminosities of UV photons, but their nuclear fuel was exhausted after 2–3 million years, resulting in a huge supernova or in collapse to a black hole. The heavy elements which were dispersed by the first supernovae in the surrounding gas, enabled the enriched gas to cool more effectively and fragment into lower mass stars. Simple calculations indicate that a carbon or oxygen enrichment of merely $<10^{-3}$ of the solar abundance is sufficient to allow solar mass stars to form. These second-generation "low-metallicity" stars are long-lived and could in principle be discovered in the halo of the Milky Way galaxy, providing fossil record of the earliest star formation episode in our cosmic environment.

Advances in computing power have made possible detailed numerical simulations of how the first stars formed. These simulations begin in the early universe, in which dark matter and gas are distributed uniformly, apart from tiny variations in density and temperature that are statistically distributed according to the patterns observed in the CMB. In order to span the vast range of scales needed to simulate an individual star within a cosmological context, the adopted codes zoom in repeatedly on the densest part of the first collapsing cloud that is found within the simulated volume. The simulation follows gravity, hydrodynamics, and chemical processes in the primordial gas, and resolves a scale that is >10 orders of magnitudes smaller than that of the simulated box. In state-of-the-art simulations, the resolved scale is approaching the scale of the proto-star. The simulations have established that the first stars formed within halos containing $\sim 10^5 M_\odot$ in total mass, and indicate that the first stars most likely weighed tens to hundreds of solar masses each.

To estimate *when* the first stars formed we must remember that the first 100, 000 solar mass halos collapsed in regions that happened to have a particularly high density enhancement very early on. There was initially only a small abundance of such regions in the entire universe, so a simulation that is limited to a small volume is unlikely to find such halos until much later. Simulating the entire universe is well beyond the capabilities of current simulations, but analytical models predict that the first observable star in the universe probably formed 30 million years after the Big Bang, less than a quarter of one percent of the Universe's total age of 13.7 billion years.

Although stars were extremely rare at first, gravitational collapse increased the abundance of galactic halos and star formation sites with time (Fig. 2). Radiation from the first stars is expected to have eventually dissociated all the molecular hydrogen in the intergalactic medium, leading to the domination of a second generation of larger galaxies where the gas cooled via radiative transitions in atomic hydrogen and helium. Atomic cooling occurred in halos of mass above $\sim 10^8 M_\odot$ (Oh and Haiman 2002), in which the infalling gas was heated above 10,000 K and became ionized. The first galaxies to form through atomic cooling are expected to have formed around redshift 45, and such galaxies were likely the main sites of star formation by the time reionization began in earnest (for a recent review, see Bromm and Yoshida 2011). As the IGM was heated above 10,000 K by reionization, its pressure jumped and prevented the gas from accreting into newly forming halos below $\sim 10^9 M_\odot$. The first Milky-Way-sized halo $M = 10^{12} M_\odot$ is predicted to have formed 400 million years after the Big Bang, but such halos have become typical galactic hosts only in the last 5 billion years.

Hydrogen is the most abundant element in the Universe, The prominent Lyman-α spectral line of hydrogen (corresponding to a transition from its first excited level to its ground state) provides an important probe of the condensation of primordial gas into the first galaxies. Existing searches for Lyman-α emission have discovered galaxies robustly out to a redshift $z \sim 7$ with some unconfirmed candidate galaxies out to $z \sim 10$. The spectral break owing to Lyman-α absorption by the IGM allows to identify high-redshifts galaxies photometrically. Existing observations provide only a preliminary glimpse into the formation of the first galaxies.

Within the next decade, NASA plans to launch an infrared space telescope (*JWST*; Fig. 5) that will image some of the earliest sources of light (stars and black holes) in the Universe (Gardner et al. 2006). In parallel, there are several initiatives to construct large-aperture infrared telescopes on the ground with the same goal in mind.

The next generation of ground-based telescopes will have a diameter of 20–30 m (Fig. 6). Together with *JWST* (which will not be affected by the atmospheric background) they will be able to image and make spectral studies of the early galaxies. Given that these galaxies also create the ionized bubbles around them by their UV emission, during reionization the locations of galaxies should correlate with bubbles within the neutral hydrogen. Within a decade it should be possible to explore the environmental influence of individual galaxies by using these telescopes in combination with 21-cm probes of reionization.

Fig. 5 TA full scale model of the James Webb Space Telescope (JWST), the successor to the Hubble Space Telescope. JWST includes a primary mirror 6.5 m in diameter, and offers instrument sensitivity across the infrared wavelength range of 0.6–28 μm which will allow detection of the first galaxies. The size of the Sun shield (the large flat screen in the image) is 22×10 m (72×29 ft). The telescope will orbit 1.5 million kilometers from Earth at the Lagrange L2 point. **Image credit:** JWST/NASA (http://www.jwst.nasa.gov/)

2.4 Gamma-Ray Bursts: Probing the First Stars One Star at a Time

So far, to learn about diffuse IGM gas pervading the space outside and between galaxies, astronomers routinely study its absorption signatures in the spectra of distant quasars, the brightest long-lived astronomical objects. Quasars' great luminosities are believed to be powered by accretion of gas onto black holes weighing up to a few billion times the mass of the Sun that are situated in the centers of massive galaxies. As the surrounding gas spirals in toward the black hole sink, the viscous dissipation of heat makes the gas glow brightly into space, creating a luminous source visible from afar.

Over the past decade, an alternative population of bright sources at cosmological distances was discovered, the so-called afterglows of *Gamma-Ray Bursts* (GRBs). These events are characterized by a flash of high-energy (>0.1 MeV) photons, typically lasting 0.1–100 s, which is followed by an afterglow of lower-energy photons over much longer timescales (see Bloom 2011, and references therein). The afterglow peaks at X-ray, UV, optical and eventually radio wavelengths on time scales of minutes, hours, days, and months, respectively. The central engines of GRBs are believed to be associated with the compact remnants (neutron stars or stellar-mass black holes) of massive stars. Their high luminosities make them detectable out to the edge of the visible Universe. GRBs offer the opportunity

Fig. 6 Artist's conception of the designs for three future giant telescopes that will be able to probe the first generation of galaxies from the ground: the European Extremely Large Telescope (EELT, left), the Giant Magellan Telescope (GMT, middle), and the Thirty Meter Telescope (TMT, right). **Image credits:** http://www.eso.org/sci/facilities/eelt/,http://www.gmto.org/, and http://www.tmt.org/

to detect the most distant (and hence earliest) population of massive stars, the so-called Population III (or Pop III), one star at a time (e.g., Lamb and Reichart 2000; Bromm and Loeb 2006). In the hierarchical assembly process of halos that are dominated by cold dark matter (CDM), the first galaxies should have had lower masses (and lower stellar luminosities) than their more recent counterparts. Consequently, the characteristic luminosity of galaxies or quasars is expected to decline with increasing redshift. GRB afterglows, which already produce a peak flux comparable to that of quasars or starburst galaxies at $z \sim 1$–2, are therefore expected to outshine any competing source at the highest redshifts, when the first dwarf galaxies formed in the Universe.

GRBs, the electromagnetically-brightest explosions in the Universe, should be detectable out to redshifts $z > 10$. High-redshift GRBs can be identified through infrared photometry, based on the Lyman-α break induced by absorption of their spectrum at wavelengths below $1.216\,\mu\text{m}\,[(1+z)/10]$. Follow-up spectroscopy of high-redshift candidates can then be performed on a 10-m-class telescope. GRB afterglows offer the opportunity to detect stars as well as to probe the metal enrichment level of the intervening IGM. Recently, the *Swift* satellite has detected a GRB originating at $z \simeq 8.2$, thus demonstrating the viability of GRBs as probes of the early Universe.

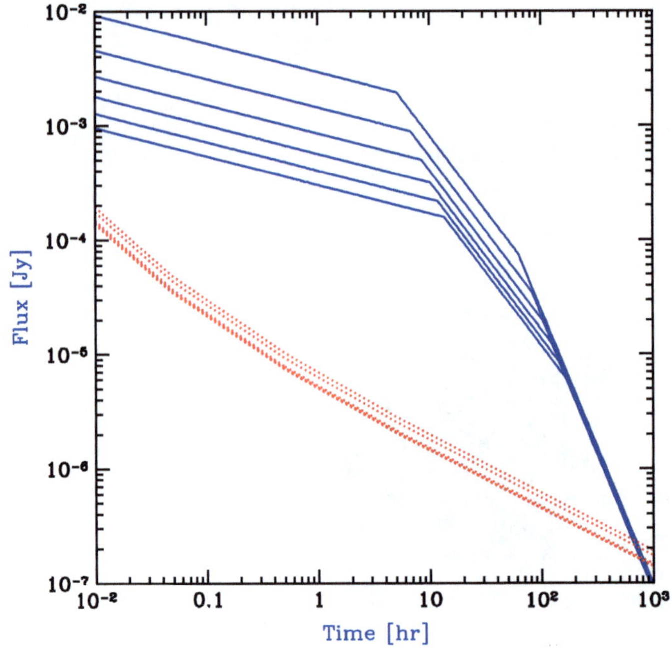

Fig. 7 GRB afterglow flux as a function of time since the γ-ray trigger in the observer frame. The flux (solid curves) is calculated at the redshifted Lyman-α wavelength. The dotted curves show the planned detection threshold for the *James Webb Space Telescope* (*JWST*), assuming a spectral resolution $R = 5,000$ with the near infrared spectrometer, a signal to noise ratio of 5 per spectral resolution element, and an exposure time equal to 20 % of the time since the GRB explosion. Each set of curves shows a sequence of redshifts, namely $z = 5, 7, 9, 11, 13$, and 15, respectively, from top to bottom. **Figure credit:** Barkana and Loeb (2004)

Another advantage of GRBs is that the GRB afterglow flux at a given observed time lag after the γ-ray trigger is not expected to fade significantly with increasing redshift, since higher redshifts translate to earlier times in the source frame, during which the afterglow is intrinsically brighter. For standard afterglow lightcurves and spectra, the increase in the luminosity distance with redshift is compensated by this cosmological time-stretching effect as shown in Fig. 7.

GRB afterglows have smooth (broken power-law) continuum spectra unlike quasars which show strong spectral features (such as broad emission lines or the so-called "blue bump") that complicate the extraction of IGM absorption features. In particular, the extrapolation into the spectral regime marked by the IGM Lyman-α absorption during the epoch of reionization is much more straightforward for the smooth UV spectra of GRB afterglows than for quasars with an underlying broad Lyman-α emission line. However, the interpretation may be complicated by the presence of damped Lyman-α absorption by dense neutral hydrogen in the immediate environment of the GRB within its host galaxy. Since GRBs originate from the dense environment of active star formation, such damped absorption is expected and indeed has been seen, including in the most distant GRB at $z = 8.2$.

2.5 Supermassive Black Holes

The fossil record in the present-day Universe indicates that every bulged galaxy hosts a supermassive black hole (BH) at its center. This conclusion is derived from a variety of techniques which probe the dynamics of stars and gas in galactic nuclei. The inferred BHs are dormant or faint most of the time, but occasionally flash in a short burst of radiation that lasts for a small fraction of the age of the Universe. The short duty cycle accounts for the fact that bright quasars are much less abundant than their host galaxies, but it begs the more fundamental question: *why is the quasar activity so brief?* A natural explanation is that quasars are suicidal, namely the energy output from the BHs regulates their own growth.

Supermassive BHs make up a small fraction, $<10^{-3}$, of the total mass in their host galaxies, and so their direct dynamical impact is limited to the central star distribution where their gravitational influence dominates. Dynamical friction on the background stars keeps the BH close to the center. Random fluctuations in the distribution of stars induces a Brownian motion of the BH. This motion can be described by the same Langevin equation that captures the motion of a massive dust particle as it responds to random kicks from the much lighter molecules of air around it. The characteristic speed by which the BH wanders around the center is small, $\sim(m_\star/M_{\rm BH})^{1/2}\sigma_\star$, where $m_\star$ and $M_{\rm BH}$ are the masses of a single star and the BH, respectively, and $\sigma_\star$ is the stellar velocity dispersion. Since the random force fluctuates on a dynamical time, the BH wanders across a region that is smaller by a factor of $\sim(m_\star/M_{\rm BH})^{1/2}$ than the region traversed by the stars inducing the fluctuating force on it.

The dynamical insignificance of the BH on the global galactic scale is misleading. The gravitational binding energy per rest-mass energy of galaxies is of order $\sim(\sigma_\star/c)^2 < 10^{-6}$. Since BH are relativistic objects, the gravitational binding energy of material that feeds them amounts to a substantial fraction its rest mass energy. Even if the BH mass amounts to a fraction as small as $\sim10^{-4}$ of the baryonic mass in a galaxy, and only a percent of the accreted rest-mass energy is deposited into the gaseous environment of the BH, this slight deposition can unbind the entire gas reservoir of the host galaxy. This order-of-magnitude estimate explains why quasars may be short lived. As soon as the central BH accretes large quantities of gas so as to significantly increase its mass, it releases large amounts of energy and momentum that could suppress further accretion onto it. In short, the BH growth might be *self-regulated*.

The principle of *self-regulation* naturally leads to a correlation between the final BH mass, $M_{\rm bh}$, and the depth of the gravitational potential well to which the surrounding gas is confined. The latter can be characterized by the velocity dispersion of the associated stars, $\sim\sigma_\star^2$. Indeed a correlation between $M_{\rm bh}$ and $\sigma_\star^4$ is observed in the present-day Universe. If quasars shine near their Eddington limit as suggested by observations of low and high-redshift quasars, then a fraction of $\sim$5–10 % of the energy released by the quasar over a galactic dynamical time needs to be captured in the surrounding galactic gas in order for the BH growth to be self-regulated.

With this interpretation, the $M_{\rm bh}$–$\sigma_\star$ relation reflects the limit introduced to the BH mass by self-regulation; deviations from this relation are inevitable during episodes of BH growth or as a result of mergers of galaxies that have no cold gas in them. A physical scatter around this upper envelope could also result from variations in the efficiency by which the released BH energy couples to the surrounding gas.

Various prescriptions for self-regulation were sketched in the literature. These involve either energy or momentum-driven winds, with the latter type being a factor of $\sim v_c/c$ less efficient. The quasar remains active during the dynamical time of the initial gas reservoir, $\sim 10^7$ years, and fades afterwards due to the dilution of this reservoir. The BH growth may resume if the cold gas reservoir is replenished through a new merger. Following early analytic work, extensive numerical simulations demonstrated that galaxy mergers do produce the observed correlations between black hole mass and spheroid properties. Because of the limited resolution near the galaxy nucleus, these simulations adopt a simple prescription for the accretion flow that feeds the black hole. The actual feedback in reality may depend crucially on the geometry of this flow and the physical mechanism that couples the energy or momentum output of the quasar to the surrounding gas.

The inflow of cold gas towards galaxy centers during the growth phase of the BH would naturally be accompanied by a burst of star formation. The fraction of gas that is not consumed by stars or ejected by supernova-driven winds, will continue to feed the BH. It is therefore not surprising that quasar and starburst activities co-exist in Ultra Luminous Infrared Galaxies, and that all quasars show broad metal lines indicating pre-enrichment of the surrounding gas with heavy elements.

The upper mass of galaxies may also be regulated by the energy output from quasar activity. This would account for the fact that cooling flows are suppressed in present-day X-ray clusters, and that massive BHs and stars in galactic bulges were already formed at $z \sim 2$. In the cores of cooling X-ray clusters, there is often an active central BH that supplies sufficient energy to compensate for the cooling of the gas. The primary physical process by which this energy couples to the gas is still unknown.

The quasars discovered so far at $z \sim 6$ mark the early growth of the most massive BHs and galactic spheroids. The BHs powering these bright quasars possess a mass of a few billion solar masses. A quasar radiating at its Eddington limiting luminosity, $L_E = 1.4 \times 10^{47}$ erg s$^{-1}(M_{\rm bh}/10^9 M_\odot)$, with a radiative efficiency, $\epsilon_{\rm rad} = L_E/\dot{M}c^2$, for converting accreted mass into radiation, would grow exponentially in mass as a function of time t, $M_{\rm bh} = M_{\rm seed}\exp\{t/t_E\}$ from its initial seed mass $M_{\rm seed}$, on a time scale, $t_E = 4.1 \times 10^7$ year$(\epsilon_{\rm rad}/0.1)$. Thus, the required growth time in units of the Hubble time $t_{\rm hubble} = 10^9$ year$[(1+z)/7]^{-3/2}$ is

$$\frac{t_{\rm growth}}{t_{\rm hubble}} = 0.7 \left(\frac{\epsilon_{\rm rad}}{10\,\%}\right) \left(\frac{1+z}{7}\right)^{3/2} \ln\left(\frac{M_{\rm bh}/10^9 M_\odot}{M_{\rm seed}/100 M_\odot}\right). \tag{29}$$

The age of the Universe at $z \sim 6$ provides just sufficient time to grow a BH with $M_{\rm bh} \sim 10^9 M_\odot$ out of a stellar mass seed with $\epsilon_{\rm rad} = 10\,\%$. The growth time is shorter for smaller radiative efficiencies or a higher seed mass.

2.6 A New Frontier: Gravitational Waves from Black Hole Mergers

When two galaxies collide, their cores migrate by dynamical friction to the center of mass of the merged galaxy. If the galaxy is rich in gas (as expected for high-redshift galaxies), the orbit of the two black holes tightens on a timescale that is much shorter than the age of the Universe. The final phase of binary coalescence is driven by the emission of gravitational waves. The emitted waves could be detected by new observatories which are currently being planned or constructed.

As long as the binary separation is much larger than its Schwarzschild radius, the emitted gravitational wave luminosity can be derived in the post-Newtonian approximation. For two black holes on a circular orbit, the luminosity is

$$L_{\rm GW} = \frac{32}{5}\frac{G^4}{c^5}\frac{M^3\mu^2}{a^5}, \tag{30}$$

where a is the semi-major axis of the binary, $M = (M_1 + M_2)$ and $\mu = M_1 M_2/M$, with M_1 and M_2 being the masses of the binary members. The loss of energy to the emitted waves leads to a decrease in the binary separation a and an eventual coalescence of the two black holes over a time,

$$t_{\rm GW} = \frac{5}{256}\frac{c^5}{G^3}\frac{a^4}{M^2\mu}. \tag{31}$$

Supermassive binaries with comparable mass members merge in less than a Hubble time once their separation shrinks to $a < 10^{3.5} r_{\rm Sch}$ (where $r_{\rm Sch} = 2GM/c^2 = 3 \times 10^{11}{\rm cm}(M/10^6 M_\odot)$) or once their relative orbital velocity $v = (GM/a)^{1/2} > 10^{-2}c = 3 \times 10^3$ km s^{-1}.

Future gravitational wave observatories will be sensitive to the gravitational wave amplitude. To an order of magnitude, the observed wave amplitude from an equal mass binary with a Schwarzschild radius $r_{\rm Sch}$ and an orbital velocity v is given by, $h \sim (1+z)(r_{\rm Sch}/d_L)(v^2/c^2)$, where d_L is the luminosity distance to the binary. Since the signal amplitude only declines as (distance)$^{-1}$ rather than (distance)$^{-2}$ as for electromagnetic detectors which respond to photon flux, the first generation of sensitive gravitational wave observatories will already be able to find sources at cosmological distances.

More accurately, in a reference frame centered on the solar system's barycenter, the gravitational wave amplitude in its two polarization states is given by,

$$h_+ = \frac{2\mathcal{M}_z^{5/3}[\pi f_{\rm obs}]^{2/3}}{d_L}\left[1 + (\hat{L}\cdot\hat{n})^2\right]\cos[2\Phi(t)];\, h_x = -\frac{4\mathcal{M}_z^{5/3}[\pi f]^{2/3}(\hat{L}\cdot\hat{n})}{d_L}\sin[2\Phi(t)]; \tag{32}$$

where the so-called "redshifted chirp mass" $\mathcal{M}_z \equiv (1+z)\mu^{3/5}/M^{2/5}$ sets the rate at which the binary shrinks, determining the "chirp" of their orbital frequency $P = 2\pi/\sqrt{GM/a^3}$. The precise orbital phase of the binary $\Phi(t)$ then depends

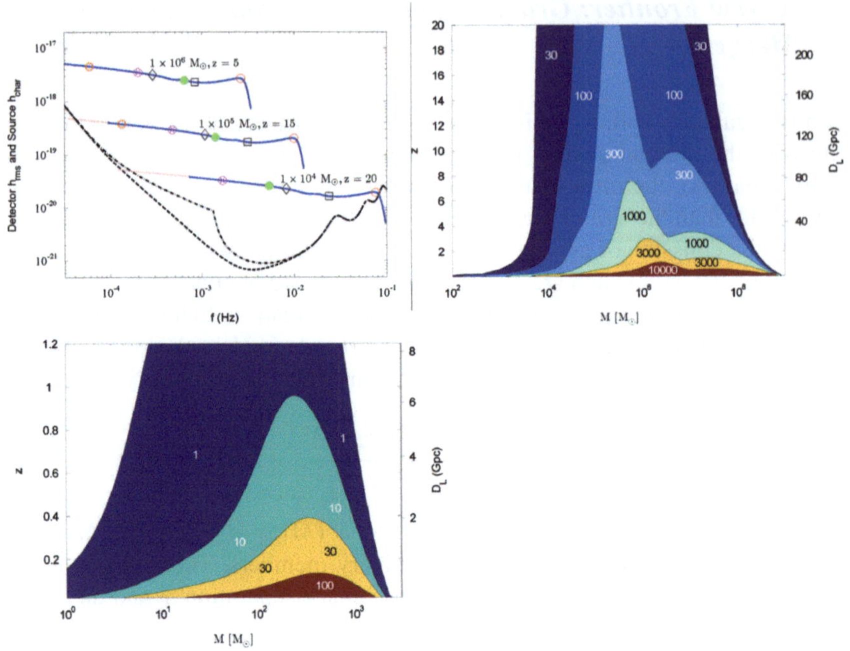

Fig. 8 Sensitivity of the future gravitational wave observatories, LISA and Advanced-LIGO to equal-mass ($M_1 = M_2 = M/2$) binaries. *Left panel:* Root-mean-square noise amplitude of LISA h_{rms} from the detector only (*dashed*) and from the detector combined with the anticipated foreground confusion (*dash-dotted*), along with the characteristic amplitudes h_{char} of three binary masses M (*solid*). The locations on each h_{char} curve correspond to the peak amplitude (*circle*), 1 h before the peak (*filled circle*), 1 day before the peak (circle with inscribed cross), and 1 month before the peak (circle with inscribed square) in the observer frame, as well as times of $25 r_{\rm Sch}/c$ (*square*) and $500 r_{\rm Sch}/c$ (*diamond*) before the peak in the source frame. *Middle panel:* Contour plot of the signal-to-noise ratio (SNR) with binary mass and redshift dependence for LISA. *Right panel:* SNR contour plot with mass and redshift dependence for Advanced-LIGO. **Figure credit:** Baker et al. (2007)

on the masses and spins of the binary members, and yields the observed wave frequency, $f_{\rm obs}(t) = [\pi]^{-1}(d\Phi/dt)$, which is $(1+z)$ times smaller than the emitted wave frequency. The unit vector $\hat{n}$ points from the solar system frame to the binary—defining the sky coordinates of the source, and the unit vector $\hat{L}$ points along the binary angular momentum—defining the binary orientation relative to the line-of-sight. The inspiral signal does not provide explicitely the cosmological redshift separately from the binary masses, but the redshift can be inferred from $d_L(z)$ (or from an electromagnetic counterpart to the gravitational wave signal). Any particular detector is sensitive to a linear combination of the two polarization signals, with coefficients that depend on the orientation of the source relative to the detector.

The sensitivity of various gravitational wave observatories is shown in Fig. 8. The Laser Interferometer Space Antenna (LISA; http://lisa.nasa.gov/) is a planned space interferometer consisting of three spacecrafts whose positions mark the vertices of an equilateral triangle 5 km on a side in an orbit around the Sun. As evident from Fig. 8, LISA will be able to detect $\sim 10^{4-7} M_{\odot}$ binaries out to arbitrary redshifts during the epoch of reionization. The next generation ground-based interferometer, Advanced-LIGO (http://www.advancedligo.mit.edu/), will be sensitive to binaries involving black hole remnants of massive Pop-III stars (with $\sim 10^{2-3} M_{\odot}$) out to $z \gg 1$.

The expected event rate of massive binary mergers can be calculated based on the halo merger rate under various assumptions about the relation between the black hole and halo masses. For reasonable assumptions, LISA is expected to detect many cosmological events per year. The actual detection of these signals would open a new window into the Universe and enable to trace the hierarchical assembly of black holes in galaxies throughout cosmic history. Since gravitational waves pass freely through all forms of matter, gravitational wave observatories might discover new populations of black hole binaries that are electromagnetically faint because of their modest mass relative to bright quasars or because they are enshrouded in gas and dust.

2.7 *The Epoch of Reionization*

Given the understanding described above of how many galaxies formed at various times, the course of reionization can be determined universe-wide by counting photons from all sources of light. Both stars and black holes contribute ionizing photons, but the early universe is dominated by small galaxies which in the local universe have central black holes that are disproportionately small, and indeed quasars are rare above redshift 6. Thus, stars most likely dominated the production of ionizing UV photons during the reionization epoch [although high-redshift galaxies should have also emitted X-rays from accreting black holes and accelerated particles in collisionless shocks]. Since most stellar ionizing photons are only slightly more energetic than the 13.6 eV ionization threshold of hydrogen, they are absorbed efficiently once they reach a region with substantial neutral hydrogen). This makes the IGM during reionization a two-phase medium characterized by highly ionized regions separated from neutral regions by sharp ionization fronts.

We can obtain a first estimate of the requirements of reionization by demanding one stellar ionizing photon for each hydrogen atom in the IGM. If we conservatively assume that stars within the early galaxies were similar to those observed locally, then each star produced $\sim 4,000$ ionizing photons per baryon. Star formation is observed today to be an inefficient process, but even if stars in galaxies formed out of only $\sim 10\,\%$ of the available gas, it was still sufficient to accumulate a small fraction (of order $0.1\,\%$) of the total baryonic mass in the universe into galaxies in order to ionize the entire IGM. More accurate estimates of the actual required fraction account for the formation of some primordial stars (which were massive, efficient ionizers, as discussed above), and for recombinations of hydrogen atoms at high redshifts and in dense regions.

From studies of quasar absorption lines at $z \sim 6$ we know that the IGM is highly ionized a billion years after the big bang. There are hints, however, that some large neutral hydrogen regions persist at these early times and so this suggests that we may not need to go to much higher redshifts to begin to see the epoch of reionization. We now know that the universe could not have fully reionized earlier than an age of 300 million years, since WMAP observed the effect of the freshly created plasma at reionization on the large-scale polarization anisotropies of the CMB and this limits the reionization redshift; an earlier reionization, when the universe was denser, would have created a stronger scattering signature that would be inconsistent with the WMAP observations. In any case, the redshift at which reionization ended only constrains the overall cosmic efficiency of ionizing photon production. In comparison, a detailed picture of reionization as it happens will teach us a great deal about the population of young galaxies that produced this cosmic phase transition. A key point is that the spatial distribution of ionized bubbles is determined by clustered groups of galaxies and not by individual galaxies. At such early times galaxies were strongly clustered even on very large scales (up to tens of Mpc), and these scales therefore dominate the structure of reionization. The basic idea is simple. At high redshift, galactic halos are rare and correspond to rare, high density peaks. As an analogy, imagine searching on Earth for mountain peaks above 5,000 m. The 200 such peaks are not at all distributed uniformly but instead are found in a few distinct clusters on top of large mountain ranges. Given the large-scale boost provided by a mountain range, a small-scale crest need only provide a small additional rise in order to become a 5,000 m peak. The same crest, if it formed within a valley, would not come anywhere near 5,000 m in total height. Similarly, in order to find the early galaxies, one must first locate a region with a large-scale density enhancement, and then galaxies will be found there in abundance.

The ionizing radiation emitted from the stars in each galaxy initially produces an isolated ionized bubble. However, in a region dense with galaxies the bubbles quickly overlap into one large bubble, completing reionization in this region while the rest of the universe is still mostly neutral. Most importantly, since the abundance of rare density peaks is very sensitive to small changes in the density threshold, even a large-scale region with a small enhanced density (say, 10 % above the mean density of the universe) can have a much larger concentration of galaxies than in other regions (e.g., a 50 % enhancement). On the other hand, reionization is harder to achieve in dense regions, since the protons and electrons collide and recombine more often in such regions, and newly-formed hydrogen atoms need to be reionized again by additional ionizing photons. However, the overdense regions end up reionizing first since the number of ionizing sources in these regions is increased so strongly. The large-scale topology of reionization is therefore inside out, with underdense voids reionizing only at the very end of reionization, with the help of extra ionizing photons coming in from their surroundings (which have a higher density of galaxies than the voids themselves). This is a key prediction awaiting observational testing.

Detailed analytical models that account for large-scale variations in the abundance of galaxies confirm that the typical bubble size starts well below a Mpc early in reionization, as expected for an individual galaxy, rises to 5–10 Mpc during the central phase (i.e., when the universe is half ionized), and then by another factor of ∼5 towards the end of reionization. These scales are given in comoving units that scale with the expansion of the universe, so that the actual sizes at a redshift z were smaller than these numbers by a factor of $(1+z)$. Numerical simulations have only recently begun to reach the enormous scales needed to capture this evolution. Accounting precisely for gravitational evolution on a wide range of scales but still crudely for gas dynamics, star formation, and the radiative transfer of ionizing photons, the simulations confirm that the large-scale topology of reionization is inside out, and that this topology can be used to study the abundance and clustering of the ionizing sources.

The characteristic observable size of the ionized bubbles at the end of reionization can be calculated based on simple considerations that only depend on the power-spectrum of density fluctuations and the redshift. As the size of an ionized bubble increases, the time it takes a 21-cm photon emitted by hydrogen to traverse it gets longer. At the same time, the variation in the time at which different regions reionize becomes smaller as the regions grow larger. Thus, there is a maximum size above which the photon crossing time is longer than the cosmic variance in ionization time. Regions bigger than this size will be ionized at their near side by the time a 21-cm photon will cross them towards the observer from their far side. They would appear to the observer as one-sided, and hence signal the end of reionization. These considerations imply a characteristic size for the ionized bubbles of ∼10 physical Mpc at $z \sim 6$ (equivalent to 70 Mpc today). This result implies that future radio experiments should be tuned to a characteristic angular scale of tens of arcminutes for an optimal detection of 21-cm brightness fluctuations near the end of reionization (see Sect. 3.2).

Existing data on the polarization anisotropies of the CMB as well as the Lyman-α forest can be used to derive a probability distribution for the hydrogen ionization fraction (x_i) as a function of redshift. Figure 9 shows this likelihood distribution in four redshift bins of interest to upcoming observations. Although there is considerable uncertainty in x_i at each redshift, it is evident from existing data that hydrogen is highly ionized by $z = 8$ (at least to $x_i > 0.8$).

To produce one ionizing photon per baryon requires a minimum comoving density of Milky-Way (so-called Population II) stars of,

$$\rho_\star \approx 1.7 \times 10^6 f_{\rm esc}^{-1}\ M_\odot\ {\rm Mpc}^{-3}, \tag{33}$$

or equivalently, a cosmological density parameter in stars of $\Omega_\star \sim 1.25 \times 10^{-5} f_{\rm esc}^{-1}$. More typically, the threshold for reionization involves at least a few ionizing photons per proton (with the right-hand-side being $\sim 10^{-6}\ {\rm cm}^{-3}$), since the recombination time at the mean density is comparable to the age of the Universe at $z \sim 10$.

For the local mass function of (Population II) stars at solar metallicity, the star formation rate per unit comoving volume that is required for balancing recombinations in an already ionized IGM, is given by

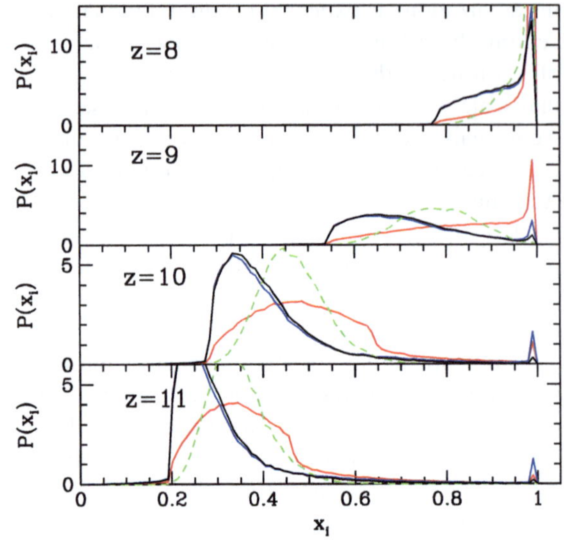

Fig. 9 Probability distribution of x_i at redshifts $z = 8, 9, 10$, and 11, based on existing data on the CMB polarization anisotropies and the Lyman-α forest. **Figure credit:** Pritchard et al. (2010)

$$\dot{\rho_\star} \approx 2 \times 10^{-3} f_{\rm esc}^{-1} C \left(\frac{1+z}{10}\right)^3 M_\odot \text{ year}^{-1} \text{ Mpc}^{-3}, \tag{34}$$

where $C = \langle n_e^2 \rangle / \langle n_e \rangle^2$ is the volume-averaged clumpiness factor of the electron density up to some threshold overdensity of gas which remains neutral. Current state-of-the-art surveys (HST WFC3/IR) are only sensitive to the bright end of the luminosity function of galaxies at $z > 6$ and hence provide a lower limit on the production rate of ionizing photons during reionization (for a recent review, see Robertson 2010).

2.8 Post-Reionization Suppression of Low-Mass Galaxies

After the ionized bubbles overlapped in each region, the ionizing background increased sharply, and the IGM was heated by the ionizing radiation to a temperature $T_{\rm IGM} > 10^4$ K. Due to the substantial increase in the IGM pressure, the smallest mass scale into which the cosmic gas could fragment, the so-called Jeans mass, increased dramatically, changing the minimum mass of forming galaxies.

Gas infall depends sensitively on the Jeans mass. When a halo more massive than the Jeans mass begins to form, the gravity of its dark matter overcomes the gas pressure. Even in halos below the Jeans mass, although the gas is initially held up by pressure, once the dark matter collapses its increased gravity pulls in some gas. Thus, the Jeans mass is generally higher than the actual limiting mass for accretion. Before reionization, the IGM is cold and neutral, and the Jeans mass plays a secondary role in limiting galaxy formation compared to cooling. After reionization, the Jeans mass is increased by several orders of magnitude due to the photoionization heating of the IGM, and hence begins to play a dominant role in limiting the formation of stars. Gas infall in a reionized and heated Universe has been investigated in a number of numerical simulations. Three dimensional numerical simulations found a significant suppression of gas infall in even larger halos ($V_c \sim 75$ km s^{-1}), but this was mostly due to a suppression of late infall at $z < 2$.

When a volume of the IGM is ionized by stars, the gas is heated to a temperature $T_{\rm IGM} \sim 10^4$ K. If quasars dominate the UV background at reionization, their harder photon spectrum leads to $T_{\rm IGM} > 2 \times 10^4$ K. Including the effects of dark matter, a given temperature results in a linear Jeans mass corresponding to a halo circular velocity of

$$V_J \approx 80 \left(\frac{T_{\rm IGM}}{1.5 \times 10^4 {\rm K}} \right)^{1/2} \ {\rm km\ s^{-1}}. \tag{35}$$

In halos with a circular velocity well above V_J, the gas fraction in infalling gas equals the universal mean of Ω_b/Ω_m, but gas infall is suppressed in smaller halos. A simple estimate of the limiting circular velocity, below which halos have essentially no gas infall, is obtained by substituting the virial overdensity for the mean density in the definition of the Jeans mass. The resulting estimate is

$$V_{\rm lim} = 34 \left(\frac{T_{\rm IGM}}{1.5 \times 10^4 {\rm K}} \right)^{1/2} \ {\rm km\ s^{-1}}. \tag{36}$$

This value is in rough agreement with the numerical simulations mentioned before.

Although the Jeans mass is closely related to the rate of gas infall at a given time, it does not directly yield the total gas residing in halos at a given time. The latter quantity depends on the entire history of gas accretion onto halos, as well as on the merger histories of halos, and an accurate description must involve a time-averaged Jeans mass. The gas content of halos in simulations is well fit by an expression which depends on the filtering mass, a particular time-averaged Jeans mass.

The reionization process was not perfectly synchronized throughout the Universe. Large-scale regions with a higher density than the mean tended to form galaxies first and reionized earlier than underdense regions. The suppression of low-mass galaxies by reionization is therefore modulated by the fluctuations in the timing of reionization. Inhomogeneous reionization imprint a signature on the power-spectrum of low-mass galaxies. Future high-redshift galaxy surveys hoping to constrain inflationary parameters must properly model the effects of reionization; conversely, they will also place new constraints on the thermal history of the IGM during reionization.

2.9 Gravitational Lensing of Faint Galaxies From the Epoch of Reionization

The chance alignment of a foreground object along the line of sight to a high redshift source could result in the magnification, distortion, and potentially splitting of the source image due the deflection of its light rays by the gravitational field of the foreground object. The probability for *gravitational lensing* grows with increasing source redshift, due to the increase in the path length of the source photons. Although the lensing probability is only of anecdotal significance of <1 % for sources at $z < 2$, its magnitude could rise by an order of magnitude and affect the statistics of bright sources during the epoch of reionization.

A simple model for the mass distribution in a galaxy assumes that its collisionless particles (stars and dark matter) possess the same isotropic velocity dispersion everywhere. Surprisingly, this simple model, labeled as a *singular isothermal sphere (SIS)*, appears to describe extremely well the dynamics of stars and gas in the cores of disk galaxies (whose rotation curve is roughly flat), as well the strong lensing properties of spheroidal galaxies.

A SIS lens has the simple property that the deflection angle $\hat{\alpha}$ is independent of the impact parameter of the light ray. The condition for multiple imaging (and hence strong lensing) is then that the source would lie inside the so-called Einstein angle, θ_E. The probability that a line-of-sight to a source at a redshift z_s passes within the cross-sectional area associated with the Einstein radius of SIS lenses $\pi\theta_E^2$ gives a lensing optical depth,

$$\tau(z_s) = \frac{16\pi^3}{H_0}\int_0^{z_s} dz \frac{D^2(1+z)^2}{(\Omega_m(1+z)^3+\Omega_\Lambda)^{1/2}}\int_0^\infty d\sigma_v \frac{dn}{d\sigma_v}\sigma_v^4, \tag{37}$$

where $(dn/d\sigma_v)d\sigma_v$ is the (redshift-dependent) comoving density of SIS halos with a one-dimensional velocity dispersion between σ_v and $\sigma_v + d\sigma_v$, and

$$\theta_E = 4\pi\,\frac{\sigma_v^2}{c^2}\,\frac{D_{ds}}{D_s}\,. \tag{38}$$

with

$$D = \frac{D_d D_{ds}}{D_s}\,. \tag{39}$$

Here the angular diameter distances between observer and lens, lens and source, and observer and source are denoted here as D_d, D_{ds}, and D_s, respectively.

In calculating the probability of lensing it is important to allow for various selection effects. Lenses magnify the observed flux, and lift sources which are intrinsically too faint to be observed over the detection threshold. At the same time, lensing increases the solid angle within which sources are observed so that their number density in the sky is reduced. If there is a large reservoir of faint sources,

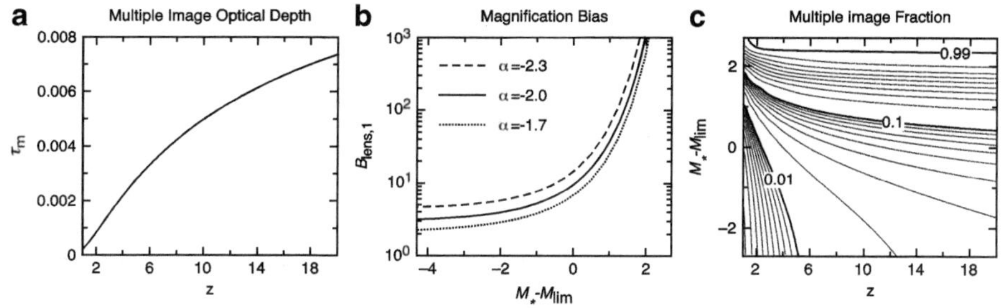

Fig. 10 Probability for multiple imaging of high redshift galaxies by an unevolving population of SIS lenses. *Panel a*: lensing probability τ as a function of source redshift. *Panel b*: magnification bias as a function of the difference between the characteristic magnitude of a galaxy $M_\star$ (assuming a Schechter luminosity function) and the limiting survey magnitude $M_{\rm lim}$. Three values of the faint-end slope of the luminosity function (labeled by α here) are shown. *Panel c*: Contours of the fraction of multiply-imaged sources as a function of source redshift and $(M_\star - M_{\rm lim})$, assuming a faint end slope of -2. **Figure credit:** Wyithe et al. (2011). See also Barkana and Loeb (2000)

the increase in source number due to the apparent brightening outweighs their spatial dilution, and the observed number of sources is increased due to lensing. This so-called magnification bias can substantially increase the probability of lensing for bright sources whose number-count function is steep. The magnification bias for sources at redshift z_s with luminosities between L and $L + dL$ is,

$$B(L) = \frac{1}{dn_s(L)/dL} \int_{\mu_{\rm min}}^{\mu_{\rm max}} \frac{d\mu}{\mu} \frac{dP}{d\mu} \frac{dn_s(L)}{dL}, \tag{40}$$

where $n_s(<L)$ is the density of sources with luminosity $<L$ and $dP/d\mu$ is the probability for magnification μ. For example, the brighter SIS image has a magnification distribution $(dP/d\mu) = 2(\mu - 1)^{-3}$ for $2 < \mu < \infty$.

A simplified model for the redshift evolution of SIS lenses is to use the mass function of dark matter halos and identify $\sigma_v = V_c/\sqrt{2}$ at the virial radius. Another simplified approach is to adopt the observed $(dn/d\sigma_v)$ at $z = 0$ and assume no evolution in the comoving density of lenses. The latter approach gives the approximate results shown in Fig. 10.

3 Probing the Diffuse Intergalactic Hydrogen

3.1 Lyman-Alpha Absorption

Resonant Lyman-α absorption has thus far proved to be the best probe of the state of the IGM. The optical depth to absorption by a uniform intergalactic medium is

$$\tau_s = \frac{\pi e^2 f_\alpha \lambda_\alpha n_{\rm H\,I}(z)}{m_e c H(z)} \qquad (41)$$
$$\approx 6.45 \times 10^5 x_{\rm H\,I} \left(\frac{\Omega_b h}{0.0315}\right)\left(\frac{\Omega_m}{0.3}\right)^{-1/2}\left(\frac{1+z}{10}\right)^{3/2},$$

where $H \approx 100h\ {\rm km\,s^{-1}\,Mpc^{-1}}\Omega_m^{1/2}(1+z)^{3/2}$ is the Hubble parameter at redshift z; $f_\alpha = 0.4162$ and $\lambda_\alpha = 1{,}216\,\text{Å}$ are the oscillator strength and the wavelength of the Lyman-α transition; $n_{\rm H\,I}(z)$ is the neutral hydrogen density at z (assuming primordial abundances); Ω_m and Ω_b are the present-day density parameters of all matter and of baryons, respectively; and $x_{\rm H\,I}$ is the average fraction of neutral hydrogen. In the second equality we have implicitly considered high redshifts.

Lyman-α absorption is thus highly sensitive to the presence of even trace amounts of neutral hydrogen. The lack of full absorption in quasar spectra then implies that the IGM has been very highly ionized during much of the history of the universe, from at most a billion years after the big bang to the present time. At redshifts approaching six, however, the optical depth increases, and the observed absorption becomes very strong. The difference between the unabsorbed expectation and the actual observed spectrum can be used to measure the amount of absorption, and thus to infer the atomic hydrogen density.

Several quasars beyond $z \sim 6.1$ show in their spectra a strong (so-called "Gunn–Peterson") trough, a blank spectral region at wavelengths shorter than Lyα at the quasar redshift (Fig. 11). The detection of Gunn–Peterson troughs indicates a rapid change in the neutral content of the IGM at $z \sim 6$, and hence a rapid change in the intensity of the background ionizing flux. However, even a small atomic hydrogen fraction of $\sim 10^{-3}$ would still produce nearly complete Lyα absorption.

While only resonant Lyα absorption is important at moderate redshifts, the damping wing of the Lyα line plays a significant role when neutral fractions of order unity are considered at $z > 6$. The scattering cross-section of the Lyα resonance line by neutral hydrogen is given by

$$\sigma_\alpha(\nu) = \frac{3\lambda_\alpha^2 \Lambda_\alpha^2}{8\pi} \frac{(\nu/\nu_\alpha)^4}{4\pi^2(\nu-\nu_\alpha)^2 + (\Lambda_\alpha^2/4)(\nu/\nu_\alpha)^6}, \qquad (42)$$

where $\Lambda_\alpha = (8\pi^2 e^2 f_\alpha / 3 m_e c \lambda_\alpha^2) = 6.25 \times 10^8\ {\rm s^{-1}}$ is the Lyα $(2p \to 1s)$ decay rate, $f_\alpha = 0.4162$ is the oscillator strength, and $\lambda_\alpha = 1{,}216\,\text{Å}$ and $\nu_\alpha = (c/\lambda_\alpha) = 2.47 \times 10^{15}$ Hz are the wavelength and frequency of the Lyα line. The term in the numerator is responsible for the classical Rayleigh scattering.

Although reionization is an inhomogeneous process, we consider here a simple illustrative case of instantaneous reionization. Consider a source at a redshift z_s beyond the redshift of reionization, $z_{\rm reion}$, and the corresponding scattering optical depth of a uniform, neutral IGM of hydrogen density $n_{\rm H,0}(1+z)^3$ between the source and the reionization redshift. The optical depth is a function of the observed wavelength $\lambda_{\rm obs}$,

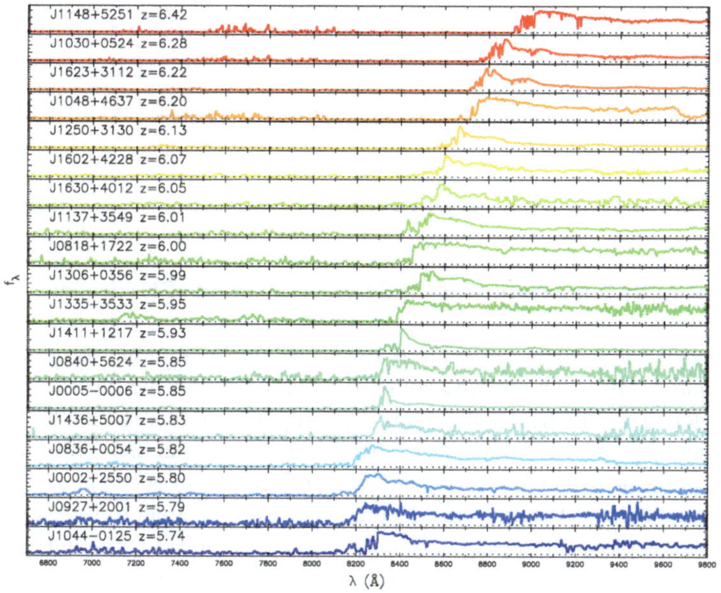

Fig. 11 Spectra of 19 quasars with redshifts $5.74 < z < 6.42$ from the *Sloan Digital Sky Survey*. For some of the highest-redshift quasars, the spectrum shows no transmitted flux shortward of the Lyman-α wavelength at the quasar redshift (the so-called "Gunn–Peterson trough"), indicating a non-negligible neutral fraction in the IGM. **Figure credit:** Fan et al. (2005)

$$\tau(\lambda_{\rm obs}) = \int_{z_{\rm reion}}^{z_s} dz \frac{cdt}{dz} n_{\rm H,0}(1+z)^3 \sigma_\alpha \left[\nu_{\rm obs}(1+z)\right], \tag{43}$$

where $\nu_{\rm obs} = c/\lambda_{\rm obs}$ and for a flat Universe with $(\Omega_m + \Omega_\Lambda) = 1$,

$$\frac{dt}{dz} = [(1+z)H(z)]^{-1} = H_0^{-1} \times \left[\Omega_m(1+z)^5 + \Omega_\Lambda(1+z)^2\right]^{-1/2}. \tag{44}$$

At wavelengths longer than Lyα at the source, the optical depth obtains a small value; these photons redshift away from the line center along its red wing and never resonate with the line core on their way to the observer. Considering only the regime in which $|\nu - \nu_\alpha| \gg \Lambda_\alpha$, we may ignore the second term in the denominator of Eq. (42). This leads to an analytical result for the red damping wing of the Gunn–Peterson trough,

$$\tau(\lambda_{\rm obs}) = \tau_s \left(\frac{\Lambda}{4\pi^2\nu_\alpha}\right) \tilde{\lambda}_{\rm obs}^{3/2} \left[I(\tilde{\lambda}_{\rm obs}^{-1}) - I([(1+z_{\rm reion})/(1+z_s)]\tilde{\lambda}_{\rm obs}^{-1})\right], \tag{45}$$

an expression valid for $\tilde{\lambda}_{\rm obs} \geq 1$, where τ_s is given in Eq. (41), and we also define

$$\tilde{\lambda}_{\rm obs} \equiv \frac{\lambda_{\rm obs}}{(1+z_s)\lambda_\alpha} \tag{46}$$

and

$$I(x) \equiv \frac{x^{9/2}}{1-x} + \frac{9}{7}x^{7/2} + \frac{9}{5}x^{5/2} + 3x^{3/2} + 9x^{1/2} - \frac{9}{2}\ln\left[\frac{1+x^{1/2}}{1-x^{1/2}}\right]. \tag{47}$$

3.2 21-cm Absorption or Emission

3.2.1 The Spin Temperature of the 21-cm Transition of Hydrogen

The ground state of hydrogen exhibits hyperfine splitting owing to the possibility of two relative alignments of the spins of the proton and the electron. The state with parallel spins (the triplet state) has a slightly higher energy than the state with anti-parallel spins (the singlet state). The 21-cm line associated with the spin-flip transition from the triplet to the singlet state is often used to detect neutral hydrogen in the local universe. At high redshift, the occurrence of a neutral pre-reionization IGM offers the prospect of detecting the first sources of radiation and probing the reionization era by mapping the 21-cm emission from neutral regions. While its energy density is estimated to be only a 1 % correction to that of the CMB, the redshifted 21-cm emission should display angular structure as well as frequency structure due to inhomogeneities in the gas density field, hydrogen ionized fraction, and spin temperature. Indeed, a full mapping of the distribution of H I as a function of redshift is possible in principle (see Furlanetto et al. 2006 for a comprehensive review).

The basic physics of the hydrogen spin transition is determined as follows. The ground-state hyperfine levels of hydrogen tend to thermalize with the CMB background, making the IGM unobservable. If other processes shift the hyperfine level populations away from thermal equilibrium, then the gas becomes observable against the CMB in emission or in absorption. The relative occupancy of the spin levels is usually described in terms of the hydrogen spin temperature T_S, defined by

$$\frac{n_1}{n_0} = 3\,\exp\left\{-\frac{T_*}{T_S}\right\}, \tag{48}$$

where n_0 and n_1 refer respectively to the singlet and triplet hyperfine levels in the atomic ground state ($n = 1$), and $T_* = 0.068$ K is defined by $k_B T_* = E_{21}$, where the energy of the 21 cm transition is $E_{21} = 5.9 \times 10^{-6}$ eV, corresponding to a frequency of 1,420 MHz. In the presence of the CMB alone, the spin states reach thermal equilibrium with $T_S = T_{\rm CMB} = 2.725(1+z)$ K on a time-scale of $T_*/(T_{\rm CMB}A_{10}) \simeq 3 \times 10^5 (1+z)^{-1}$ year, where $A_{10} = 2.87 \times 10^{-15}$ s^{-1} is the spontaneous decay rate of the hyperfine transition. This time-scale is much shorter than the age of the universe at all redshifts after cosmological recombination.

The IGM is observable when the kinetic temperature T_k of the gas differs from $T_{\rm CMB}$ and an effective mechanism couples T_S to T_k. Collisional de-excitation of the triplet level dominates at very high redshift, when the gas density (and thus the collision rate) is still high, but once a significant galaxy population forms in the universe, the spin temperature is affected also by an indirect mechanism that acts through the scattering of Lyman-α photons. Continuum UV photons produced by early radiation sources redshift by the Hubble expansion into the local Lyman-α line at a lower redshift. These photons mix the spin states via the Wouthuysen-Field process whereby an atom initially in the $n = 1$ state absorbs a Lyman-α photon, and the spontaneous decay which returns it from $n = 2$ to $n = 1$ can result in a final spin state which is different from the initial one. Since the neutral IGM is highly opaque to resonant scattering, and the Lyman-α photons receive Doppler kicks in each scattering, the shape of the radiation spectrum near Lyman-α is determined by T_k, and the resulting spin temperature (assuming $T_S \gg T_*$) is then a weighted average of T_k and $T_{\rm CMB}$:

$$T_S = \frac{T_{\rm CMB} T_k (1 + x_{\rm tot})}{T_k + T_{\rm CMB} x_{\rm tot}}, \tag{49}$$

where $x_{\rm tot} = x_\alpha + x_c$ is the sum of the radiative and collisional threshold parameters. These parameters are

$$x_\alpha = \frac{P_{10} T_\star}{A_{10} T_{\rm CMB}}, \tag{50}$$

and

$$x_c = \frac{4\kappa_{1-0}(T_k)\, n_H T_\star}{3 A_{10} T_{\rm CMB}}, \tag{51}$$

where P_{10} is the indirect de-excitation rate of the triplet $n = 1$ state via the Wouthuysen-Field process, related to the total scattering rate P_α of Lyman-α photons by $P_{10} = 4P_\alpha/27$. Also, the atomic coefficient $\kappa_{1-0}(T_k)$ is tabulated as a function of T_k. The coupling of the spin temperature to the gas temperature becomes substantial when $x_{\rm tot} > 1$; in particular, $x_\alpha = 1$ defines the thermalization rate of P_α:

$$P_{\rm th} \equiv \frac{27 A_{10} T_{\rm CMB}}{4 T_*} \simeq 7.6 \times 10^{-12} \left(\frac{1+z}{10}\right) \text{ s}^{-1}. \tag{52}$$

A patch of neutral hydrogen at the mean density and with a uniform T_S produces (after correcting for stimulated emission) an optical depth at a present-day (observed) wavelength of $21(1 + z)$ cm,

$$\tau(z) = 9.0 \times 10^{-3} \left(\frac{T_{\rm CMB}}{T_S}\right) \left(\frac{\Omega_b h}{0.03}\right) \left(\frac{\Omega_m}{0.3}\right)^{-1/2} \left(\frac{1+z}{10}\right)^{1/2}, \tag{53}$$

assuming a high redshift $z \gg 1$. The observed spectral intensity I_ν relative to the CMB at a frequency ν is measured by radio astronomers as an effective brightness temperature T_b of blackbody emission at this frequency, defined using the Rayleigh-Jeans limit of the Planck radiation formula: $I_\nu \equiv 2k_B T_b \nu^2/c^2$.

The brightness temperature through the IGM is $T_b = T_{\rm CMB}e^{-\tau} + T_S(1-e^{-\tau})$, so the observed differential antenna temperature of this region relative to the CMB is

$$T_b = (1+z)^{-1}(T_S - T_{\rm CMB})(1-e^{-\tau})$$
$$\simeq 28\,{\rm mK}\left(\frac{\Omega_b h}{0.033}\right)\left(\frac{\Omega_m}{0.27}\right)^{-1/2}\left(\frac{1+z}{10}\right)^{1/2}\left(\frac{T_S - T_{\rm CMB}}{T_S}\right), \quad (54)$$

where $\tau \ll 1$ is assumed and T_b has been redshifted to redshift zero. Note that the combination that appears in T_b is

$$\frac{T_S - T_{\rm CMB}}{T_S} = \frac{x_{\rm tot}}{1+x_{\rm tot}}\left(1 - \frac{T_{\rm CMB}}{T_k}\right). \quad (55)$$

In overdense regions, the observed T_b is proportional to the overdensity, and in partially ionized regions T_b is proportional to the neutral fraction. Also, if $T_S \gg T_{\rm CMB}$ then the IGM is observed in emission at a level that is independent of T_S. On the other hand, if $T_S \ll T_{\rm CMB}$ then the IGM is observed in absorption at a level that is enhanced by a factor of $T_{\rm CMB}/T_S$. As a result, a number of cosmic events are expected to leave observable signatures in the redshifted 21-cm line, as discussed below in further detail.

Figure 12 illustrates the mean IGM evolution for three examples in which reionization is completed at different redshifts, namely $z = 6.47$ (thin curves), $z = 9.76$ (medium curves), and $z = 11.76$ (thick curves). The top panel shows the global evolution of the CMB temperature $T_{\rm CMB}$ (dotted curve), the gas kinetic temperature T_k (dashed curve), and the spin temperature T_S (solid curve). The middle panel shows the evolution of the ionized gas fraction and the bottom panel presents the mean 21 cm brightness temperature, T_b.

3.2.2 A Handy Tool for Studying Cosmic Reionization

The prospect of studying reionization by mapping the distribution of atomic hydrogen across the universe using its prominent 21-cm spectral line has motivated several teams to design and construct arrays of low-frequency radio telescopes; the Low Frequency Array (http://www.lofar.org/), the Murchison Wide-Field Array (http://www.mwatelescope.org/), PAPER (http://arxiv.org/abs/0904.1181), GMRT (http://arxiv.org/abs/0807.1056), 21CMA (http://21cma.bao.ac.cn/), and ultimately the Square Kilometer Array (http://www.skatelescope.org) will search over the next decade for 21-cm emission or absorption from $z \sim 6.5$–15, redshifted and observed today at relatively low frequencies which correspond to wavelengths of 1.5–4 m.

The idea is to use the resonance associated with the hyperfine splitting in the ground state of hydrogen. While the CMB spectrum peaks at a wavelength of 2 mm, it provides a still-measurable intensity at meter wavelengths that can be used as the bright background source against which we can see the expected 1 % absorption

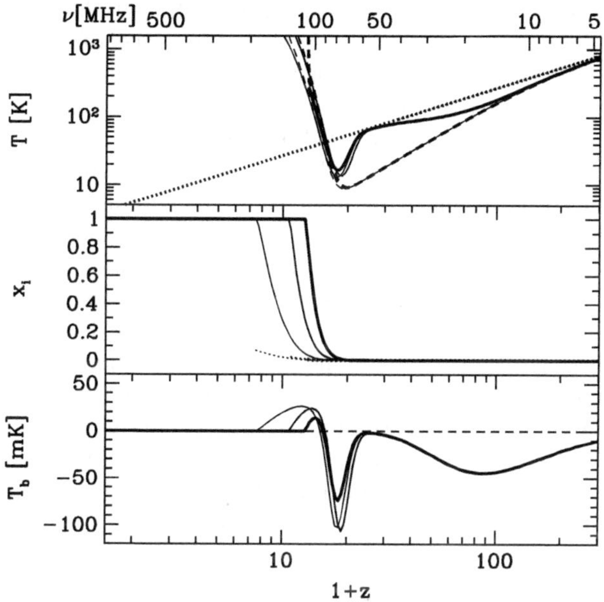

Fig. 12 *Top panel:* Evolution with redshift z of the CMB temperature T_{CMB} (*dotted curve*), the gas kinetic temperature T_k (*dashed curve*), and the spin temperature T_S (*solid curve*). *Middle panel:* Evolution of the gas fraction in ionized regions x_i (*solid curve*) and the ionized fraction outside these regions (due to diffuse X-rays) x_e (*dotted curve*). *Bottom panel:* Evolution of mean 21 cm brightness temperature T_b. The horizontal axis at the top provides the observed photon frequency at the different redshifts shown at the bottom. Each panel shows curves for three models in which reionization is completed at different redshifts, namely $z = 6.47$ (*thin curves*), $z = 9.76$ (*medium curves*), and $z = 11.76$ (*thick curves*). **Figure credit:** Pritchard and Loeb (2008)

by neutral hydrogen along the line of sight. The hydrogen gas produces 21-cm absorption if its spin temperature is colder than the CMB and excess emission if it is hotter. Since the CMB covers the entire sky, a complete three-dimensional map of neutral hydrogen can in principle be made from the sky position of each absorbing gas cloud together with its redshift z. Different observed wavelengths slice the Universe at different redshifts, and ionized regions are expected to appear as cavities in the hydrogen distribution, similar to holes in swiss cheese. Because the smallest angular size resolvable by a telescope is proportional to the observed wavelength, radio astronomy at wavelengths as large as a meter has remained relatively undeveloped. Producing resolved images even of large sources such as cosmological ionized bubbles requires telescopes which have a kilometer scale. It is much more cost-effective to use a large array of thousands of simple antennas distributed over several kilometers, and to use computers to cross-correlate the measurements of the individual antennas and combine them effectively into a single large telescope. The new experiments are being placed mostly in remote sites, because the cosmic wavelength region overlaps with more mundane terrestrial telecommunications.

In approaching redshifted 21-cm observations, although the first inkling might be to consider the mean emission signal in the bottom panel of Fig. 12, the signal is orders of magnitude fainter than foreground synchrotron emission from relativistic electrons in the magnetic field of our own Milky Way as well as other galaxies (see Fig. 13). Thus cosmologists have focused on the expected characteristic variations in T_b, both with position on the sky and especially with frequency, which signifies redshift for the cosmic signal. The synchrotron foreground is expected to have a smooth frequency spectrum, and so it is possible to isolate the cosmological signal by taking the difference in the sky brightness fluctuations at slightly different frequencies (as long as the frequency separation corresponds to the characteristic size of ionized bubbles). The 21-cm brightness temperature depends on the density of neutral hydrogen. As explained in the previous subsection, large-scale patterns in the reionization are driven by spatial variations in the abundance of galaxies; the 21-cm fluctuations reach $\sim$5 mK (root mean square) in brightness temperaturen on a scale of 10 comoving Mpc. While detailed maps will be difficult to extract due to the foreground emission, a statistical detection of these fluctuations is expected to be well within the capabilities of the first-generation experiments now being built. Current work suggests that the key information on the topology and timing of reionization can be extracted statistically.

While numerical simulations of reionization are now reaching the cosmological box sizes needed to predict the large-scale topology of the ionized bubbles, they do this at the price of limited small-scale resolution (see Fig. 14). These simulations cannot yet follow in any detail the formation of individual stars within galaxies, or the feedback that stars produce on the surrounding gas, such as photo-heating or the hydrodynamic and chemical impact of supernovae, which blow hot bubbles of gas enriched with the chemical products of stellar nucleosynthesis. Thus, the simulations cannot directly predict whether the stars that form during reionization are similar to the stars in the Milky Way and nearby galaxies or to the primordial $100 M_{\odot}$ stars. They also cannot determine whether feedback prevents low-mass dark matter halos from forming stars. Thus, models are needed that make it possible to vary all these astrophysical parameters of the ionizing sources and to study the effect on the 21-cm observations.

The theoretical expectations presented here for reionization and for the 21-cm signal are based on rather large extrapolations from observed galaxies to deduce the properties of much smaller galaxies that formed at an earlier cosmic epoch. Considerable surprises are thus possible, such as an early population of quasars or even unstable exotic particles that emitted ionizing radiation as they decayed. In any case, the forthcoming observational data in 21-cm cosmology should make the next few years a very exciting time.

At high redshifts prior to reionization, spatial perturbations in the thermodynamic gas properties are linear and can be predicted precisely (see Sect. 2.1). Thus, if the gas is probed with the 21-cm technique then it becomes a promising tool of fundamental, precision cosmology, able to probe the primordial power spectrum of density fluctuations imprinted in the very early universe, perhaps in an era of cosmic inflation. The 21-cm fluctuations can be measured down to the smallest scales

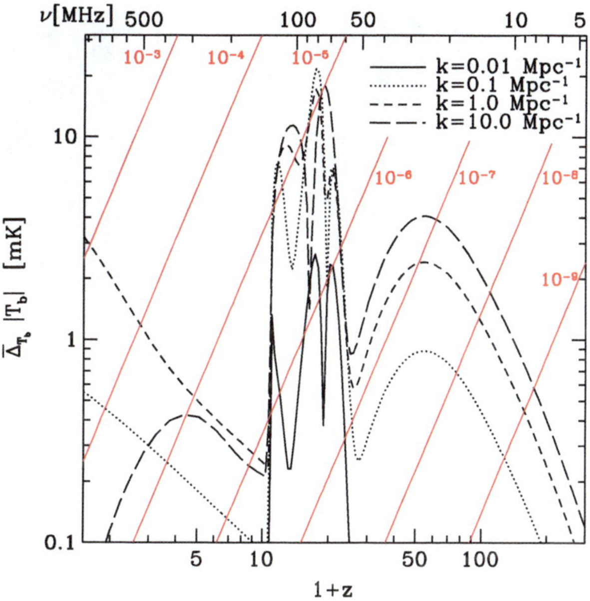

Fig. 13 Predicted redshift evolution of the angle-averaged amplitude of the 21-cm power spectrum ($|\bar{\Delta}_{T_b}| = [k^3 P_{21-\mathrm{cm}}(k)/2\pi^2]^{1/2}$) at comoving wavenumbers $k = 0.01$ (*solid curve*), 0.1 (*dotted curve*), 1.0 (*short dashed curve*), 10.0 (*long dashed curve*), and 100.0 Mpc^{-1} (*dot-dashed curve*). In the model shown, reionization is completed at $z = 9.76$. The horizontal axis at the top shows the observed photon frequency at the different redshifts. The diagonal *straight (red) lines* show various factors of suppression for the synchrotron Galactic foreground, necessary to reveal the 21-cm signal. **Figure credit:** Pritchard and Loeb (2008)

where the baryon pressure suppresses gas fluctuations, while the CMB anisotropies are damped on small scales (through the so-called Silk damping). This difference in damping scales can be seen by comparing the baryon-density and photon-temperature power spectra. Since the 21-cm technique is also three-dimensional (while the CMB yields a single sky map), there is a much large potential number of independent modes probed by the 21-cm signal: $N_{21-\mathrm{cm}} \sim 3 \times 10^{16}$ compared to $N_{\mathrm{cmb}} \sim 2 \times 10^7$. This larger number should provide a measure of non-Gaussian deviations to a level of $\sim N_{21\mathrm{cm}}^{-1/2}$, constituting a test of the inflationary origin of the primordial inhomogeneities which are expected to possess non-Gaussian deviations $>10^{-6}$.

The 21-cm fluctuations are expected to simply trace the primordial power-spectrum of matter density perturbations (which is shaped by the initial conditions from inflation and the dark matter) either before the first population of galaxies had formed (at redshifts $z > 25$) or after reionization ($z < 6$)—when only dense pockets of self-shielded hydrogen (such as damped Lyman-α systems) survive. During the epoch of reionization, the fluctuations are mainly shaped by the topology of ionized regions, and thus depend on uncertain astrophysical details involving star formation.

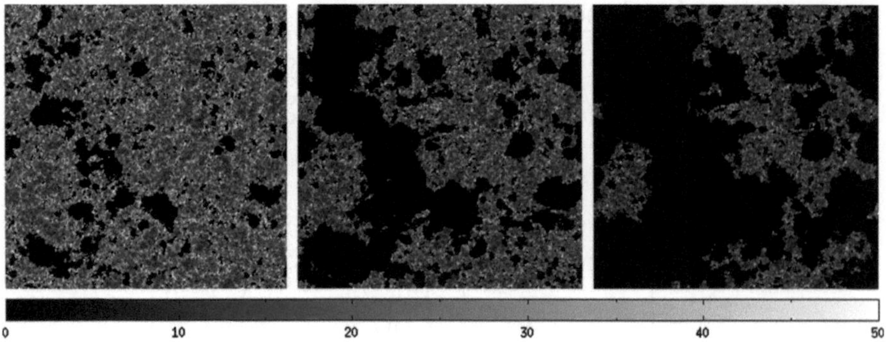

Fig. 14 Map of the fluctuations in the 21 cm brightness temperature on the sky, ΔT_b (mK), based on a numerical simulation which follows the dynamics of dark matter and gas in the IGM as well as the radiative transfer of ionizing photons from galaxies. The panels show the evolution of the signal in a slice of 140 comoving Mpc on a side, in three snapshots corresponding to the simulated volume being 25, 50, and 75 % ionized. Since neutral regions correspond to strong emission (i.e., a high T_b), the 21-cm maps illustrate the global progress of reionization and the substantial large-scale spatial fluctuations in the reionization history. **Figure credit:** Trac et al. (2009)

However, even during this epoch, the imprint of peculiar velocities (which are induced gravitationally by density fluctuations), can in principle be used to separate the implications for fundamental physics from the astrophysics.

Peculiar velocities imprint a particular form of anisotropy in the 21-cm fluctuations that is caused by gas motions along the line of sight. This anisotropy, expected in any measurement of density that is based on a spectral resonance or on redshift measurements, results from velocity compression. Consider a photon traveling along the line of sight that resonates with absorbing atoms at a particular point. In a uniform, expanding universe, the absorption optical depth encountered by this photon probes only a narrow strip of atoms, since the expansion of the universe makes all other atoms move with a relative velocity that takes them outside the narrow frequency width of the resonance line. If there is a density peak, however, near the resonating position, the increased gravity will reduce the expansion velocities around this point and bring more gas into the resonating velocity width. This effect is sensitive only to the line-of-sight component of the velocity gradient of the gas, and thus causes an observed anisotropy in the power spectrum even when all physical causes of the fluctuations are statistically isotropic. This anisotropy is particularly important in the case of 21-cm fluctuations. When all fluctuations are linear, the 21-cm power spectrum takes the form

$$P_{21-\mathrm{cm}}(\mathbf{k}) = \mu^4 P_\rho(k) + 2\mu^2 P_{\rho-\mathrm{iso}}(k) + P_{\mathrm{iso}}, \tag{56}$$

where $\mu = \cos\theta$ in terms of the angle θ between the wave-vector $\mathbf{k}$ of a given Fourier mode and the line of sight, P_{iso} is the isotropic power spectrum that would result from all sources of 21-cm fluctuations without velocity compression, $P_\rho(k)$

is the 21-cm power spectrum from gas density fluctuations alone, and $P_{\rho-\mathrm{iso}}(k)$ is the Fourier transform of the cross-correlation between the density and all sources of 21-cm fluctuations. The three power spectra can also be denoted $P_{\mu^4}(k)$, $P_{\mu^2}(k)$, and $P_{\mu^0}(k)$, according to the power of μ that multiplies each term. At these redshifts, the 21-cm fluctuations probe the infall of the baryons into the dark matter potential wells. The power spectrum shows remnants of the photon-baryon acoustic oscillations on large scales, and of the baryon pressure suppression on small scales.

Once stellar radiation becomes significant, many processes can contribute to the 21-cm fluctuations. The contributions include fluctuations in gas density, temperature, ionized fraction, and Lyα flux. These processes can be divided into two broad categories: The first, related to *"physics"*, consists of probes of fundamental, precision cosmology, and the second, related to *"astrophysics"*, consists of probes of stars. Both categories are interesting—the first for precision measures of cosmological parameters and studies of processes in the early universe, and the second for studies of the properties of the first galaxies. However, the astrophysics depends on complex non-linear processes (collapse of dark matter halos, star formation, supernova feedback), and must be cleanly separated from the physics contribution, in order to allow precision measurements of the latter. As long as all the fluctuations are linear, the anisotropy noted above allows precisely this separation of the *fundamental physics* from the *astrophysics* of the 21-cm fluctuations. In particular, the $P_{\mu^4}(k)$ is independent of the effects of stellar radiation, and is a clean probe of the gas density fluctuations. Once non-linear terms become important, there arises a significant mixing of the different terms; in particular, this occurs on the scale of the ionizing bubbles during reionization.

The 21-cm fluctuations are affected by fluctuations in the Lyman-α flux from stars, a result that yields an indirect method to detect and study the early population of galaxies at $z \sim 20$. The fluctuations are caused by biased inhomogeneities in the density of galaxies, along with Poisson fluctuations in the number of galaxies. Observing the power-spectra of these two sources would probe the number density of the earliest galaxies and the typical mass of their host dark matter halos. Furthermore, the enhanced amplitude of the 21-cm fluctuations from the era of Lyα coupling improves considerably the practical prospects for their detection. Precise predictions account for the detailed properties of all possible cascades of a hydrogen atom after it absorbs a photon. Around the same time, X-rays may also start to heat the cosmic gas, producing strong 21-cm fluctuations due to fluctuations in the X-ray flux.

In difference from interferometric arrays, single dipole experiments which integrate over most of the sky, can search for the global (spectral) 21-cm signal shown in Fig. 12. Examples of such experiments are CoRE or EDGES (http://www.haystack.mit.edu/ast/arrays/Edges/). Rapid reionization histories which span a redshift range $\Delta z < 1$ can already be ruled out, and the current constraints might be improved provided that local foregrounds (see Fig. 15) can be well modelled by low-order polynomials in frequency (Fig. 16). Observations in the frequency range 50–100 MHz can potentially constrain the Lyman-α and X-ray emissivity of the first stars forming at redshifts $z \sim 15$–25, as illustrated in Fig. 17.

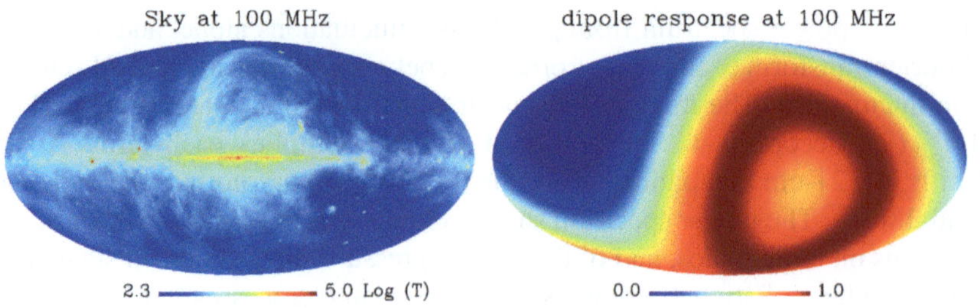

Fig. 15 *Left panel:* Radio map of the sky at 100 MHz. *Right panel:* Ideal dipole response averaged over 24 h. **Figure credits:** Pritchard and Loeb (2010a); de Oliveira-Costa et al. (2008)

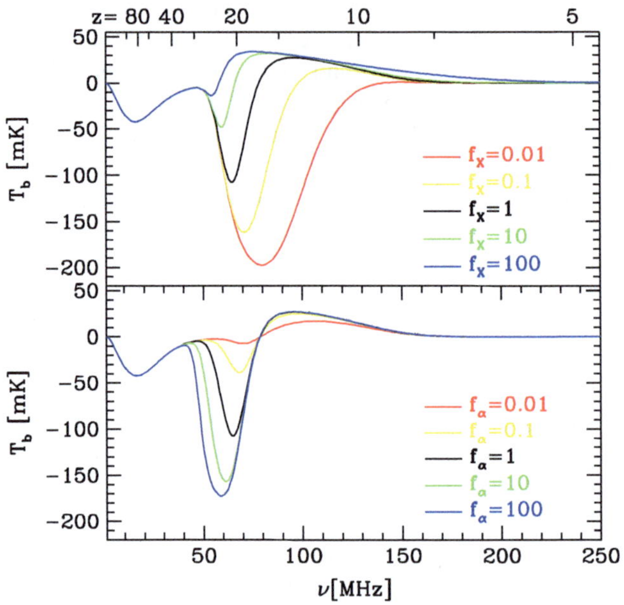

Fig. 16 Dependence of global 21-cm signal on the X-ray (*top panel*) and Lyman-α (*bottom panel*) emissivity of stars. Each case depicts examples with the characteristic emissivity reduced or increased by a factor of up to 100. **Figure credit:** Pritchard and Loeb (2010a)

4 Epilogue

The initial conditions of our Universe can be summarized on a single sheet of paper. Yet the Universe is full of complex structures today, such as stars, galaxies and groups of galaxies. This chapter discussed the standard theoretical model for how complexity emerged from the simple initial state of the Universe through the action of gravity. In order to test and inform the related theoretical calculations, large-aperture telescopes and arrays of radio antennae are currently being designed and constructed.

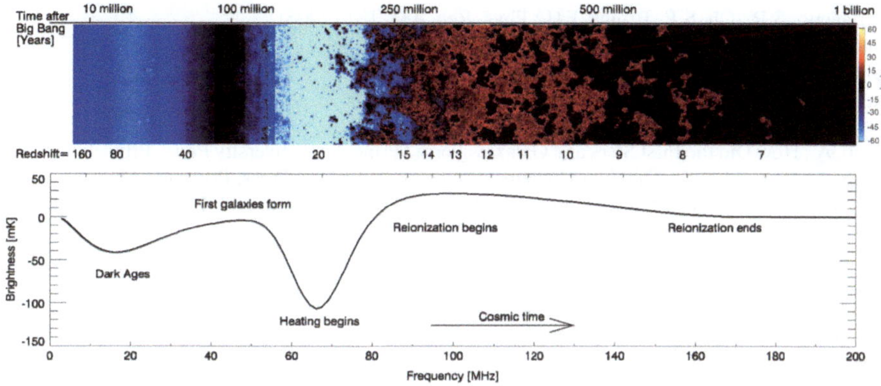

Fig. 17 Schematic illustration of the correspondence between the evolution of structure in intergalactic hydrogen (*upper panel*) and the sky-averaged spectrum (global 21-cm signal in the *lower panel*). Here *blue* denotes absorption and *red* denotes emission of 21-cm radiation relative to the cosmic microwave background. **Figure credit:** Pritchard and Loeb (2010b)

The actual transition from simplicity to complexity has not been observed as of yet. The simple initial conditions were already traced in maps of the microwave background radiation, but the challenge of detecting the first generation of galaxies defines one of the exciting frontiers in the future of cosmology. Once at hand, the missing images of the infant Universe might potentially surprise us and revise our current ideas.

For a comprehensive exposition of the topics covered here, we refer the reader to Loeb (2010) and Loeb and Furlanetto (2012).

Acknowledgements I thank my collaborators on the topics covered by this chapter: Dan Babich, Rennan Barkana, Volker Bromm, Steve Furlanetto, Zoltan Haiman, Joey Munoz, Jonathan Pritchard, Hy Trac, Stuart Wyithe, and Matias Zaldarriaga.

References

Baker, J., et al.: Phys. Rev. D **75**, 4024 (2007)
Barkana, R., Loeb, A.: ApJ **531**, 613 (2000)
Barkana, R., Loeb, A.: Phys. Rep. **349**, 125 (2001)
Barkana, R., Loeb, A.: ApJ **601**, 64 (2004)
Barkana, R., Loeb, A.: Rep. Prog. Phys. **70**, 627 (2007)
Bloom, J.S.: What are Gamma-ray Bursts? Princeton University Press, Princeton (2011)
Bromm, V., Larson, R.B.: ARA&A **42**, 79 (2004)
Bromm, V., Loeb, A.: ApJ **642**, 382 (2006)
Bromm, V., Yoshida, N., Hernquist, L., McKee, C.F.: Nature **459**, 49 (2009)
Bromm, V., Yoshida, N.: ARA&A **49**, 373 (2011)
de Oliveira-Costa, A., et al.: MNRAS **388**, 247 (2008)
Fan, X., et al.: AJ **125**, 1649 (2005)

Furlanetto, S.R., Oh, S.P., Briggs, F.H.: Phys. Rep. **433**, 181 (2006)
Gardner, J.P., et al.: Space Sci. Rev. **123**, 485 (2006)
Glover, S.C.O.: Space Sci. Rev. **117**, 445 (2005)
Komatsu, E., et al.: ApJS **192**, 18 (2011)
Lamb, D.Q., Reichart, D.E.: ApJ **536**, 1 (2000)
Loeb, A.: How Did the First Stars and Galaxies Form? Princeton University Press, Princeton (2010)
Loeb, A., Furlanetto, S.: The First Galaxies. Princeton University Press, Princeton (2012)
Mo, H., Van den Bosch, F., White, S.D.M.: Galaxy Formation and Evolution. Cambridge University Press, Cambridge (2010)
Oh, S.P., Haiman, Z.: ApJ **569**, 558 (2002)
Pritchard, J.R., Loeb, A.: Phys. Rev. D **78**, 3511 (2008)
Pritchard, J.R., Loeb, A.: Phys. Rev. D **82**, 23006 (2010a)
Pritchard, J.R., Loeb, A.: Nature **468**, 772 (2010b)
Pritchard, J.R., Loeb, A., Wyithe, J.S.B.: MNRAS **408**, 57 (2010)
Robertson, B.E., et al.: Nature **468**, 49 (2010)
Sheth, R.K., Tormen, G.: MNRAS **329**, 61 (2002)
Trac, H., Cen, R., Loeb, A.: ApJ **689**, L81 (2009)
Wyithe, J.S.B., et al.: Nature **469**, 7329 (2011)

The Epoch of Reionization

Saleem Zaroubi

Abstract The Universe's *dark ages* end with the formation of the first generation of galaxies. These objects start emitting ultraviolet radiation that carves out ionized regions around them. After a sufficient number of ionizing sources have formed, the ionized fraction of the gas in the Universe rapidly increases until hydrogen becomes fully ionized. This period, during which the cosmic gas went from neutral to ionized, is known as the Universe's Epoch of Reionization . The Epoch of Reionization is related to many fundamental questions in cosmology, such as properties of the first galaxies, physics of (mini-)quasars, formation of very metal-poor stars and a slew of other important research topics in astrophysics. Hence uncovering it will have far reaching implications on the study of structure formation in the early Universe. This chapter reviews the current observational evidence for the occurrence of this epoch, its key theoretical aspects and main characteristics, and finally the various observational probes that promise to uncover it. A special emphasis is put on the redshifted 21 cm probe, the various experiments that are currently being either built or designed, and what we can learn from them about the Epoch of Reionization.

1 Introduction

The formation of the first galaxies marks a major transition in the evolution of structure in the Universe. These same galaxies with their zero metallicity Population III stars and black hole driven sources (e.g., mini-quasars, x-ray binaries, etc.) transformed the intergalactic medium from neutral to ionized. This process, known as the Epoch of Reionization (EoR), is the central topic discussed in this chapter.

S. Zaroubi (✉)
Kapteyn Astronomical Institute, Landleven 12, 9747AD Groningen, The Netherlands

Physics Department, The Technion, Haifa 32000, Israel
e-mail: saleem@astro.rug.nl

T. Wiklind et al. (eds.), *The First Galaxies*, Astrophysics and Space Science Library 396,
DOI 10.1007/978-3-642-32362-1_2, © Springer-Verlag Berlin Heidelberg 2013

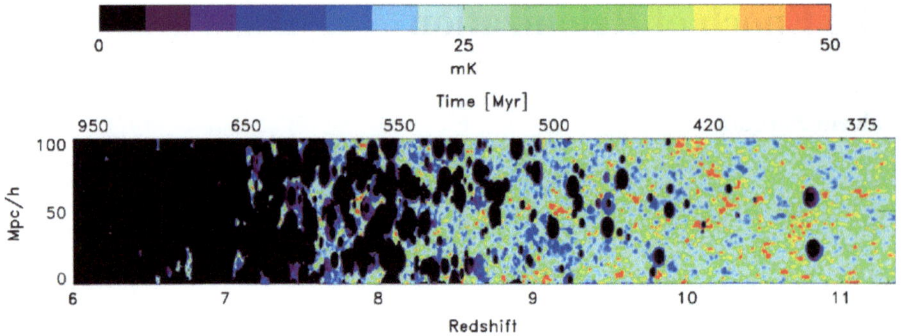

Fig. 1 This figure shows a slice through redshift of the 21 cm radiation in which the reionization process progresses through the volume of a cosmological simulation with radiative transfer (Thomas et al. 2009)

As mentioned in chapter #1 [by A. Loeb in this book], about 400,000 years after the Big Bang, the Universe's density decreased enough so that the temperature fell below 3,000 K, allowing ions and electrons to (re)combine into neutral hydrogen and helium—the fraction of heavier elements was negligible. Immediately afterwards, photons decoupled from baryons and the Universe became transparent, leaving a relic signature known as the cosmic microwave background (CMB) radiation. This event ushered the Universe into a period of darkness, known as the *dark ages*.

The *dark ages* ended about 400 million years later, when the first galaxies formed and start emitting ionizing radiation. Initially during the EoR, the intergalactic medium (IGM) is neutral except in regions surrounding the first objects. However, as this reionization progresses, an evolving patchwork of neutral (H I) and ionized hydrogen (H II) regions unfolds. After a sufficient number of UV-radiation emitting objects formed, the temperature and the ionized fraction of the gas in the Universe increase rapidly until eventually the ionized regions permeate to fill the whole Universe (Barkana and Loeb 2001; Loeb and Barkana 2001; Bromm and Larson 2004; Ciardi and Ferrara 2005; Choudhury and Ferrara 2006; Furlanetto et al. 2006; Morales and Wyithe 2010).

The current constraints strongly suggest that the EoR roughly occurs within the redshift range of $z \sim [6 - 15]$. Figure 1 shows a space-redshift slice of a simulation of the progression of reionization with time and how it appears in 21 cm brightness temperature, which is proportional to the density of neutral hydrogen (see Sect. 4). At high redshifts most of the gas is neutral, hence, the signal is mostly sensitive to cosmological density fluctuations, whereas at lower redshifts ionization bubbles start to appear until they fill the whole Universe (Barkana and Loeb 2001).

The EoR is a watershed epoch in the history of the Universe. Prior to it, the formation and evolution of structure was dominated by dark matter alone, while baryonic matter played a marginal role. The EoR marks the transition to an era in which the role of cosmic gas in the formation and evolution of structure became prominent and, on small scales, even dominant.

The details of the reionization scenario I have laid out are yet to be clarified. For example, it is not known what controls the formation of the first objects and how much ionizing radiation they produce, or how the ionization bubbles expand into the intergalactic medium and what they ionize first, high-density or low-density regions?. The answer to these questions and many others that arise in the context of studying the EoR needs knowledge of fundamental issues in cosmology, galaxy formation, quasars and the physics of very metal poor stars; all including foremost research in topics in modern astrophysics. Substantial theoretical and observational efforts are currently dedicated to understanding the physical processes that trigger this epoch and govern its evolution, and ramifications on subsequent structure formation (c.f., Barkana and Loeb 2001; Bromm and Larson 2004; Ciardi and Ferrara 2005; Choudhury and Ferrara 2006; Furlanetto et al. 2006). However, despite the pivotal role played by the EoR in cosmic history, observational support for the proposed scenarios is very scarce, and when available, is indirect and model dependent.

In principle, there are many different ways to observationally probe the EoR. In this contribution, I mainly focus on the redshifted 21cm emission line from neutral hydrogen at high redshifts. This is one of the most promising techniques for studying the *dark ages* and the EoR. To date, there are a number of telescopes dedicated to measure this faint radiation. In the short term, these consist of: The Low Frequency Array (LOFAR), the Murchison Widefield Array (MWA), Precision Array to Probe Epoch of Reionization (PAPER) and Giant Metrewave Radio Telescope (GMRT), while, on a somewhat longer time scales the Square Kilometer Array (SKA). One of the most challenging tasks in studying the EoR is to extract and identify the cosmological signal from the data and interpret it correctly. This is because the detectable signal in the frequency range relevant to the EoR is composed of a number of components—the cosmological EoR signal, extragalactic and Galactic foreground, ionospheric distortions, instrumental response and noise—each with its own physical origin and statistical properties.

Figure 2 shows a sketch of the likely evolution of reionization from the neutral hydrogen point of view. The figure emphasizes the other non-cosmological effects that are seen with the 21 cm experiments, e.g., foreground, ionosphere and instrumental effects. The radio antennas seen at the bottom are LOFAR's Low Band Antennas.

In this chapter I discuss various observational and theoretical aspects of the Epoch of Reionization. In Sect. 2 the current observational scene is reviewed, specifically focusing on the CMB data and the Lyman α forest spectra. In Sects. 3 and 4, we discuss, respectively, the physics of the reionization process and the 21 cm line transition and how it could be used to probe reionization. The redshifted 21 cm experiments their potentials and the challenges are discussed in Sect. 5. Extraction and quantification of the information stored in the redshifted 21 cm data using various statistics is discussed in Sect. 6. This chapter concludes with a brief summary (Sect. 7).

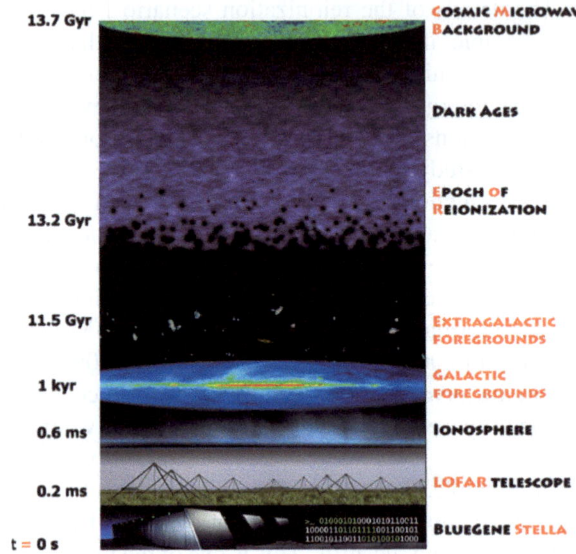

Fig. 2 This figure shows a sketch of the likely development of the EoR. About 500,000 years after the Big Bang ($z \sim 1,000$) hydrogen recombined and remained neutral for a few hundred million years during the *dark ages*. At a redshift, $z \sim 15$, the first stars, galaxies and quasars began to form, heating and reionizing the hydrogen gas. The neutral IGM can be observed with LOFAR up to $z \approx 11.5$ through its redshifted 21cm spin-flip transition. However, many atmospheric, galactic and extra-galactic emission contaminate the 21 cm signal

2 Observational Evidence for Reionization

To date, the majority of observations related to the EoR provide weak and model dependent constraints on reionization. However, there are currently a number of observations which could impose strong constraints on reionization models, as discussed below. It should be noted however that none of these observations constrains the EoR evolution in detail.

2.1 *The Lyman α Forest at $z \approx 2.5 - 6.5$*

The state of the intergalactic medium (IGM) can be studied through the analysis of the Lyman-α forest. This is an absorption phenomenon seen in the spectra of background quasi-stellar objects (QSOs). The history of this field goes back to 1965 when a number of authors (Gunn and Peterson 1965; Scheuer 1965) predicted that an expanding Universe, homogeneously filled with gas, will produce an absorption trough due to neutral hydrogen, known as the Gunn–Peterson trough, in the spectra of distant QSOs bluewards of the Lyman-α emission line of the quasar. That is, the quasar flux will be absorbed at the UV resonance line frequency of 1,215.67 Å. Gunn and Peterson (1965) found such a spectral region of reduced flux, and used this measurement to put upper limits on the amount of intergalactic neutral hydrogen. The large cross-section for the Lyman α absorption makes this technique very powerful for studying gas in the intergalactic medium.

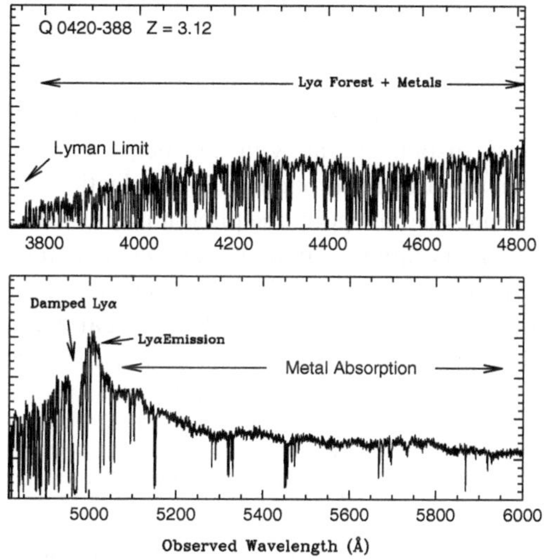

Fig. 3 High resolution spectrum of the $z = 3.12$ quasar Q0420-388 obtained with the Las Campanas echelle spectrograph by J. Bechtold and S. A. Shectman. The two panels cover the whole wavelength range of the spectrum. The Lyman α forest is clearly indicated in the upper panel of the figure, bluewards of the quasar rest frame Lyman α emission feature. Remember, at the rest frame this feature should have a wavelength of 1,215.67 Å but since it is redshifted by a factor $1+z$ it appears at a wavelength of about 5,000 Å. The figure is courtesy of Jill Bechtold and appeared in Bechtold (2003)

In the last 15 years two major advances occurred. The first was the development of high-resolution echelle spectrographs on large telescopes (e.g., HIRES on the Keck and UVES on the Very Large Telescope) that provided data of unprecedented quality. The second was the emergence of a theoretical paradigm within the context of cold dark matter (CDM) cosmology that accounts for all the features seen in these systems (e.g. Bi et al. 1992; Cen et al. 1994; Hernquist et al. 1996; Machacek et al. 2000; Miralda-Escudé et al. 1996; Theuns et al. 1998; Zhang et al. 1995, 1997). According to this paradigm, the absorption is produced by volume filling photoionized gas that contains most of the baryons at redshifts at $z \sim 3-6$ and resides in mildly non-linear overdensities.

Figure 3 shows a typical example of the Lyman α forest seen in the spectrum of the $z = 3.12$ quasar Q0420-388. An interesting feature of such spectra is the density of weak absorbing lines which increase with redshift due to the expansion of the Universe. In fact, at redshifts above 4, the density of the absorption features become so high that it is hard to define them as separate absorption features. Instead, one sees only the flux in between the absorption minima which appears as if they are emission rather than absorption lines.

The Lyman α forest has turned out to be a treasure trove for studying the intergalactic medium and its properties in both low and high density regions. In particular, it is very sensitive to the neutral hydrogen column density and hence, to the neutral fraction as a function of redshift along the line of sight. In the following, we demonstrate how one could constrain the neutral fraction of hydrogen from the forest and what the values obtained from the data are. For a review on the Lyman α forest the reader is referred to Rauch (1998).

We need to calculate the optical depth for absorption of Lyman α photons. A photon emitted by a distant quasar with an energy higher than 10.196 eV is continuously redshifted as it travels through the intergalactic medium until it reaches the observer. At some intermediate point the photon is redshifted to around 1,216 Å in the rest-frame of the intervening medium, which may contain neutral hydrogen. It can then excite the Lyman α transition and be absorbed. Let us consider a particular line of sight from the observer to the quasar. The optical depth τ_α of a photon is related to the probability of the photon's transmission $e^{-\tau_\alpha}$. At a given observed frequency, ν_0, the Lyman α optical depth is given by

$$\tau_\alpha(\nu_0) = \int_O^Q n_{\rm HI}\sigma_\alpha dl/(1+z), \tag{1}$$

where l is the comoving radial coordinate of some intermediate point along the line of sight, z is the redshift and $n_{\rm HI}$ is the proper number density of neutral hydrogen at that point. The limits of the integration, O and Q, are the comoving distance between the observer and the quasar, respectively. The Lyman α absorption cross section is denoted by σ_α. It is a function of the frequency of the photon, ν, with respect to the rest-frame of the intervening H I at position l. The cross section is peaked when ν is equal to the Lyman α frequency ν_α. The frequency ν is related to the observed frequency ν_0 by $\nu = \nu_0(1+z)$, where $1+z$ is the redshift factor due to the uniform Hubble expansion alone at the same position. Note that for the sake of simplicity here we ignore peculiar velocity effects.

Using $dl = cdt/a$, where a is the Hubble scale factor and t is the proper time and Friedmann equation for a flat Universe with cosmological constant, we have,

$$\tau_\alpha = \int \sigma_\alpha(\nu) n_{\rm HI} \frac{cH_0^{-1}dz}{(1+z)\sqrt{\Omega_m(1+z)^3+\Omega_\Lambda}}. \tag{2}$$

This optical depth should also depend on the Lyman α line profile function but here we assume that it is basically a δ-function centered at the frequency ν. Considering $n_{\rm HI} = n_H x_{\rm HI}$, where $x_{\rm HI}$ is the neutral fraction of hydrogen, and integrating over this equation, one obtains the following result:

$$\frac{n_{\rm HI}}{n_H} = x_{\rm HI} \approx 10^{-4}\Omega_m^{1/2} h(1+z)^{\frac{3}{2}}\tau_\alpha. \tag{3}$$

Since the Lyman α features mostly show mild absorption probability ($\tau_\alpha \lesssim 1$) this equation clearly implies that at the mean density of the Universe at τ_α of about one the ionized fraction is on the order of 10^{-4}. Therefore, the fact that we observe the Lyman α forest at all means that the Universe is highly ionized at least until $z \approx 6$. This is the most reliable and robust evidence that the Universe has in fact reionized.

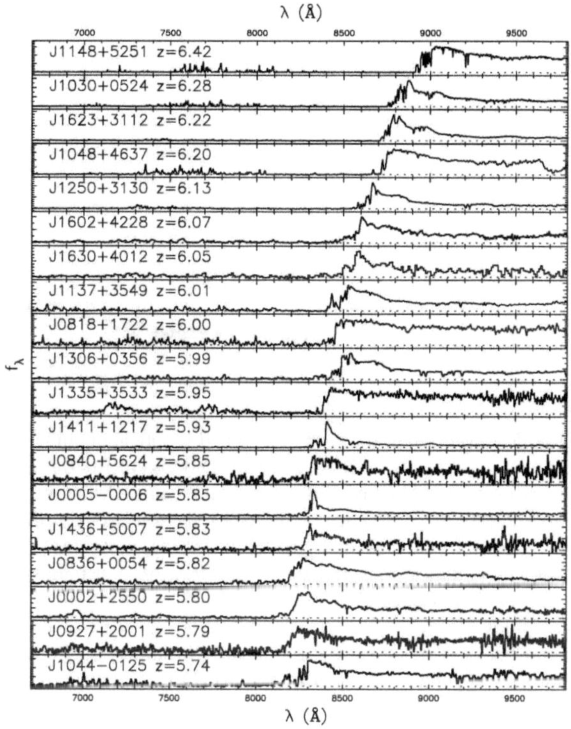

Fig. 4 Spectra for high redshift SDSS quasars. The Gunn–Peterson trough bluewards of the QSO Lyman α emission that is clearly apparent in the highest redshift ones indicates that the Universe has become somewhat more neutral at these redshifts. A similar behavior is also seen bluewards of the QSO Lyman β region of the same spectra. The actual amount of increase in neutral hydrogen implied by these spectra is not clear (Fan et al. 2006)

Another important evidence relevant for reionization comes from high resolution spectroscopy of high redshift Sloan Digital Sky Survey (SDSS) quasars (Fan et al. 2003, 2006). The SDSS has discovered about 19 QSOs with redshifts around 6 that are powered by black holes with masses on the order of $10^9 M_\odot$. In a follow up observations with 10 meter class telescopes (Fan et al. 2003, 2006) were able to obtain high resolution spectra of these objects.

Figure 4 shows the spectra of these high redshift quasars (Fan et al. 2003, 2006). Notice the complete absence of structure that some of these spectra exhibit bluewards of the quasar Lyman α restframe emission, especially those with redshift $z \gtrsim 6$. This is normally attributed to an increase in τ_α as a result of the decrease in the ionized fraction of the Universe. Notice also, that although the trend with redshift is clear, it is by no means monotonic. For example, quasar $J1411 + 3533$ at $z = 5.93$ shows an "emptier" trough relative to quasars $J0818 + 1722$ at $z = 6$. Such trend might be indicating a more patchy ionization of the IGM at such redshifts.

Figure 5 shows the effective Lyman α or Gunn–Peterson optical depth, τ_{GP}^{eff}, as a function of redshift as estimated from the joint optical depths of Lyman α, β and γ. From this plot it is clear that the increase in the optical depth as a function of redshift is much larger than expected (shown in the dashed line) from passive redshift evolution of the density of the Universe.

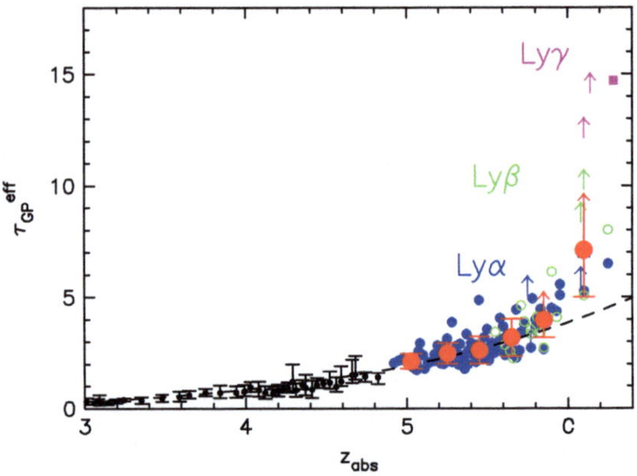

Fig. 5 Evolution of the Lyman α, β and γ optical depth from the high redshift Sloan quasars. The Lyman β and Lyman γ restframe wavelengths are 1,026 Å and 972.5 Å, respectively. The Lyman β measurements are converted to Lyman α Gunn–Peterson optical depth using a conversion factor that reflects the difference in the cross section between the two transitions, which is a factor of 5.27 lower in the case of Lyman β (see Lidz et al. 2002; Cen and McDonald 2002). The values in the two highest redshift bins are lower limits, since they both contain complete Gunn–Peterson troughs. The *dashed line* shows a redshift evolution of $\tau_\alpha \approx (1+z)^{4.3}$. At $z > 5.5$, the best-fit evolution has $\tau_\alpha \approx (1+z)^{>10.9}$, indicating an accelerated evolution. The *large filled symbols* with error bars are the average and standard deviation of the optical depth at each redshift. The sample variance also increases rapidly with redshift. Figure taken from Fan et al. (2006)

The interpretation of the increase in the optical depth at $z \gtrsim 6.3$ has been the subject of some debate. All authors agree that this is a sign of an increase in the Universe's neutral fraction at high redshifts, marking the tail end of the reionization process. The controversy is centered on the question of by how much the neutral fraction increases. Some authors (Wyithe and Loeb 2004a,b; Mesinger and Haiman 2004) have argued that the size of the so call Near Zone ionized by the quasar itself and set redwards of the Gunn–Peterson trough indicates that the neutral fraction around the SDSS high redshift quasars is $\approx$10 %. More recently is had been suggested that the variations seen across various SDSS quasars indicate that the ionization state of the IGM at these redshifts changes significantly across different sightlines (Mesinger and Furlanetto 2009). However, given the intense radiation field around these quasars, it is not possible to put general constraints on the neutral fraction of the IGM from quasars at redshift below 6.5 (see e.g., Bolton and Haehnelt 2007; Wyithe et al. 2008; Maselli et al. 2009, 2007). Moreover, recently and with the discovery of the redshift $z = 7.1$ QSO ULAS J1120 + 0641 Mortlock et al. (2011) by the UKIDSS survey Lawrence et al. (2007) it has been argued that this quasar's Near Zone gives a clear evidence for an increase in the neutral fraction of hydrogen in the IGM at $z = 7.1$ (Mortlock et al. 2011; Bolton et al. 2011). Note however that his conclusion relies on one quasar and might change as more of such quasars at $z \gtrsim 7$ are discovered.

There are more things that we can learn about reionization from the Lyman α forest that we will discuss later. But to summarize, the main conclusion from the

Lyman α optical depth measurements is that the Universe is highly ionized at redshifts below 6 (as seen in Fig. 4), while at about z = 6.3, the its neutral fraction increases, forming the tail end of the reionization process (see Fig. 5) .

2.2 The Thomson Scattering Optical Depth for the Cosmic Microwave Background (CMB) Radiation

This is a very evolved topic, discussed and reviewed by many authors (e.g., Peebles and Yu 1970; Sunyaev and Zeldovich 1972; Bond and Efstathiou 1984; Ma and Bertschinger 1995; Hu and White 1997; Aghanim et al. 2008). Here, I give a general review of the constraints provided by the CMB on reionization. The CMB provides important information relevant to the history of reionization. It is known that the Universe has indeed recombined and became largely neutral at $z \approx 1,100$. If recombination had been absent or substantially incomplete, the resulting high density of free electrons would imply that photons could not escape Thomson scattering until the density of the Universe dropped much further. This scattering would inevitably destroy the correlations at subhorizon angular scales seen in the CMB data (see e.g., Hu 1995; Sugiyama 1995).

In order to calculate the effect of reionization on CMB photons, a function is often defined called the visibility function,[1]

$$g(\eta) = -\dot{\tau} e^{-\tau(\eta)}, \tag{4}$$

where $\eta \left(\equiv \int dt/a\right)$ is the conformal time, a is the scale factor of the Universe and $\dot{\tau}$ is the derivative of the optical depth with respect to η. The optical depth for Thomson scattering is given by $\tau(\eta) = -\int_{\eta}^{\eta_0} d\eta \dot{\tau} = \int_{\eta}^{\eta_0} d\eta a(\eta) n_e \sigma_T$, where η_0 is the present time, n_e is the electron density and σ_T is the Thomson cross section. The visibility function gives the probability density that a photon had scattered out of the line of sight between η and $\eta + \mathrm{d}\eta$. The influence of reionization on the CMB temperature fluctuations is obtained by integrating Eq. (4) along each sightline to estimate the temperature fluctuation suppression due to the EoR. The suppression probability turns out to be roughly proportional to $1 - e^{-\tau}$ (Zaldarriaga 1997). Since the amount of suppression in the measured power spectrum is small, the optical depth for Thomson scattering must be small too (Page et al. 2007). The left hand panel in Fig. 6 shows the influence of increasing the value of τ, the Thomson optical depth, on the CMB temperature fluctuation power spectrum. The right hand panel shows the reionization history of the Universe assumed in the left panel. Since in this case a sudden global reionization is assumed, there is one to one correspondence between the optical depth for Thomson scattering and the redshift of reionization.

[1]Notice that this is a different "visibility" than the one used in radio interferometry which we discuss in Sect. 5.

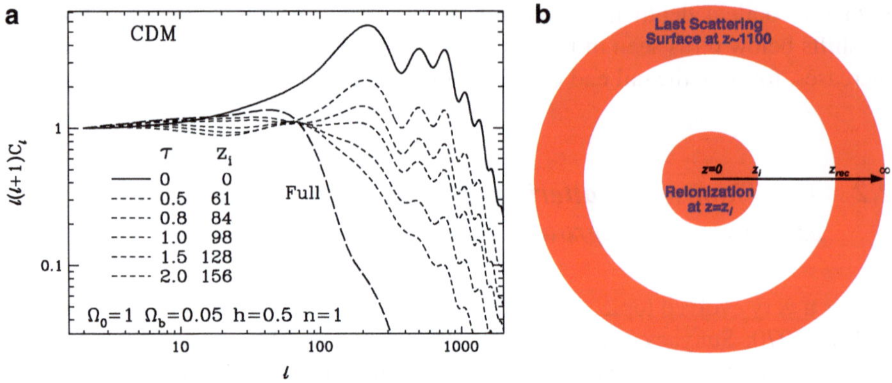

Fig. 6 Left hand panel (**a**): The influence of reionization on the CMB temperature angular power spectrum. Reionization damps anisotropy power as $e^{-2\tau}$ under the horizon (diffusion length) at last scattering. The models here are fully ionized $x_e = 1.0$ out to a reionization redshift z_i. Notice that with high optical depth, fluctuations at intermediate scales are regenerated as the fully ionized (*long-dashed*) model shows. This figure is taken from Wayne Hu's PhD thesis Hu (1995). Right panel (**b**) shows the assumed reionization history used. It is obvious that since we are considering a uniform and sudden reionization model, a change in the reionization redshift, z_i, will translate uniquely to an optical depth for Thomson scattering

Further information can be obtained from observations of CMB via the polarization power spectrum. The polarization of the CMB emerges naturally from the Cold Dark Matter paradigm which stipulates that small fluctuations in the early universe grow, through gravitational instability, into the large scale structure we see today (Bond and Efstathiou 1984; Hu and White 1997; Kamionkowski et al. 1997; Zaldarriaga and Seljak 1997). Since, the temperature anisotropies observed in the CMB are the result of primordial fluctuations, they would naturally polarize the CMB anisotropies. The degree of linear polarization of the CMB photons at any scale reflects the quadrupole anisotropy in the plasma when they last scattered at that same scales. From this argument it is clear that the amount of polarization at scales larger than the horizon scale at the last scattering surface should fall down since there is no more coherent quadrupole contribution due to the lack of causality. This is shown in the sketch presented in the left hand panel in Fig. 7. The largest scale at which a primordial quadrupole exists is the scale of the horizon at recombination, which roughly corresponds to 1°. Therefore, any polarization signature on scales larger than the horizon scale provides a clear evidence for Thomson scattering at later stages where the horizon scale is equivalent to the scale on which polarization has been detected.

Furthermore, the polarized fraction of the temperature anisotropy must be small, normally one order of magnitude smaller than the anisotropy in the temperature. This is simply because these photons mush have passed through an optically thin plasma, otherwise they would not have reached us but they would have scattered and destroyed the sub-horizon (i.e., below 1°) correlation in the CMB, contrary to what we observe (see e.g., Sugiyama 1995).

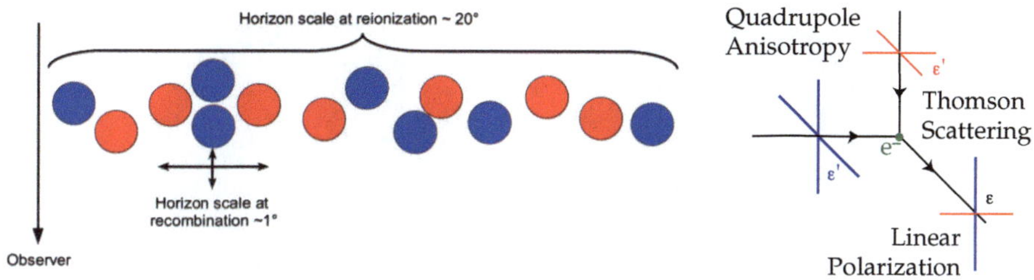

Fig. 7 *Left hand panel*: A sketch that shows why the CMB polarization is sensitive to the quadrupole momentum of temperature fluctuations. *Right hand panel*: Thomson scattering of radiation with quadrupole anisotropy generates linear polarization. The *blue and red lines* represent cold and hot radiation

The dependence Thomson scattering differential cross section on polarization is expressed as

$$\frac{d\sigma_T}{d\Omega} = \frac{e^4}{m_e^2 c^4} |\boldsymbol{\epsilon} \cdot \boldsymbol{\epsilon}'|^2 \tag{5}$$

where e and m_e are the electron charge and mass and $\boldsymbol{\epsilon} \cdot \boldsymbol{\epsilon}'$ is the angle between the incident and scattered photons. The right hand panel of Fig. 7 shows how the Thomson scattering produces polarization of the CMB photons. If the CMB photons scatter later due to reionization and the incident radiation has a quadrupole moment, then it will be scattered in a polarized manner on the scale roughly equivalent to the horizon scale at the redshift of scattering. That is why the scale at which the large scale polarization is detected gives information about the reionization redshift.

The polarization field of the CMB photons is usually described in terms of the so called "electric" (E) and "magnetic" (B) components which can be derived from a scalar or vector field. The harmonics of an E-mode have $(-1)^\ell$ parity on the sphere, whereas those of B-mode have $(-1)^{\ell+1}$ parity. Under parity transformation, i.e., $\hat{n} \rightarrow -\hat{n}$, the E-mode thus remains unchanged for even ℓ, whereas the B-mode changes sign and vice versa. Figure 8 illustrates such (a)symmetry under parity transformation for the simple case of $\ell = 2$, $m = 0$ (Hu and White 1997).

Various physical processes lead to different effects on the CMB polarization. Most of these effects are expected to produce E mode polarization patterns on the CMB. However, gravitational waves in the primordial signal and gravitational lensing of the CMB on its way to us produce a B mode polarization patterns. A large scale E mode polarization signal could only be caused by the process of reionization. The main reason for this is that large scale polarization could not be caused by causal effects on the last scattering surface which has a 1° scale whereas reionization, which occurs much later, has no such restriction. Figure 9 shows the measured and predicted CMB angular power and cross-power spectra from the WMAP 3rd year data. The existence of large scale correlation in the E-mode is a strong indication that the Universe became ionized around redshift $z \approx 10$. The argument in essence is mostly geometric, namely it has to do with the scale of the E-mode power spectrum as well as the line of sight distance to the onset of the reionization front along

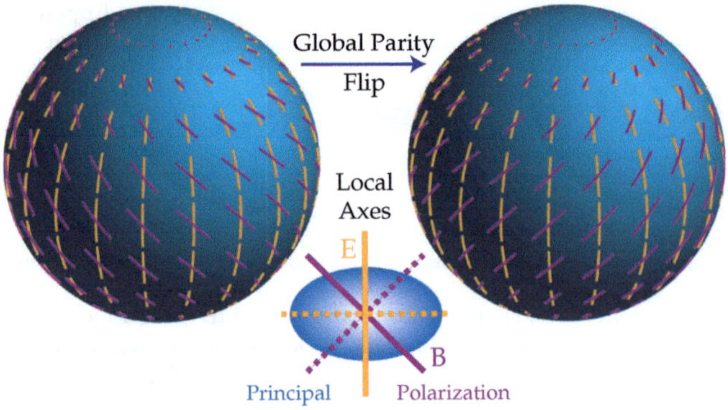

Fig. 8 The E and B polarization modes are distinguished by their behavior under parity transformation. The local distinction is that the E mode is aligned with the principle axis of polarization whereas the B mode is 45° crossed with it (this figure is taken from Hu and White 1997)

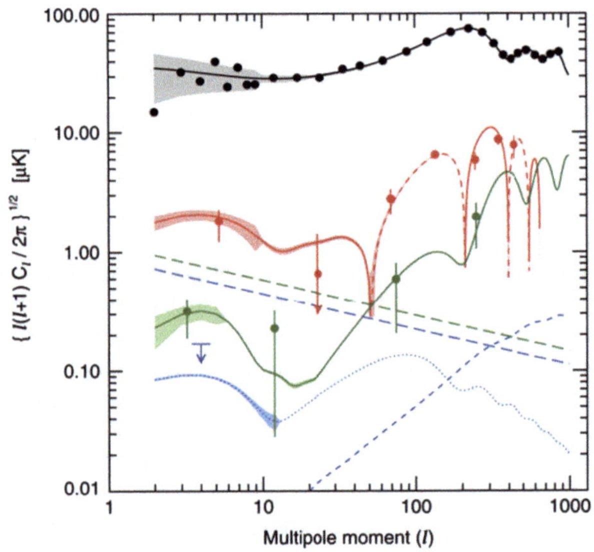

Fig. 9 The temperature and E-mode polarization power and cross-power spectra as measure by the WMAP satellite (Page et al. 2007). Plots of signal for TT (*black*), TE (*red*), and EE (*green*) for the best-fit model. The *dashed line* for TE indicates areas of anticorrelation. For more details about this figure we refer the reader to the (Page et al. 2007). Notice the excess power on large scales caused by reionization seen in the TE and EE power spectra

a given direction. Some authors have also argued that one can have somewhat more detailed constraints on reionization from the exact shape of the CMB E-mode polarization large scale bump (Holder et al. 2003; Lewis et al. 2006; Mortonson and Hu 2008). Unfortunately however, the large cosmic variance at large scales limits the amount of possible information one can extract. Still, the Planck surveyor is expected to be able to retrieve some of the large scale bump shape.

From Fig. 9 one can also deduce the optical depth for Thomson scattering, τ, caused by the scattering of the CMB photons off free electrons released by reionization to be 0.087 ± 0.017 (Dunkley et al. 2009). This could be turned into a constraint on the global reionization history through the integral,

$$\tau = \int_0^{z_{dec}} \sigma_T n_e \frac{c H_0^{-1} dz}{(1+z)\sqrt{\Omega_m (1+z)^3 + \Omega_\Lambda}}. \tag{6}$$

Here z_{dec} is the decoupling redshift, σ_T is the Thomson cross section, μ is the mean molecular weight and n_e is the electron density. This formula works for the optical depth along each sight line but also for the mean electron density, i.e., mean reionization history, of the Universe.

An important point to notice here is that, in order to turn τ into a measurement of the reionization redshift, one needs a model for n_e as a function of redshift. Hence, one has to be careful when using the reionization redshift given by CMB papers as in most cases a gradual reionization is assumed. Sudden reionization gives a one to one correspondence between the measured optical depth and the reionization redshift, e.g., the WMAP measurement optical depth implied $z_i = 11.0 \pm 1.4$.

However, sudden reionization is very unlikely and most models predict a more gradual evolution of the electron density as a function of redshift. Furthermore, in such scenarios the redshift of reionization is not clearly defined, therefore authors refer instead to the redshift at which half of the IGM volume is ionized, $z_{x_{HI}=0.5}$. Obviously, in the case of sudden reionization the two redshifts coincide, $z_i = z_{x_{HI}=0.5}$. It is also important to notice that in the case of sudden reionization the WMAP measured Thomson optical depth does not imply that the redshift at which half the IGM is ionized is the same as z_i and in most cases one obtains $z_{x_{HI}=0.5} < z_i$ (Thomas et al. 2009).

The patchy nature of the reionization process will also leave an imprint at arcminute scales on the CMB sky. Such an imprint will be mostly caused by the reionization bubbles that form during the EoR. However, the strength of the reionization signal at small scales is found to be smaller than that caused by gravitation lensing and is very hard to extract unless the experiment has a very high signal-to-noise at such small scales (Doré 2007).

2.3 *The Intergalactic Medium at* $z \lesssim 6$

There are a number of other observations that put somewhat less certain constraints on reionization. Those constraints come mostly from detailed analysis of high resolution Lyman α forest data and from the observation of high redshift Lyman break galaxies. Here we present the two "strongest" of those constraints.

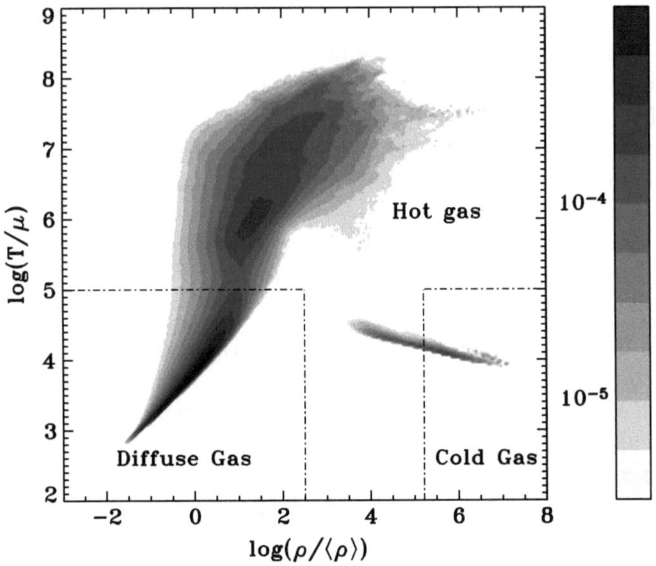

Fig. 10 The different baryon phases in the $\rho - T$ diagram. *Gray* contours show a mass-weighted histogram: the baryon mass fraction at a given density and temperature. Each region corresponds to a given phase (diffuse background, hot, or cold gas) (Rasera and Teyssier 2006)

2.3.1 IGM Temperature Evolution

Another constraint on the reionization history comes from studying the thermal history of the IGM. Due to its low density, the intergalactic medium cooling time is long and retains some memory of when and how it was last heated, namely, reionized. Hence, measuring the IGM temperature at a certain redshift ($\gtrsim$3.5) allows us to reconstruct, under certain assumptions, its thermal history up to the reionization phase where the IGM has been substantially heated. Such a measurement has been carried out by a number of authors using high resolution Lyman α forest data, especially using the very low column density absorption lines. The width of these absorption features carries information about the temperature of the underlying IGM. This temperature obviously varies with density and with other parameters like the background UV flux. Based on both theoretical arguments (Hui and Gnedin 1997) and on numerical simulations (Theuns et al. 1998) in the linear and quasilinear regime, the temperature–density relation follows the simple power law,

$$T = \bar{T}\left(\frac{\rho}{\bar{\rho}}\right)^{\gamma-1}, \tag{7}$$

where $\bar{T}$ is the temperature of the IGM at the mean density of the Universe and γ is the adiabatic power law index. Figure 10 shows the so called phase diagram, i.e., the relation between the temperature and density, obtained from a cosmological hydrodynamical simulation (Rasera and Teyssier 2006). The relation between the density and temperature at the low density end of the diagram, marked as diffuse background, follows a power law. The hot phase at intermediate densities where cooling is not efficient, is driven by shock heating. At high densities, cooling becomes very efficient and drives the gas temperature. At high redshifts more than 90 % of the gas is in the diffuse phase.

Given the validity of Eq. (7) at low densities, it is meaningful to define an IGM temperature as the gas temperature at the mean density, $\bar{T}$. Such a measurement has been performed by a number of authors at $z \approx 3-4$ (Lidz et al. 2010; Schaye et al. 2000; Theuns et al. 2002; Zaldarriaga 2002) and recently at $z \approx 6$ by Bolton et al. (2010).

The usefulness of this temperature to constrain the reionization history was first realized by Theuns et al. (2002), Hui and Haiman (2003) who used the measured temperature around redshift 3 to set $z \approx 9$ as an upper limit for the reionization process. Bolton et al. (2010) have recently confirmed these findings with higher redshift quasars. That is, the measured temperatures of the IGM at redshift $z \approx 3$ and $z \approx 6$ are too high for the bulk of reionization to have occurred at redshift $\gtrsim 10$.

After reionization, the evolution of the IGM mean temperature $\bar{T}$ is given by

$$\frac{1}{\bar{T}}\frac{d\bar{T}}{dt} - \frac{1}{\mu}\frac{d\mu}{dt} = -2H + \frac{\mu \Delta_\epsilon}{\frac{3}{2}k_B\bar{T}}, \tag{8}$$

where H is the Hubble parameter, k_B is the Boltzmann constant, μ is the mean molecular weight, and Δ_ϵ is the effective radiative cooling rate (in units of ergs $g^{-1}s^{-1}$). Δ_ϵ is negative (positive) for net cooling (heating) and includes photoelectric heating and cooling via recombination, excitation, inverse Compton scattering, collisional ionization, and bremsstrahlung. Without cooling/heating processes the cooling rate is set by adiabatic cooling, namely, Hubble expansion. This equation enables us to calculate the temperature evolution as a function of redshift. Measuring the IGM temperature at a given redshift will allow us to extrapolate back in time until we reach a temperature of 6×10^4K which is the temperature at which hydrogen ionizes. Figure 11 demonstrates this procedure (Theuns et al. 2002).

Obviously, the weak point of this argument is the assumption that one knows the cooling/heating function of the IGM at every redshift up to the time of reionization. Still, this is a useful argument and certainly any model for the reionization history would have to explain the temperature we measure at lower redshifts.

2.3.2 Number of Ionizing Photons per Baryon

Another constraint that comes mostly from the Lyman α forest but also from the recently discovered galaxies at $z \gtrsim 7$ is the number of ionizing photons per baryon. Using physically motivated assumptions for the mean free path of ionizing photons, Bolton and Haehnelt (2007) turned the measurement of the photoionization rate into an estimate of the ionizing emissivity. They showed that the inferred ionizing emissivity in comoving units, is nearly constant over the redshift range 2–6 and

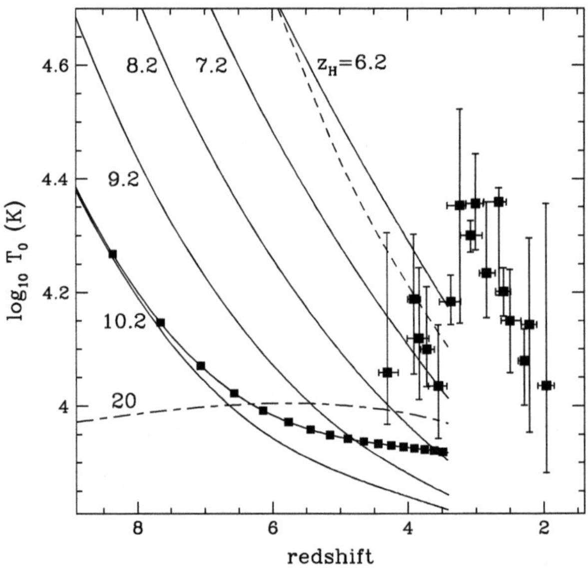

Fig. 11 Temperature evolution of the IGM above redshift 3.4. The *solid curves* indicate the evolution of the temperature at the mean density for various H I reionization redshifts z_H, as indicated. The temperature after hydrogen reionization is assumed to be $T_0 = 6 \times 10^4$ K, and the hydrogen photoionization rate is $\Gamma = 10^{-13}s^{-1}$ ($\Gamma = 10^{-14}s^{-1}$, *short-dashed curve*). The He II photoionization rate is adjusted so that the He III abundance is $x_{\rm He\,III} \approx 0.1$ at $z = 3.5$. The *solid curve* connecting the *filled squares* indicates $z_H = 10.2$ and a higher He II photoionization rate, $x_{\rm He\,III}(z = 3.5) = 0.6$. Finally, the *long-dashed curves* has $z_H = 20$ but a still higher He II photoionization rate, $x_{\rm He\,III}(z = 3.5) = 0.95$. If He is mostly singly ionized at $z \approx 3.5$, then the rapid decrease in T_0 after reionization places an upper limit of $z_H < 9$ on the redshift of hydrogen reionization. The *filled squares* with error bars show the measured IGM temperature as function of redshift. This figure taken from Theuns et al. (2002)

corresponds to $1.5 - 3$ photons emitted per hydrogen atom over a time interval corresponding to the age of the Universe at $z = 6$. Completion of reionization at or before $z = 6$ requires therefore, either an emissivity which rises towards higher redshifts or one which remains constant but is dominated by sources with a rather hard spectral index, e.g., mini-quasars.

With the installation of the WFC3 camera aboard the Hubble Space Telescope, searches for high redshift galaxies at $z = 6 - 10$ have improved dramatically. In particular, a number of authors (Oesch et al. 2010; Bouwens et al. 2010; Bunker et al. 2010; McLure et al. 2010) have reported detection of very high redshifts galaxies using the Lyman-break drop-out technique. The most striking result of these studies is the low number of galaxies found beyond redshift $\approx$6, making it very hard for these galaxies to ionize the Universe. This conclusion depends however on assuming a luminosity function for galaxies at these redshifts, a function that is very poorly known. More surprising is the very steep drop in the number of galaxies at redshift $z \approx 9$ (Bouwens et al. 2011) which makes it even harder to explain reionization with such galaxies.

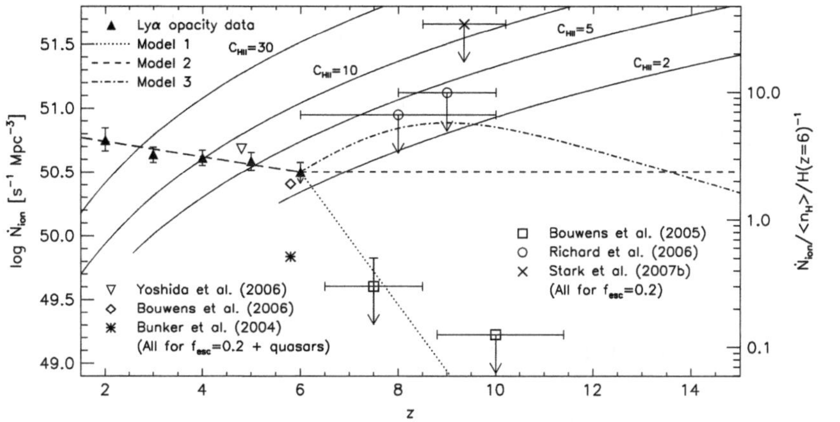

Fig. 12 Observational constraints on the emission rate of ionizing photons per comoving Mpc, $\dot{N}_{\rm ion}$, as a function of redshift. The scale on the right-hand vertical axis corresponds to the number of ionizing photons emitted per hydrogen atom over the Hubble time at $z = 6$. The *filled triangles* give an estimate of $\dot{N}_{\rm ion}$ based on the constraints obtained from the Lyman α effective optical depth from Bolton et al. (2005) . The *inverted triangle* at $z = 5$ and the *diamond* and star at $z = 6$ correspond to estimates of $\dot{N}_{\rm ion}$ based on the Lyman limit emissivities of LBGs and quasars. The data have been slightly offset from their actual redshifts for clarity. An escape fraction of $f_{\rm esc} = 0.2$ has been assumed in this instance. At $z > 6$, the *open squares and circles* are derived from the upper limits on the comoving star formation rate per unit volume inferred by Bouwens et al. (2005); Richard et al. (2006), respectively. The *cross* is derived from the number density of Lyman α emitters estimated by Stark et al. (2007). Three simple models for the evolution of $\dot{N}_{\rm ion}$ are also shown as the *dotted, short dashed and dot-dashed lines*. The *solid lines* correspond to the emission rate of ionizing photons per unit comoving volume, $\dot{N}_{\rm rec}$, needed to keep the IGM ionized for various H II clumping factors. This figure is taken from Bolton and Haehnelt (2007), see also Calverley et al. (2011)

The last two observational findings have led some authors to claim that the reionization is photon starved, i.e., has a low number of ionizing photons per baryon, which results in a very slow and extended reionization process (Bolton and Haehnelt 2007; Calverley et al. 2011). Figure 12 shows the number density of ionizing photons (left-hand vertical axis) and number of ionizing photons per baryon (right-hand vertical axis) as a function of redshift. The number of ionizing photons per baryon at redshift 6 is of the order of 2. More recent results deduced from Lyman-break galaxies are consistent with this figure and show an even lower ratio of ionizing photons per baryon at higher redshifts.

2.4 Other Observational Probes

In addition to the probes that we discussed so far, there are a large number of other observational probes that could potentially add valuable input to the reionization models. Examples of such probes are cosmic infrared and soft x-ray backgrounds (Dijkstra et al. 2004), Lyman α emitters (Ouchi et al. 2009), high redshift QSOs (Mortlock et al. 2011) and GRBs (Bromm and Loeb 2006), metal abundance at high redshift (Rudie et al. 2012), etc. However, such probes currently provide very limited constraints on the EoR.

In the coming chapters we will focus on the very large effort currently made to measure the diffuse neutral hydrogen in the IGM as a function of redshift up to $z \gtrsim 11$ using the redshifted 21 cm emission line. This probe will give the most direct and detailed evidence on the reionization process.

3 The Reionization Process

The inflationary process that occurred very early in the Universe has created the initial tiny fluctuations in matter density field. The high density peaks in these fluctuations field are the seeds around which galaxies form. The formation process is initially driven by gravitational instability alone but later gas physics, cooling, heating, radiation processes and feedback effects play an important role as well (Mo et al. 2010; Peebles 1993). The first galaxies form when primordial gas (H I and He I) condenses within dark matter potential wells which leads to radiative cooling driven mostly by the Lyman α line transition (Dijkstra et al. 2006; Haiman et al. 2000; Latif et al. 2011a,b,c; Partridge and Peebles 1967) and, probably, by H_2 cooling. To date Lyman α emission has been observed in many high redshift galaxies (Kashikawa et al. 2006; Murayama et al. 2007; Ouchi et al. 2009, 2010). This gas condenses further to form the first stars and black holes which in turn produce radiation that starts ionizing the Universe. The efficiency with which these objects produce ionizing radiation is subject to many different physical processes and assumptions (see e.g., Ciardi and Ferrara 2005). Since this book's topic is the first galaxies, the reader is referred to the other chapters in this volume for detailed discussion of how the first radiation emitting objects form and how efficient are they in producing ionizing radiation.

An important unknown in these galaxies is the so called escape fraction, namely the fraction of ionizing radiation that escapes the galaxy into the IGM. It is these ionizing photons that are relevant to the Universe's reionization. Determining the escape fraction of ionizing radiation observationally is very difficult especially at high redshifts where the available information is very limited. Nevertheless, such observations have been carried out by a number of authors (Giallongo et al. 2002; Inoue et al. 2005; Iwata et al. 2009; Shapley et al. 2006; Steidel et al. 2001) where the measured fraction is found to be between 0.1 and 0.5. Theoretical prediction of the escape fraction is also difficult. Early studies have assumed idealized smooth galaxies (Dove and Shull 1994; Dove et al. 2000; Ricotti and Shull 2000; Wood and Loeb 2000) but later studies have simulated more realistic galaxies (see e.g., Ciardi et al. 2002). Each of these studies have considered different set up and different sources but all conclude that the escape fraction of radiation is roughly in the range of 0.1–0.5.

The most accepted picture of how reionization unfolds is simple. The first radiation-emitting objects ionize their immediate surroundings, forming bubbles that expand until their ionizing photons are consume by the neutral IGM. As the number of radiating sources increases, so do the number and size of the ionization bubbles, which eventually spread to fill the whole Universe. However, most of the details of this scenario are yet to be clarified. For example: what controls the formation of the first objects and how much ionizing radiation do they produce? How do the bubbles expand into the intergalactic medium and what do they ionize first, high-density or low density regions? The answer to these important questions and many others touch upon many fundamental questions in cosmology, galaxy formation, quasars activity and the physical properties of very metal poor stars (Barkana and Loeb 2001; Bromm and Larson 2004; Ciardi and Ferrara 2005; Choudhury and Ferrara 2006; Furlanetto et al. 2006; Morales and Wyithe 2010).

To ionize hydrogen one needs photons with energy of 13.6 eV or higher meaning the reionization of the Universe requires ultraviolet photons. A crucial question is which sources in the Universe provide the UV photons needed to ionize the Universe and maintain it in that state. Obvious candidates are the first stars (so called Population III stars), second generation stars (Population II stars) and (mini)quasars which are objects powered by intermediate mass black holes ($10^{3-6} M_{\odot}$). There are other candidate sources of reionization, like decaying or self-annihilating dark matter particles or decaying cosmic strings. However, the constraints on such objects make it unlikely that they could reionize the Universe by themselves (Chen and Kamionkowski 2004; Kasuya and Kawasaki 2004; Mapelli and Ferrara 2005; Mapelli et al. 2006; Natarajan and Schwarz 2010; Padmanabhan and Finkbeiner 2005; Ripamonti et al. 2007; Zhang et al. 2006).

Massive black holes powering quasars convert mass to radiation extremely efficiently. They produce a large amount of UV and X-ray radiation above the ionization threshold. In fact, one of the main discoveries of the last decade is that quasars, powered by very large black holes with masses in excess of $10^9 M_{\odot}$, already existed at redshift above 7 [e.g., QSO ULAS J1120 + 0641 (Mortlock et al. 2011) from the UKIDSS survey (Lawrence et al. 2007)]. How these black holes managed to accumulate so much mass in such a short time is a puzzle in its own right (Mortlock et al. 2011; Bolton et al. 2011). However, the mass distribution of the massive black holes in the early Universe is unknown, rendering the role played by quasars during reionization very uncertain.

Population III stars formed from the primordial mix of the elements and thus only contain hydrogen and helium. This composition makes them very different from present-day stars. In order for a star to form, the initial proto-star has to radiate some of the energy gained by gravitational contraction, or the collapse will rapidly halt as the cloud reaches hydrostatic equilibrium. Population III stars are poor radiators until the cloud from which they form reaches high temperatures. This causes them to be very massive, and hence, they are very efficient and abundant sources of UV photons, yet are very short lived. Theoretically, these objects could have reionized the Universe but our knowledge of them, including the question of whether they existed in sufficient numbers, is very uncertain.

The first stars' early metal enrichment was likely the dominant effect that brought about the transition from Population III to Population II star formation. Recent numerical simulations of collapsing primordial objects with masses of $\approx 10^6 M_\odot$, have shown that the gas has to be enriched with heavy elements to a minimum level of $Z_{crit} \approx 10^{-4} Z_\odot$, in order to have any effect on the dynamics and fragmentation properties of the system. Normal, low-mass (Population II) stars are hypothesized to form only out of gas with metallicity $Z \geq Z_{crit}$. Thus, the characteristic mass scale for star formation is expected to be a function of metallicity, with a discontinuity at Z_{crit} where the mass scale changes by about two orders of magnitude. The redshift where this transition occurs has important implications for the early growth of cosmic structure, and the resulting observational signature includes the extended nature of reionization (see the review by Ciardi and Ferrara 2005).

Most studies of reionization have focused on stars as being the primary source (Mellema et al. 2006; Abel et al. 2000, 2002; Bromm et al. 2002; Yoshida et al. 2003). Due to the deficiency of hard photons in the spectral energy distributions (SEDs) of these "first stars", heating due to these objects is limited in extent (Thomas and Zaroubi 2008). On the other hand, miniquasars (miniqsos), characterized by central black hole masses $<10^6 M_\odot$, have also been considered as an important contributor to reionization (Madau et al. 1997; Ricotti and Ostriker 2004a,b; Nusser 2005; Furlanetto and Loeb 2002; Furlanetto et al. 2004; Wyithe and Loeb 2004a; Thomas and Zaroubi 2008). Ionization aspects of the miniquasar radiation have been explored by several authors (Madau et al. 1997; Ricotti and Ostriker 2004a,b; Thomas and Zaroubi 2008; Thomas et al. 2009; Zaroubi and Silk 2005). Thomas and Zaroubi (2008) have shown that although the ionization pattern around miniqsos is similar to that of stellar-type sources, the heating due to the presence of hard photons in miniqsos is very different. The reason being is that stars produce thermal radiation that is mostly in the UV range, which is very efficient in ionization, but once it is absorbed by H I, the energy left will be too small to be converted to heat effectively. On the other hand black hole powered sources have hard x-ray photons as their spectral energy distribution (SED) follows a power law (typically assumed to be -1). Such x-ray photons have lower bound-free cross section relative to UV photons but once they are absorbed, their leftover energy is very large and can easily be converted to heat. Also, x-ray photons penetrate much deeper into the IGM and can heat it up much further from the source than UV radiation.

Miniqsos heat the surrounding IGM well beyond their ionization front (Thomas and Zaroubi 2008; Chuzhoy et al. 2006). Several authors (e.g., Madau et al. 1997; Nusser 2005; Zaroubi et al. 2007 have shown the importance of heating the IGM with respect to the observability of the redshifted 21 cm radiation in either emission or absorption. Figure 13 shows the ionization and heating patterns around a number of stars (upper panels) and miniqsos (lower panel). The mass of the stars and black-holes are indicated next to the lines, and their SEDs are assumed to be thermal or to have a power law dependence on the photon energy, $\propto E^{-1}$, respectively. The calculation here is spherically symmetric and assumes a single object forming in the IGM (Zaroubi et al. 2007; Thomas and Zaroubi 2008). The ionization pattern

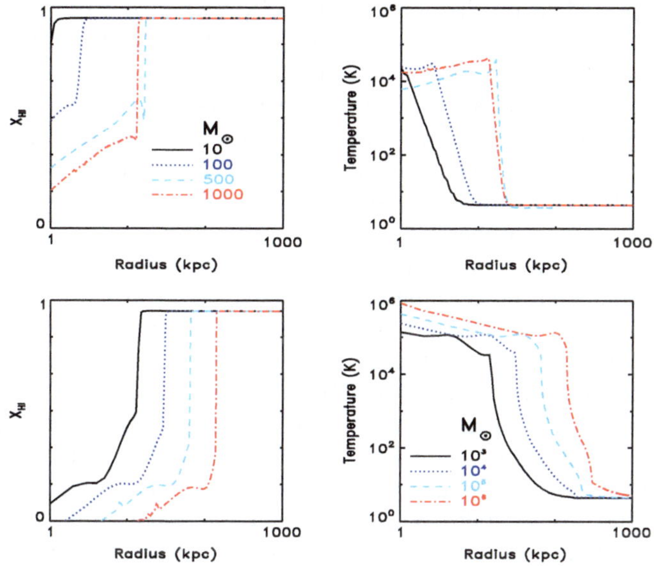

Fig. 13 This figure shows the ionization and heating profile around a single star and black hole forming in the IGM assuming spherical symmetry. The *upper panels* show the stars case whereas the *lower panels* show the black holes case. The *left hand panels* show the neutral fraction of H I as a function of distance from the star and the *right hand panels* show the gas temperature as a function of distance from the source (Thomas and Zaroubi 2008)

around stars and black holes are very similar, they both show a very abrupt increase in H I with a clear ionization front (see e.g. Kramer and Haiman 2009; Thomas and Zaroubi 2008; Zaroubi et al. 2007). Of course the radius at which such front is seen depends on the mass of the star or the black hole but the pattern is the same (see the left hand side panels of Fig. 13). The heating profile around the two types of sources, on the other hand, is different since in power law sources (miniqsos) the radiation can penetrate the neutral gas and reach large distances from the sources (see right hand panels of Fig. 13). This high energy radiation is efficient in heating the IGM gas through secondary electrons (Shull and van Steenberg 1985) (see discussion later) whereas UV radiation is efficient in ionizing the gas but has little energy left to heat too much and can not penetrate the neutral gas as far as x-ray radiation does.

We have seen that unlike stars, x-ray source a (e.g., miniquasars) have an additional property of heating the IGM to a large extent and through secondary Lyman α radiation making the neutral IGM visible to a 21-cm experiment. However, some authors (e.g., Dijkstra et al. 2004; Salvaterra et al. 2005) argue that miniquasars alone can not reionize the Universe as they will produce far more soft X-ray background radiation than currently observed (Moretti et al. 2003; Sołtan 2003) while simultaneously satisfying the WMAP3 polarisation results (Page et al. 2007; Spergel et al. 2007). It should be noted, however, that these calculations have been carried out assuming specific models for the evolution of black hole mass density and spectral energy distributions of UV/X-ray radiation of the miniquasars, whereas one can easily construct other models in which the discrepancy is not so severe (Zaroubi et al. 2007; Ripamonti et al. 2008).

Some authors (Kashlinsky et al. 2005) have claimed a detection of excess IR background radiation and argued that it provides evidence for stars being the primary source of reionization. This too has been subject to controversy because of the sensitivity of the result to the subtraction of the contaminants, e.g., Zodiacal light, within the same waveband (Cooray et al. 2007).

Although uncertainty looms about the sources that resided during the *dark ages*, it is conceivable from observations of our Universe up to redshifts of 6.5, that sources of reionization could have been a mixture of both stellar and miniquasar sources. Implementing radiative transfer that includes both ionizing and hard X-ray photons has been difficult and, as a result, most 3-D radiative transfer schemes restrict themselves to ionization due to stars (Benson et al. 2006; Ciardi et al. 2001; Gnedin and Abel 2001; Mellema et al. 2006; Mesinger and Furlanetto 2007; Nakamoto et al. 2001; Pawlik and Schaye 2008; Ritzerveld et al. 2003; Razoumov and Cardall 2005; Susa 2006; Whalen and Norman 2006; Zahn et al. 2007). In Ricotti and Ostriker (2004a), a "semi" hybrid model of stars and quasars like the one hinted above used, albeit in sequential order instead of a simultaneous implementation. That is, pre-ionization due to quasars has been invoked between $7 \leq z \leq 20$, after which stars reionize the Universe at redshift 7.

Given the numerical cost of the full 3D radiative transfer schemes, exploring a large parameter space for models of reionization, is not feasible. Such an exploration is needed in order to understand the various physical effects introduced by each such parameter. It is also needed to help interpret the available data. A number of authors have been pursuing "quick-and-dirty" methods to simulate the reionization process. These schemes can include very rough methods that use the initial density field to produce a reionization cube without the need for cosmological N-body and hydro simulations, such as 21cmFAST (Mesinger et al. 2011; Zahn et al. 2007, 2011) and SimFast21 (Santos et al. 2010). They also include more accurate (yet still fast) methods like BEARS (Thomas and Zaroubi 2008; Thomas et al. 2009; Thomas and Zaroubi 2011) that use N-body and hydro simulations but reduces the numerical cost by restricting the ionization bubbles around the radiation sources to be spherical.

4 The Redshifted 21 cm as a Probe of the EoR

In recent years it has become clear that the 21 cm line can be used to probe the neutral IGM prior to and during the reionization process. This hyperfine transition line of atomic hydrogen (in the ground state) arises due to the interaction between the electron and proton spins (Hogan and Rees 1979; Scott and Rees 1990; Madau et al. 1997). The excited triplet state is a state in which the spins are parallel whereas the spins at the lower (singlet) state are antiparallel. The 21 cm line is a forbidden line for which the probability for a spontaneous $1 \rightarrow 0$ transition is given by the Einstein A coefficient that has the value of $A_{10} = 2.85 \times 10^{-15}\text{s}^{-1}$. Such an extremely small value for Einstein-A corresponds to a lifetime of the triplet state of 1.1×10^7 years for spontaneous emission. Despite its low decay rate, the 21 cm transition line is one of the most important astrophysical probes, simply due to the vast amounts of hydrogen in the Universe (Ewen and Purcell 1951; van de Hulst 1945; Muller and Oort 1951) as well as the efficiency of collisions and Lyman-α radiation in pumping the line and establishing the population of the triplet state (Wouthuysen 1952; Field 1958). In this chapter I will describe the basic physics behind this transition, especially what decides its intensity.

4.1 *The 21 cm Spin and Brightness Temperatures*

The intensity of the 21 cm radiation is controlled by one parameter, the so called spin temperature, T_{spin}. This temperature is defined through the equation,

$$\frac{n_1}{n_0} = 3\exp(-T_*/T_{spin}), \tag{9}$$

where n_1 and n_0 are the number densities of electrons in the triplet and singlet states of the hyperfine level respectively, and $T_* = 0.0681$ K is the temperature corresponding to the 21 cm wavelength. The spin temperature is therefore, merely a shorthand for the ratio between the occupation number of the two hyperfine levels. This ratio establishes the intensity of the radiation emerging from a cloud of neutral hydrogen. Of course, in the measurement of such radiation one has to take into account the level of background being transmitted through a given cloud as well as the amount of absorption and emission within the cloud. Namely, one has to use the equation of radiative transfer.

In the following derivation I follow the description in Rybicki and Lightman (1986). The radiative transfer equation is normally written in terms of the *brightness* (or *specific intensity*) of the radiation I_ν. This quantity is defined as the intensity per differential frequency element in the form, $I_\nu = \frac{dI}{d\nu}$, where ν is the frequency. The intensity has the dimensions of ergs s^{-1} cm^{-2} sr^{-1} Hz^{-1}, namely, it quantifies the energy carried by radiation traveling along a given direction, per unit area, frequency, solid angle, and time. The radiative transfer equation for thermally emitting material at temperature T can be written in terms of the optical depth for absorption as,

$$\frac{dI_\nu}{d\tau_\nu} = -I_\nu + B_\nu(T), \tag{10}$$

where τ_ν is the optical depth for absorption through the cloud at a given frequency and B_ν is the Planck function.

In radio astronomy the intensity I_ν is often expressed by its equivalent *brightness temperature*, $T_b(\nu)$. This is convenient because at the Rayleigh–Jeans low energy

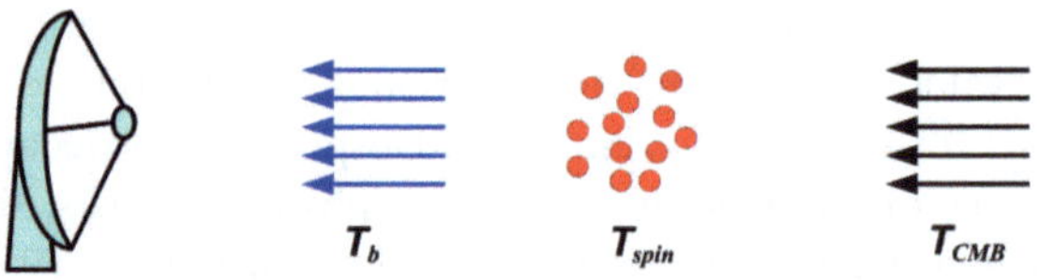

Fig. 14 A cartoon that shows that set up of the various components relevant to radiative transfer problem at hand starting from the background (CMB) radiation, going through a certain cloud with temperature T_{spin} and emerging with a temperature T_b that is measured by our telescopes

limit, the relation between the brightness temperature and specific intensity is given by,

$$T_b(\nu) \approx I_\nu c^2/2k_B\nu^2, \tag{11}$$

where c is the speed of light and k_B is the Boltzmann's constant. Expressing the radiative transfer equation (10) in terms of the brightness temperature gives it a particularly simple form,

$$\frac{dT_b}{d\tau_\nu} = -T_b + T_{CMB}, \tag{12}$$

where I substituted the CMB temperature for the background temperature. Solving Eq. (12) yields the temperature of the emergent radiation at frequency ν,

$$T_b(\nu) = T_{spin}(1 - e^{-\tau_\nu}) + T_{CMB}(\nu)e^{-\tau_\nu}, \tag{13}$$

where $T_{spin} = T_b(0)$ is the brightness temperature in the absorbing cloud (see Fig. 14). Notice that for the background radiation the factor $\exp(-\tau_\nu)$ gives the transmission probability of the background radiation whereas the $1 - \exp(-\tau_\nu)$ factor gives the emission probability of 21 cm photons from within the cloud. Therefore, in order to determine the brightness temperature, one needs to know the optical depth for absorption, τ_ν, and the spin temperature, T_{spin}, in the optically thin regime relevant to our case. Notice that in the case in which $T_{spin} = T_{CMB}$ the brightness temperature gives exactly the CMB temperature. This is simply because in such a case there is a prefect balance between the absorption and emission at every frequency. Therefore, the measurement in such a case does not reveal anything interesting about the intervening cloud, the subject we are interested in here.

I will first start with calculating the 21 cm optical depth. The hyperfine transition of atomic hydrogen is an ideal transition to be described by Einstein coefficients and their relations. The 21 cm radiation incident on the atom can cause $0 \rightarrow 1$ transitions (absorptions) and $1 \rightarrow 0$ transitions (induced emissions) corresponding to Einstein coefficient B_{01} and B_{10} respectively. The probabilities are given by,

$$I_\nu B_{01} = \frac{g_1}{g_0} B_{10} I_\nu, \tag{14}$$

and

$$I_\nu B_{10} = A_{10} \frac{\lambda^2 I_\nu}{2h\nu_{10}}, \tag{15}$$

respectively (Rybicki and Lightman 1986). Here $\nu_{10} = 1,420.4$ MHz is the frequency of the 21 cm transition.

The 21 cm line absorption cross section is given by

$$\sigma_\nu \equiv \sigma_{01}\phi(\nu) = \frac{3c^2 A_{10}}{8\pi\nu^2}\phi(\nu), \tag{16}$$

where $\phi(\nu)$ is the line profile defined so that $\int \mathrm{d}\nu\, \phi(\nu) = 1$ and has units of time.

The optical depth of a cloud of hydrogen is then:

$$\tau_\nu = \int \mathrm{d}\ell\, \sigma_{01}\, (1 - e^{-E_{10}/k_B T_{spin}})\, \phi(\nu)\, n_0 \tag{17}$$

$$\approx \sigma_{01} \left(\frac{h\nu}{k_B T_{spin}}\right) \left(\frac{N_{\mathrm{HI}}}{4}\right) \phi(\nu), \tag{18}$$

where N_{HI} is the column density of H I and $\mathrm{d}\ell$ is a line element within the cloud. The factor of 4 connecting n_0 and H I accounts for the fraction of atoms in the hyperfine singlet state. The second factor in Eq. (17) with E_{10} accounts for stimulated emission. The approximate form in Eq. (18) assumes uniformity throughout the cloud.

We now substitute for $\phi(\nu)$ and N_{HI} using cosmological quantities. In general, the line shape $\phi(\nu)$ includes natural, thermal, turbulent and velocity broadening, as well as bulk motion (which increases the effective Doppler spread). Velocity broadening is the most important effect in the IGM. Hubble expansion of the gas results in velocity broadening of a region of linear dimension ℓ will be $\Delta v \sim \ell H(z)$ so that $\phi(\nu) \sim c/(\ell H(z)\nu)$. The column density along such a segment depends on the neutral fraction x_{HI} of hydrogen, so $N_{H\,I} = \ell x_{H\,I} n_H(z)$ (Furlanetto et al. 2006). A more exact solution of Eq. (17) yields an expression for the 21 cm optical depth of the diffuse IGM,

$$\tau_{\nu_0} = \frac{3}{32\pi} \frac{hc^3 A_{10}}{k_B T_{spin} \nu_0^2} \frac{x_{HI} n_H}{(1+z)\,(\mathrm{d}v_\parallel/\mathrm{d}r_\parallel)} \tag{19}$$

$$\approx 0.0092\,(1+\delta)\,(1+z)^{3/2} \frac{x_{H\,I}}{T_{spin}} \left[\frac{H(z)/(1+z)}{\mathrm{d}v_\parallel/\mathrm{d}r_\parallel}\right], \tag{20}$$

where in the second relation, T_{spin} is in degrees Kelvin. Here the factor $(1+\delta)$ is the fractional overdensity of baryons and $\mathrm{d}v_\parallel/\mathrm{d}r_\parallel$ is the gradient of the proper velocity along the line of sight, including both the Hubble expansion and the peculiar velocity (Kaiser 1987). In the second line, we have substituted the velocity $H(z)/(1+z)$ appropriate for the uniform Hubble expansion at high redshifts.

Next we need to calculate the spin temperature and substitute in Eq. (13). In his seminal papers, George Field (1958, 1959a), used the quasi-static approximation to calculate the spin temperature, T_{spin}, as a weighted average of the CMB temperature, T_{CMB}, the gas kinetic temperature, T_{kin}, and the temperature related to the existence of ambient Lyman-α photons, T_α (Wouthuysen 1952; Field 1959a). For almost all interesting cases, one can safely assume that $T_{kin} = T_\alpha$ (Field 1958; Furlanetto et al. 2006; Madau et al. 1997; Morales and Wyithe 2010).

Three competing processes determine T_{spin}: (1) absorption of CMB photons (as well as stimulated emission); (2) collisions with other hydrogen atoms, free electrons, and protons; and (3) scattering of Lyman α photons through excitation and deexcitation. Hence, the spin temperature could be recast as (Field 1958):

$$T_{spin} = \frac{T_{CMB} + y_{kin} T_{kin} + y_\alpha T_{kin}}{1 + y_{kin} + y_\alpha}, \tag{21}$$

where y_{kin} and y_α are the kinetic and Lyman-α coupling terms, respectively. It is important to note that for the 21 cm radiation to be observed, it has to attain a different temperature than that of the CMB background (Field 1958, 1959a,b; Hogan and Rees 1979; Wouthuysen 1952). The form I use here for Eq. (21) is the original form used in the George Field's 1958 paper (Field 1958), whereas some authors use a form that relates the inverse of the various temperatures. Both ways are of course equivalent but one needs to be careful with the definitions of the coupling coefficients in each case.

The kinetic coupling term y_{kin} is due to collisional excitations of the 21 cm transitions. The Lyman-α coupling term y_α is due to the so called Lyman-α pumping mechanism, also known as the Wouthyusen-Field effect, which is produced by photo-exciting the hydrogen atoms to their Lyman transitions (Field 1958, 1959a; Wouthuysen 1952). The coupling factors y_{kin} and y_α depend on the rate of collisional and Lyman α pumping within the H I cloud. A number of authors have calculated these rates in detail (Allison and Dalgarno 1969; Liszt 2001; Smith 1966; Wild 1952; Zygelman 2005). In the case of first stars, the Wouthyusen-Field effect will depend on the intensity of the Lyman α photons produced by these sources. Collisions on the other hand are somewhat more complicated since it is normally done through the so called secondary electrons which are released by the ionization of an H I atom by an x-ray photon. An electron with such high energy will lose it to the rest of the IGM through collisions. This energy will in general be divided between collisional excitation, collisional ionization and heating (Furlanetto and Furlanetto 2007; Furlanetto and Stoever 2010; Shull and van Steenberg 1985; Valdés et al. 2010).

Since decoupling mechanisms can influence the spin temperature in different ways, it is important to explore the decoupling issue for various types of ionization sources. For instance, stars decouple the spin temperature mainly through radiative Lyman α pumping whereas x-ray sources (e.g., mini-quasars) decouple it through a combination of collisional excitation and heating (Chuzhoy et al. 2006; Zaroubi et al. 2007), both produced by the energetic secondary electrons ejected due to x-ray photons (Shull and van Steenberg 1985). The difference in the spin temperature decoupling patterns of the two, will eventually help disentangle the nature of the first ionization sources (Thomas and Zaroubi 2008; Pritchard and Furlanetto 2007).

Collisions could also be induced by Compton scattering of the CMB photons off the residual free electrons in the IGM gas. This process is dominant at high redshifts $z \gtrsim 200$ and keeps the gas temperature equal to that of the CMB. However, it is not efficient enough at lower redshifts to heat the gas, it is still sufficient to couple the spin temperature to the gas down to $z \approx 100$. In fact, one can show that the global spin temperature evolves in an intricate fashion bouncing back and forth between the gas (kinetic) temperature and the CMB temperature based on which heating/excitations mechanism is dominant.

Figure 15 shows the expected global evolution of the spin temperature as a function of redshift. The blue solid line represents T_{CMB}, which drops as $1 + z$. The green line shows the gas temperature as a function of redshift. At $z \gtrsim 200$, the gas temperature is still coupled to the CMB due to Compton scattering of the background photons off residual electrons leftover from the recombination era. At redshift ~200, however, the gas decouples from the CMB radiation and starts adiabatically cooling as a function of the redshift squared, $(1 + z)^2$, until the first objects start forming and heating up the gas at redshift below 30. The spin temperature (shown by the red lines) has a somewhat more complicated behavior. At $z \gtrsim 100$ it is coupled to the gas temperature due to collisional coupling caused by residual electrons leftover from recombination. At $z \approx 100$ the efficiency of collisional coupling to the gas drops due to the Hubble expansion. At this stage, the spin temperature starts veering towards T_{CMB} until it is completely dominated by it. At lower redshifts the first astrophysical objects that heat and ionize the IGM couple T_{spin} to the gas. Here, broadly speaking, there are two possible histories, one in which T_{spin} couples to the gas as it heats up once it obtains a temperature greater than T_{CMB} (red solid line). In the other possible evolution the spin temperature couples to the gas much before the kinetic temperature exceeds that of the CMB (red dashed line) (Baek et al. 2009; Pritchard and Loeb 2008; Thomas and Zaroubi 2011). In the former case the 21 cm radiation, after decoupling from the CMB at $z \lesssim 30$, is seen only in emission, whereas in the latter case it is seen initially in absorption and only at later stages in emission.

Currently all attempts to measure the redshifted 21 cm emission from the IGM are focused on the redshift range $6 \lesssim z \lesssim 12$. This is due to a number of reasons that are related to the limitations posed by the ionosphere and the background noise (see Sect. 5 for more detail). In this range of redshifts the spin temperature is expected to be set by the astrophysics of the first objects in the Universe, namely, gas physics, feedback, etc., which often involve very complicated and poorly understood processes. However, observing the spin temperature of the Universe within the redshift window around $z \approx 50 - 100$ will mostly probe the cosmological density field (Loeb and Zaldarriaga 2004). Such a measurement could provide a vast amount of information about the pristine Universe that, given the span of its redshift coverage, could potentially exceed that of the CMB data. Unfortunately however,

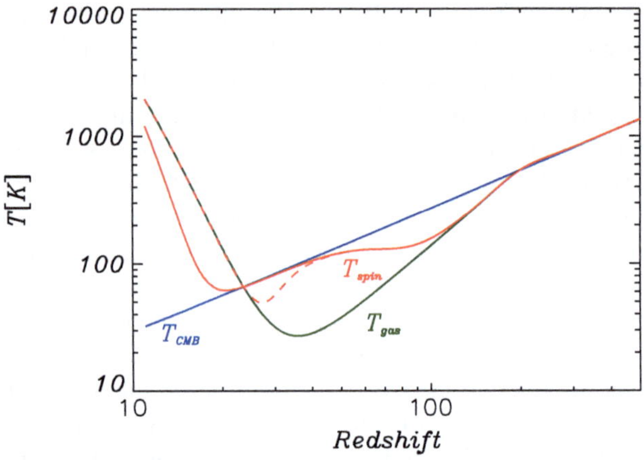

Fig. 15 The global evolution of the CMB (*blue line*), gas (*green line*) and spin (*red solid line and red dashed line*) temperatures as a function of redshift. The CMB temperature evolves steadily as $1 + z$ whereas the gas and spin temperatures evolve in a more complicated manner (see text for detail)

the ionosphere at such frequencies $\nu \lesssim 30$ MHz poses insurmountable hurdles that render such attempts futile. This has led some authors to propose setting up radio telescopes at these very low frequencies on the moon (see e.g., Lazio et al. 2009).

4.2 *The Differential Brightness Temperature*

As we mentioned above the measured quantity in radio astronomy is the brightness temperature, or more accurately the so called differential brightness temperature $\delta T_b \equiv T_b - T_{CMB}$ which reflects the fact the only meaningful brightness temperature measurement insofar as the IGM is concerned is when it deviates from T_{CMB}. In order to get this quantity one should substitute the various components into Eq. (13). Such a substitution and rearrangement yields, (Field 1958, 1959a; Madau et al. 1997; Ciardi and Madau 2003),

$$\delta T_b = 28\mathrm{mK}\,(1+\delta)\,x_{\mathrm{H}\,I}\left(1-\frac{T_{CMB}}{T_{spin}}\right)\left(\frac{\Omega_b h^2}{0.0223}\right)\sqrt{\left(\frac{1+z}{10}\right)\left(\frac{0.24}{\Omega_m}\right)}\left[\frac{H(z)/(1+z)}{\mathrm{d}v_\parallel/\mathrm{d}r_\parallel}\right], \tag{22}$$

where h is the Hubble constant in units of 100 km s^{-1}Mpc^{-1}, δ is the mass density contrast, $x_{\mathrm{H}\,I}$ is the neutral fraction, and Ω_m and Ω_b are the mass and baryon densities in units of the critical density. Note that the three quantities, δ, $x_{\mathrm{H}\,I}$ and T_{spin}, are all functions of 3D position. The term $(T_{spin} - T_{CMB})/T_{spin}$ can obtain a maximum of $+1$ for $T_{spin} \gg T_{CMB}$, i.e., in the emission case. It has no such bound for the case of $T_{spin} \ll T_{CMB}$ and can be very negative in the absorption case.

Equation (22) shows that the differential brightness temperature is composed of a mixture of cosmology dependent and astrophysics dependent terms. This makes the equation a complex yet also a very informative one. This is simply because at different stages in the evolution of this field δT_b is dominated by different contributions. For example, at high redshifts and before significant ionization takes place, i.e. $x_{\mathrm{HI}} \approx 1$, everywhere the brightness temperature is proportional to the density fluctuations making its measurement an excellent probe of cosmology. However, at low redshifts ($z \lesssim 7$) a significant fraction of the Universe is expected to be ionized and the measurement is dominated by the contrast between the neutral and ionized regions, hence, probing the astrophysical source of ionization (see e.g., Iliev et al. 2008; Thomas et al. 2009). Here I assumed that $T_{spin} \gg T_{CMB}$ at all redshifts. Figure 1, which have discussed before, shows a typical distribution of the differential brightness temperature. The figures taken from the simulations of Thomas et al. (2009).

Most radiative transfer simulations assume that the spin temperature is much larger than the CMB temperature, namely the term $\left(1 - T_{CMB}/T_{spin}\right)$ in Eq. (22) is unity. As Fig. 15 shows, this is a good assumption at the later stages of reionization, however, it is probably not valid at the early stages. Modeling this effect is somewhat complex and requires radiative transfer codes that capture the Lyman-α line formation and multifrequency effects, especially those coming from energetic photons (Baek et al. 2009; Mesinger et al. 2011; Pritchard and Furlanetto 2007; Thomas and Zaroubi 2011).

Here we show the evolution of the brightness temperature for three reionization histories: (1) With reionization, excitation and heating dominated by power law sources (miniqsos with x-rays); (2) dominated by thermal (stellar) sources; (3) dominated by a mixture of the aforementioned two types of sources. To create a contiguous observational cube or "frequency cube" (right ascension (RA) $\times$ declination (DEC) $\times$ redshift), the RA and DEC slices, taken from individual snapshots at different redshifts (or frequency), are stacked and interpolated smoothly to create a reionization history. This datacube is then convolved with the point spread function of the LOFAR telescope to simulate the mock data cube of the redshifted 21-cm signal as seen by LOFAR. For further details on creating this cube, refer to Thomas et al. (2009), Thomas and Zaroubi (2011).

As expected, the signatures (both visually and in terms of the r.m.s) of the three scenarios (Fig. 16) are markedly different. In the miniqso-only scenario, reionization proceeds extremely quickly and the Universe is almost completely ($x_{\mathrm{H\,II}} >= 0.95$) reionized by around redshift 7. The case in which stars are the only source sees reionization end at a redshift of 6. Also in this case, compared to the previous one, reionization proceeds in a rather gradual manner. The hybrid model, as explained previously, is in between the previous two scenarios.

In the models shown here, the transition from the absorption dominated brightness temperature to the emission dominated one occurs at relatively low redshifts. The transition redshift depends sensitively on the assumptions made in each case.

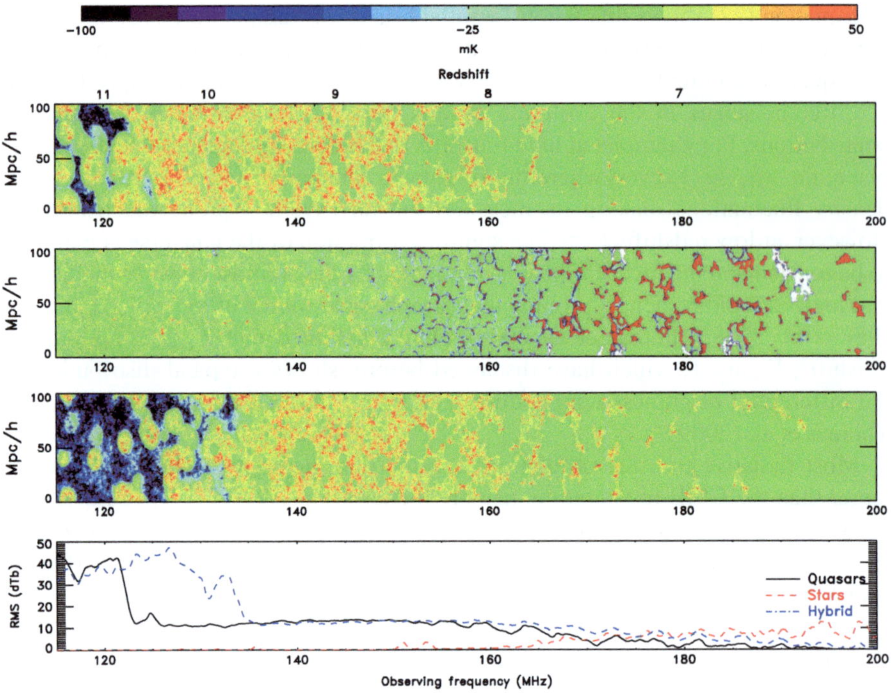

Fig. 16 *Contrasting reionization histories*: From the top, reionization histories (δT_b in mK as a function of frequency or redshift) are plotted for miniqso, stellar and hybrid sources, respectively. The bottom panel plots the r.m.s of δT_b as a function of redshift/frequency for all the three cases. This figure is taken from Thomas and Zaroubi (2011)

Other authors have explored such effects and conclude that the transition occurs at much higher redshifts (see e.g., the models in Mesinger et al. 2011; Santos et al. 2010).

The δT_b in Fig. 16 is calculated based on the effectiveness of the radiation flux, produced by the source, in decoupling the CMB temperature (T_{CMB}) from the spin temperature (T_s). This flux, both in spatial extent and amplitude, is obviously much larger in the case of miniqsos compared to that of stars, resulting in a markedly higher brightness temperature in both the miniqso-only and hybrid models when compared to that of the stars. However, we know that stars themselves produce Lyman α radiation in their spectrum. Apart from providing sufficient Lyman α flux to their immediate surroundings, this radiation builds up as the Universe evolves into a strong background (Ciardi and Madau 2003), potentially filling the Universe with enough Lyman α photons to couple the spin temperature to the kinetic temperature everywhere. It has to be noted that the results we are discussing here are extremely model dependent and any changes to the parameters can influence the results significantly.

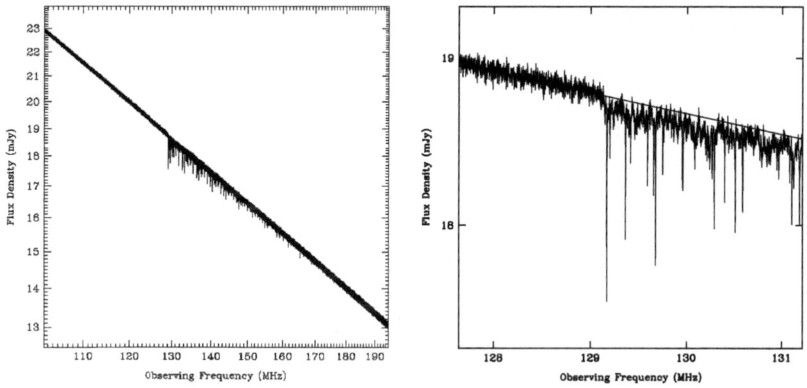

Fig. 17 *Left hand panel:* A simulated spectrum from 100 to 200 MHz of a source with S_{120} = 20 mJy at $z = 10$ using the Cygnus A spectral model and assuming H I 21cm absorption by the IGM. Thermal noise has been added using the specifications of the SKA and assuming 10 days integration with 1 kHz wide spectral channels. *Right hand panel:*The same as the left panel, but showing an expanded view of the spectral region around the frequency corresponding to the redshift H I 21cm line at the source redshift (129 MHz). The *solid line* is the Cygnus A model spectrum without noise or absorption. Figure taken from Carilli et al. (2002)

4.3 *The 21 cm Forest at High z*

Finally, I will conclude this section by discussing a very different aspect of the redshifted 21 cm radiation, and that is the case of the 21 cm forest. Very bright radio sources might exist at high redshifts. In such a case, the emission from these sources is expected to be resonantly absorbed by the neutral IGM and form a system of absorption features just like the Lyman α forest seen in the spectra of distant quasars. Such absorption features are called the 21 cm forest and they were first investigated by Carilli et al. (2002) and subsequently by other authors (Carilli et al. 2004; Furlanetto and Loeb 2002; Furlanetto 2006; Mack and Wyithe 2011; Xu et al. 2009). The discovery of such systems will provide very valuable information about the reionization process and the IGM's physical properties during the EoR which will be largely independent of calibration errors (see Sect. 5). Currently, we know of no very bright high redshift sources, but with the imminent availability of highly sensitive radio telescopes like LOFAR and SKA, the prospects for detecting such sources are very promising.

Figure 17 shows a simulated spectrum at 1 kHz resolution of a $z = 10$ radio source with a flux density of 20 mJy at an observing frequency of 120 MHz (S_{120}). The implied luminosity density at a rest frame frequency of 151 MHz is then $P_{151} = 2.5 \times 10^{35}$ erg s^{-1} Hz^{-1}. The left hand panel of Fig. 17 shows a spectrum covering a large frequency range (100–200 MHz, or HI 21cm redshifts from 13 to 6), whereas the right hand panel shows an expanded view of the frequency range corresponding to the HI 21cm line at the source redshift (129 MHz). At 129 MHz the spectrum shows a 1 % drop due to the diffuse neutral IGM. See Carilli et al. (2002) for detail.

5 The Redshifted 21 cm Observation

In Sect. 4 we discussed the cosmological 21 cm signal and showed that it is expected to be on the order of ≈ 10 mK. However, the detectable signal in the frequency range that corresponds to the epoch of reionization is composed of a number of components each with its own physical origin and statistical properties. These components are: (1) the 21 cm signal coming from the high redshift Universe. (2) galactic and extra-galactic foregrounds, (3) ionospheric influences, (4) telescope response effects (5) Radio frequency interference (RFI) (Offringa et al. 2010a,b) and (6) thermal noise (see Fig. 2). Obviously, the challenge of the experiments in the low frequency regime is to distill the cosmological signal out of this complicated mixture of influences. This will depend crucially on the ability to calibrate the data very accurately so as to correct for the ever changing ionospheric effects and variation of the instrument response with time.

Currently, there are two types of redshifted 21 cm experiments that are attempting to observe the EoR. The first type are experiments that measure the global (mean) radio signal at the frequency range of $\nu = [100 - 200]$ MHz averaged over the whole sky (hemisphere) as a function of frequency. In this radio signal the 21 cm radiation from the EoR is hidden. The expected measurement should show an increase of the intensity at higher redshifts due to the increase in the neutral fraction of H I. In particular, if the reionization process occurred rapidly such a measurement should exhibit a step-like jump in the mean brightness temperature at the redshift of reionization z_i (in this case z_i is well defined). This type of measurement is cheap and relatively easy to perform. However, given the amount of foreground contamination, especially from our Galaxy, radio frequency interference (RFI), noise and calibration errors as well as the limited amount of information contained in the data (mean intensity as a function of redshift), such experiments are in reality very hard to perform. One of those experiments, EDGES (Bowman et al. 2008), has recently reported a lower limit on the duration of the reionization process to be $\Delta z > 0.06$, thus providing a very weak constraint on reionization as most realistic simulations predict that this process occurs over a much larger span of redshift (Bowman and Rogers 2010).

The second type of experiment is interferometric experiments carried out in the frequency range of $\approx 100 - 200$ MHz, corresponding to a redshift range of $\approx 6 - 12$. This type of experiment is considered more promising. The main reason for this is that these experiments allow better control of what is being measured and contain a huge amount of information so as to allow a much more accurate calibration of the instrument. Furthermore, radio interferometers are more diverse instruments that can be used to study many scientific topics besides the EoR, which makes them appealing for a much wider community. Having said that, however, one should note that the cost involved in building and running such facilities is much higher than for the global signal experiments.

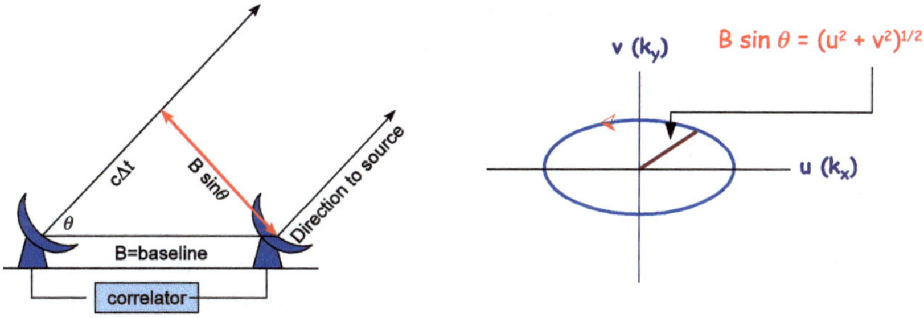

Fig. 18 *Left hand panel*: A sketch showing the basic principle of radio interferometry, the delay time Δt between the two antennas is set by the direction of the observed object on the sky. *Right hand panel*: The projection of the baseline on the sky gives the uv point measured at time, t. The rotation of earth produces a track in the uv-plane which completes half of the drawn track after 12 h. The other half is produced due to the fact that the intensity is a real function, so its Fourier transform has a complex symmetry. The width of the uv-track is set by the size of the station, i.e., the larger the station the thicker its tracks

5.1 Radio Interferometry and the Calibration Problem

Interferometers measure the spatial correlation of the electric field vector emanating from a distant source in the sky, $\mathbf{E}(\mathbf{R}, t)$, located at position $\mathbf{R}$ and measured at time t. The sketch presented in the left hand panel of Fig. 18 shows the basic principle of interferometry. The two stations (dishes) receive a wavefront from a distant source and the receivers are timed to account for the difference in the pathway to the two stations which obviously depends on the source location on the sky. The signals measured at the two stations, taken with the appropriate time difference Δt, are then cross-correlated (see for example Taylor et al. 1999; Thompson et al. 2001).

The measured spatial correlation of the electric field between two interferometric elements (stations) i and j is called the "visibility" and is approximately given by Taylor et al. (1999), Thompson et al. (2001):

$$V_\nu(\mathrm{u}, \mathrm{v}) = \int A(l, m; \nu) I_\nu(l, m) e^{i(\mathrm{u}l + \mathrm{v}m)} dl\, dm, \tag{23}$$

where A is the normalized station response pattern and I_ν is the observed intensity at frequency ν. The coordinates l and m are the projections (direction cosines) of the source in terms of the baseline in units of wavelength. As a side note, here we ignore the effect of the Earth's curvature, the so called w-projection. From this equation it is clear that the observed visibility is basically the Fourier transform of the intensity measured at the coordinates u and v. Notice that coordinates u and v depend on the baseline and its direction relative to the source position (see the right hand panel of Fig. 18). Therefore, the coordinates u and v produced by a given baseline vary with time due to Earth's rotation and will create an arc in the uv plane that completes half of the drawn track after 12 h as seen in the right hand panel of Fig. 18.

The other half is produced due the fact that the intensity is a real function. The width of the uv-track is set by the size of the station, such that large stations produce thick tracks. I will discuss the issue of uv coverage in more detail below. One also should note that the coordinates u and v are a function of wavelength, namely their value will change as a function of frequency, which one has to take into account when combining or comparing results from different frequencies.

In the interferometric visibilities there always exist errors introduced by the sky, the atmosphere (e.g. troposphere and ionosphere), the instrument (e.g. beam-shape, frequency response, receiver gains etc.) and by Radio Frequency Interference (RFI). The process of estimating and reducing the errors in these measurements is called calibration and is an essential step before understanding the measured data. Calibration normally involves knowing very well the position and intensities of the bright sources within and without the field of view of the radio telescope and using them to correct for the ionospheric and instrumental effects introduced into the data (Hamaker et al. 1996; Kazemi et al. 2011; Pearson and Readhead 1984; Yatawatta et al. 2009). This is similar to the adaptive optics techniques used in the optical regime except that here one needs to account for the variations in polarization of the radiation as well as in its total intensity.

Since most current instruments are composed of simple dipoles as their fundamental elements which have a polarized response (preferred x and y direction), the main danger in insufficient calibration lies in the possible leakage of polarized components into total intensity, thereby severely polluting the signal (see e.g., Jelić et al. 2010). That is to say, since the cosmological signal is not expected to be polarized, if the polarized response of the instrument is not very well understood and taken into account it will mix some of the polarization that exists in the Galactic foregrounds (see Sect. 5.5) with the cosmological signal and create a spurious signal that can not be distinguished from the cosmological signal. Hence, a very accurate calibration of these instruments is absolutely needed. Another issue one needs to deal with is that of the Radio Frequency interference, but we will not discuss it here and refer the reader instead to the papers by Offringa et al. (2010a,b).

5.2 Current and Future EoR Experiments

Currently, there are a number of new generation radio telescopes, GMRT,[2] LOFAR,[3] MWA,[4] 21CMA[5] and PAPER,[6] that plan to capture the lower redshift part of the δT_b evolution ($z \lesssim 12$). Unfortunately, however, none of these experiments has enough

[2]Giant Metrewave Telescope, http://gmrt.ncra.tifr.res.in

[3]Low Frequency Array, http://www.lofar.org

[4]Murchinson Widefield Array, http://www.mwatelescope.org/

[5]21 Centimeter Array, http://21cma.bao.ac.cn/

[6]Precision Array to Probe EoR, http://astro.berkeley.edu/~dbacker/eor

signal-to-noise to provide images of the EoR as it evolves with redshift. Instead, they are all designed to detect the signal statistically. In what follows I will focus on LOFAR more than the other telescopes, simply because this is the instrument I know best, but the general points I will make are applicable to the other telescopes as well.

The LOw Frequency ARray (LOFAR) is a European telescope built mostly in the Netherlands and has two observational bands, a low band and a high band covering the frequency range of 30–85 MHz and 115–230 MHz, respectively. The high band array is expected to be sensitive enough to measure the redshifted 21 cm radiation coming from the neutral IGM within the redshift range of $z = 11.4$ (115 MHz) to $z = 6$ (203 MHz), with a resolution of 3–4 arcmin and a typical field of view of $\sim$120 square degrees (with 5 beams) and a sensitivity on the order of 80 mK per resolution element for a 1 MHz frequency bandwidth. At frequencies below the FM band, probed by the low band array, the LOFAR sensitivity drops significantly and the sky noise increases so dramatically (roughly like $\approx \nu^{-2.6}$) that detection of H I signals at these frequencies is beyond the reach of LOFAR (Harker et al. 2010; Jelić et al. 2008; Labropoulos et al. 2009) and all other current generation telescopes for that matter. Figure 19 shows an artistic impression of the LOFAR telescope and its spread over Europe (left hand panel). The right hand panel shows the inner most center of the core located in the north of the Netherlands and shows the two types of stations used in the array.

In the future, SKA[7] (Carilli et al. 2004) will significantly improve on the current instruments in two major ways. Firstly, it will have at least an order of magnitude higher signal-to-noise which will allow us to actually image the reionization process. It will also give us access to the Universe's *dark ages* up to redshifts as high as $z \approx 30$ (assuming lowest frequency of about 50 MHz), hence, providing crucial information about cosmology which none of the current telescopes is able to probe. Thirdly, SKA will have a resolution better by a factor of few, at least, relative to the current telescopes (Zaroubi 2010). These three advantages—sensitivity, resolution and frequency coverage—will not only improve on the understanding we gain with current telescopes but give the opportunity to address a host of fundamental issues that current telescopes will not be able to address at all. Here, I give a few examples: (1) due to the limited resolution and poor signal-to-noise of current telescopes, the nature of the ionizing sources is expected to remain poorly constrained; (2) the mixing between the astrophysical effects and the cosmological evolution is severe during the EoR but much less so during the *dark ages*, an epoch beyond the reach of the current generation of telescopes, but within SKA's reach; (3) at redshifts larger than 30, the 21 cm could potentially provide very strong constraints, potentially much more so than the CMB, on the primordial non-gaussianity of the cosmological density field, which is essential in order to distinguish between theories of the very early Universe (e.g., between different inflationary models).

---

[7] Square Kilometer Array, http://www.skatelescope.org/

Fig. 19 *Left hand panel*: An artists impression of the layout of the LOFAR telescope over Western Europe. For the EoR, only the central part of the telescope is relevant. (courtesy of Peter Prijs). *Right hand panel*: The very central area of LOFAR. This circular area is know as the superterp and is the heart of the LOFAR core. The high-band array stations (covered in black blastic sheets) are clearly seen in this picture. In between one can also see the Low-Band Array antennas

5.3 Station Configuration and uv Coverage

In principle, Fourier space measurement and real space measurement are equivalent. However, this is only true if one has a perfect coverage of both spaces. In reality, each baseline will cover a certain line in the so called uv plane which needs to be convolved with the width of the track (see right hand panel of Fig. 18). The combination of all the tracks of the array produces the uv coverage of the interferometer. The low frequency arrays must be configured so that they have a very good uv coverage. This is crucial to the calibration effort of the data where a filled uv plane is important for obtaining precise Local (Nijboer et al. 2006) and Global (Smirnov and Noordam 2004) Sky models (LSM/GSM; i.e. catalogues of the brightest, mostly compact, sources in and outside of the beam, i.e. local versus global). It is also crucial for the ability to accurately fit for the foregrounds (Harker et al. 2009; Jelić et al. 2008) and to the measurement of the EoR signal power spectrum (Bowman et al. 2006; Harker et al. 2010; Hobson and Maisinger 2002; Santos et al. 2005).

The uv coverage of an interferometric array depends on the layout of the stations (interferometric elements), their number and size as well as on the integration time, especially, when the number of stations is not large enough to have a good instantaneous uv coverage.

For a given total collecting area one can achieve a better uv coverage by having smaller elements (stations). For example LOFAR has chosen to have large stations resulting in about $\approx 10^3$ baselines in the core area. Such a small number of baselines needs about 5–6 h of integration time per field in order to fill the uv plane (using the Earth's rotation). In comparison, MWA which has roughly 1/3 of the total collecting area of LOFAR but chose to have smaller stations with about $\approx 10^5$ baselines resulting in an almost instantaneous full uv coverage.

The decision on which strategy to follow has to do with a number of considerations that include the ability to store the raw visibilities, hence, allowing for a better calibration and an acceptable noise level for both the foreground extraction needs as well the power spectrum measurement (see the following sections; Sects. 5.4 and 5.5). A compromise between these issues as well the use of the telescopes for scientific projects other than the EoR is what drives the decision on the specific layout of the antennas.

5.4 Noise Issues

In the low frequency regime the random component of the noise, i.e., the thermal noise, is set by two effects: the sky noise and the receiver noise. At frequencies ν below ≈160 MHz the sky is so bright that the dominant source of noise is the sky itself, whereas at higher frequencies the receiver noise starts to be more important. The combination of the two effects is normally written in terms of the so called system temperature, T_{sys}. One can show that the thermal noise level for a given visibility, i.e., uv point, is,

$$\Delta V(u,v) \approx \frac{2k_B T_{sys}}{\epsilon dA\sqrt{Bt}}, \tag{24}$$

where ϵ is the efficiency of the system, dA is the station area, B is the bandwidth and t is the observation time (see e.g., Morales 2005). This expression is simple to understand in that the more one observes—either in terms of integration time, frequency bandwidth or station collecting area—the less uncertainty one has. Obviously, if the signal we are after is well localized in either time, space or frequency the relevant noise calculation should take that into account.

In order to calculate the noise in the 3D power spectrum, the main quantity we are after, one should remember that the frequency direction in the observed datacube can be mapped one-to-one with the redshift, which in turn can be easily translated to distance, whereas the u and v coordinates are in effect Fourier space coordinates. Therefore, to calculate the power spectrum one should first Fourier transform the data cube along the frequency direction. Following Morales' work (Morales 2005) I will call the new Fourier space coordinate η (with $d\eta$ resolution), which together with u and v defines the Fourier space vector $\mathbf{u} = \{u, v, \eta\}$. From this, one can calculate the noise contribution to the power spectrum at a given $|\mathbf{u}|$,

$$P_{noise}(|\mathbf{u}|) \approx 2N_{beam}^{-1} N_{cell}^{-1/2} \left(\frac{2k_B T_{sys}}{\epsilon dA d\eta}\right)^2 \frac{1}{Bn(|\mathbf{u}|)t}, \tag{25}$$

where N_{beam} is the number of simultaneous beams that could be measured, N_{cell} is the number of independent Fourier samplings per annulus and $n(|\mathbf{u}|)$ is the number of baselines covering this annulus (Morales 2005). Note that $n(|\mathbf{u}|)$ is proportional

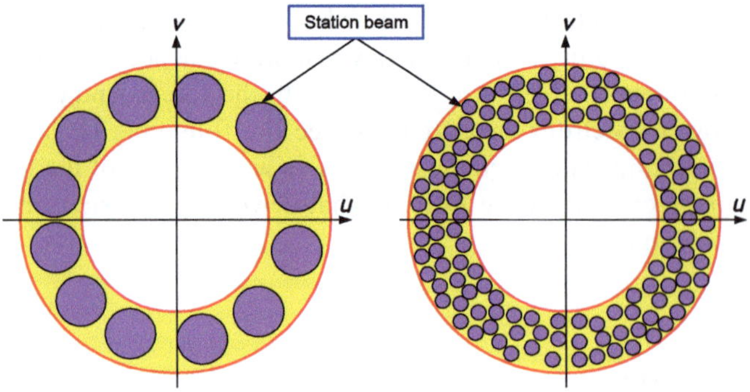

Fig. 20 This figure shows how two different experiments might sample an annulus in the uv plane. The size of uv point is given by the station (interferometric element) size where a larger station (*left hand panel*) has a larger footprint relative to the smaller station case (*right hand panel*) in the uv plane; the footprint is shown by the purple circles. Even though the sampled area in the two cases might be the same, the fact that smaller stations sample the annulus more results in an increased accuracy in their estimation of the power spectrum

to the square of the number of stations, hence, $n(|\mathbf{u}|)dA^2$ is proportional to the square of the total collecting area of the array regardless of the station size. This means that in rough terms the noise power spectrum measurement does not depend only on the total collecting area, bandwidth and integration time, it also depends the number of stations per annulus. This is easy to understand as follows. The power in a certain Fourier space annulus is given by the variance of the measured visibilities in the annulus which carries uncertainty proportional to the inverse square root of the number of points. This point is demonstrated in Fig. 20 (Morales 2005; Morales and Hewitt 2004; Zaroubi 2010).

5.5 The Foregrounds

The foregrounds in the frequency regime (40–200 MHz) are very bright and dominate the sky. In fact the amplitude of the foreground contribution, T_{sky}, at 150MHz is about 4 orders of magnitude larger than that of the expected signal. However, since we are considering radio interferometers the important part of the foregrounds is that of the fluctuations and not the mean signal, which reduces the ratio between them and the cosmological signal to about 2–3 orders of magnitude, which is still a formidable obstacle to surmount.

The most prominent foreground is the synchrotron emission from relativistic electrons in the Galaxy: this source of contamination contributes about 75 % of the foregrounds. Other sources that contribute to the foregrounds are radio galaxies, galaxy clusters, resolved supernovae remnants and free–free emission,

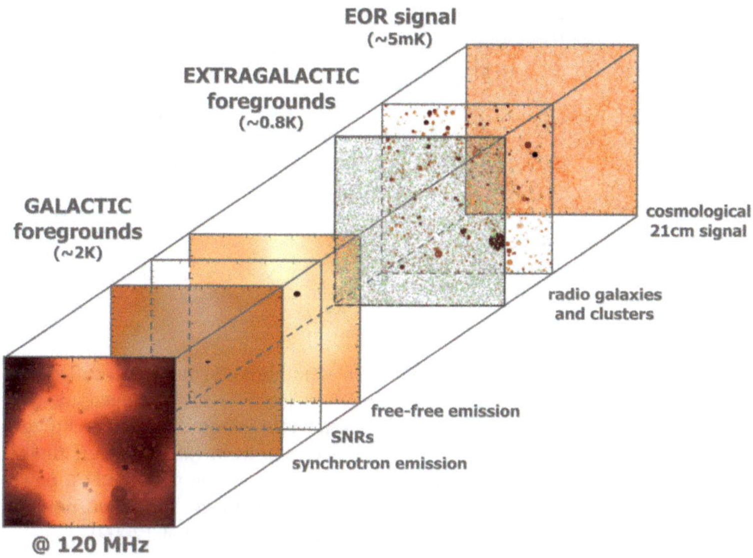

Fig. 21 A figure showing the various cosmological and galactic components that contribute to the measured signal at a given frequency. The slices are color coded with different color tables owing to the vast difference in the range of brightness temperature in each component. The figure also shows the rms of the galactic foregrounds, extra galactic foregrounds and cosmological signal

which together provide 25 % of the foreground contribution (see e.g., Shaver et al. 1999). Figure 21 shows the simulated foreground contribution at 120 MHz taking into account all the foreground sources mentioned.

Observationally, the regime of frequencies relevant to the EoR is obviously not very well explored. There are several all-sky maps of the total Galactic diffuse radio emission at different frequencies and angular resolutions. The 150 MHz map of Landecker and Wielebinski (1970) is the only all-sky map in the frequency range (100 – 200 MHz) relevant for the EoR experiments, but has only 5° resolution.

In addition to current all-sky maps, a number of recent dedicated observations have given estimates of Galactic foregrounds in small selected areas. For example, Ali et al. (2008) have used 153 MHz observations with GMRT to characterize the visibility correlation function of the foregrounds. Rogers and Bowman (2008) have measured the spectral index of the diffuse radio background between 100 and 200 MHz. Pen et al. (2009) have set an upper limit to the diffuse polarized Galactic emission.

Recently, a comprehensive program was initiated by the LOFAR-EoR collaboration to directly measure the properties of the Galactic radio emission in the frequency range relevant for the EoR experiments. The observations were carried out using the Low Frequency Front Ends (LFFE) on the WSRT radio telescope. Three different fields were observed. The first field was a highly polarized region

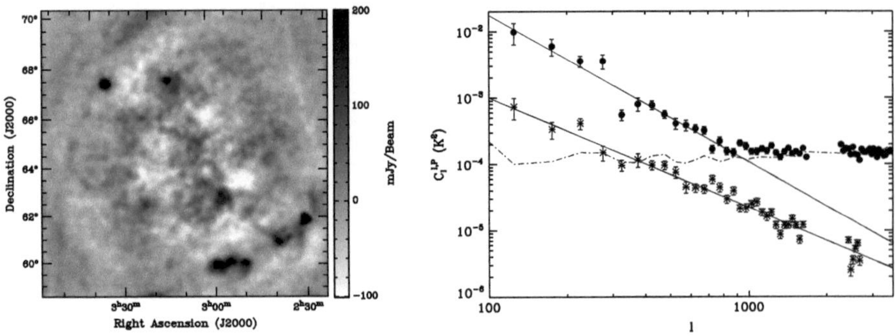

Fig. 22 *Left hand panel*: Stokes I map of the Galactic emission in the so-called Fan region, at Galactic coordinates $l = 137°$ and $b = +8°$ in the 2nd Galactic quadrant (Brouw and Spoelstra 1976; Haverkorn et al. 2003). The conversion factor is from flux (Jansky) to temperature is 1 Jy $\text{beam}^{-1} = 105.6$ K. *Right hand panel*: power spectrum (*filled circles*: total intensity; *asterisks*: polarized intensity) of the Galactic emission in Fan region with the best power-law fit. The plotted 1 σ error bars only account for the statistical errors. Power spectra are computed in the inner $6° \times 6°$ square of the map. This figure is taken from Bernardi et al. (2009)

known as the Fan region in the 2nd Galactic quadrant at a low Galactic latitude of $\sim$10° (Bernardi et al. 2009). This field is not ideal for measuring the EoR but it is a good field to learn from about calibration issues and about the influence of strong polarization.

The second field was a very cold region in the Galactic halo ($l \sim 170°$) around the bright radio quasar 3C196, and the third was a region around the North Celestial Pole (NCP) (Bernardi et al. 2010). In the Fan region fluctuations of the Galactic diffuse emission were detected at 150 MHz for the first time (see Fig. 22). The fluctuations were detected both in total and polarized intensity, with an rms of 14 K (13 arcmin resolution) and 7.2 K (4 arcmin resolution) respectively (Bernardi et al. 2009). Their spatial structure appeared to have a power law behavior with a slope of -2.2 ± 0.3 in total intensity and -1.65 ± 0.15 in polarized intensity (see Fig. 22). Note that, due to its strong polarized emission, the "Fan region" is not a representative part of the high Galactic latitude sky.

The foregrounds in the context of the EoR measurements have been studied theoretically by various authors (Shaver et al. 1999; Di Matteo et al. 2002, 2004; Cooray 2004; Santos et al. 2005; Jelić et al. 2008; Gleser et al. 2008; Wilman et al. 2008; de Oliveira-Costa et al. 2008; Sun et al. 2008; Waelkens et al. 2009; Sun and Reich 2009; Bowman et al. 2009). The first comprehensive simulation of the EoR foregrounds was carried out by Jelić et al. (2008) focusing on the LOFAR-EoR project. The Jelić et al. model takes into account the Galactic diffuse synchrotron & free–free emission, synchrotron emission from Galactic supernova remnants and extragalactic emission from radio galaxies and clusters, both in total intensity and polarization. The simulated foreground maps, in their angular and frequency characteristics, are similar to the maps expected from the LOFAR-EoR experiment (see Fig. 21).

One major problem faced when considering the LOFAR-EoR data is disentangling the desired cosmological signal from the foreground signals. Even though the foregrounds are very prominent they are very smooth along the frequency direction (Shaver et al. 1999; Jelić et al. 2008, 2010; Bernardi et al. 2009, 2010), as opposed to the cosmological signal that fluctuates along the same direction. Hence, the separation of the two is, at a first glance, very simple. One would fit a smooth function to the data along the frequency direction and subtract it to obtain the desired signal. In reality, however, things are much more complicated as the existence of thermal noise and systematic errors due to calibration imperfections make the extraction much harder. In addition, the foregrounds are partially polarized, with a complicated structure along the frequency direction. The confluence of this with the ionospheric distortions and the polarized instrumental response makes it imperative to calibrate the data very accurately over a very wide field in order to obtain a very high dynamic range of observations. These factors make the fitting non-trivial, that might result in either under-fitting or over-fitting the signal. In the former case the deduced signal retains a large contribution of the foregrounds and produce a spurious "signal". Whereas in the over-fitting case one fits out the foregrounds and some of the signal resulting in an underestimation of the cosmological signal.

The simplest method for foreground removal in total intensity as a function of frequency is a polynomial fitting performed on the log-log scale which reduces to a power law in the first order case (Jelić et al. 2008). However, one should be careful in choosing the order of the polynomial to perform the fitting. If the order of the polynomial is too small, the foregrounds will be under-fitted and the EoR signal could be dominated and corrupted by the fitting residuals, while if the order of the polynomial is too big, the EoR signal could be fitted out. Arguably, it would be better to fit the foregrounds non-parametrically, i.e., allowing the data to determine their shape rather than selecting some functional form in advance and then fitting its parameters (see Harker et al. 2009).

After foreground subtraction from the EoR observations, the residuals will be dominated by instrumental noise, i.e., the level of the noise is expected to be an order of magnitude larger than the EoR signal (assuming 300 h of observation with LOFAR). Thus, general statistical properties of the noise should be determined and used to statistically detect the cosmological 21 cm signal, e.g., the variance of the EoR signal over the image, σ^2_{EoR}, as a function of frequency (redshift) obtained by subtracting the variance of the noise, $\sigma^2_{\mathrm{noise}}$, from that of the residuals, $\sigma^2_{\mathrm{residuals}}$. It has been shown the such statistical detection of the EoR signal using the fiducial model of the LOFAR-EoR experiment is possible (Jelić et al. 2008) (see Fig. 23). Similar results by using different statistics are the skewness of the one-point distribution of brightness temperature of the EoR signal, measured as a function of observed frequency (Harker et al. 2009), and the power spectrum of variations in the intensity of redshifted 21 cm radiation from the EoR (Harker et al. 2010).

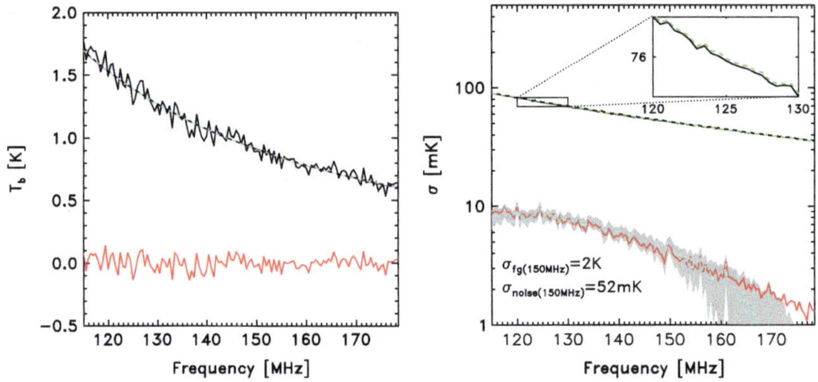

Fig. 23 This figure shows the ability to statistically extract the EoR signal from the foregrounds. Please notice the difference in the vertical axis units between the two panels. *Left hand panel*: One line of sight (one pixel along frequency) of the LOFAR-EoR data maps (*black solid line*), smooth component of the foregrounds (*dashed black line*), fitted foregrounds (*dashed cyan line*) and residuals (*red solid line*) after taking out of the foregrounds. *Right hand panel*: Statistical detection of the EoR signal from the LOFAR-EoR data maps that include diffuse components of the foregrounds and realistic instrumental noise ($\sigma_{noise}(150\text{ MHz}) = 52\text{ mK}$). *Black dashed line* represents standard deviation (σ) of the noise as a function of frequency, *cyan dashed line* is the σ of the residuals after taking out smooth component due to the foregrounds and the *red solid line* the σ of the original EoR signal. The *grey shaded surface* represents the 90 % of detected EoR signals from 1,000 independent realisations of the noise, while the *cyan dashed line* is the mean of the detected EoR signal. Note that the y-axis is in logarithmic scale (Jelić et al. 2008)

6 The Statistics of the Observed Cosmological Signal

6.1 The 21 cm Power Spectrum

One of the main goals of the EoR projects is to measure the power spectrum of variations in the intensity of redshifted 21 cm radiation from the EoR (Barkana and Loeb 2005; Bharadwaj and Ali 2004; Morales and Hewitt 2004; Morales 2005; Zaldarriaga et al. 2004). As shown in Eq. (22) the power spectrum depends on a number of astrophysical and cosmological quantities. The sensitivity with which this power spectrum can be estimated depends on the level of thermal noise (25), sample variance and systematic errors arising from the extraction process, in particular from the subtraction of foreground contamination. In the LOFAR case (Harker et al. 2009, 2010; Jelić et al. 2008, 2010; Labropoulos et al. 2009), for example, the extraction process is modeled using realistic simulations of the cosmological signal, the foregrounds and the noise. In doing so we estimate the sensitivity of the LOFAR EoR experiment to the redshifted 21 cm power spectrum. Detection of emission from the EoR should be possible within 300 h of observation with a single station beam. Integrating for longer, and synthesizing multiple station beams within

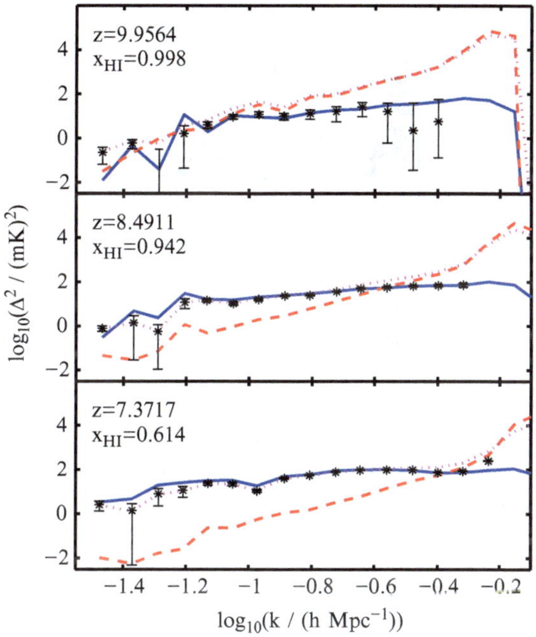

Fig. 24 Power spectra of the cosmic signal (*blue solid line*), the noise (*red dashed line*), the residuals (*magenta dotted line*) and the extracted signal (black points with error bars) at three different redshifts. Here the assumption is that, like in the real experiment, the uv coverage is frequency-dependent. Furthermore, the field is assumed to have been observed for 300 h per frequency channel with a single station beam and the foreground fitting is done using the so called Wp method but performed in Fourier space (Harker et al. 2010)

the primary (tile) beam, will then enable us to extract progressively more accurate estimates of the power at a greater range of scales and redshifts (see Fig. 24 taken from Harker et al. 2010).

6.2 *High Order Statistics*

Given the nature of the reionization process the expected signal is non-Gaussian, hence using high order statistics to characterize the data can reveal information that the power spectrum does not include. The left hand panel of Fig. 25 shows the Probability Density Function (PDF) of the brightness temperature at four different redshifts; the PDF is clearly non Gaussian in all four cases. Therefore, high order moments, like the skewness, as a function of redshift could be a useful tool for signal extraction in the presence of realistic overall levels of foregrounds and noise. Harker et al. (2009), (see also Gleser et al. 2006; Ichikawa et al. 2010) has shown that the cosmological signal, under generic assumptions, has a very characteristic pattern in the skewness as a function of redshift (the right hand panel of Fig. 25). At sufficiently high redshifts the signal is controlled by the cosmological density fluctuations which, in the linear regime, are Gaussian. At lower redshifts, and as nonlinearity becomes important, the signal starts getting a slightly positive skewness. As the ionization bubbles begin to show up the skewness starts veering towards 0 until it crosses to the negative side when the weight of the ionized bubbles

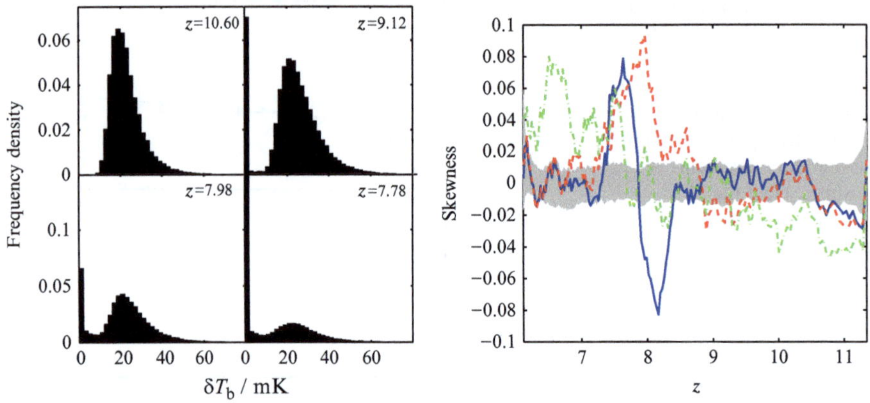

Fig. 25 *Left hand panel*: The distribution of δT_b in a certain cosmological simulation of reionization (Iliev et al. 2008) at four different redshifts, showing how the PDF evolves as reionization proceeds. Note that the y-axis scale in the top two panels is different from that in the bottom two panels. The delta-function at $\delta T_b = 0$ grows throughout this period while the rest of the distribution retains a similar shape. The bar for the first bin in the bottom-right hand panel has been cut off; approximately 58 % of points are in the first bin at $z = 7.78$ (Harker et al. 2009). *Right hand panel*: Skewness of the fitting residuals from data cubes with uncorrelated noise, but in which the residual image has been denoised by smoothing at each frequency before calculating the skewness. The three lines correspond to results from three different simulations (Thomas et al. 2009; Iliev et al. 2008). Each line has been smoothed with a moving average (boxcar) filter with a span of nine points. The *grey, shaded area* shows the errors, estimated using 100 realizations of the noise (see Harker et al. 2009)

becomes more important than the high density outliers—note high density outliers are likely to ionize first—but the distribution is still dominated by the density fluctuations. At lower redshifts the bubbles dominate the PDF and the neutral areas become the "new" outliers giving rise to a sharp positive peak to the skewness. At redshift around 6 the instrument noise, assumed to be Gaussian, dominates driving the skewness again towards zero. Exploiting this characteristic behavior might allow us to pick up the cosmological signal with this high order statistic.

6.3 Cross-Correlating the LOFAR-EoR Data with Other Data Sets

Given the challenges and uncertainties involved in measuring the redshifted 21cm signal from the EoR, it is vital to corroborate this result with other probes of the EoR. Namely, other astrophysical data that probe the EoR signal. The list of such data is long but here we focus on the most promising two such probes: CMB maps and high redshift galaxy catalogs.

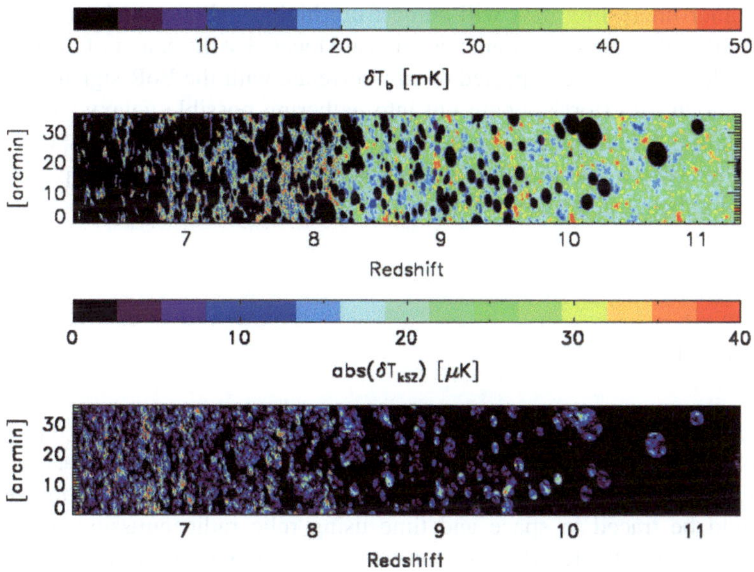

Fig. 26 A slice through a simulated reionization history of the cosmological 21cm signal (*top panel*) and the so called kinematic Sunyaev-Zeldovich effect which reflects the effect of reionization on the CMB photons (*bottom panel*) as time evolves (Ostriker and Vishniac 1986; Sunyaev and Zeldovich 1972, 1980, 1981). The expected anti-correlation between the two phenomena is patently clear. Unfortunately however, the CMB data stores no redshift (time) information and the actual map that one gets is the integral of the lower panel over redshift (time), which will make the anti-correlation more challenging to detect. This figure is taken from Jelić et al. (2010)

The CMB photons are scattered by the free electrons released during the reionization process. This scattering produces anti-correlation between the CMB signal and the EoR through a number of physical processes. This has been studied in recent years by a number of groups (Alvarez et al. 2006; Adshead and Furlanetto 2008; Cooray 2004; Jelić et al. 2008, 2010; Salvaterra et al. 2005; Tashiro et al. 2008; Tashiro at al. 2010; Tashiro et al. 2011). Figure 26 shows a slice through reionization history of the 21cm signal (top panel) and the so called kinematic Sunyaev–Zeldovich effect which reflects the effect of reionization on the CMB photons (bottom panel) as time evolves (Ostriker and Vishniac 1986; Sunyaev and Zeldovich 1972, 1980, 1981). Notice the clear anti-correlation between the two maps. However, whereas the 21cm data store the redshift information in them the actual CMB data do not, instead they are sensitive to an integral of the map in the bottom panel over time (redshift). In both cases many instrumental, foreground and background effects might hamper the anti-correlation analysis.

Of special interest here are the CMB data expected to be collected by the Planck satellite, which will have a resolution comparable to the LOFAR core and, since it is an all-sky survey, will probe the same regions as the LOFAR-EoR project.

The other interesting data will come from high z galaxies and quasars. These objects harbor the sources of ionization and are located at the center of the ionization bubbles. Hence, they are expected to anti-correlate with the EoR signal. Currently, a large amount of effort is being put into gathering possible galaxy candidates at high redshifts, e.g., Lyman-α emitters and Z, Y and J-bands drop outs, etc. Recent studies have shown that the anti-correlation signal will be detectable provided the number of galaxies in the regions with EoR observations is significant (e.g. see Lidz et al. 2009).

7 Summary

The EoR, which starts about 400 million years after the Big-Bang, represents a major phase transition for hydrogen. Due to the formation of the first astrophysical sources, hydrogen in this epoch transforms from fully neutral to fully ionized. The EoR could be traced in space and time using relic radio emission that will be observed by the LOFAR radio telescope starting from the end of this year.

The EoR is determined by how and when the Universe started forming astrophysical objects and how the ionizing radiation from these objects permeates and fills the intergalactic medium. The EoR is related to many fundamental questions in cosmology, properties of the first (mini-)quasars, formation of very metal-poor stars and a slew of other important research topics in astrophysics. Hence uncovering it will have far reaching implications on the study of structure formation in the early Universe.

Currently, there are only few observational constraints on the epoch of reionization. The CMB temperature and polarization data obtained by the WMAP satellite allow measurement of the total Thomson scattering of the primordial CMB photons off intervening free electrons produced by the epoch of reionization along the line of sight. They show that the CMB intensity has only been damped by $\sim$9 %, indicating that the Universe was mostly neutral for 400 million years and then ionized. However, the Thomson scattering measurement is an integral constraint telling us little about the sources of reionization, its duration or how it propagated to fill the whole Universe.

Another constraint comes from specific features in the spectra of distant quasars, known as the Lyman α forest. These features, which are due to neutral hydrogen, indicate two important facts about reionization. First, hydrogen in the recent Universe is highly ionized, only 1 part in 10,000 being neutral. Second, the neutral fraction of hydrogen in the distant Universe suddenly increases at redshift 6.5, i.e., about 900 million years after the Big Bang, demarcating the end of the reionization process. Despite these data providing strong constraints on the ionization state of the Universe at redshifts below 6.5, they say very little about the reionization process itself. Another couple of constraints come also from the Lyman α forest systems, IGM temperature and the number of ionizing photons per baryon, suggesting the bulk of the reionization process occurs at late redshifts $z \approx [6-9]$.

A whole slew of possible constraints currently discussed in the literature are either very controversial, very weak or, as is often the case, both. Most are very interesting and exciting, but can be investigated reliably only with a new generation of instruments such as the James Webb Space Telescope, replacing the Hubble Space Telescope in the next decade.

The imminent availability of observations of redshifted 21 cm radiation from the Universe's *dark ages* and the EoR will be one of the most exciting developments in the study of cosmology and galaxy and structure formation in recent years. Currently, there are a number of instruments that are designed to measure this radiation. In this contribution I have argued that despite the many difficulties that face such measurements they will provide a major breakthrough in our understanding of this crucial epoch. In particular current radio telescopes, such as LOFAR, will be able to provide us with the global history of the EoR progression, the fluctuations power spectrum during the EoR, etc., up to $z \approx 11$. These measurements will usher the study of the high redshift Universe into a new era which will bridge, at least in part, the large gap that currently exists in observation between the very high redshift Universe ($z \approx 1,100$) as probed by the CMB and the low redshift Universe ($z \lesssim 6$).

Although the current generation of telescopes have a great promise they will also have limitations. For example they have neither the resolution, the sensitivity nor the frequency coverage to address many fundamental issues, like the nature of the first sources. Crucially, they will not provide a lot of information about the *dark ages* which is only accessible through very low frequencies in the range of $40-120$ ($z \approx 35-11$). Fortunately, in the future SKA can improve dramatically on the current instruments in terms of sensitivity, redshift coverage and resolution.

The next decade will be extremely exciting for studying the high redshift Universe, especially as these radio telescopes gradually come online, starting with LOFAR, GMRT and MWA. They promise to resolve many of the puzzles we have today pertaining to the formation and evolution of the first objects cosmology, and the physical processes in the high redshift intergalactic medium.

Acknowledgements I would like to thank Geraint Harker, Stephen Rafter and Rajat M. Thomas for careful reading of the manuscript. Many of the results shown here have been obtained in collaboration with the members of the LOFAR EoR project whose contribution I would like to acknowledge. I would like also thank the editors of this book for giving me the opportunity to write this chapter.

References

Abel, T., Bryan, G.L., Norman, M.L.: The formation and fragmentation of primordial molecular clouds. ApJ **540**, 39–44 (2000)

Abel, T. Bryan, G.L. Norman, M.L.: The formation of the first star in the Universe. Science **295**, 93–98 (2002)

Adshead, P.J., Furlanetto, S.R.: Reionization and the large-scale 21-cm cosmic microwave background cross-correlation. MNRAS **384**, 291–304 (2008)

Aghanim, N., Majumdar, S., Silk, J.: Secondary anisotropies of the CMB. Rep. Progr. Phys. **71**(6), 066902–+ (2008)

Ali, S.S., Bharadwaj, S., Chengalur, J.N.: Foregrounds for redshifted 21-cm studies of reionization: Giant Meter Wave Radio Telescope 153-MHz observations. MNRAS **385**, 2166–2174 (2008)

Allison A.C., Dalgarno A.: Spin change in collisions of hydrogen atoms. ApJ **158**, 423–425 (1969)

Alvarez, M.A., Komatsu, E., Doré, O., Shapiro, P.R.: The cosmic reionization history as revealed by the cosmic microwave background doppler-21 cm correlation. ApJ **647**, 840–852 (2006)

Baek, S., Di Matteo, P., Semelin, B., Combes, F., Revaz, Y.: The simulated 21 cm signal during the epoch of reionization: full modeling of the Ly-α pumping. A&A **495**, 389–405 (2009)

Barkana R., Loeb, A.: In the beginning: the first sources of light and the reionization of the universe. PhysRep **349**, 125–238 (2001)

Barkana, R., Loeb, A.: A method for separating the physics from the astrophysics of high-redshift 21 centimeter fluctuations. ApJL **624**, L65–L68 (2005)

Bechtold, J.: Quasar absorption lines. In: Pérez-Fournon, I., Balcells, M., Moreno-Insertis, F., Sánchez, F. (eds.) Galaxies at High Redshift, pp. 131–184 (2003)

Benson, A.J., Sugiyama, N., Nusser, A., Lacey, C.G.: The epoch of reionization. MNRAS **369**, 1055–1080 (2006)

Bernardi, G., de Bruyn, A.G., Brentjens, M.A., Ciardi, B., Harker, G., Jelić, V., Koopmans, L.V.E., Labropoulos, P., Offringa, A., Pandey, V.N., Schaye, J., Thomas, R.M., Yatawatta, S., Zaroubi S.: Foregrounds for observations of the cosmological 21 cm line. I. First Westerbork measurements of Galactic emission at 150 MHz in a low latitude field. A&A **500**, 965–979 (2009)

Bernardi, G., de Bruyn, A.G., Harker, G., Brentjens, M.A., Ciardi, B., Jelić, V., Koopmans, L.V.E., Labropoulos, P., Offringa, A., Pandey, V.N., Schaye, J., Thomas, R.M., Yatawatta, S., Zaroubi S.: Foregrounds for observations of the cosmological 21 cm line. II. Westerbork observations of the fields around 3C 196 and the North Celestial Pole. A&A **522**, A67+ (2010)

Bharadwaj S., Ali, S.S.: The cosmic microwave background radiation fluctuations from HI perturbations prior to reionization. MNRAS **352**, 142–146 (2004)

Bi, H.G., Boerner, G., Chu, Y.: An alternative model for the Ly-alpha absorption forest. A&A **266**, 1–5 (1992)

Bolton, J.S., Becker, G.D., Wyithe, J.S.B., Haehnelt, M.G., Sargent, W.L.W.: A first direct measurement of the intergalactic medium temperature around a quasar at $z = 6$. MNRAS 771 (2010)

Bolton, J.S., Haehnelt, M.G.: The observed ionization rate of the intergalactic medium and the ionizing emissivity at z ¿= 5: evidence for a photon-starved and extended epoch of reionization. MNRAS **382**, 325–341 (2007)

Bolton, J.S., Haehnelt, M.G., Viel, M., Springel, V.: The Lyman α forest opacity and the metagalactic hydrogen ionization rate at z 2-4. MNRAS **357**, 1178–1188 (2005)

Bolton, J.S., Haehnelt, M.G., Warren, S.J., Hewett, P.C., Mortlock, D.J., Venemans, B.P., McMahon, R.G., Simpson C.: How neutral is the intergalactic medium surrounding the redshift $z = 7.085$ quasar ULAS J1120+0641?. MNRAS **416**, L70–L74 (2011)

Bond J.R., Efstathiou G.: Cosmic background radiation anisotropies in universes dominated by nonbaryonic dark matter. ApJL **285**, L45–L48 (1984)

Bouwens, R.J., Illingworth, G.D., Labbe, I., Oesch, P.A., Trenti, M., Carollo, C.M., van Dokkum, P.G., Franx, M., Stiavelli, M., González, V., Magee, D., Bradley, L.: A candidate redshift z 10 galaxy and rapid changes in that population at an age of 500Myr. Nature **469**, 504–507 (2011)

Bouwens, R.J., Illingworth, G.D., Oesch, P.A., Stiavelli, M., van Dokkum, P., Trenti, M., Magee, D., Labbé, I., Franx, M., Carollo, C.M., Gonzalez V.: Discovery of z 8 galaxies in the Hubble Ultra Deep Field from Ultra-Deep WFC3/IR observations. ApJL **709**, L133–L137 (2010)

Bouwens, R.J., Illingworth, G.D., Thompson, R.I., Franx, M.: Constraints on z 10 galaxies from the Deepest Hubble Space Telescope NICMOS fields. ApJL **624**, L5–L8 (2005)

Bowman, J.D., Morales, M.F., Hewitt, J.N.: The sensitivity of first-generation epoch of reionization observatories and their potential for differentiating theoretical power spectra. ApJ **638**, 20–26 (2006)
Bowman, J.D., Morales, M.F., Hewitt, J.N.: Foreground contamination in interferometric measurements of the redshifted 21 cm power spectrum. ApJ **695**, 183–199 (2009)
Bowman, J.D., Rogers, A.E.E.: A lower limit of $\Delta z > 0.06$ for the duration of the reionization epoch. Nature **468**, 796–798 (2010)
Bowman, J.D., Rogers, A.E.E., Hewitt, J.N.: Toward empirical constraints on the global redshifted 21 cm brightness temperature during the epoch of reionization. ApJ **676**, 1–9 (2008)
Bromm, V., Coppi, P.S., Larson, R.B.: The formation of the first stars. I. The primordial star-forming cloud. ApJ **564**, 23–51 (2002)
Bromm, V., Larson, R.B.: The first stars. ARA&A **42**, 79–118 (2004)
Bromm, V., Loeb, A.: High-redshift gamma-ray bursts from Population III progenitors. ApJ **642**, 382–388 (2006)
Brouw, W.N., Spoelstra, T.A.T.: Linear polarization of the galactic background at frequencies between 408 and 1411 MHz. Reductions. A&AS **26**, 129–+ (1976)
Bunker, A.J., Wilkins, S., Ellis, R.S., Stark, D.P., Lorenzoni, S., Chiu, K., Lacy, M., Jarvis, M.J., Hickey, S.: The contribution of high-redshift galaxies to cosmic reionization: new results from deep WFC3 imaging of the Hubble Ultra Deep Field. MNRAS **409**, 855–866 (2010)
Calverley, A.P., Becker, G.D., Haehnelt, M.G., Bolton, J.S.: Measurements of the ultraviolet background at 4.6 ¡ z ¡ 6.4 using the quasar proximity effect. MNRAS **412**, 2543–2562 (2011)
Carilli, C.L., Furlanetto, S., Briggs, F., Jarvis, M., Rawlings, S., Falcke, H.: Probing the dark ages with the square kilometer array. New Astron. Rev. **48**, 1029–1038 (2004)
Carilli, C.L., Gnedin, N., Furlanetto, S., Owen, F.: Observations of HI 21-cm absorption by the neutral IGM during the epoch of re-ionization with the square kilometer array. New Astron. Rev. **48**, 1053–1061 (2004)
Carilli, C.L., Gnedin, N.Y., Owen, F.: H I 21 centimeter absorption beyond the epoch of reionization. ApJ **577**, 22–30 (2002)
Cen, R., McDonald, P.: Evolution of the ionizing radiation background and star formation in the aftermath of cosmological reionization. ApJ **570**, 457–462 (2002)
Cen, R., Miralda-Escudé, J., Ostriker, J.P., Rauch, M.: Gravitational collapse of small-scale structure as the origin of the Lyman-alpha forest. ApJL **437**, L9–L12 (1994)
Chen X., Kamionkowski, M.: Particle decays during the cosmic dark ages. PhRvD **70**(4), 043502–+ (2004)
Choudhury, T.R., Ferrara, A.: Physics of cosmic reionization. ArXiv Astrophysics e-prints (2006)
Chuzhoy, L., Alvarez, M.A., Shapiro, P.R.: Recognizing the first radiation sources through their 21 cm signature. ApJL **648**, L1–L4 (2006)
Ciardi, B., Bianchi, S., Ferrara, A.: Lyman continuum escape from an inhomogeneous interstellar medium. MNRAS **331**, 463–473 (2002)
Ciardi, B., Ferrara, A.: The first cosmic structures and their effects. Space Sci. Rev. **116**, 625–705 (2005)
Ciardi, B., Ferrara, A., Marri, S., Raimondo, G.: Cosmological reionization around the first stars: Monte Carlo radiative transfer. MNRAS **324**, 381–388 (2001)
Ciardi, B., Madau, P.: Probing beyond the epoch of hydrogen reionization with 21 centimeter radiation. ApJ **596**, 1–8 (2003)
Cooray, A.: Cross-correlation studies between CMB temperature anisotropies and 21cm fluctuations. PhRvD **70**(6), 063509–+ (2004)
Cooray, A., Sullivan, I., Chary, R.-R., Bock, J.J., Dickinson, M., Ferguson, H.C., Keating, B., Lange, A., Wright, E.L.: IR background anisotropies in Spitzer GOODS images and constraints on first galaxies. ApJL **659**, L91–L94 (2007)
de Oliveira-Costa, A., Tegmark, M., Gaensler, B.M., Jonas, J., Landecker, T.L., Reich, P.: A model of diffuse Galactic radio emission from 10 MHz to 100 GHz. MNRAS **388**, 247–260 (2008)
Di Matteo, T., Ciardi, B., Miniati, F.: The 21-cm emission from the reionization epoch: extended and point source foregrounds. MNRAS **355**, 1053–1065 (2004)

Di Matteo, T., Perna, R., Abel, T., Rees, M.J.: Radio foregrounds for the 21 centimeter tomography of the neutral intergalactic medium at high redshifts. ApJ **564**, 576–580 (2002)
Dijkstra, M., Haiman, Z., Loeb, A.: A limit from the x-ray background on the contribution of quasars to reionization. ApJ **613**, 646–654 (2004)
Dijkstra, M., Haiman, Z., Spaans, M.: Lyα Radiation from collapsing Protogalaxies. I. Characteristics of the emergent spectrum. ApJ **649**, 14–36 (2006)
Doré, O., Holder, G., Alvarez, M., Iliev, I.T., Mellema, G., Pen, U.-L., Shapiro, P.R.: Signature of patchy reionization in the polarization anisotropy of the CMB. PhRvD **76**(4), 043002–+ (2007)
Dove, J.B., Shull, J.M.: Photoionization of the diffuse interstellar medium and galactic halo by OB associations. ApJ **430**, 222–235 (1994)
Dove, J.B., Shull, J.M., Ferrara, A.: The escape of ionizing photons from OB associations in disk galaxies: radiation transfer through superbubbles. ApJ **531**, 846–860 (2000)
Dunkley, J., Komatsu, E., Nolta, M.R., Spergel, D.N., Larson, D., Hinshaw, G., Page, L., Bennett, C.L., Gold, B., Jarosik, N., Weiland, J.L., Halpern, M., Hill, R.S., Kogut, A., Limon, M., Meyer, S.S., Tucker, G.S., Wollack, E., Wright, E.L.: Five-year Wilkinson microwave anisotropy probe observations: likelihoods and parameters from the WMAP data. ApJS **180**, 306–329 (2009)
Ewen, H.I., Purcell, E.M.: Observation of a line in the galactic radio spectrum: radiation from galactic hydrogen at 1,420 Mc./sec. Nature **168**, 356–+ (1951)
Fan, X. et al.: A survey of z>5.7 quasars in the sloan digital sky survey. II. Discovery of three additional quasars at z>6. AJ **125**, 1649–1659 (2003)
Fan, X. et al.: A survey of z>5.7 quasars in the sloan digital sky survey. IV. Discovery of seven additional quasars. AJ **131**, 1203–1209 (2006)
Field, G.B.: Excitation of the hydrogen 21-cm line. Proc. IRE **46**, 240–+ (1958)
Field, G.B.: The spin temperature of intergalactic neutral hydrogen. ApJ **129**, 536–+ (1959)
Field, G.B.: The time relaxation of a resonance-line profile. ApJ **129**, 551–+ (1959)
Furlanetto, S.R.: The 21-cm forest. MNRAS **370**, 1867–1875 (2006)
Furlanetto, S.R., Furlanetto, M.R.: Spin exchange rates in proton-hydrogen collisions. MNRAS **379**, 130–134 (2007)
Furlanetto, S.R., Loeb, A.: The 21 centimeter forest: radio absorption spectra as probes of minihalos before reionization. ApJ **579**, 1–9 (2002)
Furlanetto, S.R., Oh, S.P., Briggs, F.H.: Cosmology at low frequencies: The 21 cm transition and the high-redshift Universe. PhysRep **433**, 181–301 (2006)
Furlanetto, S.R., Stoever, S.J.: Secondary ionization and heating by fast electrons. MNRAS **404**, 1869–1878 (2010)
Furlanetto, S.R., Zaldarriaga, M., Hernquist, L.: The growth of H II regions during reionization. ApJ **613**, 1–15 (2004)
Giallongo, E., Cristiani, S., D'Odorico, S., Fontana, A.: A low upper limit to the lyman continuum emission of two galaxies at z =3. ApJL **568**, L9–L12 (2002)
Gleser, L., Nusser, A., Benson, A.J.: Decontamination of cosmological 21-cm maps. MNRAS **391**, 383–398 (2008)
Gleser, L., Nusser, A., Ciardi, B., Desjacques, V.: The morphology of cosmological reionization by means of Minkowski functionals. MNRAS **370**, 1329–1338 (2006)
Gnedin, N.Y., Abel, T.: Multi-dimensional cosmological radiative transfer with a Variable Eddington Tensor formalism. New Astron. **6**, 437–455 (2001)
Gunn, J.E., Peterson, B.A.: On the density of neutral hydrogen in intergalactic space. ApJ **142**, 1633–1641 (1965)
Haiman, Z., Spaans, M., Quataert, E.: Lyα cooling radiation from high-redshift halos. ApJL **537**, L5–L8 (2000)
Hamaker, J.P., Bregman, J.D., Sault, R.J.: Understanding radio polarimetry. I. Mathematical foundations. A&AS **117**, 137–147 (1996)
Harker, G., Zaroubi, S., Bernardi, G., Brentjens, M.A., de Bruyn, A.G., Ciardi, B., Jelić, V., Koopmans, L.V.E., Labropoulos, P., Mellema, G., Offringa, A., Pandey, V.N., Pawlik, A.H.,

Schaye, J., Thomas, R.M., Yatawatta, S.: Power spectrum extraction for redshifted 21-cm Epoch of Reionization experiments: the LOFAR case. MNRAS **405**, 2492–2504 (2010)
Harker, G., Zaroubi, S., Bernardi, G., Brentjens, M.A., de Bruyn, A.G., Ciardi, B., Jelić, V., Koopmans, L.V.E., Labropoulos, P., Mellema, G., Offringa, A., Pandey, V.N., Schaye, J., Thomas, R.M., Yatawatta, S.: Non-parametric foreground subtraction for 21-cm epoch of reionization experiments. MNRAS **397**, 1138–1152 (2009)
Harker, G.J.A., Zaroubi, S., Thomas, R.M., Jelić, V., Labropoulos, P., Mellema, G., Iliev, I.T., Bernardi, G., Brentjens, M.A., de Bruyn, A.G., Ciardi, B., Koopmans, L.V.E., Pandey, V.N., Pawlik, A.H., Schaye, J., Yatawatta, S.: Detection and extraction of signals from the epoch of reionization using higher-order one-point statistics. MNRAS **393**, 1449–1458 (2009)
Haverkorn, M., Katgert, P., de Bruyn, A.G.: Multi-frequency polarimetry of the Galactic radio background around 350 MHz. I. A region in Auriga around l = 161 deg, b = 16 deg. A&A **403**, 1031–1044 (2003)
Hernquist, L., Katz, N., Weinberg, D.H., Miralda-Escudé, J.: The lyman-alpha forest in the cold dark matter model. ApJL **457**, L51+ (1996)
Hobson, M.P., Maisinger, K.: Maximum-likelihood estimation of the cosmic microwave background power spectrum from interferometer observations. MNRAS **334**, 569–588 (2002)
Hogan, C.J., Rees, M.J.: Spectral appearance of non-uniform gas at high Z. MNRAS **188**, 791–798 (1979)
Holder, G.P., Haiman, Z., Kaplinghat, M., Knox, L.: The reionization history at high redshifts. II. Estimating the optical depth to Thomson scattering from cosmic microwave background polarization. ApJ **595**, 13–18 (2003)
Hu, W.: Wandering in the background: A CMB explorer. ArXiv Astrophysics e-prints (1995)
W. Hu White, M.: A CMB polarization primer. New Astron. **2**, 323–344 (1997)
Hui, L., Gnedin, N.Y.: Equation of state of the photoionized intergalactic medium. MNRAS **292**, 27–+ (1997)
Hui, L., Haiman, Z.: The thermal memory of reionization history. ApJ **596**, 9–18 (2003)
Ichikawa, K., Barkana, R., Iliev, I.T., Mellema, G., Shapiro, P.R.: Measuring the history of cosmic reionization using the 21-cm probability distribution function from simulations. MNRAS **406**, 2521–2532 (2010)
Iliev, I.T., Mellema, G., Pen, U.-L., Bond, J.R., Shapiro, P.R.: Current models of the observable consequences of cosmic reionization and their detectability. MNRAS **384**, 863–874 (2008)
Inoue, A.K., Iwata, I., Deharveng, J.-M., Buat, V., Burgarella, D.: VLT narrow-band photometry in the Lyman continuum of two galaxies at z 3. Limits to the escape of ionizing flux. A&A **435**, 471–482 (2005)
Iwata, I., Inoue, A.K., Matsuda, Y., Furusawa, H., Hayashino, T., Kousai, K., Akiyama, M., Yamada, T., Burgarella, D., Deharveng, J.-M.: Detections of lyman continuum from star-forming galaxies at z 3 through subaru/suprime-cam narrow-band imaging. ApJ **692**, 1287–1293 (2009)
Jelić, V., Zaroubi, S., Aghanim, N., Douspis, M., Koopmans, L.V.E., Langer, M., Mellema, G., Tashiro, H., Thomas, R.M.: A cross-correlation study between the cosmological 21 cm signal and the kinetic Sunyaev-Zel'dovich effect. MNRAS **402**, 2279–2290 (2010)
Jelić, V., Zaroubi, S., Labropoulos, P., Bernardi, G., de Bruyn, A.G., Koopmans, L.V.E.: Realistic simulations of the Galactic polarized foreground: consequences for 21-cm reionization detection experiments. MNRAS **409**, 1647–1659 (2010)
Jelić, V., Zaroubi, S., Labropoulos, P., Thomas, R.M., Bernardi, G., Brentjens, M.A., de Bruyn, A.G., Ciardi, B., Harker, G., Koopmans, L.V.E., Pandey, V.N.B, Schaye, J., Yatawatta, S.: Foreground simulations for the LOFAR-epoch of reionization experiment. MNRAS **389**, 1319–1335 (2008)
Kaiser, N.: Clustering in real space and in redshift space. MNRAS **227**, 1–21 (1987)
Kamionkowski, M., Kosowsky, A., Stebbins, A.: A probe of primordial gravity waves and vorticity. Phys. Rev. Lett. **78**, 2058–2061 (1997)
Kashikawa, N., Shimasaku, K., Malkan, M.A., Doi, M., Matsuda, Y., Ouchi, M., Taniguchi, Y., Ly, C., Nagao, T., Iye, M., Motohara, K., Murayama, T., Murozono, K., Nariai, K., Ohta, K.,

Okamura, S., Sasaki, T., Shioya, Y., Umemura, M.: The end of the reionization epoch probed by Lyα emitters at z = 6.5 in the Subaru Deep Field. ApJ **648**, 7–22 (2006)
Kashlinsky, A., Arendt, R.G., Mather, J., Moseley, S.H.: Tracing the first stars with fluctuations of the cosmic infrared background. Nature **438**, 45–50 (2005)
Kasuya, S., Kawasaki, M.: Early reionization by decaying particles and cosmic microwave background radiation. PhRvD **70**(10), 103519–+ (2004)
Kazemi, S., Yatawatta, S., Zaroubi, S., Lampropoulos, P., de Bruyn, A.G., Koopmans, L.V.E., Noordam, J.: Radio interferometric calibration using the SAGE algorithm. MNRAS **414**, 1656–1666 (2011)
Kramer, R.H., Haiman, Z.: Probing re-ionization with quasar spectra: the impact of the intrinsic Lyman α emission line shape uncertainty. MNRAS **400**, 1493–1511 (2009)
Labropoulos, P., Koopmans, L.V.E., Jelic, V., Yatawatta, S., Thomas, R.M., Bernardi, G., Brentjens, M., de Bruyn, G., Ciardi, B., Harker, G., Offringa, A., Pandey, V.N., Schaye, J., Zaroubi, S.: The LOFAR EoR data model: (I) Effects of noise and instrumental corruptions on the 21-cm reionization signal-extraction strategy. ArXiv e-prints (2009)
Landecker, T.L., Wielebinski, R.: The galactic metre wave radiation: A two-frequency survey between declinations $+25^o$ and -25° and the preparation of a map of the whole sky. Aust. J. Phys. Astrophys. Suppl. **16**, 1–+ (1970)
Latif, M.A., Schleicher, D.R.G., Spaans, M., Zaroubi, S.: Lyman alpha emission from the first galaxies: Implications of UV backgrounds and the formation of molecules. A&A **532**, A66 (2011)
Latif, M.A., Schleicher, D.R.G., Spaans, M., Zaroubi, S.: Lyman α emission from the first galaxies: signatures of accretion and infall in the presence of line trapping. MNRAS **413**, L33–L37 (2011)
Latif, M.A., Zaroubi, S., Spaans, M.: The impact of Lyman α trapping on the formation of primordial objects. MNRAS **411**, 1659–1670 (2011)
Lawrence, A., Warren, S.J., Almaini, O., Edge, A.C., Hambly, N.C., Jameson, R.F., Lucas, P., Casali, M., Adamson, A., Dye, S., Emerson, J.P., Foucaud, S., Hewett, P., Hirst, P., Hodgkin, S.T., Irwin, M.J., Lodieu, N., McMahon, R.G., Simpson, C., Smail, I., Mortlock, D., Folger, M.: The UKIRT Infrared Deep Sky Survey (UKIDSS). MNRAS **379**, 1599–1617 (2007)
Lazio, J., Carilli, C., Hewitt, J., Furlanetto, S., Burns, J.: The lunar radio array (LRA). In Society of Photo-Optical Instrumentation Engineers (SPIE) Conference Series, vol. 7436 of Society of Photo-Optical Instrumentation Engineers (SPIE) Conference Series (2009)
Lewis, A., Weller, J., Battye, R.: The cosmic microwave background and the ionization history of the Universe. MNRAS **373**, 561–570 (2006)
Lidz, A., Faucher-Giguère, C.-A., Dall'Aglio, A., McQuinn, M., Fechner, C., Zaldarriaga, M., Hernquist, L., Dutta, S.: A measurement of small-scale structure in the 2.2 ¡= z ¡= 4.2 Lyα forest. ApJ **718**, 199–230 (2010)
Lidz, A., Hui, L., Zaldarriaga, M., Scoccimarro, R.: How neutral is the intergalactic medium at z 6? ApJ **579**, 491–499 (2002)
Lidz, A., Zahn, O., Furlanetto, S.R., McQuinn, M., Hernquist, L., Zaldarriaga, M.: Probing reionization with the 21 cm galaxy cross-power spectrum. ApJ **690**, 252–266 (2009)
Liszt, H.: The spin temperature of warm interstellar H I. A&A **371**, 698–707 (2001)
Loeb, A., Barkana, R.: The first light. ARA&A **39**, 19–130 (2001)
Loeb, A., Zaldarriaga, M.: Measuring the small-scale power spectrum of cosmic density fluctuations through 21cm tomography prior to the epoch of structure formation. Phys. Rev. Lett. **92**(21), 211301–+ (2004)
Ma, C.-P., Bertschinger, E.: Cosmological perturbation theory in the synchronous and conformal newtonian gauges. ApJ **455**, 7–+ (1995)
Machacek, M.E., Bryan, G.L., Meiksin, A., Anninos, P., Thayer, D., Norman, M., Zhang, Y.: Hydrodynamical simulations of the Lyα forest: model comparisons. ApJ **532**, 118–135 (2000)
Mack, K.J., Wyithe, J.S.B.: Detecting the redshifted 21cm forest during reionization. ArXiv e-prints (2011)
Madau, P., Meiksin, A., Rees, M.J.: 21 centimeter tomography of the intergalactic medium at high redshift. ApJ **475**, 429–+ (1997)

Mapelli, M., Ferrara, A.: Background radiation from sterile neutrino decay and reionization. MNRAS **364**, 2–12 (2005)
Mapelli, M., Ferrara, A., Pierpaoli, E.: Impact of dark matter decays and annihilations on reionization. MNRAS **369**, 1719–1724 (2006)
Maselli, A., Ferrara, A., Gallerani, S.: Interpreting the transmission windows of distant quasars. MNRAS **395**, 1925–1933 (2009)
Maselli, A., Gallerani, S., Ferrara, A., Choudhury, T.R.: On the size of HII regions around high-redshift quasars. MNRAS **376**, L34–L38 (2007)
McLure, R.J., Dunlop, J.S., Cirasuolo, M., Koekemoer, A.M., Sabbi, E., Stark, D.P., Targett, T.A., Ellis, R.S.: Galaxies at z = 6-9 from the WFC3/IR imaging of the Hubble Ultra Deep Field. MNRAS **403**, 960–983 (2010)
Mellema, G., Iliev, I.T., Alvarez, M.A., Shapiro, P.R.: C^2-ray: A new method for photon-conserving transport of ionizing radiation. New Astron. **11**, 374–395 (2006)
Mesinger, A., Furlanetto, S.: Efficient simulations of early structure formation and reionization. ApJ **669**, 663–675 (2007)
Mesinger, A., Furlanetto, S.: The inhomogeneous ionizing background following reionization. MNRAS **400**, 1461–1471 (2009)
Mesinger, A., Furlanetto, S., Cen, R.: 21cmFAST: a fast, semi-numerical simulation of the high-redshift 21-cm signal. MNRAS **411**, 955–972 (2011)
Mesinger, A., Haiman, Z.: Evidence of a cosmological Strömgren surface and of significant neutral hydrogen surrounding the quasar SDSS J1030+0524. ApJL **611**, L69–L72 (2004)
Miralda-Escudé, J., Cen, R., Ostriker, J.P., Rauch, M.: The Ly alpha forest from gravitational collapse in the cold dark matter + lambda model. ApJ **471**, 582–+ (1996)
Mo, H., van den Bosch, F.C., White, S.: Galaxy Formation and Evolution. Cambridge University Press (2010)
Morales, M.F.: Power spectrum sensitivity and the design of epoch of reionization observatories. ApJ **619**, 678–683 (2005)
Morales, M.F., Hewitt, J.: Toward epoch of reionization measurements with wide-field radio observations. ApJ **615**, 7–18 (2004)
Morales, M.F., Wyithe, J.S.B.: Reionization and cosmology with 21-cm fluctuations. ARA&A **48**, 127–171 (2010)
Moretti, A., Campana, S., Lazzati, D., Tagliaferri, G.: The resolved fraction of the cosmic x-ray background. ApJ **588**, 696–703 (2003)
Mortlock, D.J., Warren, S.J., Venemans, B.P., Patel, M., Hewett, P.C., McMahon, R.G., Simpson, C., Theuns, T., Gonzáles-Solares, E.A., Adamson, A., Dye, S., Hambly, N.C., Hirst, P., Irwin, M.J., Kuiper, E., Lawrence, A., Röttgering, H.J.A.: A luminous quasar at a redshift of z = 7.085. Nature **474**, 616–619 (2011)
Mortonson, M.J., Hu, W.: Model-independent constraints on reionization from large-scale cosmic microwave background polarization. ApJ **672**, 737–751 (2008)
Muller, C.A., Oort, J.H.: Observation of a Line in the galactic radio spectrum: the interstellar hydrogen line at 1,420 Mc./sec., and an estimate of galactic rotation. Nature **168**, 357–358 (1951)
Murayama, T., Taniguchi, Y., Scoville, N.Z., Ajiki, M., Sanders, D.B., Mobasher, B., Aussel, H., Capak, P., Koekemoer, A., Shioya, Y., Nagao, T., Carilli, C., Ellis, R.S., Garilli, B., Giavalisco, M., Kitzbichler, M.G., Le Fèvre, O., Maccagni, D., Schinnerer, E., Smolčić, V., Tribiano, S., Cimatti, A., Komiyama, Y., Miyazaki, S., Sasaki, S.S., Koda, J., Karoji, H.: Lyα emitters at redshift 5.7 in the COSMOS field. ApJS **172**, 523–544 (2007)
Nakamoto, T., Umemura, M., Susa, H.: The effects of radiative transfer on the reionization of an inhomogeneous universe. MNRAS **321**, 593–604 (2001)
Natarajan, A., Schwarz, D.J.: Distinguishing standard reionization from dark matter models. PhRvD **81**(12), 123510–+ (2010)
Nijboer, R.J., Noordam, J.E., Yatawatta, S.B.: LOFAR self-calibration using a local sky model. In: Gabriel, C., Arviset, C., Ponz, D., Enrique, S. (eds.) Astronomical Data Analysis Software and Systems XV, vol. 351 of Astronomical Society of the Pacific Conference Series, p. 291 (2006)

Nusser, A.: The spin temperature of neutral hydrogen during cosmic pre-reionization. MNRAS **359**, 183–190 (2005)
Oesch, P.A., Bouwens, R.J., Illingworth, G.D., Carollo, C.M., Franx, M., Labbé, I., Magee, D., Stiavelli, M., Trenti, M., van Dokkum, P.G.: z 7 galaxies in the HUDF: first epoch WFC3/IR results. ApJL **709**, L16–L20 (2010)
Offringa, A.R., de Bruyn, A.G., Biehl, M., Zaroubi, S., Bernardi, G., Pandey, V.N.: Post-correlation radio frequency interference classification methods. MNRAS **405**, 155–167 (2010)
Offringa, A.R., de Bruyn, A.G., Zaroubi, S., Biehl, M.: A LOFAR RFI detection pipeline and its first results. ArXiv e-prints, (2010)
Ostriker, J.P., Vishniac, E.T.: Generation of microwave background fluctuations from nonlinear perturbations at the ERA of galaxy formation. ApJL **306**, L51–L54 (1986)
Ouchi, M., Ono, Y., Egami, E., Saito, T., Oguri, M., McCarthy, P.J., Farrah, D., Kashikawa, N., Momcheva, I., Shimasaku, K., Nakanishi, K., Furusawa, H., Akiyama, M., Dunlop, J.S., Mortier, A.M.J., Okamura, S., Hayashi, M., Cirasuolo, M., Dressler, A., Iye, M., Jarvis, M.J., Kodama, T., Martin, C.L., McLure, R.J., Ohta, K., Yamada, T., Yoshida, M.: Discovery of a giant Lyα emitter near the reionization epoch. ApJ **696**, 1164–1175 (2009)
Ouchi, M., Shimasaku, K., Furusawa, H., Saito, T., Yoshida, M., Akiyama, M., Ono, Y., Yamada, T., Ota, K., Kashikawa, N., Iye, M., Kodama, T., Okamura, S., Simpson, C., Yoshida, M.: Statistics of 207 Lyα emitters at a redshift near 7: Constraints on reionization and galaxy formation models. ApJ **723**, 869–894 (2010)
Padmanabhan, N., Finkbeiner, D.P.: Detecting dark matter annihilation with CMB polarization: Signatures and experimental prospects. PhRvD **72**(2), 023508–+ (2005)
Page, L., Hinshaw, G., Komatsu, E., Nolta, M.R., Spergel, D.N., Bennett, C.L., Barnes, C., Bean, R., Doré, O., Dunkley, J., Halpern, M., Hill, R.S., Jarosik, N., Kogut, A., Limon, M., Meyer, S.S., Odegard, N., Peiris, H.V., Tucker, G.S., Verde, L., Weiland, J.L., Wollack, E., Wright, E.L.: Three-Year Wilkinson Microwave Anisotropy Probe (WMAP) observations: polarization analysis. ApJS **170**, 335–376 (2007)
Partridge, R.B., Peebles, P.J.E.: Are young galaxies visible? ApJ **147**, 868–+ (1967)
Pawlik, A.H., Schaye, J.: TRAPHIC – radiative transfer for smoothed particle hydrodynamics simulations. MNRAS **389**, 651–677 (2008)
Pearson, T.J., Readhead, A.C.S.: Image formation by self-calibration in radio astronomy. ARA&A **22**, 97–130 (1984)
Peebles, P.J.E.: Principles of Physical Cosmology. Princeton University Press (1993)
Peebles, P.J.E., Yu, J.T.: Primeval adiabatic perturbation in an expanding Universe. ApJ **162**, 815–+ (1970)
Pen, U.-L., Chang, T.-C., Hirata, C.M., Peterson, J.B., Roy, J., Gupta, Y., Odegova, J., Sigurdson, K.: The GMRT EoR experiment: limits on polarized sky brightness at 150 MHz. MNRAS **399**, 181–194 (2009)
Pritchard, J.R., Furlanetto, S.R.: 21-cm fluctuations from inhomogeneous X-ray heating before reionization. MNRAS **376**, 1680–1694 (2007)
Pritchard, J.R., Loeb, A.: Evolution of the 21cm signal throughout cosmic history. PhRvD **78**(10), 103511–+ (2008)
Rasera, Y., Teyssier, R.: The history of the baryon budget. Cosmic logistics in a hierarchical universe. A&A **445**, 1–27 (2006)
Rauch, M.: The lyman alpha forest in the spectra of QSOs. ARA&A **36**, 267–316 (1998)
Razoumov, A.O., Cardall, C.Y.: Fully threaded transport engine: new method for multi-scale radiative transfer. MNRAS **362**, 1413–1417 (2005)
Richard, J., Pelló, R., Schaerer, D., Le Borgne, J.-F., Kneib, J.-P.: Constraining the population of 6 ¡z ¡ 10 star-forming galaxies with deep near-IR images of lensing clusters. A&A **456**, 861–880 (2006)
Ricotti, M., Ostriker, J.P.: Reionization, chemical enrichment and seed black holes from the first stars: is Population III important? MNRAS **350**, 539–551 (2004)
Ricotti, M., Ostriker, J.P.: X-ray pre-ionization powered by accretion on the first black holes - I. A model for the WMAP polarization measurement. MNRAS **352**, 547–562 (2004)

Ricotti, M., Shull, J.M.: Feedback from galaxy formation: escaping ionizing radiation from galaxies at high redshift. ApJ **542**, 548–558 (2000)

Ripamonti, E., Mapelli, M., Ferrara, A.: The impact of dark matter decays and annihilations on the formation of the first structures. MNRAS **375**, 1399–1408 (2007)

Ripamonti, E., Mapelli, M., Zaroubi, S.: Radiation from early black holes - I. Effects on the neutral intergalactic medium. MNRAS **387**, 158–172 (2008)

Ritzerveld, J., Icke, V., Rijkhorst, E.-J.: Triangulating radiation: radiative transfer on unstructured grids. ArXiv Astrophysics e-prints (2003)

Rogers, A.E.E., Bowman, J.D.: Spectral index of the diffuse radio background measured from 100 to 200 MHz. AJ **136**, 641–648 (2008)

Rudie, G.C., Steidel, C.C., Trainor, R.F., Rakic, O., Bogosavljevic, M., Pettini, M., Reddy, N., Shapley, A.E., Erb, D.K., Law, D.R.: The gaseous environment of high-z galaxies: precision measurements of neutral hydrogen in the circumgalactic medium of z 2-3 galaxies in the keck baryonic structure survey. ApJ **750**, 67 (2012)

Rybicki, G.B., Lightman, A.P.: Radiative Processes in Astrophysics. Wiley-VCH (Weinheim) (1986)

Salvaterra, R., Ciardi, B., Ferrara, A., Baccigalupi, C.: Reionization history from coupled cosmic microwave background/21-cm line data. MNRAS **360**, 1063–1068 (2005)

Salvaterra, R., Haardt, F., Ferrara, A.: Cosmic backgrounds from miniquasars. MNRAS **362**, L50–L54 (2005)

Santos, M.G., Cooray, A., Knox, L.: Multifrequency analysis of 21 centimeter fluctuations from the era of reionization. ApJ **625**, 575–587 (2005)

Santos, M.G., Ferramacho, L., Silva, M.B., Amblard, A., Cooray, A.: Fast large volume simulations of the 21-cm signal from the reionization and pre-reionization epochs. MNRAS **406**, 2421–2432 (2010)

Schaye, J., Theuns, T., Rauch, M., Efstathiou, G., Sargent, W.L.W.: The thermal history of the intergalactic medium. MNRAS **318**, 817–826 (2000)

Scheuer, P.A.G.: A sensitive test for the presence of atomic hydrogen in intergalactic space. Nature **207**, 963–+ (1965)

Scott, D., Rees, M.J.: The 21-cm line at high redshift: a diagnostic for the origin of large scale structure. MNRAS **247**, 510–+ (1990)

Shapley, A.E., Steidel, C.C., Pettini, M., Adelberger, K.L., Erb, D.K.: The direct detection of lyman continuum emission from star-forming galaxies at z 3. ApJ **651**, 688–703 (2006)

Shaver, P.A., Windhorst, R.A., Madau, P., de Bruyn, A.G.: Can the reionization epoch be detected as a global signature in the cosmic background? A&A **345**, 380–390 (1999)

Shull, J.M., van Steenberg, M.E.: X-ray secondary heating and ionization in quasar emission-line clouds. ApJ **298**, 268–274 (1985)

Smirnov, O.M., Noordam, J.E.: The LOFAR global sky model: some design challenges. In: Ochsenbein, F., Allen, M.G., & Egret, D. (eds.) Astronomical Data Analysis Software and Systems (ADASS) XIII, vol. 314 of Astronomical Society of the Pacific Conference Series, pp. 18–+, (2004)

Smith, F.J.: Hydrogen atom spin-change collisions. Planet. Space Sci. **14**, 929–+ (1966)

Sołtan, A.M.: The diffuse X-ray background. A&A **408**, 39–42 (2003)

Spergel, D.N., Bean, R., Doré, O., Nolta, M.R., Bennett, C.L., Dunkley, J., Hinshaw, G., Jarosik, N., Komatsu, E., Page, L., Peiris, H.V., Verde, L., Halpern, M., Hill, R.S., Kogut, A., Limon, M., Meyer, S.S., Odegard, N., Tucker, G.S., Weiland, J.L., Wollack, E., Wright, E.L.: Three-year Wilkinson Microwave Anisotropy Probe (WMAP) observations: implications for cosmology. ApJS **170**, 377–408 (2007)

Stark, D., Ellis, R., Richard, J.: The case for an abundant population of feeble Lyman-alpha emitting galaxies at z ¿ 8. In American Astronomical Society Meeting Abstracts, volume 38 of Bulletin of the American Astronomical Society, pp. 143.02–+ (2007)

Steidel, C.C., Pettini, M., Adelberger, K.L.: Lyman-continuum emission from galaxies at Z = 3.4. ApJ **546**, 665–671 (2001)

Sugiyama, N.: Cosmic background anisotropies in cold dark matter cosmology. ApJS **100**, 281–+ (1995)
Sun, X.H., Reich, W.: Simulated square kilometre array maps from Galactic 3D-emission models. A&A **507**, 1087–1105 (2009)
Sun, X.H., Reich, W., Waelkens, A., Enßlin, T.A.: Radio observational constraints on Galactic 3D-emission models. A&A **477**, 573–592 (2008)
Sunyaev, R.A., Zeldovich, I.B.: The velocity of clusters of galaxies relative to the microwave background - The possibility of its measurement. MNRAS **190**, 413–420 (1980)
Sunyaev, R.A., Zeldovich, I.B.: Intergalactic gas in clusters of galaxies, the microwave background, and cosmology. Astrophys. Space Phys. Rev. **1**, 1–60 (1981)
Sunyaev, R.A., Zeldovich, Y.B.: The observations of relic radiation as a test of the nature of x-ray radiation from the clusters of galaxies. Comments Astrophys. Space Phys. **4**, 173–+ (1972)
Susa, H.: Smoothed particle hydrodynamics coupled with radiation transfer. PASJ **58**, 445–460 (2006)
Tashiro, H., Aghanim, N., Langer, M., Douspis, M., Zaroubi, S.: The cross-correlation of the CMB polarization and the 21-cm line fluctuations from cosmic reionization. MNRAS **389**, 469–477 (2008)
Tashiro, H., Aghanim, N., Langer, M., Douspis, M., Zaroubi, S., Jelic, V.: Detectability of the 21-cm CMB cross-correlation from the epoch of reionization. MNRAS **402**, 2617–2625 (2010)
Tashiro, H., Aghanim, N., Langer, M., Douspis, M., Zaroubi, S., Jelić, V.: Second order cross-correlation between kinetic Sunyaev-Zel'dovich effect and 21-cm fluctuations from the epoch of reionization. MNRAS 638–+ (2011)
Taylor, G.B., Carilli, C.L., Perley, R.A. (eds.): In: Synthesis Imaging in Radio Astronomy II, vol. 180 of Astronomical Society of the Pacific Conference Series (1999)
Theuns, T., Leonard, A., Efstathiou, G., Pearce, F.R., Thomas, P.A.: P^3M-SPH simulations of the Lyalpha forest. MNRAS **301**, 478–502 (1998)
Theuns, T., Schaye, J., Zaroubi, S., Kim, T.-S., Tzanavaris, P., Carswell, B.: Constraints on reionization from the thermal history of the intergalactic medium. ApJL **567**, L103–L106 (2002)
Theuns, T., Zaroubi, S., Kim, T.-S., Tzanavaris, P., Carswell, R.F.: Temperature fluctuations in the intergalactic medium. MNRAS **332**, 367–382 (2002)
Thomas, R.M., Zaroubi, S.: Time-evolution of ionization and heating around first stars and miniqsos. MNRAS **384**, 1080–1096 (2008)
Thomas, R.M., Zaroubi, S.: On the spin-temperature evolution during the epoch of reionization. MNRAS **410**, 1377–1390 (2011)
Thomas, R.M., Zaroubi, S., Ciardi, B., Pawlik, A.H., Labropoulos, P., Jelić, V., Bernardi, G., Brentjens, M.A., de Bruyn, A.G., Harker, G.J.A., Koopmans, L.V.E., Mellema, G., Pandey, V.N., Schaye, J., Yatawatta, S.: Fast large-scale reionization simulations. MNRAS **393**, 32–48 (2009)
Thompson, A.R., Moran, J.M., Swenson, G.W. Jr.: Interferometry and Synthesis in Radio Astronomy, 2nd edn. (2001)
Valdés, M., Evoli, C., Ferrara, A.: Particle energy cascade in the intergalactic medium. MNRAS **404**, 1569–1582 (2010)
van de Hulst, H.C.: Nederlands Tijdschrift voor Natuuurkunde **11**, 210–221 (1945)
Waelkens, A., Jaffe, T., Reinecke, M., Kitaura, F.S., Enßlin, T.A.: Simulating polarized Galactic synchrotron emission at all frequencies. The Hammurabi code. A&A **495**, 697–706 (2009)
Whalen, D., Norman, M.L.: A multistep algorithm for the radiation hydrodynamical transport of cosmological ionization fronts and ionized flows. ApJS **162**, 281–303 (2006)
Wild, J.P.: The radio-frequency line spectrum of atomic hydrogen and its applications in astronomy. ApJ **115**, 206–+ (1952)
Wilman, R.J., Miller, L., Jarvis, M.J., Mauch, T., Levrier, F., Abdalla, F.B., Rawlings, S., Klöckner, H.-R., Obreschkow, D., Olteanu, D., Young, S.: A semi-empirical simulation of the extragalactic radio continuum sky for next generation radio telescopes. MNRAS **388**, 1335–1348 (2008)
Wood, K., Loeb, A.: Escape of ionizing radiation from high-redshift galaxies. ApJ **545**, 86–99 (2000)

Wouthuysen, S.A.: On the excitation mechanism of the 21-cm (radio-frequency) interstellar hydrogen emission line. AJ **57**, 31–32 (1952)

Wyithe, J.S.B., Bolton, J.S., Haehnelt, M.G.: Reionization bias in high-redshift quasar near-zones. MNRAS **383**, 691–704 (2008)

Wyithe, J.S.B., Loeb, A.: A characteristic size of 10Mpc for the ionized bubbles at the end of cosmic reionization. Nature **432**, 194–196 (2004)

Wyithe, J.S.B., Loeb, A.: A large neutral fraction of cosmic hydrogen a billion years after the Big Bang. Nature **427**, 815–817 (2004)

Xu, Y., Chen, X., Fan, Z., Trac, H., Cen, R.: The 21 cm forest as a probe of the reionization and the temperature of the intergalactic medium. ApJ **704**, 1396–1404 (2009)

Yatawatta, S., Zaroubi, S., de Bruyn, G., Koopmans, L., Noordam, J.: Radio interferometric calibration using The SAGE algorithm. In: Digital Signal Processing Workshop and 5th IEEE Signal Processing Education Workshop, 2009. DSP/SPE 2009. IEEE 13th (2009)

Yoshida, N., Abel, T., Hernquist, L., Sugiyama, N.: Simulations of early structure formation: primordial gas clouds. ApJ **592**, 645–663 (2003)

Zahn, O., Lidz, A., McQuinn, M., Dutta, S., Hernquist, L., Zaldarriaga, M., Furlanetto, S.R.: Simulations and analytic calculations of Bubble growth during hydrogen reionization. ApJ **654**, 12–26 (2007)

Zahn, O., Mesinger, A., McQuinn, M., Trac, H., Cen, R., Hernquist, L.E.: Comparison of reionization models: radiative transfer simulations and approximate, seminumeric models. MNRAS **414**, 727–738 (2011)

Zaldarriaga, M.: Polarization of the microwave background in reionized models. PhRvD **55**, 1822–1829 (1997)

Zaldarriaga, M.: Searching for fluctuations in the intergalactic medium temperature using the Lyα forest. ApJ **564**, 153–161 (2002)

Zaldarriaga, M., Furlanetto, S.R., Hernquist, L.: 21 centimeter fluctuations from cosmic gas at high redshifts. ApJ **608**, 622–635 (2004)

Zaldarriaga, M., Seljak, U.: All-sky analysis of polarization in the microwave background. PhRvD **55**, 1830–1840 (1997)

Zaroubi, S.: Probing the epoch of reionization with low frequency arrays. In: Torchinsky, S.A., van Ardenne, A., van den Brink-Havinga, T., van Es, A., Faulkner, A.J. (eds.) Widefield Science and Technology for the SKA, pp. 75–+ (2010)

Zaroubi, S., Silk, J.: LOFAR as a probe of the sources of cosmological reionization. MNRAS **360**, L64–L67 (2005)

Zaroubi, S., Thomas, R.M., Sugiyama, N., Silk, J.: Heating of the intergalactic medium by primordial miniquasars. MNRAS **375**, 1269–1279 (2007)

Zhang, L., Chen, X., Lei, Y.-A., Si, Z.-G.: Impacts of dark matter particle annihilation on recombination and the anisotropies of the cosmic microwave background. PhRvD **74**(10), 103519–+ (2006)

Zhang, Y., Anninos, P., Norman, M.L.: A multispecies model for hydrogen and helium absorbers in Lyman-Alpha forest clouds. ApJL **453**, L57+ (1995)

Zhang, Y., Anninos, P., Norman, M.L., Meiksin, A.: Spectral analysis of the Ly alpha forest in a cold dark matter cosmology. ApJ **485**, 496–+ (1997)

Zygelman, B.: Hyperfine level-changing collisions of hydrogen atoms and tomography of the dark age universe. ApJ **622**, 1356–1362 (2005)

The First Stars

Simon Glover

Abstract The first stars to form in the Universe—the so-called Population III stars—bring an end to the cosmological Dark Ages, and exert an important influence on the formation of subsequent generations of stars and on the assembly of the first galaxies. Developing an understanding of how and when the first Population III stars formed and what their properties were is an important goal of modern astrophysical research. In this review, I discuss our current understanding of the physical processes involved in the formation of Population III stars. I show how we can identify the mass scale of the first dark matter halos to host Population III star formation, and discuss how gas undergoes gravitational collapse within these halos, eventually reaching protostellar densities. I highlight some of the most important physical processes occurring during this collapse, and indicate the areas where our current understanding remains incomplete. Finally, I discuss in some detail the behaviour of the gas after the formation of the first Population III protostar. I discuss both the conventional picture, where the gas does not undergo further fragmentation and the final stellar mass is set by the interplay between protostellar accretion and protostellar feedback, and also the recently advanced picture in which the gas does fragment and where dynamical interactions between fragments have an important influence on the final distribution of stellar masses.

S. Glover (✉)
Institut für Theoretische Astrophysik, Zentrum für Astronomie der Universität Heidelberg, Albert-Ueberle-Str. 2, 69120 Heidelberg, Germany
e-mail: sglover@ita.uni-heidelberg.de

T. Wiklind et al. (eds.), *The First Galaxies*, Astrophysics and Space Science Library 396, DOI 10.1007/978-3-642-32362-1_3,

1 Formation of the First Star-Forming Minihalo

1.1 The Jeans Mass and the Filter Mass

In the currently dominant ΛCDM paradigm, gravitationally bound objects form in a hierarchical fashion, with the smallest, least massive objects forming first, and larger objects forming later through a mixture of mergers and accretion. The mass scale of the least massive objects to form from dark matter is set by free-streaming of the dark matter particles, and so depends on the nature of these particles. However, in most models, this minimum mass is many orders of magnitude smaller than the mass of even the smallest dwarf galaxies (Green et al. 2005). More relevant for the formation of the first stars and galaxies is the mass scale of the structures (frequently referred to as dark matter "minihalos") within which the baryonic component of matter, the gas, can first cool and collapse.

A lower limit on this mass scale comes from the theory of the growth of small density perturbations in an expanding universe (see e.g. Barkana and Loeb 2001). From the analysis of the linearized equations of motion, one can identify a critical length scale, termed the Jeans length, that marks the boundary between gravitationally stable and gravitationally unstable regimes. The Jeans length is given (in physical units) by

$$\lambda_{\mathrm{J}} = c_{\mathrm{s}} \sqrt{\frac{\pi}{G\rho_0}}, \tag{1}$$

where c_{s} is the sound speed in the unperturbed intergalactic medium and ρ_0 is the cosmological background density. Perturbations on scales $\lambda > \lambda_{\mathrm{J}}$ are able to grow under the influence of their own self-gravity, while those with $\lambda < \lambda_{\mathrm{J}}$ are prevented from growing by thermal pressure. We can associate a mass scale with λ_{J} by simply taking the mass within a sphere of radius $\lambda_{\mathrm{J}}/2$ (Barkana and Loeb 2001):

$$M_{\mathrm{J}} = \frac{4\pi}{3}\rho_0 \left(\frac{\lambda_{\mathrm{J}}}{2}\right)^3. \tag{2}$$

This mass, termed the Jeans mass, describes the minimum mass that a perturbation must have in order to be gravitationally unstable.

In the simplest version of this analysis, the value used for the sound speed in the equations for the Jeans length and Jeans mass is the instantaneous value; i.e. to determine λ_{J} and M_{J} at a redshift z, we use the value of c_s at that redshift. In this approximation, the Jeans mass is given in the high redshift limit (where the gas temperature is strongly coupled to the cosmic microwave background (CMB) temperature by Compton scattering) by the expression (Barkana and Loeb 2001)

$$M_{\mathrm{J}} = 1.35 \times 10^5 \left(\frac{\Omega_m h^2}{0.15}\right)^{-1/2} \mathrm{M}_{\odot}, \tag{3}$$

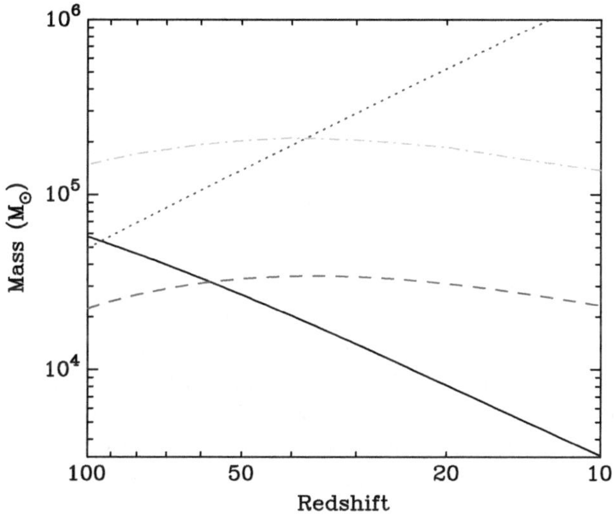

Fig. 1 Evolution with redshift of the Jeans mass (*solid line*), the filter mass computed in the limit where the relative streaming velocity between gas and dark matter is zero (*dashed line*) and the filter mass computed assuming a streaming velocity $v = \sigma_{\rm vbc}$ (*dash-dotted line*). Also plotted is the critical minihalo mass, $M_{\rm crit}$, required for efficient H_2 cooling (*dotted line*). The estimate of the filter mass in the no streaming limit comes from Naoz and Barkana (2007), who account for a number of effects not treated in the original Gnedin and Hui (1998) formulation, while the estimate of $M_{\rm F}$ in the streaming case comes from Tseliakhovich et al. (2011). The value of $M_{\rm crit}$ was computed using Eq. (38)

where $\Omega_{\rm m}$ is the dimensionless cosmological matter density parameter, and h is the value of the Hubble constant in units of $100\,{\rm km\,s^{-1}\,Mpc^{-1}}$. In the low redshift limit (where the coupling between radiation and matter is weak and the gas temperature evolves adiabatically), the Jeans mass is given instead by

$$M_{\rm J} = 5.18 \times 10^3 \left(\frac{\Omega_m h^2}{0.15}\right)^{-1/2} \left(\frac{\Omega_b h^2}{0.026}\right)^{-3/5} \left(\frac{1+z}{10}\right)^{3/2} {\rm M_\odot}, \quad (4)$$

where $\Omega_{\rm b}$ is the dimensionless cosmological baryon density parameter. The evolution of $M_{\rm J}$ with redshift is also illustrated in Fig. 1.

A more careful treatment of the growth of linear density perturbations accounts for the fact that the sound speed, the Jeans length and potentially also the Jeans mass may all change significantly during the time it takes for a perturbation to grow into the non-linear regime. Gnedin and Hui (1998) showed that in this case, the appropriate mass scale separating the gravitationally stable and gravitationally unstable regimes is a form of time-averaged Jeans mass that they denote as the "filter mass", $M_{\rm F}$. This is given in physical units by

$$M_{\mathrm{F}} = \frac{4\pi}{3}\rho_0 \left(\frac{\lambda_{\mathrm{F}}}{2}\right)^3, \tag{5}$$

where the filter wavelength λ_{F} is given in the high redshift limit by (Gnedin 2000)

$$\lambda_{\mathrm{F}}^2 = \frac{3}{1+z}\int_z^{\infty} \lambda_{\mathrm{J}}^2 \left[1 - \left(\frac{1+z}{1+z'}\right)^{1/2}\right] \mathrm{d}z'. \tag{6}$$

It is possible to improve further on this analysis by accounting for spatial variations in the sound speed (Barkana and Loeb 2005; Naoz and Barkana 2005) and by properly accounting for the separate rates of growth of the dark matter and baryonic perturbations in the high redshift limit in which the gas is mechanically coupled to the CMB by Compton scattering (Naoz and Barkana 2007). The net result is to somewhat lower the filter mass in comparison with the predictions of Eqs. (5)–(6). Comparing the resulting filter mass with the Jeans mass (Fig. 1), we see that the filter mass can be a factor of a few smaller than the Jeans mass at high redshift, but that for redshifts below $z \sim 50$, the filter mass is the larger of the two mass scales.

Another complication was recently pointed out by Tseliakhovich and Hirata (2010). They show that prior to the recombination epoch, the strong coupling between gas and radiation leads to the gas developing a non-zero velocity relative to the dark matter. While the gas and radiation are coupled, the sound-speed in the gas is approximately $c/\sqrt{3}$, where c is the speed of light, and the relative velocity between gas and dark matter is highly subsonic. Once the gas and radiation decouple, however, the sound-speed of the gas decreases enormously, becoming $\sim$6 km s^{-1} at the end of the recombination epoch. Tseliakhovich and Hirata (2010) show that at the same time, the RMS velocity of the gas relative to the dark matter is about 30 km s^{-1}, implying that the gas is moving supersonically with respect to the dark matter. The coherence length of the supersonic flow is of the order of the Silk damping scale (Silk 1968), i.e. several comoving Mpc, and so on the much smaller scales corresponding to the formation of the first star-forming minihalos, the gas can be treated as being in uniform motion with respect to the dark matter. Tseliakhovich and Hirata (2010) also show that the relative velocity between gas and dark matter acts to suppress the growth of small-scale structure in both components, and that because this effect is formally a second-order term in cosmological perturbation theory, it was not included in previous studies based on linear perturbation theory.

In a follow-up study, Tseliakhovich et al. (2011) improve on the Tseliakhovich and Hirata (2010) analysis by accounting for spatial variations in the sound speed, and study the effect that the relative velocity between the gas and the dark matter has on the size of the filter mass. The magnitude of the relative velocity v is randomly distributed with a Gaussian probability distribution function (PDF) with total variance σ_{vbc}^2, i.e.

$$P_{\mathrm{vbc}}(v) = \left(\frac{3}{2\pi\sigma_{\mathrm{vbc}}^2}\right)^{3/2} 4\pi v^2 \exp\left(-\frac{3v^2}{2\sigma_{\mathrm{vbc}}^2}\right). \tag{7}$$

Tseliakhovich et al. (2011) show that for a relative velocity $v = \sigma_{\mathrm{vbc}}$ (i.e. a one sigma perturbation), the effect of the relative velocity between gas and dark matter is to increase M_{F} by roughly an order of magnitude, as illustrated in Fig. 1. Higher sigma perturbations lead to even greater increases in M_{F}, but Tseliakhovich et al. (2011) show that the global average case (obtained by computing M_{F} for a range of different v and then integrating over the PDF given in Eq. (7)) is very similar to the one sigma case. Numerical studies of the effects of these streaming velocities (see e.g. Stacy et al. 2011; Greif et al. 2011b) have generally confirmed this result, although these studies still disagree somewhat regarding the influence of the streaming velocities on minihalos with masses greater than the revised M_{F}.

Nevertheless, even the most careful version of this analysis only tells us the mass scale of the first gravitationally bound structures to have a significant gas content, which is merely a lower limit on the mass scale of the first *star-forming* minihalos. The reason for this is that for stars to form within a minihalo, it is not enough that the gas be gravitationally bound; it must also be able to cool efficiently. In order for the gas within a minihalo to dissipate a large fraction of its gravitational binding energy—a necessary condition if pressure forces are not to halt the gravitational collapse of the gas (Hoyle 1953; Rees 1976; Rees and Ostriker 1977)—it must be able to radiate this energy away. The timescale over which this occurs is known as the cooling time, and is defined as

$$t_{\mathrm{cool}} = \frac{1}{\gamma - 1} \frac{n_{\mathrm{tot}} k T}{\Lambda}, \tag{8}$$

where n_{tot} is the total number density of particles, γ is the adiabatic index, k is Boltzmann's constant, T is the gas temperature and Λ is the radiative cooling rate per unit volume. If the cooling time of the gas is longer than the Hubble time, then it is very unlikely that the minihalo will survive as an isolated object for long enough to form stars. Instead, it is far more likely that it will undergo a major merger with another dark matter halo of comparable or larger mass before any of its gas has cooled significantly, since major mergers occur, on average, approximately once per Hubble time (Lacey and Cole 1993). Therefore, to determine the minimum mass of a star-forming minihalo, we must first understand how cooling occurs within primordial gas, a topic that we explore in the next section.

1.2 Cooling and Chemistry in Primordial Gas

At high temperatures ($T \sim 10^4$ K and above), primordial gas can cool efficiently through the collisional excitation of excited electronic states of atomic hydrogen, atomic helium, and singly-ionized helium. However, it is relatively easy to show that most of the gas within a minihalo with $M \sim M_{\rm F}$ will have a temperature significantly below 10^4 K. If we assume that the gas within the minihalo relaxes into a state of virial equilibrium, such that the total potential energy W and total kinetic energy K are related by $W = -2K$, then we can use this fact to define a virial temperature for the minihalo (Barkana and Loeb 2001)

$$T_{\rm vir} = \frac{\mu m_{\rm p} v_{\rm c}^2}{2k}, \quad (9)$$

where μ is the mean molecular weight, $m_{\rm p}$ is the proton mass, and v_c is the circular velocity of the minihalo. This can be rewritten in terms of the redshift z and the mass M of the minihalo as

$$T_{\rm vir} = 1.98 \times 10^4 \left(\frac{\mu}{0.6}\right) \left(\frac{M}{10^8 h^{-1}\,{\rm M}_{\odot}}\right)^{2/3} \left[\frac{\Omega_m}{\Omega_m(z)} \frac{\Delta_c}{18\pi^2}\right]^{1/3} \left(\frac{1+z}{10}\right) {\rm K}, \quad (10)$$

where $\Omega_m(z)$ is the dimensionless cosmological density parameter evaluated at redshift z and $\Delta_c = 18\pi^2 + 82d - 39d^2$, with $d = \Omega_m(z) - 1$ (Bryan and Norman 1998). In the standard ΛCDM cosmology, $\Omega_m(z) \simeq 1$ at $z > 6$, and hence the term in square brackets reduces to $\Omega_m^{1/3}$. If we rearrange Eq. (10) and solve for the mass $M_{\rm atom}$ of a cloud that has a virial temperature $T_{\rm vir} = 10^4$ K and that can therefore cool via atomic excitation, we find that

$$M_{\rm atom} = 5 \times 10^7 h^{-1} \left(\frac{\mu}{0.6}\right)^{-3/2} \Omega_m^{-1/2} \left(\frac{1+z}{10}\right)^{-3/2} {\rm M}_{\odot}, \quad (11)$$

significantly larger than our estimates for $M_{\rm J}$ and $M_{\rm F}$ above. Minihalos with masses close to $M_{\rm J}$ or $M_{\rm F}$ will therefore have virial temperatures much less than 10^4 K, placing them in the regime where molecular coolants dominate.

In primordial gas, by far the most abundant and hence most important molecule is molecular hydrogen, H_2. The chemistry of H_2 in primordial gas has been reviewed in a number of different studies (see e.g. Abel et al. 1997; Galli and Palla 1998; Stancil et al. 1998; Glover and Abel 2008), and so we only briefly discuss it here. Direct formation of H_2 by the radiative association of two hydrogen atoms is highly forbidden (Gould and Salpeter 1963), and so at low densities, most H_2 forms via the reaction chain (McDowell 1961; Peebles and Dicke 1968)

$$\rm H + e^- \rightarrow H^- + \gamma, \quad (12)$$

$$\rm H^- + H \rightarrow H_2 + e^-, \quad (13)$$

with a minor fraction forming via the reaction chain (Saslaw and Zipoy 1967)

$$\rm H + H^+ \rightarrow H_2^+ + \gamma, \quad (14)$$

$$\rm H_2^+ + H \rightarrow H_2 + H^+. \quad (15)$$

In warm gas, H_2 can be destroyed by collisional dissociation (see e.g. Martin et al. 1998)

$$H_2 + H \rightarrow H + H + H, \tag{16}$$

$$H_2 + H_2 \rightarrow H + H + H_2, \tag{17}$$

or by charge transfer with H^+ (see e.g. Savin et al. 2004)

$$H_2 + H^+ \rightarrow H_2^+ + H, \tag{18}$$

but at low temperatures there are no collisional processes that can efficiently remove it from the gas.

When the fractional ionization of the gas is low, the rate at which H_2 forms is limited primarily by the rate at which H^- ions form via reaction (12), as any ions that form are rapidly converted to H_2 by associative detachment with atomic hydrogen (reaction (13)). If the fractional ionization is large, on the other hand, then many of the H^- ions formed by reaction (12) do not survive for long enough to form H_2, but instead are destroyed by mutual neutralization with H^+ ions:

$$H^+ + H^- \rightarrow H + H. \tag{19}$$

The ratio of the rates of reactions (13) and (19) is given by $k_{(13)}n_{\rm H}/k_{(19)}n_{\rm H^+}$, where $n_{\rm H}$ is the number density of atomic hydrogen, $n_{\rm H^+}$ is the number density of protons, and $k_{(13)}$ and $k_{(19)}$ are the rate coefficients for reactions (13) and (19), respectively. Mutual neutralization therefore becomes significant whenever $n_{\rm H^+}/n_{\rm H} \geq k_{(13)}/k_{(19)}$. Although the value of $k_{(13)}/k_{(19)}$ is temperature dependent, the temperature dependence is weak if one uses the best available determinations of the rate coefficients (Kreckel et al. 2010 for reaction (13), Stenrup et al. 2009 for reaction (19)[1]), and $k_{(13)}/k_{(19)} \sim 0.03$ to within 50 % for all temperatures $100 < T < 10^4$ K. If we compare this value with the residual fractional ionization of the intergalactic medium (IGM) at this epoch, $x \sim 2 \times 10^{-4}$ (Schleicher et al. 2008), we see that mutual neutralization is unimportant within the very first star-forming minihalos. It becomes important once larger minihalos, with virial temperatures $T_{\rm vir} \sim 10^4$ K or above, begin to form, as in these minihalos, substantial collisional ionization of the gas can occur, leading to an initial fractional ionization much higher than the residual value in the IGM. It also becomes an important process within the "fossil" HII regions left behind by the first generation of massive stars (Oh and Haiman 2003; Nagakura and Omukai 2005; Glover et al. 2006; Kreckel et al. 2010).

Although H_2 is by far the most abundant primordial molecule, it is actually not a particularly efficient coolant. The H_2 molecule has no dipole moment, and so

---

[1]A group lead by X. Urbain at the Université Catholique de Louvain has recently made new experimental measurements of the rate of this reaction at low temperatures, but at the time of writing, the results of this work remain unpublished.

dipole transitions between its excited rotational and vibrational levels are forbidden. Although radiative transitions between levels do occur, they are quadrupole transitions and the associated transition rates are small. In addition, application of the Pauli exclusion principle to the hydrogen molecule shows that it must have two distinct states, distinguished by the nuclear spin of the two hydrogen nuclei: para-hydrogen, in which the nuclear spins are parallel, and which must have an even value for the rotational quantum number J, and ortho-hydrogen, which has anti-parallel nuclear spins and an odd value for J. Radiative transitions between ortho-hydrogen and para-hydrogen involve a change in orientation of the spin of one of the nuclei and are therefore strongly forbidden. As a result, the least energetic rotational transition of H_2 that has any significant probability of occurring is the transition between the $J = 2$ and $J = 0$ rotational levels in the vibrational ground-state of para-hydrogen. This transition has an associated energy $E_{20}/k \simeq 512$ K. The H_2 molecule therefore has large energy separations between the ground state and any of the accessible excited rotational or vibrational states[2], and has only weak radiative transitions between these states.

These features of the H_2 molecule have two important consequences. First, it becomes a very inefficient coolant at temperatures $T \ll E_{20}/K$, as it becomes almost impossible to collisionally populate any of the excited states. The minimum temperature that can be reached solely with H_2 cooling depends somewhat on the H_2 abundance and the time available for cooling, but typically $T_{\rm min} \sim 150$-200 K. Second, its rotational and vibrational levels reach their local thermodynamic equilibrium (LTE) level populations at a relatively low density, $n_{\rm crit} \sim 10^4\ {\rm cm}^{-3}$. This means that at densities $n \gg n_{\rm crit}$, the H_2 cooling rate scales only linearly with density and the cooling time due to H_2 becomes independent of density. Since other important timescales, such as the free-fall collapse time of the gas, continue to decrease with increasing density, the implication is that H_2 becomes an increasingly ineffective coolant as one moves to higher densities.

For these reasons, primordial molecules or molecular ions that do not share these drawbacks have attracted a certain amount of attention. In an early study, Lepp and Shull (1984) suggested that deuterated hydrogen, HD, and lithium hydride, LiH, may both be significant coolants in primordial gas. More recently, work by Yoshida et al. (2007) has suggested that H_2^+ may be an important coolant in some circumstances, while Glover and Savin (2006) show that H_3^+ is also worthy of attention. In practice, the only one of these molecules or ions that has proved to be important is HD. Detailed modelling of the chemistry of lithium in primordial gas (e.g. Stancil et al. 1996; Mizusawa et al. 2005) has shown that LiH is efficiently destroyed by the reaction

$$\mathrm{LiH} + \mathrm{H} \rightarrow \mathrm{Li} + \mathrm{H_2}, \qquad (20)$$

---

[2]For comparison, note that the energy separation between the $J = 0$ and $J = 1$ rotational levels of CO is roughly 5 K.

and that only a small fraction of the available lithium (which itself has an abundance of only 5×10^{-10} relative to hydrogen; see Cyburt et al. 2008) is ever incorporated into LiH. Cooling from the molecular ion H_2^+ was re-examined by Glover and Savin (2009), who showed that the collisional excitation rates cited by Galli and Palla (1998) and used as a basis for the fits given in Yoshida et al. (2007) were a factor of ten too large, and that if the correct rates are used, H_2^+ cooling is no longer important. Finally, Glover and Savin (2009) also examined the possible role played by H_3^+ cooling in considerable detail, but found that even if one makes optimistic assumptions regarding its formation rate and collisional excitation rate, it still contributes to the total cooling rate at the level of only a few percent, and hence at best is a minor correction term.

These studies leave HD as the only viable alternative to H_2 as a coolant of primordial gas. HD has a small, but non-zero dipole moment, giving it radiative transition rates that are somewhat larger than those of H_2, resulting in a critical density $n_{\rm crit} \sim 10^6\,{\rm cm}^{-3}$. Unlike H_2, it is not separated into ortho and para states, and so the lowest energy transition accessible from the ground state is the $J = 1$ to $J = 0$ rotational transition, with an energy $E_{10}/k = 128$ K. Although the cosmological ratio of deuterium to hydrogen is small [$\rm D/H = (2.49 \pm 0.17) \times 10^{-5}$; Cyburt et al. 2008], the ratio of HD to H_2 can be significantly boosted in low temperature gas by chemical fractionation. The reaction

$$\rm H_2 + D^+ \rightarrow HD + H^+ \tag{21}$$

that converts H_2 into HD is exothermic and so proceeds rapidly at all temperatures, while the inverse reaction

$$\rm HD + H^+ \rightarrow H_2 + D^+ \tag{22}$$

is endothermic and so proceeds very slowly at low temperatures. In equilibrium, these two reactions produce an HD-to-H_2 ratio given by

$$\frac{x_{\rm HD}}{x_{\rm H_2}} = 2 \exp\left(\frac{464}{T}\right) [\rm D/H], \tag{23}$$

where [D/H] is the cosmological D:H ratio. Together, these factors render HD a much more effective coolant than H_2 in low temperature gas.

In practice, for HD cooling to take over from H_2 cooling, the gas must already be fairly cold, with $T \sim 150$ K (Glover 2008), and temperatures this low are typically not reached during the collapse of the first star-forming minihalos, meaning that HD remains a minor coolant (Bromm et al. 2002). However, there are a number of situations, typically involving gas with an enhanced fractional ionization, in which HD cooling does become significant (see e.g. Nakamura and Umemura 2002; Nagakura and Omukai 2005; Johnson and Bromm 2006; Yoshida et al. 2007; McGreer and Bryan 2008; Greif et al. 2008; Kreckel et al. 2010).

1.3 The Minimum Mass Scale for Collapse

The relative simplicity of the chemistry discussed in the previous section allows one to construct a very simple model that captures the main features of the evolution of the H_2 fraction within low density gas falling into a dark matter minihalo. We start by assuming that radiative recombination is the only process affecting the electron abundance, and writing the rate of change of the electron number density as

$$\frac{dn_e}{dt} = -k_{rec} n_e n_{H^+}, \tag{24}$$

where n_e is the number density of electrons, n_{H^+} is the number density of protons, and k_{rec} is the recombination coefficient. If we assume that ionized hydrogen is the only source of free electrons, implying that $n_e = n_{H^+}$, and that the temperature remains roughly constant during the evolution of the gas, then we can solve for the time evolution of the electron fraction:

$$x = \frac{x_0}{1 + k_{rec} n t x_0}, \tag{25}$$

where $x \equiv n_e/n$, n is the number density of hydrogen nuclei, and x_0 is the initial value of x. We next assume that all of the H_2 forms via the H^- pathway, and that mutual neutralization of H^- with H^+ (reaction (19)) is the only process competing with associative detachment (reaction (13)) for the H^- ions. In this case, we can write the time evolution of the H_2 fraction, $x_{H_2} \equiv n_{H_2}/n$, as

$$\frac{dx_{H_2}}{dt} = k_{(12)} x n_H \, p_{AD}, \tag{26}$$

where $k_{(12)}$ is the rate coefficient of reaction (12), the formation of H^- by the radiative association of H and e^-, and p_{AD} is the probability that any given H^- ion will be destroyed by associative detachment rather than by mutual neutralization. Given our assumptions above, this can be written as

$$p_{AD} = \frac{k_{(13)} n_H}{k_{(13)} n_H + k_{(19)} n_{H^+}}, \tag{27}$$

where $k_{(13)}$ is the rate coefficient for reaction (13) and $k_{(19)}$ is the rate coefficient for reaction (19). If we again assume that $n_e = n_{H^+}$, and in addition assume that $n_H \simeq n$, then the expression for p_{AD} can be simplified to

$$p_{AD} = \left(1 + \frac{k_{(19)}}{k_{(13)}} x\right)^{-1}. \tag{28}$$

Substituting this into Eq. (26), we obtain

$$\frac{\mathrm{d}x_{\mathrm{H}_2}}{\mathrm{d}t} = k_{(12)} x n_{\mathrm{H}} \left(1 + \frac{k_{(19)}}{k_{(13)}} x\right)^{-1}. \tag{29}$$

If the initial fractional ionization $x_0 \ll k_{(13)}/k_{(19)}$, then the term in parentheses is of order unity and this equation has the approximate solution

$$x_{\mathrm{H}_2} = \frac{k_{(12)}}{k_{\mathrm{rec}}} \ln\left(1 + k_{\mathrm{rec}} n x_0 t\right), \tag{30}$$

$$= \frac{k_{(12)}}{k_{\mathrm{rec}}} \ln\left(1 + t/t_{\mathrm{rec}}\right), \tag{31}$$

where $t_{\mathrm{rec}} = 1/(k_{\mathrm{rec}} n x_0)$ is the recombination time. The growth of the H_2 fraction is therefore logarithmic in time, with most of the H_2 forming within the first few recombination times. In the more complicated case in which x_0 is comparable to or larger than $k_{(13)}/k_{(19)}$, but still significantly less than unity (so that $n_{\mathrm{H}} \sim n$), the H_2 fraction is given instead by

$$x_{\mathrm{H}_2} = \frac{k_{(12)}}{k_{\mathrm{rec}}} \ln\left(\frac{1 + x_0 k_{(19)}/k_{(13)} + t/t_{\mathrm{rec}}}{1 + x_0 k_{(19)}/k_{(13)}}\right). \tag{32}$$

From this analysis, we see that the main factor determining the final H_2 abundance is the ratio $k_{(12)}/k_{\mathrm{rec}}$, since for times of the order of a few recombination times, the logarithmic term in Eq. (32) is of order unity, implying that the final H_2 abundance is at most a factor of a few times $k_{(12)}/k_{\mathrm{rec}}$. If we use the simple power-law fits to $k_{(12)}$ and k_{rec} given by Hutchins (1976), namely $k_{(12)} = 1.83 \times 10^{-18} T^{0.8779}\ \mathrm{cm^3\,s^{-1}}$ and $k_{\mathrm{rec}} = 1.88 \times 10^{-10} T^{-0.644}\ \mathrm{cm^3\,s^{-1}}$, then we can write the ratio of the two rate coefficients as

$$\frac{k_{(12)}}{k_{\mathrm{rec}}} \simeq 10^{-8} T^{1.5219}. \tag{33}$$

The amount of H_2 produced is a strong function of temperature, but is of the order of a few times 10^{-3} for temperatures of a few thousand Kelvin. We see therefore that the formation of H_2 via H^- never results in a gas dominated by H_2, as the H_2 abundance always remains much smaller than the abundance of atomic hydrogen.

Given this simple model for the amount of H_2 that will form in the gas, the obvious next step is to compare this to the amount of H_2 that is required to cool the gas efficiently. In order to determine the H_2 fraction necessary to significantly cool gas with a temperature T within some specified fraction of the Hubble time—say 20 % of t_{H}—we can simply equate the two timescales, and solve for the H_2 fraction. Using our previous definition of the cooling time, we have

$$\frac{1}{\gamma - 1} \frac{n_{\mathrm{tot}} k T}{\Lambda_0(T) n_{\mathrm{H}_2}} = 0.2 t_{\mathrm{H}}, \tag{34}$$

where we have assumed that H_2 is the dominant coolant and have written the cooling rate per unit volume in terms of Λ_0, the cooling rate per H_2 molecule. Rearranging this equation, using the fact that when the H_2 fraction and the ionization level are low, $\gamma = 5/3$ and $n_{\rm tot} = (1 + 4x_{\rm He})n$, where $x_{\rm He}$ is the fractional abundance of helium (given by $x_{\rm He} = 0.083$ for primordial gas), we obtain

$$x_{\rm H_2,req} = 1.38 \times 10^{-15} \frac{T}{\Lambda_0(T)} t_{\rm H^{-1}}. \tag{35}$$

In the high-redshift limit where $t_{\rm H} \simeq H_0^{-1} \Omega_m^{-1/2} (1+z)^{-3/2}$, this becomes

$$x_{\rm H_2,req} = 5.2 \times 10^{-32} \frac{T}{\Lambda_0(T)} \left(\frac{1+z}{10}\right)^{3/2}, \tag{36}$$

where we have used values for the cosmological parameters taken from Komatsu et al. (2011). Collisions of H_2 with a number of different species contribute to Λ_0, as explored in Glover and Abel (2008), but in the earliest minihalos, the dominant contributions come from collisions with H and He. Λ_0 is therefore given to a good approximation by

$$\Lambda_0 = \Lambda_H n_{\rm H} + \Lambda_{\rm He} n_{\rm He}. \tag{37}$$

Simple fits for the values of $\Lambda_{\rm H}$ and $\Lambda_{\rm He}$ as a function of temperature can be found in Glover and Abel (2008).

An illustration of the likely size of $x_{\rm H_2,req}$ is given in Fig. 2. In this Figure, we plot $x_{\rm H_2,req}$ as a function of temperature, evaluated for three different redshifts: $z = 20$, 30 and 40. In computing these values, we have assumed that the mean density of the gas in the minihalo is given by $\bar{\rho} = \Delta_c \rho_{b,0}$, where ρ_0 is the cosmological background density of baryons. In the Figure, we also show the actual H_2 fraction produced in the gas, $x_{\rm H_2,act}$, as a function of temperature at times equal to 1, 5 and 10 recombination times, and where we have taken $x_0 \ll k_{(13)}/k_{(19)}$.

Figure 2 demonstrates that the amount of H_2 produced in the gas is a strongly increasing function of temperature, while the amount required to bring about efficient cooling of the gas is a strongly decreasing function of temperature. This means that for any given choice of comparison time t and redshift z, we can identify a critical temperature $T_{\rm crit}$, such that gas with $T > T_{\rm crit}$ will cool within a small fraction of a Hubble time, while gas with $T < T_{\rm crit}$ will not. Moreover, because $x_{\rm H_2,act}$ and $x_{\rm H_2,req}$ are both steep functions of temperature, but are relatively insensitive to changes in t or z, the value of $T_{\rm crit}$ that we obtain is also relatively insensitive to our choices for t or z. We find that $T_{\rm crit} \sim 1{,}000$ K, and that at this temperature, the H_2 fraction required to provide efficient cooling lies somewhere between a few times 10^{-4} and 10^{-3} (c.f. Tegmark et al. 1997, who come to a similar conclusion using a very similar argument). If we convert this critical virial temperature into a corresponding critical minihalo mass using Eq. (10), we find that

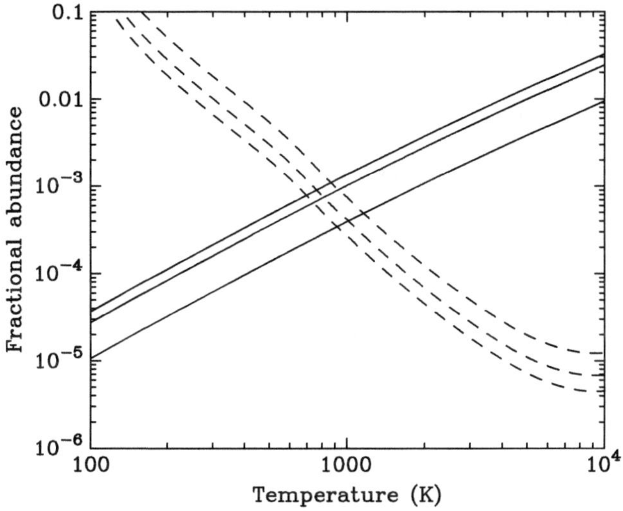

Fig. 2 Comparison of the fractional abundance of H_2 produced with our simple toy model for the chemistry (*solid lines*) versus the quantity of H_2 required in order to cool the gas within 20 % of a Hubble time (*dashed lines*). From bottom to top, the solid lines correspond to the H_2 fraction produced at times $t = 1, 5$ and $10\ t_{\rm rec}$, respectively, where $t_{\rm rec}$ is the recombination time, and the dashed lines correspond to the H_2 fraction required at redshifts $z = 40, 30$ and 20, respectively. We see that the minimum temperature that the gas must have in order to be able to cool within a fraction of a Hubble time—indicated by the point at which the lines cross—is relatively insensitive to our choices for t and z

$$M_{\rm crit} \simeq 6 \times 10^5 h^{-1} \left(\frac{\mu}{1.2}\right)^{-3/2} \Omega_m^{-1/2} \left(\frac{1+z}{10}\right)^{-3/2} \ {\rm M}_\odot. \tag{38}$$

This mass scale is illustrated by the dotted line in Fig. 1. At high redshift, it is smaller than the filter mass scale corresponding to $v_{\rm bc} = \sigma_{\rm vbc}$, demonstrating that at these redshifts, it is the streaming of the gas with respect to the dark matter that is the main process limiting the formation of Population III stars. Below a redshift of around 40, however, $M_{\rm crit}$ becomes the larger mass scale, implying that at these lower redshifts, there will be a population of small minihalos that contain a significant gas fraction, but that do not form stars, because their gas is unable to cool in less than a Hubble time. These small starless minihalos may be important sinks for ionizing photons during the epoch of reionization (Haiman et al. 2001).

To conclude our discussion of the first star-forming minihalos, we should mention one potentially important effect not taken into account in the analysis above. This is the influence of ongoing minor mergers and accretion on the thermal balance of the gas. Although major mergers occur only once per Hubble time, on average, minor mergers occur far more frequently, and act to stir up the gas, thereby heating it and lengthening the time required for it to cool. This phenomenon was noted by Yoshida et al. (2003) in their cosmological simulations of the formation of the first star-forming minihalos. Yoshida et al. (2003) show that in spite of the approximations made in its derivation, Eq. (38) gives a reasonable guide to the minimum mass of the minihalos that contain gas that can cool effectively. However, they also find that there are some minihalos with $M > M_{\rm crit}$ in which the gas does not cool. They show that these minihalos have higher mass accretion rates than minihalos of the same mass in which cooling does occur, and hence ascribe the suppression of cooling to the effects of dynamical heating by the ongoing accretion and minor mergers. This effect was also treated more recently by Wang and Abel (2008), who show that it can be included into the simple thermal model described above by the addition of a heating term describing the effects of mergers and accretion. They show that if one writes this heating term as

$$\Gamma = \frac{k}{\gamma - 1} \frac{\mathrm{d}T_{\rm vir}}{\mathrm{d}t}, \tag{39}$$

then one can relate the rate of change of the virial temperature to the mass growth rate of the minihalo in a relatively simple fashion.

2 Gravitational Collapse and the Formation of the First Protostar

As the analysis in the previous section has shown, gas in minihalos with virial temperatures greater than about 1,000 K (corresponding to masses $M \sim 10^6\ \mathrm{M_\odot}$) can form enough H_2 to cool within a small fraction of a Hubble time. This reduces the pressure and allows the gas to collapse further under the influence of its own self-gravity. As it does so, the value of the Jeans mass decreases. Many early studies of the formation of primordial stars assumed that as the Jeans mass decreases and the gas becomes more gravitationally unstable, it begins to undergo hierarchical gravitational fragmentation in a manner similar to that envisaged by Hoyle (1953), with the result that at any given moment, the mean fragment mass is approximately equal to the local Jeans mass (see Glover 2005, for a historical summary of these models). In this picture, one could predict the final mass of the first stars simply by studying the evolution of the Jeans mass. Moreover, since the minimum Jeans mass reached during the collapse can be estimated with reasonable accuracy on purely thermodynamical grounds (Rees 1976; Low and Lynden-Bell 1976), in this view of Population III star formation, the dynamics of the gas is of secondary importance. Around ten years ago, however, it first became possible to model the coupled chemical, dynamical and thermal evolution of the gas within a primordial minihalo using high resolution 3D numerical simulations (Abel et al. 2000, 2002; Bromm et al. 1999, 2002). These studies showed that the picture outlined above is wrong: the gas does not undergo hierarchical fragmentation, and so one cannot predict the masses of the first stars simply by studying the evolution of the Jeans mass. These high resolution numerical simulations, and the many that have followed them (e.g. Yoshida et al. 2006; O'Shea and Norman 2007; McGreer and Bryan 2008, to name but a few), have for the first time given us a clear picture of exactly how gravitational collapse proceeds within one of these early minihalos. In the next section, we will discuss the sequence of events that occur as we follow the collapse from the minihalo scale all the way down to the scale of a single Population III protostar. Following that, in Sects. 2.2 and 2.3 we discuss two of the main uncertainties remaining in our model for the formation of the first Pop. III protostar: the role played by heating and ionization arising from dark matter self-annihilation (Sect. 2.2) and the role played by magnetic fields (Sect. 2.3).

2.1 *Thermal and Chemical Evolution of the Gas During Collapse*

2.1.1 Initial Collapse

As gas falls into the minihalo from the intergalactic medium, it is shock-heated to a temperature close to $T_{\rm vir}$. In the post-shock gas, the electron fraction decreases due to radiative recombination, but at the same time H_2 forms, primarily via reactions (12) and (13). As we have already seen, the H_2 fraction evolves logarithmically with time, with most of the H_2 forming within the first few recombination times. As the H_2 fraction increases, so does its ability to cool the gas, and so the gas temperature slowly decreases, reducing the pressure and allowing the gas to collapse to the centre of the minihalo.

At this point, the evolution of the gas depends upon how much H_2 it has formed. There are two main outcomes, and which one occurs within a given minihalo depends primarily on the initial ionization state of the gas.

The low ionization case

During the formation of the very first Population III stars (also known as Population III.1, to use the terminology introduced by Tan and McKee 2008), the initial fractional ionization of the gas is the same as the residual ionization in the intergalactic medium, i.e. $x_0 \sim 2 \times 10^{-4}$. In this case, the amount of H_2 that forms in the gas is typically enough to cool it to a temperature of $T \sim 200$ K but not below. At this temperature, chemical fractionation has already increased the HD/H_2 ratio by a factor of 20 compared to the cosmic deuterium-to-hydrogen ratio, and as a consequence, HD is starting to become an important coolant. However, the amount of HD that forms in the gas is not enough to cool it significantly below 200 K (Bromm et al. 2002), and H_2 continues to dominate the cooling and control the further evolution of the gas. In this scenario, the collapse of the gas is greatly slowed once its temperature reaches 200 K and its density reaches a value of around

10^4 cm^{-3}, corresponding to the critical density n_{crit}, at which the rotational and vibrational level populations of H_2 approach their local thermodynamic equilibrium (LTE) values. At densities higher than this critical density, the H_2 cooling rate per unit volume scales only linearly with n (compared to a quadratic dependence, $\Lambda_{H_2} \propto n^2$ at lower densities), while processes such as compressional heating continue to increase more rapidly with n. As a result, the gas temperature begins to increase once the density exceeds n_{crit}.

Gas reaching this point in the collapse enters what Bromm et al. (2002) term a "loitering" phase, during which cold gas accumulates in the centre of the halo but only slowly increases its density. This loitering phase ends once the mass of cold gas that has accumulated exceeds the local value of the Bonnor-Ebert mass (Bonnor 1956; Ebert 1955), given in this case by (Abel et al. 2002)

$$M_{BE} \simeq 40 T^{3/2} n^{-1/2} \, M_{\odot}, \tag{40}$$

which for $n \sim 10^4$ cm^{-3} and $T \sim 200$ K yields $M_{BE} \sim 1,000\,M_{\odot}$.[3] Once the mass of cold gas exceeds M_{BE}, its collapse speeds up again, and becomes largely decoupled from the larger-scale behaviour of the gas. The next notable event to occur in the gas is the onset of three-body H_2 formation, which is discussed in the next section.

The high ionization case

If the initial fractional ionization of the gas is significantly higher than the residual fraction in the IGM, then a slightly different chain of events can occur. A larger initial fractional ionization implies a shorter recombination time, and hence a logarithmic increase in the amount of H_2 formed after a given physical time. An increase in the H_2 fraction allows the gas to cool to a slightly lower temperature, and hence boosts the HD abundance in two ways: the lower temperature increases the HD/H_2 ratio produced by fractionation, and the H_2 fraction itself is larger, so any given HD/H_2 ratio corresponds to a higher HD abundance than in the low ionization case. If the addition ionization allows enough H_2 to be produced to cool the gas to $T \sim 150$ K (which requires roughly a factor of three more H_2 than is required to reach 200 K), then chemical fractionation increases the HD abundance to such an extent that it takes over as the dominant coolant (Glover 2008). This allows the gas to cool further, in some cases reaching a temperature as low as the CMB temperature, T_{CMB} (e.g. Nakamura and Umemura 2002; Nagakura and Omukai 2005; Johnson and Bromm 2006; Yoshida et al. 2007; McGreer and Bryan 2008; Kreckel et al. 2010).

[3] Discussions of Population III star formation often refer to the cold clump of gas at the centre of the minihalo as a "fragment", and speak of M_{BE} as the "fragmentation mass scale", but in the case of the very first generation of star-forming minihalos, this is actually something of a misnomer, as very seldom does more than one "fragment" form in a given minihalo.

The higher critical density of HD, $n_{crit,HD} \sim 10^6$ cm^{-3}, means that the gas does not reach the loitering phase until much later in its collapse. Once the gas does reach this phase, however, its subsequent evolution is very similar to that in the low-ionization case discussed above. Cold gas accumulates at $n \sim n_{crit}$ until its mass exceeds the Bonnor-Ebert mass, which in this case is $M_{BE} \sim 40$ M$_\odot$ if $T = 100$ K and $n = 10^6$ cm^{-3}. Once the gas mass exceeds M_{BE}, the collapse speeds up again, and the gas begins to heat up. Aside from the substantial difference in the size of M_{BE}, the main difference between the evolution of the gas in this case and in the low ionization case lies in the fact that in the high ionization case, the gas reheats from $T \sim 200$ K or below to $T \sim 1,000$ K much more rapidly than in the low ionization case. As we shall see later, this period of rapid heating has a profound influence on the ability of the gas to fragment.

Several different scenarios have been identified that lead to an enhanced fractional ionization in the gas, and that potentially allow the gas to reach the HD-dominated regime. Gas within minihalos with $T_{vir} > 9000$ K will become hot enough for collisional ionization of hydrogen to supply the necessary electrons. However, as halos of this size will typically have at least one star-forming progenitor (Johnson et al. 2008), it is questionable whether many Pop. III stars will form in such minihalos, as we would expect the gas in most of them to have been enriched with metals by one or more previous episodes of star formation.

Another possibility that has attracted significant attention involves the gas in the minihalo being drawn from a "fossil" HII region, i.e. a region that was formerly ionized by a previous Population III protostar but has now recombined (see e.g. Oh and Haiman 2003; Nagakura and Omukai 2005; Yoshida et al. 2007). Many studies have shown that the volume of the IGM ionized by a single massive Pop. III star is significantly larger than the volume that is enriched by the metals produced in the supernova occurring at the end of the massive star's life (see e.g. the recent treatment by Greif et al. 2010, or Ciardi and Ferrara (2005) for a summary of earlier work). It is therefore possible that a significant number of Population III stars may form in such conditions.

A final possibility is that the required ionization can be produced by a flux of X-rays or high energy cosmic rays. Although X-ray ionization was initially favoured as a means of raising the ionization level of the gas, and hence promoting H_2 formation (Haiman et al. 2000), more recent work has shown that if one considers realistic models for the X-ray background that also account for the simultaneous growth of the soft UV background, then one finds that UV photodissociation of the H_2 is a more important effect, and hence that the growth of the radiation backgrounds almost always leads to an overall reduction in the amount of H_2 produced (Glover and Brand 2003; Machacek et al. 2003). Cosmic ray ionization may therefore prove to be the more important effect (Stacy and Bromm 2007; Jasche et al. 2007), although we still know very little about the likely size of the cosmic ray ionization rate in high redshift minihalos.

2.1.2 Three-Body H_2 Formation

Once the collapsing gas reaches a density of around 10^8–10^9 cm^{-3}, its chemical makeup starts to change significantly. The reason for this is that at these densities, the formation of H_2 via the three-body reactions (Palla et al. 1983)

$$\mathrm{H} + \mathrm{H} + \mathrm{H} \rightarrow \mathrm{H_2} + \mathrm{H}, \tag{41}$$

$$\mathrm{H} + \mathrm{H} + \mathrm{H_2} \rightarrow \mathrm{H_2} + \mathrm{H_2}, \tag{42}$$

$$\mathrm{H} + \mathrm{H} + \mathrm{He} \rightarrow \mathrm{H_2} + \mathrm{He}, \tag{43}$$

starts to become significant. These reactions quickly convert most of the hydrogen in the gas into H_2. At the same time, however, they generate a substantial amount of thermal energy: every time an H_2 molecule forms via one of these three-body reactions, its binding energy of 4.48 eV is converted into heat. A simple estimate of the relative sizes of the compressional heating rate and the three-body H_2 formation heating rate helps to demonstrate the importance of the latter during this stage of the collapse. Let us consider gas at a density $n = 10^8$ cm^{-3} that has a temperature $T = 1,000$ K, collapsing at a rate such that $\mathrm{d}n/\mathrm{d}t = n/t_{\mathrm{ff}}$, where t_{ff} is the gravitational free-fall time. In these conditions, the compressional heating rate is given by

$$\Lambda_{\mathrm{pdv}} \simeq 1.25 \times 10^{-31} n^{1/2} T, \tag{44}$$

$$= 1.25 \times 10^{-24} \ \mathrm{erg\ s^{-1}\ cm^{-3}}, \tag{45}$$

while the three-body H_2 formation heating rate has the value

$$\Lambda_{3\mathrm{b}} \simeq 3.9 \times 10^{-40} T^{-1} n^3 x_{\mathrm{H}^3}, \tag{46}$$

$$= 3.9 \times 10^{-19} x_{\mathrm{H}^3} \ \mathrm{erg\ s^{-1}\ cm^{-3}}, \tag{47}$$

where we have adopted the rate coefficient for reaction (41) given in Palla et al. (1983). Comparing the two heating rates, we see that three-body H_2 formation heating dominates unless x_{H} is very small (i.e. unless the gas is almost fully molecular). Therefore, even though the abundance of H_2, the dominant coolant during this phase of the collapse, increases by more than two orders of magnitude, the gas typically does not cool significantly, owing to the influence of this three-body H_2 formation heating. Indeed, the temperature often actually increases.

One major uncertainty that remains in current treatments of this phase of the collapse of the gas is exactly how quickly the gas becomes molecular. Although reaction (41) is the dominant source of H_2 at these densities, the rate coefficient for this reaction is poorly known, with published values differing by almost two orders of magnitude at 1,000 K, and by an even larger factor at lower temperatures (Glover 2008; Turk et al. 2011). The effects of this uncertainty have recently been studied by Turk et al. (2011). They show that it has little effect on the density profile of the gas, and only a limited effect on the temperature profile. However, it has much more significant effects on the morphology of the gas and on its velocity structure. Simulations in which a high value was used for the three-body rate coefficient show find that gas occurs more rapidly, and that the molecular gas develops a much more flattened, filamentary structure. Significant differences are also apparent in the infall velocities and the degree of rotational support. Turk et al. (2011) halt their simulations at the point at which a protostar first forms, and so do not directly address the issue of whether these differences continue to have an influence during the accretion phase, and whether the affect the ability of the gas to fragment (see Sect. 3.2 below). A follow-up study that focussed on these issues would be informative.

2.1.3 Optically-Thick Line Cooling

The next important event occurs at a density of around 10^{10} cm^{-3}, when the main rotational and vibrational lines of H_2 start to become optically thick (Ripamonti and Abel 2004). The effect of this is to reduce the efficiency of H_2 cooling, leading to a continued rise in the gas temperature. In one-dimensional simulations (e.g. Omukai and Nishi 1998; Omukai et al. 1998; Ripamonti et al. 2002), it is possible to treat optically thick H_2 cooling accurately by solving the full radiative transfer problem. These models show that although the optical depth of the gas becomes large at frequencies corresponding to the centers of the main H_2 emission lines, the low continuum opacity of the gas allows photons to continue to escape through the wings of the lines, with the result that the H_2 cooling rate is suppressed far less rapidly as the collapse proceeds than one might at first expect (see Omukai et al. 1998, for a detailed discussion of this point).

In three-dimensional simulations, solution of the full radiative transfer problem is not currently possible, due to the high computational expense, which has motivated a search for simpler approximations. There are two such approximations in current use in simulations of Population III star formation. The first of these was introduced by Ripamonti and Abel (2004). They proposed that the ratio of the optically thick and optically thin H_2 cooling rates,

$$f_\tau \equiv \frac{\Lambda_{\rm H_2,thick}}{\Lambda_{\rm H_2,thin}}, \qquad (48)$$

could be represented as a simple function of density:

$$f_\tau = \min\left[1, \left(\frac{n}{n_0}\right)^{-0.45}\right], \qquad (49)$$

where $n_0 = 8 \times 10^9$ cm^{-3}. They showed that this simple expression was a good approximation to the results of the full radiative transfer model used by Ripamonti et al. (2002), and suggested that this approximation would be useful for extending the results of three-dimensional simulations into the optically thick regime. However, they also noted that it may only be accurate while the collapse remains approximately spherical, as the one-dimensional model on which it is based assumes spherical infall.

An alternative approach was introduced by Yoshida et al. (2006). They compute escape probabilities for each rotational and vibrational line of H_2 using the standard Sobolev approximation (Sobolev 1960). In this approximation, the optical depth at line centre of a transition from an upper level u to a lower level l is written as

$$\tau_{ul} = \alpha_{ul} L_{\rm s}, \tag{50}$$

where α_{ul} is the line absorption coefficient and $L_{\rm s}$ is the Sobolev length. The absorption coefficient α_{ul} can be written as

$$\alpha_{ul} = \frac{\Delta E_{ul}}{4\pi} n_l B_{lu} \left[1 - \exp\left(\frac{-\Delta E_{ul}}{kT}\right)\right] \phi(\nu_{ul}), \tag{51}$$

where E_{ul} is the energy difference between the two levels, n_l is the number density of H_2 molecules in the lower levels, B_{lu} is the usual Einstein B coefficient, and $\phi(\nu_{ul})$ is the line profile at the centre of the line. The Sobolev length is given by

$$L_{\rm s} = \frac{v_{\rm th}}{|{\rm d}v_{\rm r}/{\rm d}r|}, \tag{52}$$

where $v_{\rm th}$ is the thermal velocity of the H_2 and $|{\rm d}v_{\rm r}/{\rm d}r|$ is the size of the velocity gradient along a given line of sight from the fluid element of interest. Given τ_{ul}, the escape probability for photons emitted in that direction then follows as

$$\beta_{ul} = \frac{1 - \exp(-\tau_{ul})}{\tau_{ul}}. \tag{53}$$

To account for the fact that the velocity gradient may differ along different lines of sight from any particular fluid element, Yoshida et al. (2006) utilize a mean escape probability given by

$$\beta = \frac{\beta_x + \beta_y + \beta_z}{3}, \tag{54}$$

where β_x, β_y and β_z are the escape probabilities along lines of sight in the x, y and z directions, respectively. Finally, once the escape probabilities for each transition have been calculated, the optically thick H_2 cooling rate can be computed from

$$\Lambda_{\rm H_2,thick} = \sum_{u,l} E_{ul} \beta_{ul} A_{ul} n_u, \tag{55}$$

where A_{ul} is the Einstein A coefficient for the transition from u to l and n_u is the population of the upper level u.

Strictly speaking, the Sobolev approximation is valid only for flows in which the Sobolev length L_s is much smaller than the characteristic length scales associated with changes in the density, temperature or chemical makeup of the gas, a requirement which is easy to satisfy when the velocity gradient is very large, but which is harder to justify in the case of Population III star formation, since the collapse speed is typically comparable to the sound-speed. Nevertheless, Yoshida et al. (2006) show that the optically thick H_2 cooling rates predicted by the Sobolev approximation are in very good agreement with those computed in the one-dimensional study of Omukai and Nishi (1998) by solution of the full radiative transfer problem.

Little work has been done on comparing these two approaches to treating optically-thick H_2 cooling. This issue was addressed briefly in Turk et al. (2011), who showed that the two approximations yielded similar values for f_τ for densities $n < 10^{15}$ cm^{-3} during the initial collapse of the gas, with differences of at most a factor of two. However, as yet no study has examined whether this good agreement persists past the point at which the first protostar forms.

2.1.4 Collision-Induced Emission

A further significant point in the collapse of the gas is reached once the number density increases to $n \sim 10^{14}$ cm^{-3}. At this density, a process known as collision-induced emission becomes important. Although an isolated H_2 molecule has no dipole moment, and can only emit or absorb radiation through quadrupole transitions, when two H_2 molecules collide[4] they briefly act as a kind of "supermolecule" with a non-zero dipole moment for the duration of the collision. This supermolecule can therefore absorb or emit radiation through dipole transitions, which have much higher transition probabilities than the quadrupole transitions available to isolated H_2. If radiation is absorbed, this process is termed collision-induced absorption; if it is emitted, then we refer to the process as collision-induced emission (CIE). A detailed discussion of the phenomenon can be found in Frommhold (1993).

Collision-induced emission can in principle occur in gas of any density, but the probability of a photon being emitted in any given collision is very small, owing to the short lifetime of the collision state ($\Delta t < 10^{-12}$ s at the temperatures relevant for Pop. III star formation; see Ripamonti and Abel 2004). For this reason, CIE becomes an important process only at very high gas densities. Another consequence of the short lifetime of the collision state is that the individual lines associated with the dipole transitions become so broadened that they actually merge into a continuum. This is important, as it means that the high opacity of the gas in the rovibrational lines of H_2 does not significantly reduce the amount of energy that can be radiated

[4] A similar process can also occur during collisions of atomic hydrogen or atomic helium with H_2, but it is the H_2-H_2 case that is the most relevant here.

away by CIE. Therefore, once the gas reaches a sufficiently high density, CIE becomes the dominant form of cooling, as pointed out by several authors (Omukai and Nishi 1998; Ripamonti et al. 2002; Ripamonti and Abel 2004).

The most detailed study of the effects of CIE cooling on the collapse of primordial gas was carried out by Ripamonti and Abel (2004). They showed that CIE cooling could actually become strong enough to trigger a thermal instability, However, the growth rate of this instability is longer than the gravitational free-fall time, meaning that it is unlikely that this process can drive fragmentation during the initial collapse of the gas.

2.1.5 Cooling due to H_2 Dissociation

The phase of the collapse dominated by CIE cooling lasts for only a relatively short period of time. The gas becomes optically thick in the continuum once it reaches a density $n \sim 10^{16}$ cm^{-3} (Omukai and Nishi 1998; Ripamonti and Abel 2004), which strongly suppresses any further radiative cooling. Once this occurs, the gas temperature rises until it reaches a point at which the H_2 begins to dissociate. At these densities, this occurs at a temperature $T \sim 3000$ K. Once this point is reached, the temperature rise slows, as most of the energy released during the collapse goes into dissociating the H_2 rather than raising the temperature. As it takes 4.48 eV of energy to destroy each H_2 molecule, this H_2 dissociation phase continues for a while. However, it comes to an end once almost all of the H_2 has been destroyed, at which point the temperature of the gas begins to climb steeply. The thermal pressure in the interior of the collapsing core rises rapidly and eventually becomes strong enough to halt the collapse. At the point at which this occurs, the size of the dense core is around 0.1 AU, its mass is around 0.01 $M_\odot$ and its mean density is of order 10^{20} cm^{-3} (Yoshida et al. 2008). It is bounded by a strong accretion shock. This pressure-supported, shock-bounded core is a Population III protostar, and its later evolution is discussed in Sect. 3 below.

2.2 *Dark Matter Annihilation*

One complication not accounted for in the models of Pop. III star formation described above is the role that may be played by dark matter annihilation. The nature of dark matter is not yet understood, but one plausible candidate is a weakly interacting massive particle (WIMP)—specifically, the lightest supersymmetric particle predicted in models based on supersymmetry. The simplest supersymmetry models predict that this WIMP has an annihilation cross-section $\langle \sigma v \rangle \sim 3 \times 10^{-26}$ cm^2, a mass within the range of 50 GeV to 2 TeV, and a cosmological density consistent with the inferred density of dark matter (Spolyar et al. 2008). The rate per unit volume at which energy is produced by dark matter annihilation can be written as $Q_{\rm ann} = \langle \sigma v \rangle \rho_X^2 / m_X$, where ρ_X is the mass density of dark matter and m_X is the mass of a single dark matter particle. For a plausible particle mass of 100 GeV, and a dark matter density equal to the cosmological background density of dark matter, this expression yields a tiny heating rate, $Q_{\rm ann} \sim 6 \times 10^{-62} (1+z)^6 \Omega_{\rm m}^2 h^4 \, {\rm erg\,cm^{-3}\,s^{-1}}$, even before one accounts for the fact that much of the annihilation energy is released in the form of energetic neutrinos or gamma-rays that couple only very weakly with the intergalactic gas. WIMP annihilation therefore plays no significant role in the evolution of the intergalactic medium while the WIMPs remain uniformly distributed (Myers and Nusser 2008). However, the ρ_X^2 density dependence of the heating rate means that it can potentially become significant in regions where the dark matter density is very high.

Spolyar et al. (2008) proposed that one situation in which the heating from dark matter annihilation could become important would occur during the formation of the very first Population III protostars. They assumed that any given star-forming minihalo would form only a single Pop. III protostar, and that this protostar would form at the center of the minihalo. As the gas collapsed at the center of the minihalo, its increasing gravitational influence would bring about a local enhancement of the dark matter density, via a process known as adiabatic contraction. The basic idea underlying this is very simple. For a collisionless particle on a periodic orbit, the quantity $\oint p dq$, where p is the conjugate momentum of coordinate q, is an adiabatic invariant, i.e. a quantity that does not vary when the gravitational potential varies, provided that the rate of change of the potential is sufficiently slow. If p represents the angular momentum of a particle on a circular orbit of radius r within some spherically symmetric mass distribution, then one can show that the quantity $rM(r)$ is constant for that particle, where $M(r)$ is the mass enclosed within r, so long as this enclosed mass changes on a timescale that is long compared to the orbital period. Spolyar et al. (2008) show that if one starts with a simple NFW profile for the dark matter (Navarro et al. 1997) and account for the effects of adiabatic contraction using a simple approach pioneered by Blumenthal et al. (1986), then one finds that for any WIMP mass less than 10 TeV, the effects of dark matter annihilation heating become significant during the collapse of the gas. Spolyar et al. (2008) identify the point at which this occurs by comparing the heating rate due to dark matter annihilation with the H_2 cooling rate. To determine a value for the latter, they make use of the simulation results of Yoshida et al. (2006) and Gao et al. (2007) and measure how the H_2 cooling rate of the gas in the central collapsing core evolves as the collapse proceeds. They show that for a 100 GeV WIMP, heating dominates at gas densities $n > 10^{13}$ cm^{-3}. Finally, they argue that once dark matter annihilation heating dominates over H_2 cooling, the gravitational collapse of the gas will come to a halt, and hence the gas will never reach protostellar densities. Instead, it will remain quasi-statically supported at a density of roughly 10^{13} cm^{-3} (for a 100 GeV WIMP), with a corresponding size scale of 17 AU, for as long as the dark matter annihilation rate remains large compared to the H_2 cooling rate. As the time required to consume all of the dark matter within a radius of 17 AU may be hundreds of millions of years, the resulting quasi-static gas distribution—dubbed a "dark star" by Spolyar et al. (2008)—could potentially survive for a very long time.

One criticism of the original Spolyar et al. (2008) model is its reliance on the Blumenthal et al. (1986) prescription for describing the effects of the adiabatic contraction of the dark matter. This prescription assumes that all of the dark matter particles move on circular orbits, which is unlikely to be the case in a realistic dark matter minihalo, and concerns have been expressed that it may yield values for the dark matter density after adiabatic contraction that are significantly higher than the true values (see e.g. Gnedin et al. 2004). For this reason, Freese et al. (2009) re-examined this issue using an alternative method for estimating the effects of adiabatic contraction, based on Young (1980). This alternative prescription does account for particles moving on radial orbits, and Freese et al. (2009) show that it predicts dark matter densities that are indeed systematically smaller than those predicted by the Blumenthal et al. (1986) prescription, but only by a factor of two. Freese et al. (2009) therefore conclude that although using the Young (1980) prescription for adiabatic contraction in place of the simpler Blumenthal et al. (1986) prescription will lead to some minor quantitative changes in the predicted outcome, the main qualitative results of the Spolyar et al. (2008) study are insensitive to this change, and one would still expect a "dark star" to form.

Another potential problem with the dark star hypothesis is the fact that it is not at all clear that the collapse of the gas will stop once the dark matter heating rate exceeds the H_2 cooling rate. For one thing, the values for the H_2 cooling rate used by Spolyar et al. (2008) do not account for the effects of the dark matter annihilation heating. If this leads to an increase in temperature, then this will also increase the H_2 cooling rate, allowing more of the energy produced by dark matter annihilation to be radiated away. It is therefore unlikely that the point in the collapse at which the dark matter annihilation heating rate exceeds the Spolyar et al. estimate for the H_2 cooling rate is marked by any sharp jump in the temperature. Instead, we would expect to find a more gradual temperature increase, at least up until the point at which collisional dissociation of the H_2 starts to occur.

Once H_2 begins to dissociate, this provides another outlet for the energy generated by dark matter annihilation. Spolyar et al. estimate that for a 100 GeV WIMP, the power generated by dark matter annihilation within the central core is $L_{\rm dm} \sim 140\,{\rm L}_{\odot}$, and the core mass is roughly $0.6\,{\rm M}_{\odot}$. The total energy stored within the core in the form of the binding energy of the H_2 molecules is roughly

$$E_{\rm H_2,bind} = 4.48{\rm eV} \times 0.76 \times \frac{0.6\,{\rm M}_{\odot}}{m_{\rm H_2}} \simeq 2.6 \times 10^{45}\,{\rm erg}, \tag{56}$$

and the time required for dark matter annihilation to produce this much energy is

$$t_{\rm dis} = \frac{E_{\rm H_2,bind}}{L_{\rm dm}} \simeq 200\,{\rm year}. \tag{57}$$

For comparison, the free-fall time at this point in the collapse is roughly 15 years. H_2 dissociation will therefore allow the collapse of the gas to continue until either the dark matter heating rate becomes large enough to destroy the H_2 in the core in much less than a dynamical time, or the compressional heating produced during the collapse becomes capable of doing the same job. In either case, it is likely that much higher core densities can be reached than was assumed in the Spolyar et al. study.

A first attempt to hydrodynamically model the formation of a "dark star" while correctly accounting for these thermodynamical effects was made by Ripamonti et al. (2010). They used the 1D, spherically symmetric hydrodynamical code described in Ripamonti et al. (2002) to model the collapse of the gas up to densities of order $10^{15}\,\mathrm{cm}^{-3}$ for a range of different WIMP masses between 1 GeV and 1 TeV. Adiabatic contraction of the dark matter was modelled using the algorithm described in Gnedin et al. (2004), and the effects of the dark matter annihilation heating and ionization were self-consistently accounted for in the chemical and thermal model. Ripamonti et al. (2010) show that even in the most extreme case that they study, the heating produced by the dark matter appears unable to halt the collapse for an extended period. After the dark matter heating rate exceeds the H_2 cooling rate, dissociation of H_2 in the core accounts for most of the "excess" energy not radiated away by the gas, allowing the collapse to continue. Once the H_2 in the core is exhausted, the temperature rises steeply, very briefly halting the collapse. However, the temperature quickly becomes large enough to allow other cooling mechanisms (e.g. H^- bound-free transitions or Lyman-α emission from atomic hydrogen) to operate, allowing the collapse to restart. Ripamonti et al. (2010) do not find any evidence for the formation of a hydrostatically supported "dark star" up to the highest densities that they study. Confirmation of this result in a 3D treatment of the collapse would be very useful.

2.3 *The Role of Magnetic Fields*

2.3.1 Initial Strength

The majority of the work that has been done on modelling the formation of the first stars assumes that magnetic fields play no role in the process, either because no magnetic field exists at that epoch, or because the strength of any field that does exist is too small to be significant. A number of mechanisms have been suggested that may generate magnetic seed fields during the inflationary epoch, the electroweak phase transition or the QCD phase transition (see Kandus et al. 2011, for a recent comprehensive review). Observational constraints (e.g. Barrow et al. 1997; Schleicher et al. 2008) limit the strength of the magnetic field at the epoch of first star formation to no more than about 1 nG (in comoving units), but it is quite possible that any primordial seed field resulting from one of these processes will actually have a much smaller strength.

An alternative source for magnetic fields within the first generation of star-forming minihalos is the so-called Biermann battery effect (Biermann 1950). In a partially ionized gas in which the gradient of electron density does not perfectly align with the gradient of electron pressure, as can happen if there is a temperature gradient that does not align with the pressure gradient, the magnetic induction equation takes the form

$$\frac{\partial \mathbf{B}}{\partial t} = \nabla \times (\mathbf{v} \times \mathbf{B}) + \frac{c \nabla p_{\mathrm{e}} \times \nabla n_{\mathrm{e}}}{n_{\mathrm{e}}^2 e}, \tag{58}$$

where $\mathbf{B}$ is the magnetic field, $\mathbf{v}$ is the velocity, n_{e} is the electron density, p_{e} is the electron pressure, and e is the charge on an electron. In the limit that $B \rightarrow 0$, the first term on the right-hand side of this equation also becomes zero, but the battery term does not. It can therefore act as the source of a magnetic field in a gas that is initially unmagnetized. An early investigation into the effectiveness of the Biermann battery during galaxy formation was made by Kulsrud et al. (1997), who considered the formation of massive galaxies and showed that the Biermann battery could generate a field of strength $B \sim 10^{-21}$ G during their assembly. More recently, Xu et al. (2008) have simulated the action of the Biermann battery during the formation of one of the first star-forming minihalos, finding that it is able to generate initial field strengths of the order of 10^{-18} G during this process.

2.3.2 Amplification

The seed fields generated by the Biermann battery, or by other processes acting in the very early Universe can be significantly amplified by flux-freezing during the gravitational collapse of the gas. If the diffusive timescale associated with ambipolar diffusion or Ohmic diffusion is long compared to the gravitational collapse timescale, then the magnetic field will be "frozen" to the gas, and will be carried along with it when the gas collapses.

In the optimal case of spherical collapse, perfect flux freezing implies that the field strength evolves with density as $B \propto \rho^{2/3}$, and hence the magnetic pressure $p_{\mathrm{mag}} = B^2/8\pi$ evolves as $p_{\mathrm{mag}} \propto \rho^{4/3}$. In comparison, the thermal pressure p_{therm} evolves as $p_{\mathrm{therm}} \propto \rho T$, and so if the temperature does not vary much during the collapse, the plasma β parameter, $\beta \equiv p_{\mathrm{therm}}/p_{\mathrm{mag}}$, evolves as $\beta \propto \rho^{-1/3}$. Therefore, if the gas is initially dominated by thermal pressure rather than magnetic pressure, it will remain so during much of the collapse, as a large change in the density is necessary to significantly alter β. In the case examined by Xu et al. (2008), the very small initial magnetic field strength means that β is initially very large, and remains so throughout the collapse, implying that the magnetic field never becomes dynamically significant. Moreover, even if we take an initial comoving field strength of 1 nG, comparable to the observational upper limit, at the mean halo density, $\beta \sim 10^4$ (assuming a halo formation redshift $z = 20$ and a virial temperature of 1,000 K), and does not become of order unity until very late in the collapse. Furthermore, if the collapse of the gas is not spherical, whether because of the effects of gravitational forces, angular momentum, or the influence of the magnetic field itself, the amplification due to flux freezing and collapse will be less than in the spherical case (see e.g. Machida et al. 2006, who find a somewhat shallower relationship in some of their models).

Therefore, for magnetic fields to play an important role in Pop. III star formation, they must either start with a field strength very close to the observational upper limit, or we must invoke an amplification process that is much more effective than the amplification that occurs due to flux freezing and gravitational collapse. One obvious possibility is amplification via some kind of dynamo process, which could bring about exponential amplification of an initially small seed field. Of particular interest is the small-scale turbulent dynamo (Kraichnan and Nagarajan 1967; Kazantsev 1968; Kulsrud and Anderson 1992). This produces a magnetic field that has no mean flux on the largest scales but that can have substantial mean flux within smaller-scale subregions. The growth rate of the magnetic field due to the turbulent dynamo is closely related to the rate of turnover of the smallest eddies. If the magnetic field is sufficiently small that it does not significantly affect the velocity field of the gas (the kinematic approximation), and if we assume that we are dealing with Kolmogorov turbulence, then Kulsrud and Zweibel (2008) show that the magnetic energy density grows exponentially, and that after a single gravitational free-fall time it is amplified by a factor $\exp(\mathrm{Re}^{1/2})$, where Re is the Reynolds number of the flow. If we assume that the driving scale of the turbulence is comparable to the size of the minihalo, and that the turbulent velocity is of the same order as the sound speed (see e.g. Abel et al. 2002), then $\mathrm{Re} \sim 10^4$–10^5, implying that the magnetic field is amplified by an enormous factor during the collapse. In practice, the field will not be amplified by as much as this analysis suggests, as the kinematic approximation will break down once the magnetic energy density becomes comparable to the kinetic energy density on the scale of the smallest eddies. Nevertheless, this simple treatment implies that the turbulent dynamo can amplify the magnetic field to a strength at which it becomes dynamically important.

Although the importance of dynamo processes during the formation of the first galaxies has been understood for a number of years (see e.g. Pudritz and Silk 1989; Beck et al. 1994; Kulsrud et al. 1997), they have attracted surprisingly little attention in studies of primordial star formation. Over the past couple of years, however, this has begun to change, with several recent studies focussing on the growth of magnetic fields during the formation of the first stars. The first of these was Schleicher et al. (2010), who studied the effectiveness of the turbulent dynamo during gravitational collapse using a simple one-zone Lagrangian model for the collapsing gas. Their model assumes that turbulence is generated by gravitational collapse on a scale of the order of the Jeans length, and that on smaller scales, the turbulent velocity scales with the length-scale l as $v_{\rm turb} \propto l^{\beta}$. Schleicher et al. (2010) study both Kolmogorov turbulence, with $\beta = 1/3$ and Burgers turbulence, with $\beta = 1/2$, and show that in both cases, amplification of a weak initial seed field occurs rapidly, and that the field reaches saturation on all but the largest scales at an early point during the collapse. Because Schleicher et al. (2010) did not solve directly for the fluid velocities, they were unable to model the approach to saturation directly. Instead, they simply followed Subramanian (1998) and assumed that the strength of the saturated field satisfies

$$B_{\rm sat} = \sqrt{\frac{4\pi\rho v_{\rm turb}^2}{{\rm Rm}_{\rm cr}}}, \tag{59}$$

where ${\rm Rm}_{\rm cr} \sim 60$ is a critical value of the magnetic Reynolds number, ${\rm Rm} = v_{\rm turb} l/\eta$ (where η is the resistivity), that must be exceeded in order for exponential growth of the field to occur (Subramanian 1998).

The main weakness of the Schleicher et al. (2010) study lies in the assumptions that it was forced to make about the nature of the turbulent velocity field. Therefore, in a follow-up study, Sur et al. (2010) used high-resolution adaptive mesh refinement simulations to directly follow the coupled evolution of the velocity field and the magnetic field within a 3D collapse model. For their initial conditions, Sur et al. (2010) took a super-critical Bonnor-Ebert sphere (Bonnor 1956; Ebert 1955) with a core density $n_{\rm c} = 10^4\ {\rm cm}^{-3}$ and a temperature $T = 300$ K. They included initial solid-body rotation, with a rotational energy that was 4 % of the total gravitational energy, and a turbulent velocity component with an RMS velocity equal to the sound speed and with an energy spectrum $E(k) \propto k^{-2}$. A weak magnetic field was also included, with an RMS field strength $B_{\rm rms} = 1$ nG, and with the same energy spectrum as the turbulence. For reasons of computational efficiency, Sur et al. (2010) did not follow the thermal and chemical evolution of the gas directly. Instead, they adopted a simple barotropic equation of state, $P \propto \rho^{1.1}$, inspired by the results of previous hydrodynamical models (e.g. Abel et al. 2002). In view of the sensitivity of the turbulent dynamo to the Reynolds number, and the fact that numerical dissipation on the grid scale limits the size of Re in any 3D numerical simulation to be substantially less than the true physical value, there is good reason to expect that the dynamo amplification rate will be sensitive to the numerical resolution of the simulation. Sur et al. (2010) therefore focussed on the effects of resolution, performing five different simulations with the same initial conditions, but with different grid refinement criteria. Starting with a model in which the refinement criterion ensures that the Jeans length is always resolved by 8 grid zones, they looked at the effects of increasing this number to 16, 32, 64 and 128 grid zones.

Sur et al. (2010) showed that in the 8 and 16 cell runs, the magnetic field strength increases with density at a slower rate than the $B \propto \rho^{2/3}$ that we would expect simply from flux freezing and roughly spherical collapse, indicating that in these runs, the turbulent dynamo does not operate. Starting with the 32 cell run, however, they found evidence for an increase in B with density that is larger than can be explained simply by compression, which they ascribe to the effects of the turbulent dynamo. They showed that as the number of grid zones used to resolve the Jeans length is increased, the rate at which the field grows also increases, and there is no sign of convergence at even their highest resolution. This resolution dependence explains why the earlier study of Xu et al. (2008) found no evidence for dynamo amplification, as their study used only 16 grid zones per Jeans length.

More recently, Federrath et al. (2011) have re-examined this issue of resolution dependence, and have shown that when the number of grid zones per Jeans length is small, the amount of turbulent energy on small scales is also significantly underestimated. The reason for this is the same as the reason for the non-operation of the turbulent dynamo: the effective Reynolds number is too small. Federrath et al. (2011) show that in gravitationally collapsing regions that undergo adaptive mesh refinement, the effective Reynolds number scales with the number of grid zones per Jeans length as $\mathrm{Re}_{\mathrm{eff}} = (N/2)^{4/3}$. Furthermore, an effective Reynolds number $\mathrm{Re}_{\mathrm{eff}} \sim 40$ is required in order for the turbulent dynamo to operate, implying that one needs $N \sim 30$ or more zones per Jeans length in order to begin resolving it, in agreement with the findings of Sur et al. (2010). It should also be noted that the operation of the turbulent dynamo in simulations of turbulence without self-gravity requires a similar minimum value for the Reynolds number (Haugen et al. 2004).

Together, these studies support the view that amplification of a weak initial magnetic field by the turbulent dynamo may indeed have occurred within the first star-forming minihalos. However, a number of important issues remain to be addressed. First, the three-dimensional studies carried out so far all adopt a simple barotropic equation of state, rather than solving self-consistently for the thermal evolution of the gas. This is a useful simplifying assumption, but may lead to incorrect dynamical behaviour, as one misses any effects due to thermal instabilities, or the thermal inertia of the gas (i.e. the fact that the cooling time is typically comparable in size to the dynamical time). Work is currently in progress to re-run some of these models with a more realistic treatment of the thermodynamics and chemistry in order to explore the effect that this has on the degree of amplification (T. Peters, private communication). Second, it will clearly be important to perform similar studies using more realistic initial conditions for the gas. Of particular concern is whether the turbulence that is generated during the gravitational collapse of gas within a primordial minihalo is similar in nature to that studied in these more idealized calculations, and if not, what influence this has on the amplification of the field. Finally, and most importantly, there is the issue of the level at which the field saturates. Exponential amplification of the field by the small-scale dynamo will occur only while the kinematic approximation holds, i.e. while the energy stored in the magnetic field is much smaller than the energy stored in the small-scale turbulent motions. Once the field becomes large, the Lorentz force that it exerts on the gas will act to resist further folding and amplification of the field. In addition, the dissipation of magnetic energy by Ohmic diffusion and ambipolar diffusion will grow increasingly important. However, it remains unclear which of these effects will be the most important for limiting the growth of the magnetic field in dense primordial gas.

2.3.3 Consequences

If a strong magnetic field can be generated by dynamo amplification, then it will affect both the thermal and the dynamical evolution of the gas. The possible dynamical effects of a strong magnetic field have been investigated by Machida et al. (2006, 2008). In a preliminary study, Machida et al. (2006) used nested-grid simulations to investigate the influence of a magnetic field on the collapse of a small, slowly-rotating primordial gas cloud. For their initial conditions, they used a supercritical Bonnor-Ebert sphere with mass 5.1×10^4 $M_\odot$, radius 6.6 pc, central density $n_c = 10^3$ cm^{-3} and an initial temperature of 250 K. They assumed that this cloud was in solid body rotation with angular velocity Ω_0 and that it was threaded by a uniform magnetic field oriented parallel to the rotation axis, with an initial field strength B_0. They performed simulations with several different values of Ω_0 and B_0, with the former ranging from 10^{-17} s^{-1} to 3.3×10^{-16} s^{-1}, and the latter from 10^{-9} G to 10^{-6} G. To treat the thermal evolution of the gas, they used a barotropic equation of state, based on the one-zone results of Omukai et al. (2005).

Machida et al. (2006) used this numerical setup to follow the collapse of the gas down to scales of the order of the protostellar radius. They showed that the magnetic field was significantly amplified by compression and flux freezing during the collapse, reaching strengths of order 6×10^5–6×10^6 G on the scale of the protostar. A very compact disk with a radius of few $R_\odot$ formed around the protostar, and in models with initial field strength $B_0 > 10^{-9}$ G, the magnetic field became strong enough to drive a hydromagnetic disk wind that ejected roughly 10 % of the infalling gas. Numerical limitations (discussed in Sect. 3 below) prevented Machida et al. (2006) from following the evolution of the system for longer than a few days after the formation of the protostar, and so it remains unclear whether an outflow would eventually be generated in the 10^{-9}G case, and whether the outflows continue to be driven as the protostar and disk both grow to much larger masses.

In a follow-up study, Machida et al. (2008) used a similar numerical setup, but examined a much broader range of values for $\beta_0 (\equiv E_{\rm rot}/|E_{\rm grav}|)$, the ratio of the initial rotational energy to the initial gravitational energy, and $\gamma_0 (\equiv E_{\rm mag}/|E_{\rm grav}|)$, the ratio of the initial magnetic energy to the initial gravitational energy. They found that the outcomes of the simulations could be classified into two main groupings. Clouds with $\beta_0 > \gamma_0$, i.e. ones which were rotationally dominated, formed a prominent disk during the collapse that then fragmented into a binary or higher order multiple system. In these simulations, no jets were seen (with the exception of a couple of model in which $\beta_0 \sim \gamma_0$). On the other hand, when $\beta_0 < \gamma_0$, i.e. when the cloud was magnetically dominated, the disk that formed was much less prominent and did not fragment, but instead an outflow was driven that removed of order 10 % of the gas that reached the disk, as in the Machida et al. (2006) study.

These results support the idea that outflows will be a natural consequence of the generation of strong magnetic fields during Population III star formation. However, it is important to note that the Machida et al. (2006, 2008) simulations only model the very earliest stages of outflow driving, on a timescale $t \ll 1$ year. The evolution of outflows on much longer timescales, and their influence on the infalling gas have not yet been studied in detail, and it is unclear to what extent one can safely extrapolate from the very limited period that has been studied.

A strong magnetic field could also have a direct impact on the thermal evolution of the gas, through the heating arising from ambipolar diffusion. The effects of this process in gravitationally collapsing gas within the first star-forming minihalos have been investigated by Schleicher et al. (2009) using a simple one-zone treatment of the gas. They assume that in the absence of ambipolar diffusion, the magnetic field strength would evolve as $B \propto \rho^{\alpha}$, where $\alpha = 0.57(M_{\rm J}/M_{\rm J,mag})^{0.0116}$ and $M_{\rm J,mag}$ is the magnetic Jeans mass (i.e. the minimum mass that a perturbation must have in order to be unstable to its own self-gravity when support against collapse is provided by a magnetic field, rather than by thermal pressure). This expression for α is an empirical fitting formula derived by Schleicher et al. (2009) from the results of Machida et al. (2006). In their treatment of the evolution of B within their one-zone models, Schleicher et al. (2009) also account for the loss of magnetic energy through ambipolar diffusion.

Another important simplification made in the Schleicher et al. (2009) model is the replacement of the full expression for the ambipolar diffusion heating rate (Pinto et al. 2008)

$$L_{\rm AD} = \frac{\eta_{\rm AD}}{4\pi} \frac{|(\nabla \times \mathbf{B}) \times \mathbf{B}|^2}{B^2}, \tag{60}$$

where $B = |\mathbf{B}|$ and $\eta_{\rm AD}$ is the ambipolar diffusion resistivity, with the simpler approximation

$$L_{\rm AD} = \frac{\eta_{\rm AD}}{4\pi} \frac{B^2}{L_B}, \tag{61}$$

where L_B is the coherence length of the magnetic field.

Schleicher et al. (2009) show that if the initial field strength is less than 0.01 nG (comoving), then ambipolar diffusion heating has almost no effect on the thermal evolution of the gas. For stronger fields, ambipolar diffusion heating leads to an increase in the gas temperature at densities between $n \sim 10^4\ {\rm cm}^{-3}$ and $n \sim 10^{10}\,{\rm cm}^{-3}$, amounting to as much as a factor of three at $n \sim 10^8\,{\rm cm}^{-3}$. However, at higher densities, three-body H_2 formation heating becomes a more important heat source than ambipolar diffusion, meaning that the temperature evolution becomes largely independent of the magnetic field strength once again. Schleicher et al. (2009) did not examine initial field strengths larger than 1 comoving nG, as these are ruled out by observational constraints, but if one considers the effects of the turbulent dynamo acting during the collapse, then it is possible that much larger fields could be generated on smaller scales, and it would be useful to revisit this issue and examine whether ambipolar diffusion heating from these smaller-scale fields can significantly affect the collapse of the gas.

Finally, one important caveat to bear in mind regarding the Machida et al. (2006, 2008) and Schleicher et al. (2009) simulations is the fact that they adopt a correlated initial magnetic field, while the field generated by the turbulent dynamo will have little or no correlation on large scales (Maron et al. 2004). The extent to which the dynamical and thermal effects of this uncorrelated field will be the same as those of a correlated field is unclear. Further investigation of this issue would be very valuable.

3 Evolution After the Formation of the First Protostar

As we saw in the last section, when it comes to the formation of the very first Population III protostar, there is broad agreement on the details of the process, with different groups, who use different numerical approaches, finding results that are in good qualitative agreement with each other. Some quantitative disagreements still exist (see e.g. Turk et al. 2011), but it is unclear to what extent these reflect real differences between numerical approaches as opposed to natural variation in the details of the collapse. The main uncertainties in this phase stem from uncertainties in the input physics, such as whether magnetic fields can become amplified to dynamically significant levels during the collapse, or whether dark matter annihilation significantly affects the outcome.

Once we move on to considering the evolution of the gas within star-forming minihalos *after* the formation of the first protostar, the situation becomes much less clear. The fundamental problem stems from the fact that although we can follow the gravitational collapse of the primordial gas down to scales as small as the protostellar radius (see e.g. Yoshida et al. 2008), the numerical timestep in an explicit hydrodynamical code becomes extremely short during this process. This is a consequence of the Courant condition, which states that for such a code to be numerically stable, the timestep must satisfy

$$\Delta t \leq \frac{\Delta x}{c_s}, \tag{62}$$

where Δx is the size of the smallest resolution element, and c_s is the sound speed of the gas.

The Courant condition implies that if we take a value of Δx small enough to adequately resolve the structure of the protostar and the gas immediately surrounding it (e.g. $\Delta x = 1 \mathrm{R}_{\odot}$), then the required timestep will be extremely small: $\Delta t \leq 7 \times 10^4$ s for $\Delta x = 1 \mathrm{R}_{\odot}$ and a sound speed of 10 km s^{-1}. This means that if we want to follow the later evolution of the protostar and the surrounding gas over a timescale of thousands of years in order to see how it grows in mass prior to reaching the main sequence, then we must use a very large number of timesteps: our simple estimate above yields a number of the order of a million. In practice, the computational expense of doing this within a three-dimensional hydrodynamical code is prohibitively large, meaning that it has so far proved impossible to study the evolution of the gas in this fashion.

Efforts to surmount this difficulty typically follow one of two approaches. One approach is simply to halt the numerical simulation at the point at which the Courant timestep becomes prohibitively small, and to model the later evolution of the protostar using a semi-analytical, or one-dimensional, fully numerical treatment. To do this, it is necessary to make some assumption about the behaviour of the gas surrounding the protostar. In general, models of this type assume that the gas does not fragment and form additional protostars, but instead is simply accreted by the existing protostar, either directly or via a protostellar accretion disk. The results obtained using this approach—what we afterwards refer to as the "smooth accretion model"—are discussed in Sect. 3.1 below.

The other approach that can be used to study the further evolution of the gas surrounding the protostar makes use of a technique developed for studies of contemporary star formation, which face a similar problem on protostellar scales. Gravitationally bound regions of gas that become smaller than some pre-selected size scale are replaced by what are usually termed sink particles (see e.g. Bate et al. 1995). These particles can accrete gas from their surroundings and continue to interact gravitationally with the surrounding gas, but allow one to neglect the very small-scale hydrodynamical flows that would otherwise force one to take very small numerical timesteps owing to the Courant condition. The great advantage of the sink particle technique is that one need make no assumption about the dynamical evolution of the gas surrounding the protostar on scales much larger than the effective size of the sink particle (the so-called accretion radius, discussed in more detail below), as one can simply continue to model this using the same numerical techniques as were used to model the initial gravitational collapse. The main disadvantage of the technique is that, strictly speaking, it represents an *ad hoc* modification of the fluid equations, with consequences that may not be entirely straightforward to predict. The modification to the solution caused by replacing dense gas with sink particles is unlikely to significantly affect the evolution of the gas on scales that are much larger than the accretion radius, but will clearly have an effect on the flow on scales close to the accretion radius. In addition, the common strategy of treating sink particles as point masses may not be appropriate when dealing with close encounters between sinks, as one misses the tidal forces acting between the gas clumps represented by the sinks.

Although sink particles have been used in studies of Population III star formation for over a decade, simulations using the correct initial conditions, and with sufficient spatial resolution and mass resolution to capture the details of the gas flow on scales close to those of individual protostars have only recently become possible. These simulations show that, contrary to the assumption made in the smooth accretion model, the gas generally fragments, rather than simply accreting onto a single, central protostar. The results obtained from studies using sink particles—afterwards referred to as the "fragmentation model" for Population III star formation—are discussed in Sect. 3.2 below.

3.1 *The Smooth Accretion Model*

3.1.1 Determining the Accretion Rate

As we have already discussed above, at the point at which the protostar forms, its mass is very small ($M \sim 0.01\ M_{\odot}$; see Yoshida et al. 2008), but it is surrounded by an infalling envelope of gas containing tens or hundreds of solar masses. If we assume that the gas in this infalling envelope does not undergo gravitational fragmentation, then it has only two possible fates—it must either be accreted by the central protostar (or protostellar binary; see e.g. Turk et al. 2009), or it must be prevented from accreting, and possibly expelled from the immediate vicinity of the protostar, by some form of protostellar feedback. This means that the mass of the protostar at the point at which it forms has very little to do with its final mass. To determine the size of the latter, we must understand the rate at which gas is accreted by the protostar, and how this process is affected by protostellar feedback.

Since protostellar feedback involves a number of different processes, many of which are complicated to model, it is easiest to start by considering models in which feedback effects are not included. As feedback acts to reduce the accretion rate, models of this type allow us to place an upper limit on the final mass of the Pop. III star.

A useful starting point is a simple dimensional analysis. Suppose that the protostar is embedded in a gravitationally unstable cloud of mass M and mean density $\langle \rho \rangle$, and that the protostellar mass $M_{*} \ll M$, so that its gravity is negligible in comparison to the self-gravity of the cloud. The timescale on which the gas cloud will undergo gravitational collapse and be accreted by the protostar is simply the free-fall collapse time, $t_{\rm ff} = \sqrt{3\pi / 32 G \langle \rho \rangle}$. Therefore, the time-averaged accretion rate will be given approximately by

$$\dot{M}_{\rm est} \sim \frac{M}{t_{\rm ff}} \sim M \sqrt{G \langle \rho \rangle}. \quad (63)$$

If the gas cloud were highly gravitationally unstable, then it would fragment rather than accreting onto a single object, so let us assume that it is only marginally unstable, i.e. that $M \sim M_{\rm J}$. In that case, since $M_{\rm J} \sim c_s^3 G^{-3/2} \rho^{-1/2}$, we can write our estimate of the time-averaged accretion rate as

$$\dot{M}_{\rm est} \sim M_{\rm J} \sqrt{G \langle \rho \rangle}, \quad (64)$$

$$\sim \frac{c_s^3}{G}. \quad (65)$$

We therefore find that the characteristic accretion rate scales as the cube of the sound speed. Moreover, since $c_s \propto T^{1/2}$, this implies that the accretion rate scales with temperature as $\dot{M} \propto T^{3/2}$.

This is an important result, because as we have already seen, the characteristic temperature of the dense, star-forming gas in a primordial minihalo is of the order of 1,000 K, far larger than the 10 K temperatures found within prestellar cores in local regions of star formation (see e.g. Bergin and Tafalla 2007). Our simple scaling argument therefore tells us that we will be dealing with far higher accretion rates in the Population III case than we are used to from studies of local star formation.

If we want to improve on this simple scaling argument and derive a more accurate figure for the accretion rate, then there are three main ways in which we can go about it. One possible approach is to construct a simplified model for the collapsing protostellar core from which an approximation to the true accretion rate can be derived analytically (or with only minor use of numerical calculations). For example, if we assume that the protostellar core is isothermal and spherically symmetric, then there is a whole family of similarity solutions that could potentially be used to describe the collapse (Hunter 1977; Whitworth and Summers 1985), including the familiar Larson-Penston solution (Larson 1969; Penston 1969), or the Shu solution (Shu 1977).

An example of this approach is given in Omukai and Nishi (1998). These authors used a spherically symmetric Lagrangian hydrodynamical code to simulate the formation of a Population III protostar, and found that prior to core formation, the gravitational collapse of the gas could be well described with a Larson-Penston similarity solution, with an entropy parameter $K = p/\rho^{\gamma} = 4.2 \times 10^{11}$ (in cgs units) and an effective adiabatic index $\gamma_{\rm eff} = 1.09$. Omukai and Nishi (1998) were unable to continue their numerical study past the point at which the protostar formed, for the reasons addressed above, but assumed that the same similarity solution would continue to apply. By making this assumption, they were therefore able to derive the following accretion rate for the protostar

$$\dot{M} = 8.3 \times 10^{-2} \left(\frac{t}{1 \text{ year}} \right)^{-0.27} \text{M}_{\odot} \text{ year}^{-1}. \tag{66}$$

In a similar study, using a more sophisticated treatment of the microphysics of the collapsing gas, Ripamonti et al. (2002) also found that the initial flow was well described as a Larson-Penston similarity solution, but derived a different accretion rate

$$\dot{M} = 6.0 \times 10^{-2} \left(\frac{t}{1 \text{year}} \right)^{-0.343} \text{M}_{\odot} \text{ year}^{-1}. \tag{67}$$

Another example of this approach comes from Tan and McKee (2004). They model the accretion flow onto a Pop. III protostar as a spherical, isentropic polytrope, and derive an accretion rate that is a function of three parameters: the entropy parameter K, the polytropic index γ_p (which, for an isentropic flow, is equal to the adiabatic index γ), and ϕ_*, a numeric parameter of order unity, which is related to the initial conditions of the flow. Tan and McKee (2004) use the results of Omukai and Nishi (1998) and Ripamonti et al. (2002) to argue that $\gamma_p = 1.1$, and use the 3D simulation results of Abel et al. (2002) to set the other two parameters to $\phi_* = 1.43$ and $K = 1.88 \times 10^{12} K'$ (in cgs units), where

$$K' = \left(\frac{T_{\rm eff}}{300 \text{ K}} \right) \left(\frac{n_{\rm H}}{10^4 \text{ cm}^{-3}} \right)^{-0.1}, \tag{68}$$

and where the effective temperature $T_{\rm eff} = P_{\rm eff}/(nk)$ accounts for the contribution made to the pressure by small-scale, subsonic turbulence in addition to the standard thermal pressure. Based on this, they then derive the following rate for the accretion of gas onto the protostar and its associated accretion disk

$$\dot{M} = 7.0 \times 10^{-2} K'^{3/2} \left(\frac{t}{1 \text{ year}} \right)^{-0.30} \text{M}_{\odot} \text{ year}^{-1}. \tag{69}$$

This can be directly compared to the other determinations of $\dot{M}$ if we assume that all of the gas reaching the accretion disk is eventually accreted by the star, which is a reasonable assumption for models that do not include the effects of gravitational fragmentation or protostellar feedback.

Instead of using simulation results to select a particular collapse model (e.g. Larson-Penston collapse) and then calculating $\dot{M}$ from the model, the second main approach used to determine $\dot{M}$ attempts to infer it from the state of the gas in the simulation at the point at which the protostar forms, using the information that the simulation provides on the density and velocity distributions of the gas. This approach was pioneered by Abel et al. (2002), who considered two simple models for the time taken for a given fluid element to accrete onto the central protostar. In the first of these models, they assumed that the time taken for the gas within a spherically-averaged shell of radius r to accrete onto the protostar was given by the ratio between the mass enclosed within the shell, $M(r)$, and the rate at which gas was flowing inward at that radius, i.e.

$$t_{\rm acc} = \frac{M(r)}{4\pi r^2 \rho(r) |v_{\rm r}(r)|}, \tag{70}$$

where $\rho(r)$ and $v_{\rm r}(r)$ are the spherically-averaged density and radial velocity in the shell. In the second model for $t_{\rm acc}$, they used an even simpler approximation, setting $t_{\rm acc}$ to the time that it would take for the gas to reach the protostar if it merely maintained its current radial velocity, i.e.

$$t_{\rm acc} = \frac{r}{v_{\rm r}}. \tag{71}$$

Abel et al. (2002) show that other than at the very earliest times, these two approaches yield very similar values for $t_{\rm acc}$, and hence very similar values for the accretion rate. This strategy has subsequently been used by many other authors to derive predicted protostellar accretion rates from their simulations (see e.g. Yoshida et al. 2006; O'Shea and Norman 2007; McGreer and Bryan 2008; Turk et al. 2011). Of particular note is the study by O'Shea and Norman (2007), who perform multiple simulations of Population III star formation using different random realizations of the cosmological density field. They find that minihalos assembling at higher redshifts form more H_2 than those assembling at lower redshifts, owing to the higher mean density of the virialized gas in the high redshift minihalos. They show that in their simulations, this leads to the gas at densities $n > 10^4 \text{ cm}^{-3}$ having significant differences in its mean temperature in the different halos. In the most H_2-rich minihalos, the dense gas can be as cold as 200 K, while in the minihalos with the least H_2, it can be as high as 1,000 K. As a result, the predicted accretion rates for the different minihalos span more than an order of magnitude, thanks to the strong scaling of $\dot{M}$ with temperature. Unfortunately, it is necessary to treat these results with a degree of caution, as the O'Shea and Norman (2007) simulations did not include the effect of three-body H_2 formation heating, which is known to have a significant influence on the temperature of the dense gas. It is unclear whether simulations that include this effect would produce dense gas with such a wide range of temperatures and accretion rates, although a study that is currently being carried out by Turk and collaborators should address this issue in the near future (M. Turk, private communication).

The third main approach used to determine the accretion rate involves measuring it directly in a simulation of the later evolution of the gas around the protostar. If we replace the protostar with a sink particle, then we can measure $\dot{M}$ simply by measuring the rate at which the sink particle mass increases. This approach was first used by Bromm and Loeb (2004), in a study of Population III star formation in which a sink particle was created once the gas density exceeded a threshold value $n_{\rm th} = 10^{12}\ {\rm cm}^{-3}$ (we will have more to say about this study below). Bromm and Loeb (2004) showed that the rate at which gas was accreted by the sink particle could be approximated as a broken power-law

$$\dot{M} = \begin{cases} 5.6 \times 10^{-2} \left(\frac{t}{1\rm year}\right)^{-0.25} \ {\rm M_\odot\ year^{-1}} & t \leq 10^3\ {\rm year} \\ \\ 6.3 \times 10^{-1} \left(\frac{t}{1\rm year}\right)^{-0.6} \ {\rm M_\odot\ year^{-1}} & t > 10^3\ {\rm year} \end{cases} \tag{72}$$

for times $t < 10^4$ year. Bromm & Loeb halted their simulation at $t \sim 10^4$ year and hence could not directly measure the evolution of $\dot{M}$ at later times, although they did consider what the final mass of the protostar would be if one simply extrapolated Eq. (72) over the three million year lifetime of a massive star.

Accretion rates have also been measured using the sink particle technique in the group of simulations carried out by Clark et al. (2011a,b), Greif et al. (2011a) and Smith et al. (2011) that find evidence for fragmentation of the gas (see Sect. 3.2 below). The accretion rates onto the individual sinks show a considerable degree of variability in these calculations, but the *total* accretion rate, i.e. the rate of change of the sum of all of the sink particle masses, evolves more smoothly with time, and is of a similar order of magnitude to the other estimates plotted above.

In Fig. 3 we compare several of these different estimates for the accretion rate. We plot three examples, derived using different techniques: a rate based on the Tan and McKee (2004) formalism, computed assuming that $K' = 1$; a rate inferred from the results of one of the adaptive mesh refinement simulations presented in Turk et al. (2011)—specifically, the simulation that was run using the Palla et al. (1983) rate coefficient for three-body H_2 formation; and a rate measured using sink particles, taken from Smith et al. (2011).

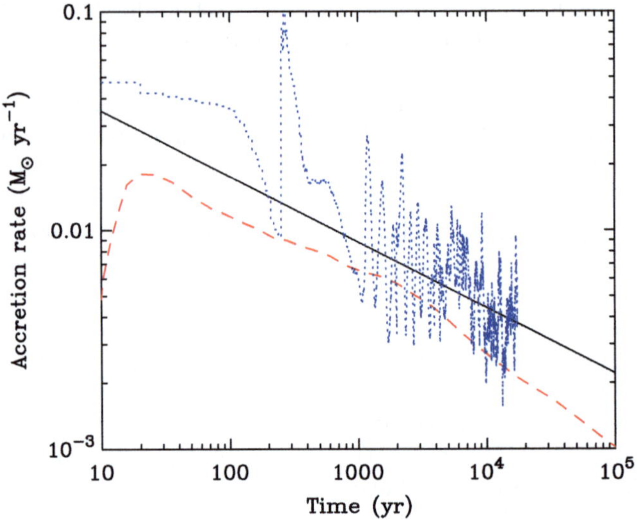

Fig. 3 Three different estimates for the accretion rate onto a Pop. III protostar, taken from Tan and McKee (2004; *solid line*), Turk et al. (2011; *dashed line*) and Smith et al. (2011; *dotted line*), as described in the text. Results from the Smith et al. (2011) simulation are only plotted for the period covered by the simulation, i.e. $t < 2 \times 10^4$ year

At very early times ($t < 100$ year), the three different techniques yield rather different estimates for $\dot{M}$, but this is primarily a consequence of the limited resolution of the numerical simulations. At later times, we see that both the Tan and McKee (2004) formalism and the Turk et al. (2011) simulation predict a similar form for the accretion rate, but disagree by about a factor of two on the normalization, which may simply indicate that our adopted value of K' is slightly too large. We also see the same general trend in the Smith et al. (2011) results, but in this case there is considerable and rapid variation in $\dot{M}$ with time. This is a result of the fragmentation of the gas in this simulation, which produces a set of sink particles that undergo chaotic N-body interactions (see Sect. 3.2 below). A similar effect is seen in simulations of protostellar accretion in present-day star-forming regions (see e.g. Stamatellos et al. 2011).

Regardless of whether $\dot{M}$ varies smoothly or erratically with time, one fact that is clear from Fig. 3 is that the protostellar accretion rate remains very large for a considerable time. This implies that the total mass of gas that is converted to stars can become fairly large after a relatively short time. For example, if we take the Tan and McKee (2004) estimate with $K' = 1$ as a guide, then we find that the total mass in stars increases with time as:

$$M_* = 0.1 \left(\frac{t}{1 \text{ year}} \right)^{0.70} \text{M}_\odot \text{ year}^{-1}. \quad (73)$$

This means that after 5×10^4 year (the Kelvin-Helmholtz relaxation time for a $100\,M_\odot$ star), we have $M_* \simeq 195\,M_\odot$, while after 2×10^6 year (the typical lifetime for an O star), we have $M_* \simeq 2575\,M_\odot$. Therefore, if the gas does not fragment and protostellar feedback is ineffective, one is led to the prediction that the resulting Population III star will be extremely massive. In practice, the gas probably does fragment (see Sect. 3.2 below), and protostellar feedback cannot be completely ignored, but even so, we would expect to be able to form massive Population III stars relatively easily.

Finally, it should be noted that so far we have considered accretion only in the standard H_2-dominated case, i.e. in a minihalo with a minimum gas temperature of around 200 K. In minihalos that reach much lower temperatures through HD cooling, the predicted accretion rates are smaller, as one would expect from the simple dimensional analysis given at the start of this section. For example, if one uses the Tan and McKee (2004) formalism to estimate the accretion rate, then Eq. (69) still applies, but the value of K' is significantly smaller. Taking $n = 10^6\ \mathrm{cm}^{-3}$ and $T_{\mathrm{eff}} = 150$ K as plausible values to substitute into Eq. (68), we find that $K' \simeq 0.3$, and hence the predicted accretion rate is roughly a factor of six smaller than in the H_2-dominated case. Values estimated from numerical simulations using the Abel et al. (2002) approach agree fairly well with this simple estimate (see e.g. Yoshida et al. 2007; McGreer and Bryan 2008).

3.1.2 Protostellar Structure and Evolution

Having established how quickly gas will be accreted by the protostar in the absence of feedback, the obvious next step is to examine how this will be modified by protostellar feedback. Before doing this, however, we must first spend a little time discussing what is known about the internal structure of Population III protostars, and how this evolves with time. This is important if we want to understand how the radius and luminosity of a given Pop. III protostar evolve, and these quantities are obviously of great importance when determining the influence of that protostar on the surrounding gas.

The internal structure of a Pop. III protostar, and how this evolves as the protostar ages and accretes matter from its surroundings was first studied in detail by Stahler et al. (1986a,b). They assume that the accretion process can be treated as a series of quasi-steady-state accretion flows onto a hydrostatic core, which is bounded by a strongly radiating accretion shock. Within the core, the standard stellar structure equations are solved. Outside of the core, the treatment depends on the optical depth of the gas. If the gas is optically thin to the radiation from the accretion shock, then the accretion flow is assumed to be in free-fall. Otherwise, a more detailed calculation is made that incorporates the effects of the radiation force on the infalling gas. The accretion shock itself is treated as a simple discontinuity.

In their initial study, Stahler et al. (1986a) began with a core mass of $0.01\,M_\odot$ and followed the growth of the protostar until its mass reached $10.5\,M_\odot$. They assumed a constant accretion rate $\dot{M} = 4.41 \times 10^{-3} \mathrm{M}_{\odot} \mathrm{year}^{-1}$, and found that for this choice of accretion rate, the evolution of the protostar could be divided into three qualitatively distinct phases.

In the first phase, which lasts until the protostellar mass $M_{*} = 0.1\ \mathrm{M}_{\odot}$, the protostar relaxes from its initial entropy profile into one consistent with the selected accretion rate. Stahler et al. (1986a) dub this a 'decay of transients' phase, and the fact that it quickly comes to an end shows that although the initial conditions used in the Stahler et al. (1986a) study are probably incorrect in detail, the flow soon loses all memory of them, and therefore any inaccuracy at this stage is unlikely to affect the later results.

Once the initial transients have died away, the protostar enters the second phase of its evolution. During this phase, its central temperature remains low ($T_{\rm c} \sim 10^{5}$ K), resulting in a high interior opacity and hence a low interior luminosity. Consequently, the evolution of the core during this phase is almost adiabatic; although the core continues to gradually contract, this contraction does not lead to any increase in the central entropy. Since the postshock entropy increases over time due to the increasing strength of the accretion shock (which is itself a natural result of the increasing protostellar mass), the core develops an off-centre distribution of entropy and temperature.

The gas surrounding the accretion shock remains optically thick throughout this period. This is a direct result of the high accretion rate, which produces a highly luminous accretion shock. This produces sufficient radiation to partially ionize the preshock gas in the vicinity of the shock, creating a structure known as a radiative precursor. The H^{-} opacity of the dense, partially ionized gas in this radiative precursor is more than sufficient to make it optically thick. Stahler et al. show that the core radius during this period evolves as

$$R_{*} = 48.1 \left(\frac{M_{*}}{\mathrm{M}_{\odot}}\right)^{0.27} \left(\frac{\dot{M}}{\dot{M}_{0}}\right)^{0.41} \mathrm{R}_{\odot}, \tag{74}$$

where $\dot{M}_{0} = 4.41 \times 10^{-3}\ \mathrm{M}_{\odot}\ \mathrm{year}^{-1}$, while the photospheric radius evolves as

$$R_{\rm p} = 66.8 \left(\frac{M_{*}}{\mathrm{M}_{\odot}}\right)^{0.27} \left(\frac{\dot{M}}{\dot{M}_{0}}\right)^{0.41} \mathrm{R}_{\odot}, \tag{75}$$

so $R_{\rm p} > R_{*}$ throughout. The strong H^{-} opacity also keeps the photospheric temperature low ($T_{\rm p} \sim 6000$ K), which prevents the protostar from being able to ionize material outside of its photosphere.

This near-adiabatic accretion phase comes to an end once the cooling time of the core, given approximately by the Kelvin-Helmholtz timescale

$$t_{\rm KH} = \frac{G M_{*}^{2}}{R_{*} L_{*}}, \tag{76}$$

becomes comparable to the accretion timescale $t_{acc} = M_*/\dot{M}$. This occurs for a core mass $M \sim 1 M_\odot$, and results in the core entering a phase of homologous collapse, while energy and entropy are transferred outwards in the form of a 'luminosity wave'. The radial position of the luminosity peak moves outwards towards the accretion shock, reaching it at about the time that the core mass has reached 8 $M_\odot$. This results in a rapid swelling of the outermost layers, which weakens the accretion shock and leads to it becoming optically thin. Stahler et al. terminate their simulation shortly afterwards, once the core mass has reached 10.5 $M_\odot$.

Stahler et al. (1986b) simulate the later stages of the evolution of a primordial protostar. Their initial protostellar core has a mass of 5 $M_\odot$, and they evolve this core forward in time, assuming that no further accretion occurs (i.e. the protostellar mass remains fixed at 5 $M_\odot$). They find that deuterium burning within the protostar begins after only 6,000 years, but that hydrogen ignition does not occur until $t = 2 \times 10^5$ year, and the protostar does not reach the zero-age main sequence (ZAMS) until $t \sim 10^6$ year.

An improved treatment of the later stages of the evolution of the protostar was made by Omukai and Palla (2001). They used a very similar setup to that in Stahler et al. (1986a), albeit with improved zero metallicity opacities, and adopted the same constant accretion rate, $\dot{M} = 4.41 \times 10^{-3}\ M_\odot\ \text{year}^{-1}$. However, unlike Stahler et al. (1986a), they initialized their simulation at the point at which the core mass was $M = 8\ M_\odot$, but did not halt the simulation once the core had grown to $10.5 M_\odot$. Instead, they continued to follow the growth of the protostar until well after hydrogen ignition. They found that deuterium burning within the core began once the core mass was $12 M_\odot$ (corresponding to a time $t = 1,000$year after the beginning of the simulation, given the assumed accretion rate), and that it was complete by the time the mass had reached 30 $M_\odot$ (corresponding to $t = 5,000$ year). Hydrogen ignition followed roughly 11,000 years later, at $t = 1.6 \times 10^4$year after the beginning of the simulation, at which time the mass of the protostar was 80 $M_\odot$. At this point, the internal luminosity of the protostar is very close to the Eddington value, which leads to the outer layers of the protostar developing oscillatory behaviour: the high luminosity leads to expansion, the expansion causes the accretion luminosity to drop, the reduced luminosity can no longer maintain the expansion, leading to contraction of the core, and the contraction raises the accretion luminosity, allowing the whole cycle to begin again. Finally, once the core mass reaches 300 $M_\odot$, at $t \sim 6.6 \times 10^4$ year, the contribution of nuclear burning to the protostellar luminosity becomes large enough to drive a final phase of expansion that is strong enough to terminate accretion onto the protostar. Omukai and Palla (2001) halt their simulation at this point.

In a follow-up study using a similar spherically-symmetric setup, Omukai and Palla (2003) performed the same analysis for a range of different values of $\dot{M}$, looking at models with $\dot{M} = (0.25, 0.5, 1.0, 2.0) \times \dot{M}_{fid}$ (where $\dot{M}_{fid}$ was the rate adopted by Stahler et al. 1986a and Omukai and Palla 2001), as well as a model using the time-dependent accretion rate predicted by Abel et al. (2002). The earliest stages of protostellar evolution are qualitatively the same in all of these models:

we see again the same sequence of adiabatic growth, propagation of a luminosity wave that triggers expansion of the outer layers, and then rapid contraction. Although some quantitative differences are apparent, significant differences in behaviour do not occur until the end of the contraction phase. At this point, the further evolution of the protostar is governed by the size of the accretion rate. For accretion rates greater than some critical value $\dot{M}_{\rm crit}$, the luminosity of the protostar becomes large enough to halt the accretion. On the other hand, for $\dot{M} < \dot{M}_{\rm crit}$, the lower accretion luminosity means that the total luminosity of the protostar remains below $L_{\rm Edd}$, and accretion continues unabated.

Omukai and Palla (2003) solve for $\dot{M}_{\rm crit}$ by equating the total luminosity of a zero-age main sequence Pop. III protostar (including accretion luminosity) with the Eddington luminosity, and find that

$$\dot{M}_{\rm crit} \simeq 4 \times 10^{-3}\,{\rm M_{\odot}\,year^{-1}}, \tag{77}$$

coincidentally close to $\dot{M}_{\rm fid}$. In principle, one would expect $\dot{M}_{\rm crit}$ to have a dependence on the current mass of the protostar, but in practice, Omukai and Palla (2003) show that this dependence is weak and may be neglected.

Finally, Omukai and Palla (2003) show that in the time-dependent accretion model, the key factor is the size of the accretion rate at the end of the contraction phase. If this is greater than $\dot{M}_{\rm crit}$, then one would expect accretion to be halted, while if it is less than $\dot{M}_{\rm crit}$ then accretion can continue. In practice, Omukai and Palla (2003) show that if one adopts the Abel et al. (2002) estimated accretion rate, then $\dot{M} < \dot{M}_{\rm crit}$, implying that accretion can continue even once the protostar reaches the zero-age main sequence.

The main limitation of the approach outlined above is the neglect of the effects of rotation. In reality, rotation can have profound effects on stellar structure and evolution, particularly for massive stars (Maeder and Meynet 2000), and it will also have a large influence on how matter reaches the protostar in the first place. The first detailed study of the pre-main sequence evolution of a Pop. III protostar to account for the effects of rotation was carried out by Tan and McKee (2004). In contrast to previous authors, they did not assume spherical symmetry. Instead, they assumed that a protostellar accretion disk would form, and fixed the size of the disk by assuming angular momentum conservation within the supersonic portion of the accretion flow. They used the polytropic accretion flow model described in the previous section to compute the accretion flow onto the disk. To solve for the disk structure, they made use of the standard theory of steady, thin viscous accretion disks (as outlined in Shakura and Sunyaev 1973), with a spatially constant viscosity parameter α. As sources for α, they considered the magnetorotational instability Balbus and Hawley (1991, 1998) and gravitational instability. With the disk structure in hand, they could then solve for the structure of the protostar itself, using a modified version of an approach developed by Nakano et al. (1995) and Nakano et al. (2000). In the zero angular momentum case, Tan and McKee (2004) show that they successfully reproduce the previous results of Stahler et al.

(1986a) and Omukai and Palla (2001, 2003). In more realistic models, Tan and McKee (2004) show that the presence of an accretion disk has little influence on the evolution of the protostar, which still evolves through the same progression of adiabatic growth, terminated by the emergence of a luminosity wave, followed by rapid contraction to the ZAMS. However, Tan and McKee (2004) do find that the photosphere surrounding the protostar behaves very differently in this case than in the spherical case. Because most of the gas accretes onto the protostar via the disk, the gas density is significantly reduced in the polar regions. Consequently, the optical depth of these regions is also significantly reduced, with the result that the flow becomes optically thin early in its evolution. For example, in the model with $f_{\rm Kep} = 0.5$, the photosphere vanishes once the protostellar mass reaches 1 $M_{\odot}$ and does not subsequently reappear. Tan and McKee (2004) argue that this may have a major influence on the effectiveness of radiative feedback from the protostar, a topic that we will return to in the next section.

3.1.3 Feedback Effects

Accretion of gas onto the protostar liberates a significant amount of energy, with most of this energy being emitted from regions close to the protostellar surface. This can be shown very simply by considering how the gravitational potential energy of a test mass changes as we move it close to a protostar of mass M_* and radius R_*. At a distance of $2R_*$, the gravitational potential energy of a fluid element with mass dM is

$$W = -\frac{GM_* dM}{2R_*}, \tag{78}$$

while at the protostellar surface it is

$$W = -\frac{GM_* dM}{R_*}. \tag{79}$$

Therefore, the amount of energy that must be dissipated by the fluid element as it moves from $2R_*$ to R_* is as large as the amount that it must have dissipated while falling in from $R \gg R_*$ to $2R_*$, or in other words, *half* of the total binding energy dissipated by the gas is dissipated while its distance from the protostellar surface is less than R_*. In addition, once the protostar reaches the main sequence, it will start generating additional energy in its own right, via nuclear fusion. The energy that is released in the vicinity of the protostar is therefore quite considerable, and it is reasonable to suppose that this will have some effect on the behaviour of the surrounding gas. It is therefore not surprising that considerable attention has been paid to the issue of protostellar feedback in the context of Pop. III star formation.

In order for the protostar to substantially reduce the rate at which matter flows onto it, it must be able to transfer a significant amount of energy and/or momentum to the infalling gas. The various mechanisms by which this can be accomplished fall under two broad headings: *mechanical feedback*, where the protostar transfers energy and momentum to some form of outflow, which subsequently transfers it to the infalling material, and *radiative feedback*, where radiation from the protostar transfers energy and momentum directly to the infalling gas.

Mechanical feedback

In the local Universe, stellar winds are an almost ubiquitous phenomenon, and play an important role in the evolution of the most massive stars (Chiosi and Maeder 1986). However, there are good reasons to expect that metal-free stars will be much less effective at driving winds than the roughly solar metallicity stars that we are familiar with in the Milky Way. Strong stellar winds are invariably radiation-driven, and at solar metallicities, the largest contribution to the radiative acceleration of the gas comes from the absorption and scattering of ultraviolet photons in the lines of the many metal atoms and ions present in the outflowing gas (Castor et al. 1975). In metal-free gas, on the other hand, the only significant sources of opacity within an outflow will come from the lines of He^+ (atomic hydrogen is typically fully ionized), and from Thomson scattering by free electrons. These provide orders of magnitude less radiative acceleration per unit luminosity than do the metal lines in a solar metallicity gas, and hence one can show that a metal-free Population III star can produce a line-driven wind only if the stellar luminosity is already very close to the Eddington limit (Kudritzki 2002).

Of course, as a Population III star evolves, it will not remain metal-free. It will start to produce carbon, nitrogen and oxygen internally once the stellar core begins to burn helium, and if the star is rotating, these elements can become well-mixed within the star (Meynet et al. 2006). This will provide an additional source of opacity in the stellar atmosphere which may allow the most massive Population III stars to produce a weak CNO-driven wind (Krtička and Kubát 2009). However, the mass-loss rate will be small, and the fraction of the stellar mass that can lost in this way is unlikely to be larger than about 1 %.

It is also possible that very massive Population III stars with luminosities close to the Eddington luminosity may produce eruptive, continuum-driven winds, similar to those we see coming from nearby luminous blue variables (LBVs) such as η Car (Smith and Owocki 2006). However, as the cause of these LBV eruptions is not yet fully understood even for nearby objects, it is difficult to say with certainty whether they will actually be produced by Pop. III stars. More work on this topic is clearly necessary.

Finally, mechanical feedback can also be generated in the form of hydrodynamical or magnetohydrodynamical jets or outflows. We have already discussed the magnetically-driven disk winds produced in the Machida et al. (2006, 2008) simulations, which are able to eject roughly 10 % of the infalling gas from the disk. Although, as we noted previously, these simulations only modelled the very earliest stages in the formation of the protostellar accretion disk, their value for the mass ejection rate is in good agreement with the predictions of a semi-analytical study of Pop. III disks and outflows carried out by Tan and Blackman (2004). If this value is correct, then it implies that the reduction in the protostellar accretion rate brought about by these outflows is small, and hence that they will not significantly limit the final stellar mass. However, one should bear in mind that their interaction with the star-forming halo on larger scales has not yet been modelled in any detail, and hence it is difficult to be certain regarding their final impact.

Radiative feedback

There are several different forms of radiative feedback that could potentially affect the accretion of gas by a Pop. III protostar. First, if the radiation is absorbed or scattered, then it will exert a force on the gas. If this force is comparable to or larger than the gravitational force acting on the gas, then it may suppress accretion onto the protostar, or even prevent it completely. Second, radiation may destroy the H_2 molecules responsible for cooling the gas. In the absence of cooling, the gas will evolve adiabatically, which again may reduce the rate at which it can be accreted. Third, the radiation may heat the gas. If radiative heating raises the gas temperature to a point at which the thermal energy of the gas exceeds the gravitational binding energy of the system, then this again will strongly suppress accretion.

In local star-forming regions, the first of these three forms of radiative feedback is believed to be the most important. Radiation pressure exerted on infalling dust grains by radiation from the protostar results in a substantial momentum transfer to the dust, and from there to the gas, since the dust and gas are strongly coupled. In spherically symmetric models, the radiative force exerted by the radiation on the dust can be strong enough to bring accretion to a complete halt (Wolfire and Cassinelli 1987). In primordial gas, there is no dust, and so this process cannot operate. However, radiation pressure can also work directly on the gas, and so it is worthwhile investigating whether this process is likely to significantly suppress accretion.

Let us start by assuming that the bolometric luminosity of the protostar is given by the Eddington luminosity

$$L_{\rm Edd} = \frac{4\pi G M_{*} c}{\kappa_{T}}, \tag{80}$$

where M_{*} is the protostellar mass, and $\kappa_T \equiv \sigma_T / m_{\rm p} \simeq 0.4\ {\rm cm}^2\ {\rm g}^{-1}$ is the opacity due to Thomson scattering for a fully ionized gas composed of pure hydrogen, with σ_T the Thomson scattering cross-section of the electron and $m_{\rm p}$ the mass of the proton. In this case, then we know from the definition of the Eddington luminosity that the radiative force exerted on a fluid element will be equal to the gravitational force exerted on it by the protostar when the opacity of the fluid element is equal to κ_T. More generally, we can write the ratio of the forces acting on the fluid element as

$$\frac{F_{\rm rad}}{F_{\rm grav}} = \frac{L_{*}}{L_{\rm Edd}} \frac{\kappa}{\kappa_{T}}, \tag{81}$$

where $F_{\rm rad}$ is the radiative force, $F_{\rm grav}$ is the gravitational force, L_* is the protostellar luminosity, and κ is the mean opacity of the fluid element. Since the protostar is unlikely to be stable if $L_* > L_{\rm Edd}$, this implies that in order for the radiative force to significantly affect the gas, it must have a mean opacity $\kappa \sim \kappa_T$ or higher. In practice, the luminosity of a Pop. III protostar before it reaches the main sequence will often be significantly less than the Eddington luminosity (see e.g. Smith et al. 2011), in which case an even higher mean opacity is required.

The mean opacity of metal-free gas has been computed by a number of authors, most recently by Mayer and Duschl (2005b). They present tabulated values for both the Rosseland mean opacity

$$\kappa_R^{-1} = \frac{\int_0^\infty (\partial B_\nu/\partial T)\kappa_\nu^{-1}\, d\nu}{\int_0^\infty (\partial B_\nu/\partial T) d\nu}, \tag{82}$$

and the Planck mean opacity

$$\kappa_P = \frac{\int_0^\infty B_\nu \kappa_\nu\, d\nu}{\int_0^\infty B_\nu d\nu}, \tag{83}$$

where κ_ν is the frequency-dependent opacity and B_ν is the Planck function. For our purposes, we are most interested in the Planck mean. Strictly speaking, this Planck mean opacity is the same as the mean opacity in Eq. (81) only if the protostar has a black-body radiation field and a photospheric temperature that is the same as the gas temperature, and in general this will not be the case. However, if the protostar is still in the pre-main sequence phase of its evolution, it will have a photospheric temperature $T_{\rm p} \sim 6000$ K (Stahler et al. 1986a) and a spectrum that does not differ too greatly from a black-body, while the temperature of the surrounding gas will typically be of the order of 1,000–2,000 K or higher (Clark et al. 2011b; Greif et al. 2011a; Smith et al. 2011). In these conditions, the error we make by using the Planck mean opacity in Eq. (81) should not be excessively large.

There are two regimes in which the tabulated values of κ_P in Mayer and Duschl (2005b) exceed κ_T. The first occurs at very high densities ($n > 10^{22}$ cm^{-3}), where $\kappa_P > \kappa_T$ for a wide range of temperatures. However, these extreme densities are only reached *within* the protostar and hence this regime is of no relevance when we are considering feedback from the protostar on the surrounding gas. The second regime in which κ_P grows to the required size is at temperatures above 8000 K, for a wide range of densities. At these temperatures, the dominant source of opacity is the scattering of photons in the Lyman series lines of hydrogen, primarily Lyman-α. The effects of Lyman-α radiation pressure in metal-free gas were considered by Oh and Haiman (2002), in the context of the formation of massive star-forming minihalos with virial temperatures $T > 10^4$ K. They argued that the Lyman-α photons produced by the cooling of the hot gas would not be important (see also Rees and Ostriker 1977), but that the Lyman-α photons produced by a massive star and its associated HII region would have a pronounced effect on the gas, and could significantly delay or even halt the inflow of the gas. However, they did not carry out a full quantitative investigation of the effects of Lyman-α radiation pressure. More recently, this issue was revisited by McKee and Tan (2008), who studied it in some detail. They found that in a rotating flow, most of the Lyman-α photons would eventually escape along the polar axis of the flow, as it is here that the optical depths are smallest. They showed that if the rotational speed of the gas were at least 10 % of the Keplerian velocity, then Lyman-α radiation pressure would be able to reverse the direction of the flow along the polar axis once the protostellar mass reached 20 $M_{\odot}$. The radiation would therefore blow out a polar cavity, allowing more Lyman-α photons to escape. This prevents the radiation pressure from rising further, and McKee & Tan argue that it never becomes large enough to significantly affect the inflow of gas from directions far away from the polar axis (e.g. from the accretion disk). For this reason, they conclude that Lyman-α radiation pressure is unlikely to be able to significantly reduce the protostellar accretion rate.

Let us now turn our attention to the second form of radiative feedback mentioned above: the photodissociation of H_2 and the consequent dramatic reduction in the cooling rate. As we have already discussed, at early times the photospheric temperature of the protostar is too low for it to produce significant quantities of far-ultraviolet radiation, and hence radiation from the protostar does not significantly affect the H_2. Once the protostar reaches the main sequence, however, it can become a significant source of far-ultraviolet radiation, provided that it has a mass greater than around 15 $M_{\odot}$ (McKee and Tan 2008). Studies by Omukai and Nishi (1999) and Glover and Brand (2001) considered the effect that this radiation would have on the H_2 surrounding the protostar, and showed that the time required to photodissociate the H_2 would be significantly less than the lifetime of the protostar. The removal of the H_2 from the gas means that it is no longer able to cool effectively at temperatures $T < 10^4$ K, and hence one would expect that as the H_2 in the accreting gas is destroyed, the gas will begin to evolve adiabatically until it reaches this temperature. McKee and Tan (2008) consider whether this switch to adiabatic evolution is sufficient to halt accretion, and conclude that it is not. If no protostar were present, then the switch to adiabatic evolution would be enough to stabilize the gas and prevent further collapse. The presence of the protostar, however, serves to destabilize the gas, allowing accretion to continue even when the evolution of the gas is fully adiabatic. McKee and Tan (2008) use the treatment of protostellar accretion introduced in Fatuzzo et al. (2004) to investigate the issue numerically, and show that an increase in the effective adiabatic increase of the gas from $\gamma_{\rm eff} = 1.1$ (which approximately characterizes the temperature evolution of the gas at $n > 10^4$ cm^{-3}; see e.g. Omukai and Nishi 1998) to $\gamma_{\rm eff} = 5/3$ reduces the accretion rate by only 20 %.

The third possible form of radiative feedback involves the heating of the surrounding gas by radiation from the protostar. If the temperature of the gas can be increased to a point at which its thermal energy exceeds its gravitational binding energy, then it will no longer be gravitationally bound to the protostar, and hence will not be accreted. A convenient way to quantify the relative importance of thermal and gravitational energy is to compare the sound-speed of the gas with the escape velocity of the system, v_{esc}: gas with $c_s > v_{esc}$ will not be gravitationally bound.

For an isolated protostar of mass M_*, we can write v_{esc} at a distance R from the protostar as:

$$v_{esc} = \sqrt{\frac{2GM_*}{R}}, \tag{84}$$

where G is the gravitational constant. If we rewrite this expression in more convenient units, we find that

$$v_{esc} \simeq 4.2 \left(\frac{M_*}{1\,M_\odot}\right)^{1/2} \left(\frac{R}{100\,\mathrm{AU}}\right)^{-1/2} \mathrm{km\,s^{-1}}. \tag{85}$$

For a primordial, fully molecular gas, $c_s = 4.2\,\mathrm{km}\,s^{-1}$ at a temperature $T \sim 3400$ K, and hence gas within 100 AU of a one solar mass protostar must be heated up to a temperature of thousands of Kelvin in order to unbind it. At larger distances, the required temperature would appear at first to be much smaller, but the reader should recall that this expression is for an *isolated* protostar, i.e. one which is not surrounded by gas. It is therefore only valid when the protostellar mass M_* is much larger than the mass of gas within a distance R of the protostar, and once we start considering scales $R \gg 100$ AU, this is unlikely to be a good approximation. If we include the influence of this gas by replacing M_* in Eq. (84) by $M_{tot} = M_* + M_{gas}$, and use the facts that prior to star formation, the mass enclosed within a sphere of radius 100 AU is roughly 5 $M_\odot$ and increases at larger distances as $M_{enc} \propto R^{0.8}$, then at distances $R > 100$ AU, we have

$$v_{esc} \simeq 9.4 \left(\frac{R}{100\,\mathrm{AU}}\right)^{-0.1} \mathrm{km\,s^{-1}}. \tag{86}$$

In other words, once we account for the mass of the infalling gas in addition to the mass of the protostar, we find that the escape velocity is of the order of 10 $\mathrm{km\,s^{-1}}$, with little dependence on the distance from the protostar. An escape velocity of this order of magnitude corresponds to a gas temperature of order 10^4 K. This immediately tells us that heating of the gas by radiation from the protostar during the pre-main sequence phase of its evolution is unlikely to significant affect the accretion rate due to the low photospheric temperature of the protostar during this phase—clearly, a protostar with an effective temperature of 6000 K will not be able to heat up distant gas to a temperature of 10,000 K. On the other hand, once the protostar reaches the main sequence, its photospheric temperature will sharply increase, and hence it may be able to heat up the surrounding gas to a much higher temperature. In particular, if the protostar is massive enough to emit a significant number of ionizing photons while on the main sequence, then it will easily be able to produce temperatures in excess of 10^4 K within the gas that it ionizes.

The idea that the formation of an HII region may strongly suppress or completely terminate protostellar accretion was discussed long ago in the context of present-day star formation (see e.g. Larson and Starrfield 1971), but has recently been re-examined by several authors in the context of primordial star formation. On large scales ($R > 0.1$ pc), the behaviour of an HII region produced by a Pop. III star is relatively simple. The radial density profile of the gas on these scales is approximately $\rho \propto R^{-2.2}$, and hence the density falls off too quickly to trap the HII region within the minihalo (see e.g. Whalen et al. 2004; Alvarez et al. 2006; Abel et al. 2007; Yoshida et al. 2007). The ionization front therefore expands rapidly, as an R-type front, with a velocity that is controlled by the rate at which ionizing photons are being produced by the star. In addition, if we are considering Pop. III star formation within one of the first star-forming minihalos, then it is easy to show that sound speed of the gas within the HII region will be higher than the escape velocity of the minihalo. Consequently, the ionized gas begins to flow out of these small minihalos, significantly reducing the mean gas density. It is therefore clear that once the HII region reaches a size of 0.1 pc or above, it will act to prevent any further infall of gas from these scales onto the protostar. However, this leaves unanswered the question of how long it takes for the HII region to expand to this scale.

In the case of steady, spherically-symmetric infall, Omukai and Inutsuka (2002) showed that in order for the HII region to avoid being trapped on scales close to the protostar, the flux of ionizing photons must exceed a critical value

$$\dot{N}_{\rm crit} = 6.4 \times 10^{52} \left(\frac{R_{\rm in}}{10\,{\rm R}_{\odot}} \right)^{-1} \left(\frac{M_{*}}{100\,{\rm M}_{\odot}} \right)^{2} \; {\rm s}^{-1}, \tag{87}$$

where $R_{\rm in}$ is the inner radius of the HII region, which we can take to be equal to the radius of the massive star. Given reasonable values for $R_{\rm in}$ and M_*, this expression yields a value for $\dot{N}_{\rm crit}$ that is much larger than the number of photons that will actually be produced by any massive star, leading Omukai and Inutsuka (2002) to conclude that the HII region would remain trapped close to the star. However, this conclusion depends crucially on the assumed spherical symmetry of the flow. In the more realistic case in which our protostar is surrounded by an accretion disk, McKee and Tan (2008) show that the HII region can expand in all directions other than those close to the midplane of the disk once the stellar mass reaches a value of around 50–100 $M_{\odot}$, where the precise value required depends on how rapidly the gas is rotating. McKee and Tan (2008) also show that the accretion disk can survive for a considerable period after the HII region has broken out, and that the protostar will stop accreting from the disk only once the rate at which gas is lost from the disk by photoevaporation exceeds the rate at which fresh gas is falling onto the disk. In their models, this occurs for $M_* \sim 140\,{\rm M}_{\odot}$, leading them to conclude that radiative feedback from the protostar on the surrounding gas cannot prevent the protostellar mass from becoming very large. Whether this conclusion will need to be modified in the case in which the accretion disk fragments remains to be seen.

3.2 *The Fragmentation Model*

3.2.1 Early Studies

The first simulations of primordial gas to make use of sink particles were the SPH simulations of Bromm et al. (1999, 2002). They studied the formation of isolated dark matter minihalos and the cooling and gravitational collapse of gas within them using a somewhat idealized set of initial conditions. At an initial redshift $z = 100$, they created a spherical, uniform density region containing both gas and dark matter, and with an initial velocity field corresponding to the Hubble expansion. The density of this spherical region was taken to be higher than the cosmological background density, and was fixed such that the region would gravitationally collapse and virialize at a specified redshift, chosen to be $z = 30$ in most of the models that they examined. Small-scale structure was introduced into the dark matter distribution by perturbing the particles slightly from their initial positions using the Zel'Dovich (1970) approximation. The amplitudes of these random perturbations were fixed such that the small-scale density structure in the dark matter would begin to evolve in the non-linear regime at the virialization redshift. Both the dark matter and the gas were also assumed to be in solid body rotation, with some specified angular velocity.

Bromm et al. examined several different choices for the halo mass and initial angular velocity of the gas, and showed that starting from these initial conditions, the gas and dark matter would initially collapse in a similar fashion, but that the gas would subsequently form H_2, dissipate energy, and sink to the center of the minihalo. In most of the cases they studied, the gas would then form a rotationally-supported disk, with a radius of order 10 pc. This disk would then break up into clumps with masses $M_{\rm cl} \sim$100–1,000 $M_{\odot}$, comparable to the Jeans mass in the disk. As these clumps were gravitationally unstable, they of course underwent gravitational collapse, and Bromm et al. (2002) therefore introduced sink particles to represent clumps that collapsed to densities greater than 10^8 cm^{-3} in order to avoid the timestep constraints discussed earlier, allowing the further evolution of the clumps to be studied.

Unfortunately, the fragmentation observed by Bromm et al. in their simulations is probably not realistic. One major problem lies in their choice of initial conditions, specifically in their use of solid-body rotation. Although the total angular momentum of the gas and dark matter in their simulations is comparable that measured for minihalos in more realistic cosmological simulations (Jang-Condell and Hernquist 2001; Davis and Natarajan 2010), their adoption of solid-body rotation leads to the gas having an incorrect radial profile for this angular momentum. This causes the collapse of the gas to be considerably more ordered than it would be in a real minihalo, leading to the formation of an over-large disk. Disks of this kind do not appear to form in simulations of small star-forming minihalos that start from more realistic cosmological initial conditions (e.g. Abel et al. 2002; Yoshida et al. 2006). A second major problem lies in the neglect of stellar feedback. Within the disk, the dynamical timescale is of the order of a million years, which is much longer than is needed for a massive Pop. III star to reach the main sequence. Therefore, if a massive star forms within the first clump to be produced within the disk, the radiation from this star may well be able to photodissociate the H_2 in the disk before a second clump can form (Omukai and Nishi 1999; Glover and Brand 2001).

The next attempt to use sink particles to study the formation of Pop. III stars was made by Bromm and Loeb (2004). The initial setup of their simulation was similar to that used by Bromm et al. (2002), but to gain improved resolution in the centre of the minihalo, they used a technique called particle splitting (Kitsionas and Whitworth 2002; Bromm and Loeb 2003). The evolution of the minihalo was followed until cold, dense gas started to build up in the centre of the halo. The simulation was then paused, and the gas within a radius of 3.1 pc of the centre of the minihalo (corresponding to roughly $3000\,M_{\odot}$ of material) was resampled using SPH particles with much smaller masses, using the resampling technique described in Bromm and Loeb (2003). The mass resolution within this resampled region was thereby improved from $M_{\rm res} = 200\ M_{\odot}$ to $M_{\rm res} = 4\ M_{\odot}$. Bromm & Loeb then restarted the simulation, and followed the further gravitational collapse of the gas within this central, higher resolution region until the gas density reached $n \sim 10^{12}\ \mathrm{cm}^{-3}$, at which point a sink particle was created. They then followed the accretion of gas onto this sink for roughly 10^4 years, as we have already described above. Bromm and Loeb (2004) found no evidence for fragmentation within the central clump of dense gas, but did note that a second dense clump formed nearby, with a final separation from the star-forming clump of roughly 0.25 pc. However, the free-fall collapse time of this clump was about 3 Myr, and so it was unclear whether it would survive for long enough to form a second star, or whether it would be destroyed by negative feedback from a massive star forming within the first clump.

3.2.2 The Importance of Turbulence

Although the Bromm and Loeb (2004) study undoubtedly represented a significant step forwards in resolution compared to Bromm et al. (2002), it still had a mass resolution which was more than two orders of magnitude greater than the actual size of a Pop. III protostar at the moment that it forms, and hence it was unable to investigate the behaviour of the gas on scales smaller than about 100 AU. The first work using sink particles that did manage to probe this regime was Clark et al. (2008). Although the main focus of their study was on the fragmentation brought about by dust cooling in low metallicity systems (see e.g. Omukai et al. 2005, Schneider et al. 2006 or Dopcke et al. 2011 for more on this topic), they also studied the behaviour of the gas in the $Z = 0$ case for the purpose of comparing it with the results on their low metallicity runs. As the initial conditions for their simulations, Clark et al. (2008) considered a uniform density cloud, with a mass of $500\ M_{\odot}$, a radius of 0.17 pc and a number density of $5 \times 10^5\ \mathrm{cm}^{-3}$. The gas within this cloud was given a low level of initial turbulence, with a turbulent energy equal to 10 % of the gravitational potential energy, and was also assumed to be rotating uniformly, with an initial rotational energy equal to 2 % of the gravitational potential energy. Two different simulations were performed, with different numbers of particles: a low resolution simulation that used only two million SPH particles, and hence had a mass resolution of 0.025 $M_{\odot}$, and a high resolution simulation that used 25 million particles, corresponding to a mass resolution of 2×10^{-3} $M_{\odot}$. Aside from the somewhat artificial initial conditions, the main simplification made in these simulations was the use of a tabulated equation of state to follow the thermal evolution of the gas. The results of the Omukai et al. (2005) one-zone model were used to derive internal energy densities and thermal pressures for the gas at a range of different densities, and this data was then used to construct a look-up table that could be used by the SPH code to compute the internal energy and pressure corresponding to a given gas density.

Clark et al. followed the collapse of the gas in their simulation down to a physical scale of less than an AU (corresponding to a gas density of over $10^{16}\,\mathrm{cm}^{-3}$). Regions collapsing to even smaller scales were replaced by sink particles, created using the standard Bate et al. (1995) prescription. Clark et al. showed that at the point at which the first sink particle formed, the radial profiles of quantities such as the gas density or the specific angular momentum were very similar to those found in previous studies of Pop. III star formation that were initialized on cosmological scales (e.g. Abel et al. 2002; Yoshida et al. 2006). They noted that at this point in the simulation, there is no sign of any fragmentation occurring, and argued that if the simulation were stopped at this point (as would be necessary if the sink particle technique were not being used), one would probably conclude that the gas would not fragment, but would merely be accreted by the protostar. However, they show that this is not what actually happens when the simulation is continued. Instead, the gas fragments, forming 25 separate protostars after only a few hundred years. Clark et al. stopped their simulations after 19 $M_{\odot}$ of gas had been incorporated into sink particles, and showed that at this point the protostars have masses ranging from 0.02 $M_{\odot}$ to 5 $M_{\odot}$, but that the mass distribution is relatively flat, with most of the mass locked up in the few most massive protostars. There is no significant difference between the mass function of sinks in the low and high resolution calculations, suggesting that fragmentation is well-resolved in both cases.

This is an intriguing result, but several reasonable concerns could be raised regarding the numerical technique adopted by Clark et al. (2008). First, the initial conditions for the gas are an idealized version of what one would find within a real star-forming minihalo, and although there are indications that the gas loses its memory of the initial conditions prior to fragmentation occurring, inevitably a few doubts remain. Second, and more importantly, the use of a tabulated equation of state represents a major simplification of the thermal evolution of the gas, and one which may make fragmentation more likely to occur. For example, this technique does not allow one to model the formation of the hot, shocked regions noted by Turk et al. (2010) in which much or all of the H_2 is dissociated, and it is likely to underestimate the temperature of gas falling in at later times, when the typical infall velocity is larger than during the initial assembly of the protostar. The Clark et al. calculation also neglects the effects of radiative feedback from the protostars, and assumes that protostars do not merge, even if they come within sub-AU distances of each other.

In a follow-up study, Clark et al. (2011a) addressed one of these concerns—the use of a tabulated equation of state—by performing simulations that replaced this with a detailed treatment of the chemistry and cooling of primordial gas. In their study, they investigated the role that low Mach number turbulence might play in triggering fragmentation in the gas by performing a set of simulations of the collapse of unstable Bonnor-Ebert spheres. They considered three initial configurations: a $1,000\ \mathrm{M}_{\odot}$ cloud with an initial temperature of 300 K; a $150\ \mathrm{M}_{\odot}$ cloud with an initial temperature of 75 K; and a $1,000\ \mathrm{M}_{\odot}$ cloud with an initial temperature of 75 K. In each case, the central density of the Bonnor-Ebert sphere was taken to be $n_{\mathrm{c}} = 10^{5}\ \mathrm{cm}^{-3}$. The first set of initial conditions were intended to correspond to the conditions that one would expect to find within one of the first star-forming minihalos, while the second set were intended to correspond to the conditions within a minihalo dominated by HD cooling. Simulations with the third set of initial conditions were run to allow the effects of lowering the temperature and lowering the total mass to be distinguished. Within the Bonnor-Ebert spheres, a turbulent velocity field was imposed, with a three-dimensional RMS velocity Δv_{turb}.

Clark et al. (2011a) did not claim that this was a completely accurate model of the physical state of the gas within a real star-forming minihalo. Instead, they treated this study as a kind of physics experiment, allowing them to investigate the effect of varying a single important parameter—the turbulent kinetic energy—without varying any of the other parameters in the problem, something that it would not be possible to do if using initial conditions derived from a cosmological simulation. For the first two setups described above, they performed four simulations, with $\Delta v_{\mathrm{turb}} = 0.1, 0.2, 0.4$ and $0.8\ c_{\mathrm{s}}$, respectively, where c_{s} was the initial sound speed. For the third setup (the large, low temperature clouds), they performed only two simulations, with $\Delta v_{\mathrm{turb}} = 0.4$ and $0.8\ c_{\mathrm{s}}$, respectively. The clouds were modelled using two million SPH particles in each case, yielding a mass resolution of $0.05\,\mathrm{M}_{\odot}$ for the $1000\,\mathrm{M}_{\odot}$ clouds and $0.0075\,\mathrm{M}_{\odot}$ for the $150\,\mathrm{M}_{\odot}$ clouds. Sink particles were created once the gas density exceeded $10^{13}\ \mathrm{cm}^{-3}$, and the sink accretion radius was 20 AU.

Clark et al. (2011a) found that fragmentation occurred in almost all of their simulated clouds. In the case of the massive, warm clouds, the only case in which fragmentation did not occur was the simulation with $\Delta v_{\mathrm{turb}} = 0.1\ c_{\mathrm{s}}$. In this simulation, the gas simply collapsed to form a single, massive protostar. In the simulations with larger turbulent energies, however, the formation of the first protostar was followed within a couple of hundred years by the fragmentation of the infalling gas and the formation of a significantly larger number of protostars. The relationship between the turbulent energy and the degree of fragmentation is not straightforward: the $\Delta v_{\mathrm{turb}} = 0.4\,c_{\mathrm{s}}$ run fragmented more than the $\Delta v_{\mathrm{turb}} = 0.2\,c_{\mathrm{s}}$, as one might expect, but the $\Delta v_{\mathrm{turb}} = 0.8\,c_{\mathrm{s}}$ run fragmented *less* than the $\Delta v_{\mathrm{turb}} = 0.4\,c_{\mathrm{s}}$ run (although still more than the $\Delta v_{\mathrm{turb}} = 0.2\ c_{\mathrm{s}}$ run). Clark et al. hypothesize that this difference in behaviour is due to the amount of angular momentum retained within the collapsing region, which in this case was larger in the $\Delta v_{\mathrm{turb}} = 0.4\ c_{\mathrm{s}}$ run than in the other runs, but note that this may not always be the case, and that a much larger series of realizations of the turbulent velocity field would be needed to make a definitive statement about the relationship between the turbulent energy and the degree of fragmentation (c.f. Goodwin et al. 2004, who come to a similar conclusion regarding present-day star formation). The total amount of mass accreted by the sinks is very similar in all four runs, and hence the runs that fragment more tend to form lower mass objects than the runs that fragment less.

In the low-mass, colder clouds, Clark et al. find a much lower degree of fragmentation, despite the fact that the initial number of Jeans masses in these clouds is the same as in the $1000\,M_{\odot}$, $T = 300$ K clouds. In this case, fragmentation occurs only in the $\Delta v_{\rm turb} = 0.2\,c_{\rm s}$ and $\Delta v_{\rm turb} = 0.8\,c_{\rm s}$ simulations, and only a small number of fragments are formed in each case. Clark et al. investigate whether this is due to the lower cloud mass by modelling the collapse of $1000\,M_{\odot}$ clouds with the same lower initial temperature, and find that although more fragmentation occurs in this case, the gas still fragments less than in the $1000\,M_{\odot}$, $T = 300$ K case. They suggest that this somewhat counterintuitive behaviour is due to the greater stiffness of the effective equation of state in the colder clouds. In both cases, the gas must heat up from its initial temperature at 10^5 cm^{-3} to a temperature of roughly 1000 K at 10^{10} cm^{-3}, and so when the initial temperature is lower, the gas must heat up more rapidly with increasing density, meaning that it has a larger effective adiabatic index. This makes it more difficult to generate the small-scale non-linear structures that are the seeds for later fragmentation, and also delays the collapse, allowing more of the turbulent energy to dissipate. A similar effect has previously been noted by Yoshida et al. (2007) and Tsuribe and Omukai (2008), and calls into question the common wisdom that minihalos in which the cooling becomes HD-dominated will inevitably form lower mass stars.

3.2.3 Models Using Cosmological Initial Conditions

In order to establish whether the fragmentation seen in the Clark et al. (2008) model was simply a consequence of the highly idealized initial conditions used in that study, several recent follow-up studies have re-examined the issue using simulations initialized on cosmological scales (i.e. scales significant larger than the virial radius of the minihalo). One of the first of these studies was carried out by Stacy et al. (2010). They first performed a medium resolution cosmological simulation, which allowed them to determine the formation site of the first minihalo large enough to cool effectively and form stars. They then used a hierarchical zoom-in procedure (Navarro and White 1994; Tormen et al. 1997; Gao et al. 2005) to improve the resolution within a region centered on this formation site, allowing them to achieve a mass resolution of $1.5\,M_{\odot}$ within the centre of the star-forming minihalo.[5] In contrast to Clark et al. (2008), the thermal and chemical evolution of the gas

[5] The value quoted here for the mass resolution of the Stacy et al. (2010) simulation assumes that 100 or more SPH particles are required to resolve gravitationally bound structures, which is the typical resolution limit adopted in studies of present-day star formation. Stacy et al. (2010) assume that only 48 SPH particles are required, and hence quote a mass resolution that is roughly a factor of two smaller.

was followed in detail during the collapse. Once the gravitationally collapsing gas reached a density of 10^{12} cm^{-3}, it was converted into a sink particle, along with all of the gas within an accretion radius $r_{acc} = 50$ AU. Stacy et al. show that following the formation of this first sink, the infalling gas collapses into a flattened disk. At a time $t = 250$ year after the formation of the first sink particle, this disk has a radius of 200 AU, but it grows with time and has reached a radius of 2,000 AU by $t = 5,000$ year, the end of the simulation. H_2 cooling allows the gas within the disk to remain at a temperature of roughly 1,000 K, and the disk soon becomes gravitationally unstable, forming a second sink particle after roughly 300 year, and a further three sinks after 4000–5000 years of evolution. At the end of the simulation, the first two sinks to form have become very massive, with masses of 43 $M_{\odot}$ and 13 $M_{\odot}$ respectively, while the three newer sinks still have masses ~ 1 $M_{\odot}$, close to the resolution limit of the simulation. Stacy et al. (2010) do not include the effects of accretion luminosity in their simulation directly, but do assess its effects during a post-processing stage. They investigate the possible effects of radiation pressure, but show that this remains unimportant within their simulation throughout the period that they simulate, in agreement with our analysis above.

The main drawback of the Stacy et al. (2010) study is their choice of mass resolution. At the point at which it fragments, the protostellar accretion disk in their simulation has a mass of roughly 35 $M_{\odot}$, and hence is resolved with only a few thousand SPH particles. This is two orders of magnitude smaller than the number of particles typically used to model gravitationally unstable accretion disks in the context of present-day star formation (see e.g. Rice et al. 2005), and it is questionable whether a few thousand particles is enough to properly model the dynamics of the disk. It is therefore possible that the results of Stacy et al. (2010) may have suffered from some degree of artificial fragmentation.

More recently, a study by Clark et al. (2011b) has dramatically improved the mass resolution used to model the build-up of a protostellar accretion disk around the first Population III protostar. Clark et al. use a similar basic strategy to Stacy et al., starting with a medium resolution cosmological simulation to identify the first star-forming minihalo, and then using a hierarchical zoom-in procedure to improve the resolution within the gas forming this minihalo. They run this zoomed-in simulation until the maximum density of the gravitationally collapsing gas reaches 10^6 cm^{-3}. At this point, they extract a spherical region containing 1000 $M_{\odot}$ of gas from the centre of the minihalo, and resimulate this region at much higher resolution, using several nested levels of particle splitting (Kitsionas and Whitworth 2002; Bromm and Loeb 2003). At the final level of splitting, the particle mass is 10^{-5} $M_{\odot}$ and the mass resolution is 10^{-3} $M_{\odot}$, several orders of magnitude better than in the Stacy et al. (2010) simulation.

In addition to the extremely high mass resolution, the other main improvement in the Clark et al. study compared to previous work is its inclusion of the effects of accretion luminosity feedback directly within the simulation. To model this, the authors start by writing the bolometric accretion luminosity produced by a given protostar as

$$L_{\rm acc} = \frac{G\dot{M}M_*}{R_*}, \tag{88}$$

where $\dot{M}$ is the accretion rate onto that protostar, M_* is the protostellar mass, and R_* is the protostellar radius. As Clark et al. attempt to model only the first few hundred years of the evolution of the gas after the formation of the first protostar, i.e. a timescale much less than the protostellar Kelvin-Helmholtz relaxation timescale, they assume that the protostars remain in the adiabatic accretion phase of their evolution, with masses and radii that are related by (Stahler et al. 1986a)

$$R_* = 26 R_\odot \left(\frac{M_*}{\rm M_\odot}\right)^{0.27} \left(\frac{\dot{M}}{10^{-3}\,{\rm M_\odot\,year^{-1}}}\right)^{0.41}. \tag{89}$$

The only remaining uncertainty is then $\dot{M}$, which can be directly measured within the simulation. Clark et al. next assume that the gas is heated by the accretion luminosity at a rate

$$\Gamma_* = \rho\kappa_{\rm P}\frac{L_{\rm acc}}{4\pi r^2}, \tag{90}$$

where ρ is the mass density, r is the distance to the protostar, and $\kappa_{\rm P}$ is the Planck mean opacity of the gas, calculated using the tabulated values given in Mayer and Duschl (2005b). This expression assumes that the gas is optically thin, and hence will tend to overestimate the heating rate.

Clark et al. model protostar formation using sink particles, which are created using the standard Bate et al. (1995) algorithm, with a density threshold $n_{\rm th} = 10^{17}\,{\rm cm^{-3}}$. The sink accretion radius was set to 1.5 AU. At the point at which the first sink particle forms, the state of the gas in the Clark et al. simulation (e.g. the density profile and the distribution of specific angular momentum) is very similar to that seen in other high resolution simulations of Pop. III star formation. However, the authors show that at later times, a protostellar accretion disk begins to build up around the central protostar, just as in the Stacy et al. study. The significantly higher resolution of the Clark et al. simulation allows them to model the build up of this disk on scales much closer to the central protostar, and to resolve the disk with a far larger number of SPH particles. The growth of the accretion disk is followed for around 100 years after the formation of the first protostar, and Clark et al. show that after around 90 years (corresponding to around 1.5 orbital periods for the disk), the accretion disk begins to fragment, forming several low-mass protostars. At the time at which it fragments, the disk contains several solar masses of gas (and hence is resolved with several hundred thousand SPH particles), compared to around $0.4\,{\rm M_\odot}$ in the central protostar, and the disk radius is a few tens of AU. The state of the disk at the onset of fragmentation is illustrated in Fig. 4.

Clark et al. argue that the reason that the protostellar accretion disk fragments is that it is unable to transfer gas onto the protostar fast enough to keep up with the rate at which fresh gas is falling onto the disk. This causes the surface density of the disk to increase, which eventually results in it becoming gravitationally unstable and fragmenting. This argument can be made somewhat more quantitative if one

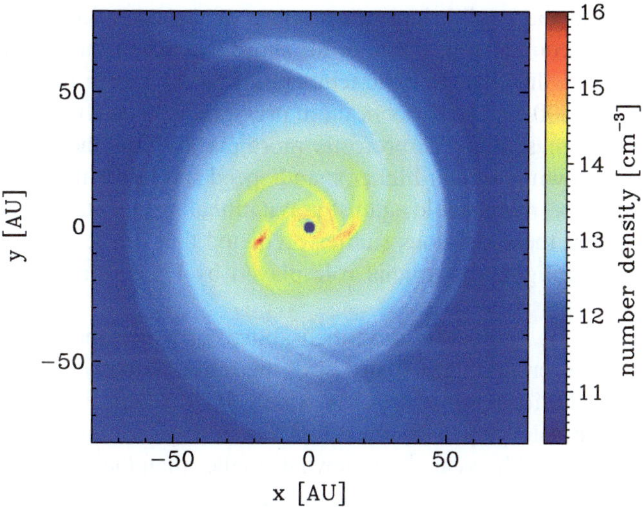

Fig. 4 The volume density of the protostellar accretion disk in the Clark et al. (2011a) simulation immediately prior to fragmentation. The 'hole' in the centre of the disk corresponds to the location of the sink particle representing the first protostar to form, and occurs because we have not accounted for the mass in the sink particle when calculating the density. The accretion disk is gravitationally unstable, and has formed several spiral arms, one of which has begun to fragment

treats the disk using the standard Shakura and Sunyaev (1973) thin disk model, and hence writes the mass flow rate through the disk at a radius r as

$$\dot{M}(r) = 3\pi\alpha c_s(r)\Sigma(r)H(r), \tag{91}$$

where α is the viscosity parameter, and $c_s(r)$, $H(r)$ and $\Sigma(r)$ are the sound speed, disk thickness and surface density, respectively, at a radius r. Clark et al. use the values of c_s, H and Σ provided by their simulation to show that $\dot{M}(r)$ is smaller than the accretion rate onto the disk for a wide range of radii, even if one adopts $\alpha = 1$ (which already presupposes that the disk is gravitationally unstable). The growth of the disk therefore appears to be unavoidable, leading to its fragmentation once portions of it develop a Toomre stability parameter $Q < 1$, where $Q \equiv c_s\kappa/\pi G\Sigma$, with κ here being the epicyclic frequency.

Previous semi-analytical studies of the structure of Pop. III accretion disks came to a somewhat different conclusion regarding their stability, predicting that $Q \gg 1$ throughout the disk (Tan and McKee 2004; Tan and Blackman 2004; Mayer and Duschl 2005a). However, these models neglected the effect of H_2 cooling, motivated by the assumption that the H_2 content of a Pop. III protostellar accretion disk is negligible (J. Tan, private communication). This assumption leaves H^- as the primary source of opacity at temperatures $T \sim 7{,}000$ K and below (Lenzuni et al. 1991; Mayer and Duschl 2005b). For gas in chemical equilibrium, the opacity of H^- decreases sharply with decreasing temperature, and consequently these models find the equilibrium temperature of the accretion disk to be high, $T \sim 6,000$ K. In comparison, Clark et al. show that when the effects of H_2 are taken into account, the characteristic temperature of the gas in the disk lies in the range of 1,500–2,000 K. The disks in these previous semi-analytical models could therefore transfer mass onto the protostar more rapidly than the disk in the Clark et al. simulation, owing to their higher sound-speed and larger thickness, but at the same time were also more stable against gravitational fragmentation. It is therefore not surprising that these previous studies predicted that the accretion disk should be stable, and highlights the crucial role played by H_2 cooling in enabling disk fragmentation.

In addition to the simulation described above, in which the value of $\dot{M}$ used to calculate the accretion luminosity was measured directly, Clark et al. also performed two additional simulations in which $\dot{M}$ was kept fixed, allowing them to investigate the role played by accretion luminosity heating. They considered cases with $\dot{M} = 10^{-3}$ $M_{\odot}$ year^{-1} (somewhat smaller than the measured value) and $\dot{M} = 10^{-2}$ $M_{\odot}$ year^{-1} (larger than the measured value), and showed that as $\dot{M}$ (and hence L_{acc}) increase, the disk becomes warmer and thicker and takes longer to fragment. However, fragmentation still occurs in every case, demonstrating that accretion luminosity heating is unable to prevent the disk from fragmenting (c.f. Krumholz 2006, who argues that it plays a crucial role in suppressing fragmentation in local star-forming systems).

One drawback of the Clark et al. (2011b) study is that they examined only a single star-forming minihalo, and although they showed that the properties of this halo (e.g. mass, spin parameter, formation redshift) were similar to those of the minihalos modelled in previous studies of Pop. III star formation, nevertheless the suspicion remains that perhaps this particular minihalo was unusual in some way. This concern was addressed by Greif et al. (2011a). They used the new moving-mesh code AREPO (Springel 2010) to study Pop. III star formation in five different minihalos, using a sink particle algorithm to allow them to follow the evolution of the gas past the point at which the first protostar formed. In all five of the systems that they modelled, they found similar behaviour to that in the Clark et al. study: an accretion disk built up around the first protostar, became gravitationally unstable, and began to fragment after only a few years. These results suggest that Clark et al. were right to claim that disk fragmentation, and the resulting formation of Pop. III binary systems, or higher order multiple systems, is an almost inevitable outcome of Population III star formation.

Greif et al. (2011a) also examined the issue of whether the objects represented by the individual sink particles would truly survive as separate protostars, or whether they would simply merge into a single massive protostar as the system evolved further. They considered two different schemes for merging sink particles. In the standard scheme, sink particles coming within a distance of 100 $R_{\odot}$ of each other were merged to form a single sink, provided that the total energy of the two-body system was negative. In an alternative model, utilizing what Greif et al. dub as "adhesive" sinks, the energy check was omitted, and sinks were always merged when within a distance of 100 $R_{\odot}$ of each other. Greif et al. justify their choice of this critical distance in two ways: first, it is also the accretion radius adopted for their sinks, meaning that the gas flow on smaller scales close to the sinks is not resolved; and second, it is roughly equal to the maximum size of a pre-main sequence protostar predicted by the models discussed in Sect. 3.1.2.

The majority of the protostars formed in the Greif et al. simulations have at least one close encounter with another protostar, but when the standard merger algorithm is used, many of these encounters result in a purely dynamical interaction, as the total energy of the protostellar pair is too large to allow them to merge. On the other hand, when the adhesive sinks are used, many of these encounters lead to mergers. Greif et al. show that although the total mass incorporated into protostars is roughly the same in both cases, the number of protostars that survive as individual objects is reduced by a factor of up to a few, and the mean protostellar mass is consequently higher. In both cases, the protostars have a relatively flat mass distribution, with most of the protostellar mass being accounted for by a small number of high-mass protostars. The protostars have a broad distribution of radial velocities, ranging from $v_{\rm rad} \sim 1$ km s^{-1} to $v_{\rm rad} \sim 100$ km s^{-1}, and in many cases the radial velocity is greater than the escape velocity of the central region of the minihalo. It is likely that this leads to a significant fraction of the Pop. III protostars escaping from the minihalo entirely, although Greif et al. do not follow their evolution for long enough to confirm this. It is possible that these protostars will accrete very little additional gas once they escape from the high density region at the centre of the minihalo (see e.g. Johnson and Khochfar 2011), in which case their final masses would be very similar to the masses that they have at the point at which they are ejected. Greif et al. show that in the standard case, a considerable number of these ejected protostars have masses $M < 1\ {\rm M_{\odot}}$. If these protostars do indeed avoid accreting further gas after their ejection, then they would have lifetimes that are comparable to the current age of the Universe. This suggests that it may be possible for some Population III stars to survive until the present day. However, when the adhesive sinks are used, the number of ejected protostars with subsolar masses is greatly reduced, demonstrating that this conclusion is highly sensitive to our treatment of protostellar mergers.

The majority of the Greif et al. simulations did not include the effects of the accretion luminosity generated by the collection of protostars. However, they did consider one case in which this was included, using the Clark et al. treatment with a fixed value for the accretion rate used in the determination of the accretion luminosity, $\dot{M} = 0.1\ {\rm M_{\odot}}$ year^{-1}. The effect of this was to puff up the disk, causing fragmentation to occur at a slightly larger distance from the initial protostar. However, despite the unrealistically high value adopted for $\dot{M}$, fragmentation still occurred and the number of protostars that formed was barely affected.

A more detailed study of the effects of accretion luminosity heating was carried out by Smith et al. (2011). They used Gadget to resimulate the central 2 pc of the minihalos simulated by Greif et al. (2011a), starting at a time prior to protostar formation at which the peak density of the gas was around 10^{9} cm^{-3}. Smith et al. (2011) evolved these systems past the time at which the first protostar formed, and used sink particles with large accretion radii ($r_{\rm acc} = 20$ AU) to allow them to follow the dynamical evolution of the system for an extended period. For each of the five minihalos, they performed simulations both with and without accretion luminosity heating. When the accretion luminosity heating was included, it was treated in the same fashion as in Clark et al. (2011b). They found that in general, the effect of the accretion luminosity heating was to delay fragmentation and reduce the number of fragments formed. However, they also showed that the effect was relatively small, and had less influence on the number of fragments formed than did the intrinsic variation in halo properties arising from their different assembly histories.

3.2.4 Open Questions

As the discussion in the previous section has shown, the past couple of years has seen a large increase in the number of simulations of Pop. III star formation that show evidence for fragmentation, suggesting that the older picture that had Pop. III stars forming in isolation with masses of $100 M_{\odot}$ or more is in need of some revision. However, many aspects of the fragmentation scenario remain unclear. Some of the most important open questions are summarized below.

- Are our treatments of optically thick H_2 cooling and accretion luminosity heating adequate?

The fragmentation of the gas observed in these simulations typically occurs at densities at which H_2 line cooling is optically thick, and hence may depend on the method used to account for the reduction that this causes in the cooling rate. At present, both of the methods in common usage represent relatively crude approximations, and it remains to be seen whether the behaviour of the gas will remain the same if a more accurate treatment is used. Similarly, the method currently used to treat the effects of accretion luminosity heating also makes a number of major simplifications that may influence the outcome of the simulations.

- What role do magnetic fields play?

If a non-negligible magnetic field can be generated by the turbulent dynamo during the gravitational collapse of the gas, as discussed in Sect. 2.3, then this may influence the evolution and stability of the disk. The presence of a magnetic field may make the disk unstable via the magnetorotational instability (Tan and Blackman 2004; Silk and Langer 2006), although the resulting mass transfer onto the protostar is unlikely to be fast enough to prevent the disk from becoming gravitationally unstable. A more important effect may be magnetic braking of the infalling gas, which could act to significantly reduce the angular momentum of the gas reaching the disk (see e.g. Hennebelle and Ciardi 2009). Whether either of these effects can significantly suppress fragmentation remains to be determined.

- How often do Population III protostars merge?

Current simulations either ignore mergers entirely (e.g. Clark et al. 2011b; Smith et al. 2011), or treat them using very simple approximations that do not properly account for the effects of tidal forces (e.g. Greif et al. 2011a). However, it is clear from the results of the Greif et al. study that the method used to treat mergers has a significant influence on the number of protostars that survive, their mass distribution and their kinematics. Improving the accuracy with which protostellar mergers are treated within this kind of simulation is therefore an important priority.

- Can we find some way to do without sink particles?

The concerns outlined above regarding the way in which protostellar mergers are treated would be greatly ameliorated if we were able to run the simulations without sink particles, as in this case we would be able to model directly how the gas behaves on scales of the order of 100 $R_\odot$. To do this, however, it will be necessary to devise some scheme for treating these very small scales that does not fall foul of the Courant time constraint discussed previously.

- How rapidly does H_2 photodissociation occur?

Fragmentation is dependent on the cooling provided by H_2 and does not occur in models of protostellar accretion disks that omit this effect (Tan and McKee 2004; Mayer and Duschl 2005a). It is therefore highly probable that fragmentation will cease once the H_2 has been photodissociated by Lyman-Werner band photons emitted from any massive protostars that form. What is not yet clear is how quickly this will occur. McKee and Tan (2008) show that the number of Lyman-Werner band photons produced by a zero-age main sequence Population III star increases sharply with increasing stellar mass, before levelling off at a value $S_{\rm lw} \sim 10^{49}$ photons s^{-1} for $M_* \sim 30\ M_\odot$. If we assume that all of these photons are absorbed by H_2 and that 20 % of these absorptions lead to dissociation (Draine and Bertoldi 1996), then the radiation from the star will photodissociate H_2 at a rate $\dot{M}_{\rm dis} = 0.1\ M_\odot\ {\rm year}^{-1}$, leading to complete removal of the H_2 within only a few hundred years. However, it is likely that many of the available photons will not be absorbed by H_2, either because they never coincide with one of the Lyman-Werner band lines, or because they escape along a direction in which most of the H_2 has already been dissociated, or because they are absorbed by atomic hydrogen (Glover and Brand 2001), and so the dissociation time for the H_2 could be significantly longer than suggested by this simple estimate, particularly once one accounts for the effects of three-body H_2 formation.

- How rapidly is the gas ionized? Does this completely suppress accretion, or simply suppress fragmentation?

As we have already discussed in Sect. 3.1.3, the most plausible mechanism for shutting off the supply of cold gas at the centre of the minihalo is the formation of an HII region whose thermal pressure is sufficient to expel most of the gas. However, with the exception of the McKee and Tan (2008) study, little work has been done on the interaction between the growth of the HII region and the evolution of the protostellar accretion disk. In particular, the issue has not yet been looked at within the context of the fragmentation model discussed above. It is therefore unclear how rapidly the HII region will grow, and whether it will immediately act to shut off accretion, or whether pockets of dense, cold gas can survive within the HII region for an extended period.

- Do any low-mass Population III stars survive until the present day?

One of the most exciting results of the Greif et al. (2011a) model is that some of the protostars that are ejected from the centre of the star-forming minihalo have masses that are below 0.8 $M_{\odot}$, and hence lifetimes that are longer than the current age of the Universe. If these protostars avoided accreting any further gas, then they could have survived until the present-day, raising the possibility of directly detecting truly metal-free stars within the Milky Way. However, as we have already discussed above, the number of protostars with sub-solar masses that are ejected from the star-forming region is very sensitive to the way in which protostellar mergers are treated, and hence is highly uncertain at present. In addition, it is possible that any Pop. III protostars that have survived until the present day have also become too polluted with metals by ongoing accretion from the ISM for us to recognize them as metal-free stars, although the best current estimates (Frebel et al. 2009; Johnson and Khochfar 2011) suggest that the effects of pollution will be very small.

4 Summary

In this review, we have focussed on three main topics: how the first star-forming minihalos come into existence and why they have the properties that they do; how gas within a representative minihalo cools, collapses and forms a protostar; and how this protostar and the massive clump of gas surrounding it subsequently evolve.

We have seen that on large scales, we now have a relatively good understanding of the physical processes involved in the formation of the first star-forming minihalos. In order for gas to accumulate within a dark matter minihalo, it must be able to overcome the effects of both gas pressure and also the large-scale streaming motion of the gas relative to the dark matter. Since this streaming motion is typically supersonic, the latter effect generally dominates, and the result is that gas is prevented from accumulating in large quantities within minihalos with masses of less than around 10^5 $M_{\odot}$. Within more massive minihalos, the gravitational force exerted by the dark matter is strong enough to overcome the effects of the gas pressure and the coherent streaming, and the gas begins to undergo gravitational collapse, reaching densities that are several hundred times higher than the cosmological background density. As the gas collapses, however, it is heated by compression and shocks. In order for the collapse to continue, the gas must be able to dissipate this energy, which it does through rotational and vibrational line emission from H_2.

In Sect. 1.2, we saw that the amount of H_2 formed within a given minihalo is a strong function of the temperature of the gas, with the final molecular fraction scaling roughly as $x_{H_2} \propto T^{3/2}$ with the temperature T. The H_2 cooling rate is also a strong function of temperature. As a result, one finds that there is a critical minihalo virial temperature, $T_{\rm crit} \sim 1,000$ K marking the division between cooler halos that do not dissipate much energy within a Hubble time, and hence which do not form stars, and warmer halos that do manage to cool and form stars. As the virial temperature of a minihalo is a simple function of its mass and redshift, one can derive a critical minihalo mass that must be exceeded in order for the gas to cool effectively. This critical mass scales approximately as $M_{\rm crit} \sim 1.6 \times 10^6 (1 + z/10)^{-3/2}\,{\rm M}_\odot$, given standard values for the cosmological parameters. Combining this constraint with that arising from coherent streaming, one finds that at redshifts $z > 40$, the minimum mass of a star-forming minihalo is set by the need to overcome the effects of the streaming, and is roughly $10^5\,{\rm M}_\odot$, while at $z < 40$, H_2 cooling is the limiting factor, and the minimum mass scale is somewhat larger.

On smaller scales, we have also developed an increasingly good understanding of how the gas evolves as it cools, undergoes runaway gravitational collapse, and forms the first protostar. As outlined in Sect. 2, the gas first passes through a "loitering" phase, during which cold gas accumulates at the centre of the minihalo. The temperature and density of the gas at this point depend on the nature of the dominant coolant. When H_2 dominates, we have $T \sim 200$ K and $n \sim 10^4\,{\rm cm}^{-3}$, while if HD dominates, then $T \sim 100$ K and $n \sim 10^6\,{\rm cm}^{-3}$. The loitering phase ends and the collapse of the gas accelerates once the mass of cold gas that has accumulated exceeds the local value of the Bonnor-Ebert mass, which is around 1000 ${\rm M}_\odot$ in the H_2-dominated case, but only 40 ${\rm M}_\odot$ in the HD-dominated case. The next major event to occur is the onset of three-body H_2 formation at $n \sim 10^8\,{\rm cm}^{-3}$ which rapidly converts most of the atomic hydrogen into H_2. The associated heat input leads to an increase in the gas temperature to $T \sim 1{,}000$–$2{,}000$ K, with the details depending to a significant extent on the rate coefficient chosen for reaction (41), which is poorly constrained at low temperatures. At $n \sim 10^{10}\,{\rm cm}^{-3}$, the gas becomes optically thick in the main H_2 cooling lines, but remains optically thin in the continuum. It can therefore continue to cool reasonably effectively at these densities, with the mean temperature only rising relatively slowly with increasing density. At $n \sim 10^{14}\,{\rm cm}^{-3}$, a new process, collision-induced emission from H_2, begins to dominate the cooling. However, this does not lead to a significant drop in the gas temperature, as the gas quickly becomes optically thick in the continuum. At densities above $n \sim 10^{16}\,{\rm cm}^{-3}$, further radiative cooling of the gas is ineffective and the only remaining process capable of slowing the temperature rise is collisional dissociation of the H_2. While the H_2 fraction in the gas remains significant, the temperature is prevented from rising much above 3000 K, but once most of the H_2 has been destroyed, the temperature in the core rises steeply, and the internal thermal pressure eventually becomes strong enough to halt the collapse. State-of-the-art simulations have followed the gravitational collapse of the gas up to this point, which we can identify as the moment at which the first true Population III protostar forms.

Nevertheless, several uncertainties remain in this picture of Pop. III star formation. As already noted, the uncertainty in the three-body H_2 formation rate limits the accuracy with which we can model the chemical and thermal evolution of the collapsing gas. In addition, current three-dimensional collapse models make use of simplified treatments of the effect of opacity on the H_2 cooling rate, and the uncertainty that this introduces into the models has not yet been properly quantified. Further uncertainty comes from two additional issues which have only recently begun to be addressed: the role played by magnetic fields, and the influence of dark matter annihilation. Although the strength of any seed magnetic field existing prior to the assembly of the first star-forming minihalos is still poorly constrained, it now seems clear that the small-scale turbulent dynamo acting during the collapse of the gas will rapidly amplify even a very weak initial field up to a point at which it could potentially become dynamically significant. However, neither the final strength of the field nor its correlation length are well constrained at present, and without a better understanding of these values it is difficult to say to what extent the magnetic field will influence the details of the collapse. The role played by heating and ionization due to dark matter annihilation is even less well understood. Simple models suggest that it may be extremely important and may result in the formation of "dark stars" supported by the energy released by dark matter annihilation rather than by nuclear fusion, but the only hydrodynamical study performed to date suggests that the influence on the collapse is small, and that dark stars do not actually form.

Finally, there remains the question of how the gas evolves after the formation of the first protostar. For much of the last decade, the leading model for this phase of the evolution of the gas has been what we have termed the "smooth accretion" model. In this model, it is assumed that the gas surrounding the newly formed protostar does not fragment, but instead simply smoothly accretes onto the protostar, primarily via a protostellar accretion disk. Considerable work has been done within the framework of this model to understand the structure of the protostar during the accretion phase, and the effect of protostellar feedback on the surrounding gas. This work has shown that any feedback occurring prior to the protostar joining the main sequence is unlikely to significantly reduce the accretion rate, and that the most plausible mechanism for terminating the accretion is photoionization of the accretion disk by ionizing radiation from the central star, implying that it must already have grown to some tens of solar masses. This model therefore predicts that Pop. III stars will generally be solitary, with only one or two forming in each minihalo, and massive, with masses $M \gg 10\,\mathrm{M}_{\odot}$.

Over the past couple of years, however, several new studies have appeared that have cast considerable doubt on the smooth accretion model. These studies have attempted to directly model the evolution of the gas as it begins to be accreted, and have shown that the accretion disk that builds up around the protostar is unstable to gravitational fragmentation even if the stabilizing effects of accretion luminosity feedback from the central protostar are taken into account. Once a few fragments have formed, the dynamical interactions between the individual fragments and between the fragments and the gas can lead to further fragmentation, and to the ejection of low-mass fragments from the system. If we assume that all of the gravitationally bound fragments form protostars, then the result of this model is the assembly of a small, extremely dense cluster of Pop. III protostars with a wide range of masses. As discussed in Sect. 3.2.4, many aspects of the fragmentation scenario remain unclear and much work remains to be done before we can hope to have a good understanding of the final protostellar mass function. Nevertheless, these results suggest that Population III star formation perhaps has far more in common with present-day star formation than has been previously recognised.

Acknowledgements The author would like to thank a large number of people with whom he has had interesting and informative discussions about the physics of Population III star formation, including T. Abel, V. Bromm, P. Clark, G. Dopcke, T. Greif, Z. Haiman, T. Hosokawa, R. Klessen, M. Norman, K. Omukai, B. O'Shea, D. Schleicher, B. Smith, R. Smith, A. Stacy, J. Tan, M. Turk, D. Whalen, and N. Yoshida. Special thanks also go to R. Smith and M. Turk for providing some of the data plotted in Fig. 3, and to P. Clark for his assistance with the production of Fig. 4. Financial support for this work was provided by the Baden-Württemberg-Stiftung via their program International Collaboration II (grant P-LS-SPII/18), from the German Bundesministerium für Bildung und Forschung via the ASTRONET project STAR FORMAT (grant 05A09VHA), and by a Frontier grant of Heidelberg University sponsored by the German Excellence Initiative.

References

Abel, T., Anninos, P., Zhang, Y., Norman, M.L.: Modeling primordial gas in numerical cosmology. New Astron. **2**, 181–207 (1997)

Abel, T., Bryan, G.L., Norman, M.L.: The formation and fragmentation of primordial molecular clouds. Astrophys. J. **540**, 39–44 (2000)

Abel, T., Bryan, G.L., Norman, M.L.: The formation of the first star in the Universe. Science **295**, 93–98 (2002)

Abel, T., Wise, J.H., Bryan, G.L.: The H II region of a primordial star. Astrophys. J. **659**, L87–L90 (2007)

Alvarez, M.A., Bromm, V., Shapiro, P.R.: The H II region of the first star. Astrophys. J. **639**, 621–632 (2006)

Balbus, S.A., Hawley, J.F.: A powerful local shear instability in weakly magnetized disks. I - Linear analysis. II - Nonlinear evolution. Astrophys. J. **376**, 214–233 (1991)

Balbus, S.A., Hawley, J.F.: Instability, turbulence, and enhanced transport in accretion disks. Rev. Mod. Phys. **70**, 1–53 (1998)

Barkana, R., Loeb, A.: In the beginning: the first sources of light and the reionization of the universe. Phys. Rep. **349**, 125–238 (2001)

Barkana, R., Loeb, A.: Probing the epoch of early baryonic infall through 21-cm fluctuations. Mon. Not. Roy. Astron. Soc. **363**, L36–L40 (2005)

Barrow, J.D., Ferreira, P.G., Silk, J.: Constraints on a primordial magnetic field. Phys. Rev. Lett. **78**, 3610–3613 (1997)

Bate, M.R., Bonnell, I.A., Price, N.M.: Modelling accretion in protobinary systems. Mon. Not. Roy. Astron. Soc. **277**, 362–376 (1995)

Beck, R., Poezd, A.D., Shukurov, A., Sokoloff, D.D.: Dynamos in evolving galaxies. Astron. Astrophys. **289**, 94–100 (1994)

Bergin, E.A., Tafalla, M.: Cold dark clouds: the initial conditions for star formation. Ann. Rev. Astron. Astrophys. **45**, 339–396 (2007)

Biermann, L.: Über den Ursprung der Magnetfelder auf Sternen und im interstellaren Raum, Zeitschrift Naturforschung A **5**, 65 (1950)

Blumenthal, G.R., Faber, S.M., Flores, R., Primack, J.R: Contraction of dark matter galactic halos due to baryonic infall. Astrophys. J. **301**, 27–34 (1986)

Bonnor, W.B.: Boyle's Law and gravitational instability. Mon. Not. Roy. Astron. Soc. **116**, 351–359 (1956)

Bromm, V., Coppi, P.S., Larson, R.B.: Forming the first stars in the Universe: The fragmentation of primordial gas. Astrophys. J. **527**, L5–L8 (1999)

Bromm, V., Coppi, P.S., Larson, R.B.: The formation of the first stars. I. The primordial star-forming cloud. Astrophys. J. **564**, 23–51 (2002)

Bromm, V., Loeb, A.: Formation of the first supermassive black holes. Astrophys. J. **596**, 34–46 (2003)

Bromm, V., Loeb, A.: Accretion onto a primordial protostar. New Astron. **9**, 353–364 (2004)

Bryan, G.L., Norman, M.L.: Statistical properties of X-Ray clusters: analytic and numerical comparisons. Astrophys. J. **495**, 80–99 (1998)

Castor, J.I., Abbott, D.C., Klein, R.I.: Radiation-driven winds in of stars. Astrophys. J. **195**, 157 (1975)

Ciardi, B., Ferrara, A.: The first cosmic structures and their effects. Space Sci. Rev. **116**, 625–705 (2005)

Chiosi, C., Maeder, A.: The evolution of massive stars with mass loss. Ann. Rev. Astron. Astrophys. **24**, 329–375 (1986)

Clark, P.C., Glover, S.C.O., Klessen, R.S.: The first stellar cluster. Astrophys. J. **672**, 757–764 (2008)

Clark, P.C., Glover, S.C.O., Klessen, R.S., Bromm, V.: Gravitational fragmentation in turbulent primordial gas and the initial mass function of Population III stars. Astrophys. J. **727**, 110 (2011a)

Clark, P.C., Glover, S.C.O., Smith, R.J., Greif, T.H., Klessen, R.S., Bromm, V.: The formation and fragmentation of disks around primordial protostars. Science **331**, 1040–1042 (2011b)

Cyburt, R.H., Fields, B.D., Olive, K.A.: An update on the big bang nucleosynthesis prediction for ^{7}Li: the problem worsens. J. Cosmol. Astropart. Phys. **11**, 012 (2008)

Davis, A.J., Natarajan, P.: Spin and structural halo properties at high redshift in a Λ cold dark matter universe. Mon. Not. Roy. Astron. Soc. **407**, 691–703 (2010)

Draine, B.T., Bertoldi, F.: Structure of stationary photodissociation fronts. Astrophys. J. **468**, 269–289 (1996)

Dopcke, G., Glover, S.C.O., Clark, P.C., Klessen, R.S.: The effect of dust cooling on low-metallicity star-forming clouds. Astrophys. J. **729**, L3 (2011)

Ebert, R.: Über die Verdichtung von H I-Gebieten. Z. Astrophys. **37**, 217–232 (1955)

Fatuzzo, M., Adams, F.C., Myers, P.C.: Generalized collapse solutions with nonzero initial velocities for star formation in molecular cloud cores. Astrophys. J. **615**, 813–831 (2004)

Federrath, C., Sur, S., Schleicher, D.R.G., Banerjee, R., Klessen, R.S.: A new jeans resolution criterion for (M)HD simulations of self-gravitating gas: application to magnetic field amplification by gravity-driven turbulence. Astrophys. J. **731**, 62 (2011)

Frebel, A., Johnson, J.L., Bromm, V.: The minimum stellar metallicity observable in the Galaxy. Mon. Not. Roy. Astron. Soc. **392**, L50–L54 (2009)

Freese, K., Gondolo, P., Sellwood, J.A., Spolyar, D.: Dark matter densities during the formation of the first stars and in dark stars. Astrophys. J. **693**, 1563–1569 (2009)

Frommhold, L.: Collision-Induced Absorption in Gases. Cambridge University Press, Cambridge (1993)

Galli, D., Palla, F.: The chemistry of the early Universe. Astron. Astrophys. **335** 403–420 (1998)

Gao, L., White, S.D.M., Jenkins, A., Frenk, C.S., Springel, V.: Early structure in ΛCDM. Mon. Not. Roy. Astron. Soc. **363**, 379–392 (2005)

Gao, L., Yoshida, N., Abel, T., Frenk, C.S., Jenkins, A., Springel, V.: The first generation of stars in the Λ cold dark matter cosmology. Mon. Not. Roy. Astron. Soc. **378**, 449–468 (2007)

Glover, S.C.O.: The formation of the first stars in the Universe. Space. Sci. Rev. **117**, 445–508 (2005)

Glover, S.C.O.: Chemistry and cooling in metal-free and metal-poor gas. In: O'Shea, B.W., Heger, A., Abel, T. (eds.) First Stars III, pp. 25–29. AIP (2008)

Glover, S.C.O., Abel, T.: Uncertainties in H_2 and HD chemistry and cooling and their role in early structure formation. Mon. Not. Roy. Astron. Soc. **388**, 1627–1651 (2008)

Glover, S.C.O., Brand, P.W.J.L.: On the photodissociation of H_2 by the first stars. Mon. Not. Roy. Astron. Soc. **321**, 385–397 (2001)
Glover, S.C.O., Brand, P.W.J.L.: Radiative feedback from an early X-ray background. Mon. Not. Roy. Astron. Soc. **340**, 210–226 (2003)
Glover, S.C.O., Savin, D.W.: H_3^+ cooling in primordial gas. Phil. Trans. Roy. Soc. Lond. A **364**, 3107–3112 (2006)
Glover, S.C.O., Savin, D.W.: Is H_3^+ cooling ever important in primordial gas? Mon. Not. Roy. Astron. Soc. **393**, 911–948 (2009)
Glover, S.C.O., Savin, D.W., Jappsen, A.-K.: Cosmological implications of the uncertainty in H^- destruction rate coefficients. Astrophys. J. **640**, 553–568 (2006)
Gnedin, N.Y.: Effect of reionization on structure formation in the Universe. Astrophys. J. **542**, 535–541 (2000)
Gnedin, N.Y., Hui, L.: Probing the Universe with the Lyalpha forest - I. Hydrodynamics of the low-density intergalactic medium. Mon. Not. Roy. Astron. Soc. **296**, 44–55 (1998)
Gnedin, O.Y., Kravtsov, A.V., Klypin, A.A., Nagai, D.: Response of dark matter halos to condensation of baryons: cosmological simulations and improved adiabatic contraction model. Astrophys. J. **616**, 16–26 (2004)
Goodwin, S.P., Whitworth, A.P., Ward-Thompson, D.: Simulating star formation in molecular cloud cores. I. The influence of low levels of turbulence on fragmentation and multiplicity. Astron. Astrophys. **414**, 633–650 (2004)
Gould, R.J., Salpeter, E.E.: The interstellar abundance of the hydrogen molecule. I. Basic processes. Astrophys. J. **138**, 393–407 (1963)
Green, A.M., Hofmann, S., Schwarz, D.J.: The first WIMPy halos. J. Cosmol. Astropart. Phys. **8**, 3 (2005)
Greif, T.H., Johnson, J.L., Klessen, R.S., Bromm, V.: The first galaxies: assembly, cooling and the onset of turbulence. Mon. Not. Roy. Astron. Soc. **387**, 1021–1036 (2008)
Greif, T.H., Glover, S.C.O., Bromm, V., Klessen, R.S.: The first galaxies: chemical enrichment, mixing, and star formation. Astrophys. J. **716**, 510–520 (2010)
Greif, T., Springel, V., White, S., Glover, S., Clark, P., Smith, R., Klessen, R., Bromm, V.: Simulations on a moving mesh: the clustered formation of Population III protostars. Astrophys. J. **737**, 75 (2011)
Greif, T., White, S., Klessen, R., Springel, V.: The delay of Population III star formation by supersonic streaming velocities. Astrophys. J. **736**; 147 (2011)
Haiman, Z., Abel, T., Madau, P.: Photon consumption in minihalos during cosmological reionization. Astrophys. J. **551**, 599–607 (2001)
Haiman, Z., Abel, T., Rees, M.J.: The radiative feedback of the first cosmological objects. Astrophys. J. **534**, 11–24 (2000)
Haugen, N.E., Brandenburg, A., Dobler, W.: Simulations of nonhelical hydromagnetic turbulence. Phys. Rev. E **70**, 016308 (2004)
Hennebelle, P., Ciardi, A.: Disk formation during collapse of magnetized protostellar cores. Astron. Astrophys. **506**, L29–L32 (2009)
Hoyle, F.: On the fragmentation of gas clouds into galaxies and stars. Astrophys. J. **118**, 513–528 (1953)
Hunter, C.: The collapse of unstable isothermal spheres. Astrophys. J. **218**, 834–845 (1977)
Hutchins, J.B.: The thermal effects of H_2 molecules in rotating and collapsing spheroidal gas clouds. Astrophys. J. **205**, 103–121 (1976)
Jang-Condell, H., Hernquist, L.: First structure formation: A simulation of small-scale structure at high redshift. Astrophys. J. **548**, 68–78 (2001)
Jasche, J., Ciardi, B., Ensslin, T.A.: Cosmic rays and the primordial gas. Mon. Not. Roy. Astron. Soc. **380**, 417–429 (2007)
Johnson, J.L., Bromm, V.: The cooling of shock-compressed primordial gas. Mon. Not. Roy. Astron. Soc. **366**, 247–256 (2006)

Johnson, J.L., Greif, T.H., Bromm, V.: Occurrence of metal-free galaxies in the early Universe. Mon. Not. Roy. Astron. Soc. **388**, 26–38 (2008)
Johnson, J.L., Khochfar, S.: Suppression of accretion on to low-mass Population III stars. Mon. Not. Roy. Astron. Soc. **413**, 1184–1191 (2011)
Kandus, A., Kunze, K.E., Tsagas, C.G.: Primordial magnetogenesis. Phys. Rep. **505**, 1–58 (2011)
Kazantsev, A.P.: Enhancement of a magnetic field by a conducting fluid. Sov. Phys. JETP **26**, 1031 (1968)
Kitsionas, S., Whitworth, A.P.: Smoothed particle hydrodynamics with particle splitting, applied to self-gravitating collapse. Mon. Not. Roy. Astron. Soc. **330**, 129–136 (2002)
Komatsu, E., et al.: Seven-year Wilkinson Microwave Anisotropy Probe (WMAP) observations: cosmological interpretation. Astrophys. J. Suppl. S. **192**, 18 (2011)
Kraichnan, R.H., Nagarajan, S.: Growth of turbulent magnetic fields. Phys. Fluids **10**, 859–870 (1967)
Kreckel, H., Bruhns, H., Čížek, M., Glover, S.C.O., Miller, K.A., Urbain, X., Savin, D.W.: Experimental results for H_2 formation from H^- and H and implications for first star formation. Science **329**, 69–71 (2010)
Krtička, J., Kubát, J.: CNO-driven winds of hot first stars. Astron. Astrophys. **493**, 585–593 (2009)
Krumholz, M.R.: Radiation feedback and fragmentation in massive protostellar cores. Astrophys. J. **641**, L45–L48 (2006)
Kudritzki, R.P.: Line-driven winds, ionizing fluxes, and ultraviolet spectra of hot stars at extremely low metallicity. I. Very massive O stars. Astrophys. J. **577**, 389–408 (2002)
Kulsrud, R.M., Anderson, S.W.: The spectrum of random magnetic fields in the mean field dynamo theory of the Galactic magnetic field. Astrophys. J. **396**, 606–630 (1992)
Kulsrud, R.M., Cen, R., Ostriker, J.P., Ryu, D.: The protogalactic origin for cosmic magnetic fields. Astrophys. J. **480**, 481–491 (1997)
Kulsrud, R.M., Zweibel, E.G.: On the origin of cosmic magnetic fields. Rep. Prog. Phys. **71**, 046901 (2008)
Lacey, C., Cole, S.: Merger rates in hierarchical models of galaxy formation. Mon. Not. Roy. Astron. Soc. **262**, 627–649 (1993)
Larson, R.B.: Numerical calculations of the dynamics of collapsing proto-star. Mon. Not. Roy. Astron. Soc. **145**, 271–295 (1969)
Larson, R.B., Starrfield, S.: On the formation of massive stars and the upper limit of stellar masses. Astron. Astrophys. **13**, 190–197 (1971)
Lenzuni, P., Chernoff, D.F., Salpeter, E.E.: Rosseland and Planck mean opacities of a zero-metallicity gas. Astrophys. J. Suppl. S. **76**, 759–801 (1991)
Lepp, S., Shull, J.M.: Molecules in the early Universe. Astrophys. J. **280**, 465–469 (1984)
Low, C., Lynden-Bell, D.: The minimum Jeans mass or when fragmentation must stop. Mon. Not. Roy. Astron. Soc. **176**, 367–390 (1976)
Machacek, M.E., Bryan, G.L., Abel, T.: Effects of a soft X-ray background on structure formation at high redshift. Mon. Not. Roy. Astron. Soc. **338**, 273–286 (2003)
Machida, M.N., Omukai, K., Matsumoto, T., Inutsuka, S.: The first jets in the Universe: protostellar jets from the first stars. Astrophys. J. **647**, L1–L4 (2006)
Machida, M.N., Matsumoto, T., Inutsuka, S.: Magnetohydrodynamics of Population III star formation. Astrophys. J. **685**, 690–704 (2008)
Maeder, A., Meynet, G.: The evolution of rotating stars. Ann. Rev. Astron. Astrophys. **38**, 143–190 (2000)
Maron, J., Cowley, S., McWilliams, J.: The nonlinear magnetic cascade. Astrophys. J. **603**, 569–583 (2004)
Martin, P.G., Keogh, W.J., Mandy, M.E.: Collision-induced dissociation of molecular hydrogen at low densities. Astrophys. J. **499**, 793–798 (1998)
Mayer, M., Duschl, W.J.: Stationary Population III accretion discs. Mon. Not. Roy. Astron. Soc. **356**, 1 (2005a)
Mayer, M., Duschl, W.J.: Rosseland and Planck mean opacities for primordial matter. Mon. Not. Roy. Astron. Soc. **358**, 614–631 (2005b)

McDowell, M.R.C.: On the formation of H2 in HI regions. Observatory **81**, 240–243 (1961)
McGreer, I.D., Bryan, G.L.: The impact of HD cooling on the formation of the first stars. Astrophys. J. **685**, 8–20 (2008)
McKee, C.F., Tan, J.C.: The formation of the first stars. II. Radiative feedback processes and implications for the initial mass function. Astrophys. J. **681**, 771–797 (2008)
Meynet, G., Ekström, S., Maeder, A.: The early star generations: the dominant effect of rotation on the CNO yields. Astron. Astrophys. **447**, 623–639 (2006)
Mizusawa, H., Omukai, K., Nishi, R.: Primordial molecular emission in Population III galaxies. Pub. Astron. Soc. Jpn. **57**, 951–967 (2005)
Myers, Z., Nusser, A.: Neutralino annihilations and the gas temperature in the dark ages. Mon. Not. Roy. Astron. Soc. **384**, 727–732 (2008)
Nagakura, T., Omukai, K.: Formation of Population III stars in fossil HII regions: significance of HD. Mon. Not. Roy. Astron. Soc. **364**, 1378–1386 (2005)
Nakamura, F., Umemura, M.: The stellar initial mass function in primordial galaxies. Astrophys. J. **569**, 549–557 (2002)
Nakano, T., Hasegawa, T., Morino, J.-I., Yamashita, T.: Evolution of protostars accreting mass at very high rates: Is Orion IRc2 a huge protostar?. Astrophys. J. **534**, 976–983 (2000)
Nakano, T., Hasegawa, T., Norman, C.: The mass of a star formed in a cloud core: theory and its application to the Orion A Cloud. Astrophys. J. **450**, 183–195 (1995)
Navarro, J.F., Frenk, C.S., White, S.D.M.: A universal density profile from hierarchical clustering. Astrophys. J. **490**, 493–508 (1997)
Navarro, J.F., White, S.D.M.: Simulations of dissipative galaxy formation in hierarchically clustering universes-2. Dynamics of the baryonic component in galactic haloes. Mon. Not. Roy. Astron. Soc. **267**, 401–412 (1994)
Naoz, S., Barkana, R.: Growth of linear perturbations before the era of the first galaxies. Mon. Not. Roy. Astron. Soc. **362**, 1047–1053 (2005)
Naoz, S., Barkana, R.: The formation and gas content of high-redshift galaxies and minihaloes. Mon. Not. Roy. Astron. Soc. **377**, 667–676 (2007)
Oh, S.P., Haiman, Z.: Second-generation objects in the Universe: radiative cooling and collapse of halos with virial temperatures above 10^4 Kelvin. Astrophys. J. **569**, 558–572 (2002)
Oh, S.P., Haiman, Z.: Fossil H II regions: self-limiting star formation at high redshift. Mon. Not. Roy. Astron. Soc. **346**, 456–472 (2003)
Omukai, K., Inutsuka, S.: An upper limit on the mass of a primordial star due to the formation of an Hii region: the effect of ionizing radiation force. Mon. Not. Roy. Astron. Soc. **332**, 59–64 (2002)
Omukai, K., Nishi, R.: Formation of primordial protostars. Astrophys. J. **508**, 141–150 (1998)
Omukai, K., Nishi, R.: Photodissociative regulation of star formation in metal-free pregalactic clouds. Astrophys. J. **518**, 64–68 (1999)
Omukai, K., Nishi, R., Uehara, H., Susa, H.: Evolution of primordial protostellar clouds—quasi-static analysis. Prog. Theor. Phys. **99**, 747–761 (1998)
Omukai, K., Palla, F.: On the formation of massive primordial stars. Astrophys. J. **589**, 677–687 (2003)
Omukai, K., Palla, F.: Formation of the first stars by accretion. Astrophys. J. **561**, L55–L58 (2001)
Omukai, K., Tsuribe, T., Schneider, R., Ferrara, A.: Thermal and fragmentation properties of star-forming clouds in low-metallicity environments. Astrophys. J. **626**, 627–643 (2005)
O'Shea, B.W., Norman, M.L.: Population III star formation in a ΛCDM Universe. I. The effect of formation redshift and environment on protostellar accretion rate. Astrophys. J. **654**, 66–92 (2007)
Palla, F., Salpeter, E.E., Stahler, S.W.: Primordial star formation - The role of molecular hydrogen. Astrophys. J. **271**, 632–641 (1983)
Peebles, P.J.E., Dicke, R.H.: Origin of the globular star clusters. Astrophys. J. **154**, 891–908 (1968)
Penston, M.V.: Dynamics of self-gravitating gaseous spheres-III. Analytical results in the free-fall of isothermal cases. Mon. Not. Roy. Astron. Soc. **144**, 425–448 (1969)

Pinto, C., Galli, D., Bacciotti, F.: Three-fluid plasmas in star formation. I. Magneto-hydrodynamic equations. Astron. Astrophys. **484**, 1–15 (2008)
Pudritz, R.E., Silk, J.: The origin of magnetic fields and primordial stars in protogalaxies. Astrophys. J. **342**, 650–659 (1989)
Rees, M.J.: Opacity-limited hierarchical fragmentation and the masses of protostars. Mon. Not. Roy. Astron. Soc. **176**, 483–486 (1976)
Rees, M.J., Ostriker, J.P.: Cooling, dynamics and fragmentation of massive gas clouds - Clues to the masses and radii of galaxies and clusters. Mon. Not. Roy. Astron. Soc. **179**, 541–559 (1977)
Rice, W.K.M., Lodato, G., Armitage, P.J.: Investigating fragmentation conditions in self-gravitating accretion discs. Mon. Not. Roy. Astron. Soc. **364**, L56–L60 (2005)
Ripamonti, E., Haardt, F., Ferrara, A., Colpi, M.: Radiation from the first forming stars. Mon. Not. Roy. Astron. Soc. **334**, 401 (2002)
Ripamonti, E., Abel, T.: Fragmentation and the formation of primordial protostars: the possible role of collision-induced emission. Mon. Not. Roy. Astron. Soc. **348**, 1019–1034 (2004)
Ripamonti, E., Iocco, F., Ferrara, A., Schneider, R., Bressan, A., Marigo, P.: First star formation with dark matter annihilation. Mon. Not. Roy. Astron. Soc. **406**, 2605–2615 (2010)
Saslaw, W.C., Zipoy, D.: Molecular hydrogen in pre-galactic gas clouds. Nature **216**, 976–978 (1967)
Savin, D.W., Krstić, P.S., Haiman, Z., Stancil, P.C.: Rate coefficient for $H^{+} + H_2(X^1\Sigma_g^+, v=0, J=0) \rightarrow H(1s) + H_2^+$ charge transfer and some cosmological implications. Astrophys. J. **606**, L167–L170 (2004)
Schleicher, D.R.G., Banerjee, R., Klessen, R.S.: Reionization: A probe for the stellar population and the physics of the early Universe. Phys. Rev. D **78**, 083005 (2008)
Schleicher, D.R.G., Galli, D., Palla, F., Camenzind, M., Klessen, R.S., Bartelmann, M., Glover, S.C.O.: Effects of primordial chemistry on the cosmic microwave background. Astron. Astrophys. **490**, 521–535 (2008)
Schleicher, D.R.G., Banerjee, R., Sur, S., Arshakian, T.G., Klessen, R.S., Beck, R., Spaans, M.: Small-scale dynamo action during the formation of the first stars and galaxies. I. The ideal MHD limit. Astron. Astrophys. **522**, A115 (2010)
Schleicher, D.R.G., Galli, D., Glover, S.C.O., Banerjee, R., Palla, F., Schneider, R., Klessen, R.S.: The influence of magnetic fields on the thermodynamics of primordial star formation. Astrophys. J. **703**, 1096–1106 (2009)
Schneider, R., Omukai, K., Inoue, A.K., Ferrara, A.: Fragmentation of star-forming clouds enriched with the first dust. Mon. Not. Roy. Astron. Soc. **369**, 1437–1444 (2006)
Shakura, N.I., Sunyaev, R.A.: Black holes in binary systems. Observational appearance. Astron. Astrophys. **24**, 337–355 (1973)
Shu, F.: Self-similar collapse of isothermal spheres and star formation. Astrophys. J. **214**, 488–497 (1977)
Silk, J.: Cosmic black-body radiation and galaxy formation. Astrophys. J. **151**, 459–471 (1968)
Silk, J., Langer, M.: On the first generation of stars. Mon. Not. Roy. Astron. Soc. **371**, 444–450 (2006)
Smith, R.J., Glover, S.C.O., Clark, P.C., Greif, T., Klessen, R.S.: The effects of accretion luminosity upon fragmentation in the early Universe. Mon. Not. Roy. Astron. Soc. **414**, 3633–3644 (2011)
Smith, N., Owocki, S.P.: On the role of continuum-driven eruptions in the evolution of very massive stars and Population III stars. Astrophys. J. **645**, L45–L48 (2006)
Sobolev, V.V.: Moving envelopes of stars. Harvard University Press, Cambridge (1960)
Spolyar, D., Freese, K., Gondolo, P.: Dark matter and the first stars: a new phase of stellar evolution. Phys. Rev. Lett. **100**, 051101 (2008)
Springel, V.: E pur si muove: Galilean-invariant cosmological hydrodynamical simulations on a moving mesh. Mon. Not. Roy. Astron. Soc. **401**, 791–851 (2010)
Stacy, A., Bromm, V.: Impact of cosmic rays on Population III star formation. Mon. Not. Roy. Astron. Soc. **382**, 229–238 (2007)

Stacy, A., Bromm, V., Loeb, A.: Effect of streaming motion of baryons relative to dark matter on the formation of the first stars. Astrophys. J. **730**, L1 (2011)

Stacy, A., Greif, T.H., Bromm, V.: The first stars: formation of binaries and small multiple systems. Mon. Not. Roy. Astron. Soc. **403**, 45–60 (2010)

Stahler, S.W., Palla, F., Salpeter, E.E.: Primordial stellar evolution - The protostar phase. Astrophys. J. **302**, 590–605 (1986a)

Stahler, S.W., Palla, F., Salpeter, E.E.: Primordial stellar evolution - The pre-main-sequence phase. Astrophys. J. **308**, 697–705 (1986b)

Stamatellos, D., Whitworth, A.P., Hubber, D.A.: The importance of episodic accretion for low-mass star formation. Astrophys. J. **730**, 32 (2011)

Stancil, P.C., Lepp, S., Dalgarno, A.: The lithium chemistry of the early Universe. Astrophys. J. **458**, 401–406 (1996)

Stancil, P.C., Lepp, S., Dalgarno, A.: The deuterium chemistry of the early Universe. Astrophys. J. **509**, 1–10 (1998)

Stenrup, M., Larson, Å, Elander, N.: Mutual neutralization in low-energy H^+ + H^- collisions: A quantum ab initio study. Phys. Rev. A **79**, 012713 (2009)

Subramanian, K.: Can the turbulent galactic dynamo generate large-scale magnetic fields? Mon. Not. Roy. Astron. Soc. **294**, 718–728 (1998)

Sur, S., Schleicher, D.R.G., Banerjee, R., Federrath, C., Klessen, R.S.: The generation of strong magnetic fields during the formation of the first stars. Astrophys. J. **721**, L134–L138 (2010)

Tan, J.C., Blackman, E.G.: Protostellar disk dynamos and hydromagnetic outflows in primordial star formation. Astrophys. J. **603**, 401–413 (2004)

Tan, J.C., McKee, C.F.: The formation of the first stars. I. Mass infall rates, accretion disk structure, and protostellar evolution. Astrophys. J. **603**, 383–400 (2004)

Tan, J.C., McKee, C.F.: Star formation at zero and very low metallicities, In: O'Shea, B.W., Heger, A., Abel, T. (eds.) First Stars III, pp. 47–62. AIP (2008)

Tegmark, M., Silk, J., Rees, M.J., Blanchard, A., Abel, T., Palla, F.: How small were the first cosmological objects? Astrophys. J. **474**, 1–12 (1997)

Tormen, G., Bouchet, F.R., White, S.D.M.: The structure and dynamical evolution of dark matter haloes. Mon. Not. Roy. Astron. Soc. **286**, 865–884 (1997)

Tseliakhovich, D., Barkana, R., Hirata, C.: Suppression and spatial variation of early galaxies and minihalos. Mon. Not. Roy. Astron. Soc. **418**, 906–915 (2011)

Tseliakhovich, D., Hirata, C.: Relative velocity of dark matter and baryonic fluids and the formation of the first structures. Phys. Rev. D **82**, 083520 (2010)

Tsuribe, T., Omukai, K.: Physical mechanism for the intermediate characteristic stellar mass in extremely metal poor environments. Astrophys. J. **676**, L45–L48 (2008)

Turk, M.J., Abel, T., O'Shea, B.: The formation of Population III binaries from cosmological initial conditions. Science **325**, 601–605 (2009)

Turk, M.J., Norman, M.L., Abel, T.: High-entropy polar regions around the first protostars. Astrophys. J. **725**, L140–L144 (2010)

Turk, M.J., Clark, P.C., Glover, S.C.O., Greif, T.H., Abel, T., Klessen, R.S., Bromm, V.: Effects of varying the three-body molecular hydrogen formation rate in primordial star formation. Astrophys. J. **726**, 55 (2011)

Wang, P., Abel, T.: Dynamical treatment of virialization heating in galaxy formation. Astrophys. J. **672**, 752–756 (2008)

Whalen, D., Abel, T., Norman, M.L.: Radiation hydrodynamic evolution of primordial H II regions. Astrophys. J. **610**, 14–22 (2004)

Whitworth, A., Summers, D.: Self-similar condensation of spherically symmetric self-gravitating isothermal gas clouds. Mon. Not. Roy. Astron. Soc. **214**, 1–25 (1985)

Wolfire, M.G., Cassinelli, J.P.: Conditions for the formation of massive stars. Astrophys. J. **319**, 850–867 (1987)

Xu, H., O'Shea, B.W., Collins, D.C., Norman, M.L., Li, H., Li, S.: The biermann battery in cosmological MHD simulations of Population III star formation. Astrophys. J. **688**, L57–L60 (2008)

Young, P.: Numerical models of star clusters with a central black hole. I - Adiabatic models. Astrophys. J. **242**, 1232–1237 (1980)

Yoshida, N., Abel, T., Hernquist, L., Sugiyama, N.: Simulations of early structure formation: primordial gas clouds. Astrophys. J. **592**, 645–663 (2003)

Yoshida, N., Omukai, K., Hernquist, L., Abel, T.: Formation of primordial stars in a ΛCDM Universe. Astrophys. J. **652**, 6–25 (2006)

Yoshida, N., Oh, S.P., Kitayama, T., Hernquist, L.: Early cosmological H II/He III egions and their impact on second-generation star formation. Astrophys. J. **663**, 687–707 (2007)

Yoshida, N., Omukai, K., Hernquist, L.: Formation of massive primordial stars in a reionized gas. Astrophys. J. **667**, L117–L120 (2007)

Yoshida, N., Omukai, K., Hernquist, L.: Protostar formation in the early Universe. Science **321**, 669–671 (2008)

Zel'Dovich, Ya. B.: Gravitational instability: An approximate theory for large density perturbations. Astron. Astrophys. **5**, 84–89 (1970)

Part II
The First Galaxies and Normal Stellar Populations

Formation of the First Galaxies: Theory and Simulations

Jarrett L. Johnson

Abstract The properties of the first galaxies are shaped in large part by the first generations of stars, which emit high energy radiation and unleash both large amounts of mechanical energy and the first heavy elements when they explode as supernovae. We survey the theory of the formation of the first galaxies in this context, focusing on the results of cosmological simulations to illustrate a number of the key processes that define their properties. We first discuss the evolution of the primordial gas as it is incorporated into the earliest galaxies under the influence of the high energy radiation emitted by the earliest stars; we then turn to consider how the injection of heavy elements by the first supernovae transforms the evolution of the primordial gas and alters the character of the first galaxies. Finally, we discuss the prospects for the detection of the first galaxies by future observational missions, in particular focusing on the possibility that primordial star-forming galaxies may be uncovered.

1 Introduction: Defining Characteristics of the First Galaxies

While the first stars are for the most part well-defined objects, the definition of the first galaxies is somewhat more ambiguous (see e.g. Bromm and Yoshida 2011). Here we adopt the common view that a galaxy must be able to host ongoing star formation, even in the face of the radiative and mechanical feedback that accompanies the formation and evolution of stars. By this definition, the formation sites of the first stars, dark matter minihalos with masses 10^5–10^6 $M_{\odot}$, are unlikely candidates for the first galaxies, as the high energy radiation emitted by young stars and the supernovae that mark their end of life can rarify and expel any dense gas from which

J.L. Johnson (✉)
Max-Planck-Institut für extraterrestrische Physik, Giessenbachstraße, 85748 Garching, Germany

Los Alamos National Laboratory, Los Alamos, NM 87545, USA
e-mail: jjohnson@mpe.mpg.de

T. Wiklind et al. (eds.), *The First Galaxies*, Astrophysics and Space Science Library 396, DOI 10.1007/978-3-642-32362-1_4,

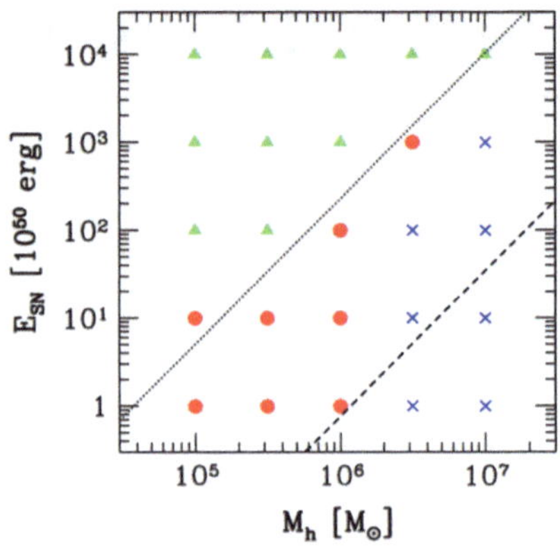

Fig. 1 The fate of host halos of the first stars, as a function of the mass $M_{\rm h}$ of the halo and the energy $E_{\rm SN}$ with which the stars explode as supernovae. Less massive halos have their gas blown away when even relatively weak (e.g. $E_{\rm SN} \sim 10^{50}$ erg) Pop III supernovae explode within them, both along with (*circles*) and in the absence of (*triangles*) the additional radiative feedback from the progenitor stars. Halos which retain their gas are shown by crosses. From Eq. (1) we can see that for the suite of halos shown here at $z \simeq 20$, those with $T_{\rm vir} \gtrsim 10^4$ K are able to retain their gas, in contrast to the less massive minihalos. From Kitayama and Yoshida (2005)

stars may form at a later time. As shown in Fig. 1, it is only somewhat larger halos, with masses 10^7–10^8 $M_\odot$, which have deep enough gravitational potential wells and enough mass to prevent the expulsion of gas after an episode of star formation (e.g. Kitayama and Yoshida 2005; Read et al. 2006; Whalen et al. 2008).

As can be inferred from this figure, one of the distinguishing characteristics of halos massive enough to host ongoing star formation, and so to host the first galaxies, is the characteristic temperature $T_{\rm vir}$ that gas reaches during their virialization. This, referred to as the virial temperature of the halo, can be derived by assuming that the absolute magnitude of the gravitational potential energy of the halo is twice its kinetic energy, which yields

$$T_{\rm vir} \simeq 4\times10^4 \left(\frac{\mu}{1.2}\right)\left(\frac{M_{\rm h}}{10^8 h^{-1} M_\odot}\right)^{\frac{2}{3}}\left(\frac{1+z}{10}\right) {\rm K}\,, \qquad (1)$$

where $M_{\rm h}$ is the mass of the halo, z is the redshift at which it collapses, and μ is the mean molecular weight of the gas in the halo, here normalized to a value appropriate for neutral primordial gas. The Hubble constant $H_0 = 100\ h\,{\rm km\,s^{-1}}$ ${\rm Mpc^{-1}}$ also appears here through h.[1] From Fig. 1 we see that the mass of halos which are large enough to host ongoing star formation, at $z \sim 20$, is $\sim 10^7$ $M_\odot$; this corresponds to

[1]Note that this formula is derived assuming a standard CDM cosmological model in which $h \simeq 0.7$ (see e.g. Barkana and Loeb 2001); as such, this formula is valid at the high redshifts (i.e. $z >> 1$) at which the first galaxies form, but must be modified at lower redshifts in order to account for a cosmological constant Λ.

a virial temperature of $T_{vir} \sim 10^4$ K. One of the reasons for this is that 10^4 K is roughly the temperature to which photoionization by stars heats the gas (see e.g. Osterbrock and Ferland 2006); thus, gas that is photoheated by stars remains bound within a halo with such a virial temperature. In turn, the presence of this gas when stars explode as supernova leads to the rapid loss of the mechanical energy in the explosion to radiation, thereby limiting the amount of gas blown out of the halo, in contrast to the case of the first supernovae in less massive minihalos (see Sect. 3.1). Also, due to the efficient cooling of atomic hydrogen at this temperature, gas can collapse into halos with $T_{vir} \geq 10^4$ K regardless of its molecular content, in contrast to the minihalos that host the first stars, into which primordial gas only collapses if it is cooled by H_2 molecules (e.g. Oh and Haiman 2002); this implies that star formation can take place even under the influence of the molecule-dissociating radiation emitted by the first stars (see Sect. 2.2).

Figure 2 shows the physical properties of an atomic cooling halo,[2] in which a first galaxy would form, at $z \sim 10$ in a cosmological simulation (see Greif et al. 2008). As shown here, much of the primordial gas that falls from the intergalactic medium (IGM) into the potential well of the halo is shock-heated to $T_{vir} \sim 10^4$ K at a physical distance of $\sim$1 kpc from the center of the halo. This distance corresponds to the virial radius r_{vir} of the halo, defined in general terms as the radius within which the average matter density is equal to the value at which virial equilibrium is established, which is $\simeq 18\pi^2$ times the mean matter density of the universe at the redshift z at which the halo forms (e.g. Barkana and Loeb 2001). For the standard ΛCDM cosmological model, this is given in physical units as

$$r_{vir} \simeq 800 h^{-1} \left(\frac{M_h}{10^8 h^{-1} M_\odot} \right)^{\frac{1}{3}} \left(\frac{1+z}{10} \right)^{-1} \text{pc} \,, \tag{2}$$

where we have normalized to values of halo mass and redshift that are typical for atomic cooling halos hosting the first galaxies. Near the virial radius a large fraction of the gas is hot ($\geq$ 500 K) and rotating about the center of the halo at nearly the circular velocity v_{circ} of the halo (Greif et al. 2008), defined as the velocity with which a body must move in order to be centripetally supported against gravity at the virial radius:

$$v_{circ} = \left(\frac{G M_h}{r_{vir}} \right)^{\frac{1}{2}} \simeq 20 \left(\frac{M_h}{10^8 h^{-1} M_\odot} \right)^{\frac{1}{3}} \left(\frac{1+z}{10} \right)^{\frac{1}{2}} \text{km s}^{-1} \,. \tag{3}$$

However, there is also a substantial portion of the infalling gas that falls to the center of the halo in cool, dense filaments and is not shock-heated to the virial temperature. These dense filaments feed cold gas into the central $\sim$100 pc of the halo, contributing to the majority of the gas the temperature of which is $<$ 500 K and which may collapse to form stars (Greif et al. 2008).

[2]Because the primordial gas can cool via emission from atomic hydrogen and collapse into halos with $T_{vir} \sim 10^4$ K, such halos are commonly referred to as 'atomic cooling' halos.

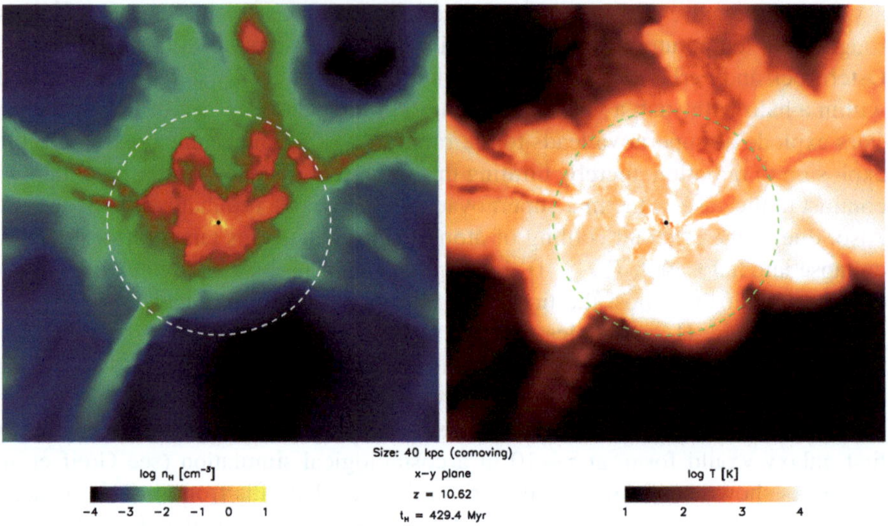

Fig. 2 The properties of the primordial gas collapsing into an atomic cooling dark matter halo at $z \simeq 10$. Shown are the hydrogen number density *(left panel)* and temperature *(right panel)*, the dashed lines denoting the virial radius $r_{\rm vir}$ at a distance of $\simeq 1$ kpc. Note that most of the gas is accreted directly from the IGM and shock-heated to the virial temperature of $T_{\rm vir} \simeq 10^4$ K, although cold accretion also becomes important as soon as gas cools in filaments and flows towards the centre of the galaxy, such as through the streams coming from the left- and right-hand sides of the panels. In contrast to the minihalos in which the first stars form, a halo with a virial temperature $T_{\rm vir} \geq 10^4$ K is massive enough and has a deep enough gravitational potential well to retain its gas even when stars formed within it explode as supernovae (see Fig. 1). Hence, such halos are strong candidates for the formation sites of the first galaxies. From Greif et al. (2008)

While the atomic cooling halo shown in Fig. 2 is a prime example of the type of halo in which the first galaxies likely formed, there are numerous physical effects that were not included in the cosmological simulation from which this halo was drawn, most notably the feedback effects of Population (Pop) III stars (see e.g. Wise and Abel 2008; Johnson et al. 2008; Greif et al. 2010; Whalen et al. 2010). The high energy radiation emitted by the first stars both ionizes the primordial gas and dissociates molecules, which are critical cooling agents. Also, many of the first stars explode as violent supernovae, which inject large amounts of mechanical energy into their host minihalos and the IGM, as well as dispersing the first heavy elements, thereby altering forever the properties of the gas from which the first galaxies form.

In this Chapter, we shall focus on how this feedback from the first generations of stars impacts the formation and evolution of the first galaxies. In Sect. 2, we briefly discuss how the cooling properties of the primordial gas, which shape the nature of Pop III star formation, are affected by the radiation emitted from the first stars and accreting black holes. In Sect. 3, we then turn to discuss how the first supernovae enrich the primordial gas with heavy elements, and how this process leads to the epoch of metal-enriched Pop II star formation. In Sect. 4, we briefly discuss the prospects for observing the first galaxies, and for finding Pop III star formation therein, using facilities such as the *James Webb Space Telescope* (JWST). Finally, in Sect. 5, we close with a summary of the results presented in this Chapter.

2 Evolution of the Primordial Gas and Formation of the First Galaxies

Being composed solely of the hydrogen, helium, and trace amounts of lithium and beryllium synthesized in the Big Bang, the primordial gas contains a limited number of coolants, chief among these H_2 at temperatures $\lesssim 10^4$ K. Because of the inefficient cooling of the gas relative to the metal-enriched[3] gas from which stars form today, it is likely that the Pop III initial mass function (IMF) is top-heavy compared to that of the stars observed in our Milky Way. A simple explanation for this is based on the mass scale at which the fragmentation of the primordial gas takes place. Known as the Jeans mass $M_{\rm J}$, this is essentially the mass at which density enhancements grow via gravity more quickly than they can be erased due to pressure gradients. To estimate $M_{\rm J}$ for a gas with a number density n and a temperature T, related to the sound speed $c_{\rm s}$ by $3k_{\rm B}T/2 = \mu m_{\rm H}\, c_{\rm s}^2/2$, we first estimate the timescale at which density enhancements grow as the free-fall time $t_{\rm ff} \simeq (G\rho)^{-\frac{1}{2}} = (G\mu m_{\rm H} n)^{-\frac{1}{2}}$, where G is Newton's constant. Then, estimating the timescale in which density enhancements are erased as the sound-crossing time $t_{\rm sc} \simeq L/c_{\rm s}$. We equate these two timescales to estimate the characteristic size $L_{\rm J}$ and mass of a gas cloud which is just massive enough to collapse under its own gravity. We thus arrive at an expression for the Jeans mass $M_{\rm J}$, given by

$$M_{\rm J} \simeq \mu m_{\rm H} n L_{\rm J}^3 \simeq 700 \left(\frac{T}{200{\rm K}}\right)^{\frac{3}{2}} \left(\frac{n}{10^4{\rm cm}^{-3}}\right)^{-\frac{1}{2}} {\rm M}_{\odot}\,, \tag{4}$$

where we have assumed $\mu = 1.2$, appropriate for neutral primordial gas, and have again normalized to quantities typical for primordial star-forming clouds. As we shall discuss in Sect. 3, the primordial gas is in general unable to cool as efficiently as metal-enriched gas, which leads in general to higher temperatures at fragmentation and so to a larger characteristic mass of the gravitationally unstable gas clouds from which stars form (e.g. Bromm and Larson 2004).

While the Jeans mass is an estimate of the mass of a collapsing gas cloud, the amount of gas that is finally incorporated into a star is also dictated by the rate at which gas accretes onto it, starting from the formation of a protostar. Thus, another reason that primordial stars are likely to be more massive than stars forming

[3] We use the common term "metals" to refer to elements heavier than helium which are produced in stars and supernovae.

from metal-enriched gas is that higher gas temperatures also translate into higher accretion rates, as can be seen by estimating the accretion rate $\dot{M}_{\rm acc}$ as a function of the temperature of the gas (see e.g. Stahler et al. 1980). Assuming that, through the action of gravity, the protostar grows by accreting from a gas cloud of mass $\simeq M_{\rm J}$, the accretion rate can be estimated as

$$\dot{M}_{\rm acc} \simeq \frac{M_{\rm J}}{t_{\rm ff}} \simeq 10^{-3} \left(\frac{T}{200{\rm K}}\right)^{\frac{3}{2}} {\rm M}_{\odot}\ {\rm year}^{-1} , \quad (5)$$

where we have again assumed $\mu = 1.2$ and normalized to the characteristic temperature of the gas from which Pop III stars form in minihalos (see e.g. Glover 2005). Therefore, it is a combination of both the relatively large reservoir of gas available in gravitationally unstable gas clouds and the relatively high accretion rates onto primordial protostars (e.g. Omukai and Palla 2003; Tan and McKee 2004; Yoshida et al. 2008) which suggests that Pop III stars are more massive than metal-enriched Pop II or Pop I stars. As both $M_{\rm J}$ and $\dot{M}_{\rm acc}$ depend strongly on the temperature of the gas, one of the central questions with regard to star formation in the first galaxies is the degree to which the gas is able to cool. In the next section, we discuss the cooling of primordial gas in the first galaxies, focusing on how it is different from the case of cooling in the minihalos in which the first stars form. Later, in Sect. 3.3, we discuss how the cooling properties of the primordial gas change when it is mixed with heavy elements and collapses to form the first Pop II stars.

2.1 Cooling of the Primordial Gas

As shown in Figs. 2 and 3, the primordial gas collapsing into atomic cooling halos is typically shock-heated to the virial temperature of $\geq 10^4$ K. In contrast to the case of Pop III star formation in minihalos, the gas at these temperatures is partially ionized, and this can have important consequences for the evolution of the gas as it collapses to form stars in the first galaxies. To see why, we note that the primary reaction sequence leading to the formation of H_2 molecules is (e.g. Galli and Palla 1998; Glover 2005)

$$e^- + {\rm H} \rightarrow {\rm H}^- + \gamma \quad (6)$$

$$ {\rm H}^- + {\rm H} \rightarrow {\rm H}_2 + e^- , \quad (7)$$

where γ denotes the emission of a photon. Whereas the primordial gas which collapses into minihalos to form the first stars has a free electron fraction $X_{\rm e} \leq 10^{-4}$, the collisional ionization of the primordial gas collapsing into atomic cooling halos can lead to an enhancement of the free electron fraction by a factor of more than an order of magnitude, as shown in Fig. 3. In turn, this leads to high rates of H_2

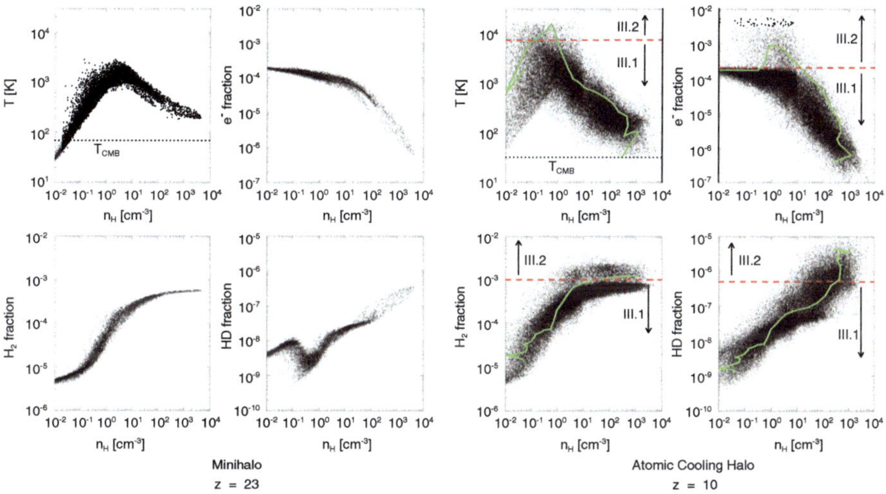

Fig. 3 The properties of the primordial gas inside a first star-forming minihalo at $z = 23$ (*left panel*) and a first galaxy-hosting halo at $z = 10$ (*right panel*), as found in a cosmological hydrodynamics simulation. The temperature, electron fraction, HD fraction, and H_2 fraction are shown as functions of number density, clockwise from top left to bottom left. *Left panel*: In the minihalo case, adiabatic collapse drives the temperature to $T \geq 10^3$ K and the density to $n \geq 1$ cm^3, where molecule formation sets in and allows the gas to cool to $\simeq$200 K. At this point, the central clump becomes Jeans-unstable and presumably collapses to form at least one massive Pop III star. *Right panel*: In the first galaxy, formed in an atomic cooling halo, a second cooling channel emerges due to an elevated electron fraction at the virial shock, where the temperature rises to $\sim 10^4$ K; this, in turn, enhances molecule formation and allows the gas to cool to the much lower temperatures, as shown by the evolutionary track of a representative parcel of gas (*green lines*). Due to the correspondingly lower Jeans mass (4) and accretion rate (5), less massive Pop III stars (Pop III.2; see Sect. 2.1) are expected to form in such a first galaxy, perhaps with a characteristic mass of the order of $\simeq 10\ M_{\odot}$. From Greif et al. (2008)

formation in atomic cooling halos, principally via the above reactions for which free electrons act as catalysts (e.g. Shapiro and Kang 1987). The net result is a generally higher H_2 fraction in the high density, central regions of atomic cooling halos than in minihalos, as is also shown in Fig. 3, and hence also to higher cooling rates due to molecular emission. Therefore, somewhat counter-intuitively, because of the higher virial temperatures of the atomic cooling halos in which the first galaxies form, the dense gas in the centers of these halos can cool more effectively than in the minihalos in which the first Pop III stars form.

In fact, the ionization of the primordial gas in atomic cooling halos results in the formation of another molecule which can be even more effective at cooling the gas than H_2: deuterium hydride (HD). With the high H_2 fraction that develops in partially ionized gas, HD forms rapidly via the following reaction:

$$D^{+} + H_2 \rightarrow HD + H^{+}. \tag{8}$$

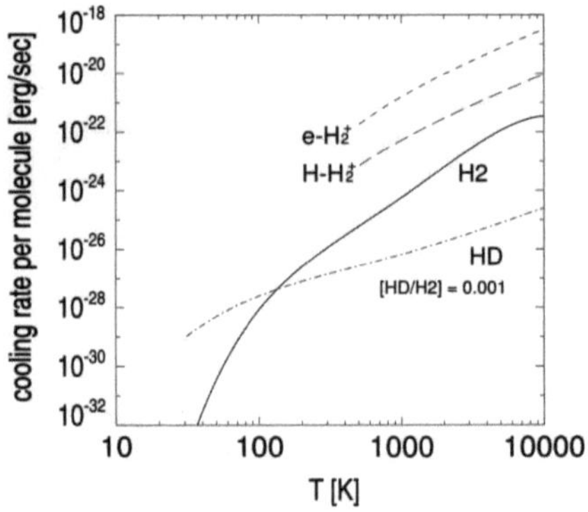

Fig. 4 Molecular cooling rates for H_2 (*solid line*), HD (*dot-dashed line*), and ${H_2}^+$ (*dashed lines*). Owing primarily to the permanent dipole moment of HD, at low temperatures the cooling rate per HD molecule is significantly higher than that per H_2 molecule. Thus, primordial gas enriched in HD is able to cool to much lower temperatures than gas containing solely H_2; indeed, as is shown in Sect. 2.1, primordial gas sufficiently enriched in HD can cool to the lowest temperature attainable by radiative cooling, that of the CMB. While the cooling rates per ${H_2}^+$ molecule (shown here for collisional excitation by free electrons and hydrogen atoms) can be very high, the low abundance of this molecule limits its importance for the thermal evolution of the primordial gas. Finally, we note that at temperatures $\sim 10^4$ K, the primordial gas is cooled primarily by recombination and resonance lines of atomic hydrogen (not shown). From Yoshida et al. (2007a)

While deuterium is less abundant in the primordial gas than hydrogen by a factor of the order of 10^{-5}, the HD molecule is able to cool to temperatures considerably lower than H_2 (e.g. Flower et al. 2000). Firstly, this owes to the fact that HD has a permanent dipole moment, allowing dipole rotational transitions, which spontaneously occur much more often than the quadrupole rotational transitions in H_2. Also, the dipole moment of HD allows transitions between rotational states of $\Delta J = \pm 1$, which are of lower energy than the $\Delta J = \pm 2$ quadrupole transitions of H_2. Thus, collisions with other particles, such as neutral hydrogen, can excite the HD molecule from the ground to the first excited rotational state ($J = 1$), from which it decays back to the ground state by a dipole transition. The photon that is emitted in the process carries away energy and thus cools the gas. Because HD can be excited to the $J = 1$ state by relatively low energy collisions, and because its subsequent radiative decay occurs quickly compared to that of H_2, the cooling rate per molecule is higher for HD than H_2 at temperatures $\lesssim 100$ K, as shown in Fig. 4.

Whereas in cosmological minihalos H_2 cooling alone can cool the gas to $\simeq$200 K, as shown in the left panel of Fig. 3, HD cooling can be so effective as to allow the primordial gas to cool to the lowest temperature that can be achieved via radiative cooling, that of the cosmic microwave background (CMB), $T_{\rm CMB} = 2.7(1+z)$ (e.g. Larson 2005; Johnson and Bromm 2006; Schneider and Omukai 2010). It is useful at this point to derive this fundamental result, as we will draw on the formalism introduced here later as well, in discussing the impact of the first heavy elements on the cooling of the primordial gas (see Sect. 3.3).

To begin, note that the frequency ν_{10} of emitted radiation for the rotational transition $J = 1 \rightarrow 0$ of HD can be expressed as

$$\frac{h\nu_{10}}{k_B} \simeq 130\,\mathrm{K}\,, \tag{9}$$

where k_B is the Boltzmann constant and h is the Planck constant. For clarity, here we shall consider the simple case in which only this transition and its reverse occur. Next, consider a finite parcel of primordial gas with a temperature T_{gas}. For simplicity, we shall assume that the density of the gas is sufficiently high to establish local thermodynamic equilibrium (LTE) level populations according to the Boltzmann distribution[4]

$$\frac{n_1}{n_0} = \frac{g_1}{g_0} e^{-\frac{h\nu_{10}}{k_B T_{gas}}}\,, \tag{10}$$

where n_i is the number density of HD molecules in the ith excited rotational state and g_i is the statistical weight of that state; specifically, here we have $g_1 = 3g_0$. Furthermore, as we are considering only transitions between the ground state and the first excited state, we shall take it that no other rotational levels are occupied. Equivalently, we take it here that $T_{gas} < 130$ K, as otherwise collisions with other particles would be sufficiently energetic to excite the molecule to higher levels. Finally, we make the assumption that $T_{gas} \geq T_{CMB}$. Thus, if we denote the specific intensity of the CMB, which is an almost perfect blackbody, at the frequency ν_{10} as $I_{\nu_{10}}$, then it follows that

$$I_{\nu_{10}} = \frac{2h\nu_{10}^3/c^2}{e^{\frac{h\nu_{10}}{k_B T_{CMB}}} - 1} \leq \frac{2h\nu_{10}^3/c^2}{e^{\frac{h\nu_{10}}{k_B T_{gas}}} - 1}\,. \tag{11}$$

Now, with the Einstein coefficients for spontaneous and stimulated emission from $J = 1 \rightarrow 0$ denoted by A_{10} and B_{10}, respectively, and that for absorption of a photon effecting the transition $J = 0 \rightarrow 1$ by B_{01}, we have the standard relations $B_{10}g_1 = B_{01}g_0$ and

$$\frac{2h\nu_{10}^3}{c^2} = \frac{A_{10}}{B_{10}}\,. \tag{12}$$

Along with Eqs. (10) and (11), these imply that

$$n_0 B_{01} I_{\nu_{10}} < n_1 A_{10} + n_1 B_{10} I_{\nu_{10}}\,. \tag{13}$$

Thus, the gas is cooled, as more energy is emitted into the CMB radiation field than is absorbed from it. The rate at which the temperature drops can be found by first expressing the energy density of the gas as

[4]Due to infrequent particle collisions at low densities, the rate of radiative deexcitations can exceed that of collisional deexcitations, leading to non-LTE level populations (see Sect. 3.3).

$$u_{\text{gas}} = \frac{3}{2} n k_{\text{B}} T_{\text{gas}} \,, \tag{14}$$

where n is the total number density of the gas particles, including all species. With this, Eq. (13) implies that, with no change in the density of the gas,

$$h\nu_{10}[n_0 B_{01} I_{\nu_{10}} - n_1 A_{10} - n_1 B_{10} I_{\nu_{10}}] = \frac{3}{2} n k_{\text{B}} \frac{dT_{\text{gas}}}{dt} \,. \tag{15}$$

Next, we take it that the ratio of the number density of HD molecules n_{HD} to the total number density of particles n in the gas is given by the constant factor

$$X_{\text{HD}} \equiv \frac{n_{\text{HD}}}{n} \simeq \frac{n_0 + n_1}{n} \simeq \frac{n_0}{n} \,. \tag{16}$$

Then using Eqs. (10)–(12) in Eq. (15), and neglecting stimulated emission for simplicity, the thermal evolution of the gas is approximately described by

$$\frac{dT_{\text{gas}}}{dt} \simeq \frac{2h\nu_{10} A_{10} X_{\text{HD}}}{k_{\text{B}}} \left(e^{-\frac{h\nu_{10}}{k_{\text{B}} T_{\text{CMB}}}} - e^{-\frac{h\nu_{10}}{k_{\text{B}} T_{\text{gas}}}} \right) \,. \tag{17}$$

It is clear from this result that if $T_{\text{CMB}} \leq T_{\text{gas}} < 130$ K, with the gas cooling only by radiative decay of the excited rotational state $J = 1$ to $J = 0$, the temperature of the gas will asymptotically approach T_{CMB}. Thus, Eq. (17) describes the fact that the CMB temperature is indeed a lower limit on the temperature to which a gas can cool via line emission only. Using the previous equation, we can estimate the timescale for reaching the CMB temperature floor as

$$\begin{aligned} t_{\text{CMB}} &\simeq \frac{1}{2A_{10} X_{\text{HD}}} \left(\frac{k_{\text{B}} T_{\text{CMB}}}{h\nu_{10}} \right)^2 \exp\left(\frac{h\nu_{10}}{k_{\text{B}} T_{\text{CMB}}} \right) \\ &\simeq (A_{10} X_{\text{HD}})^{-1} \,. \end{aligned} \tag{18}$$

Finally, we may use this timescale to define a critical HD abundance above which the gas may cool to the CMB, by demanding that the gas is able to cool faster than it is heated by compression during its collapse, which takes place roughly on the free-fall timescale. We thus require that $t_{\text{CMB}} \sim t_{\text{ff}}$, where the free-fall time is calculated at the characteristic density $n \sim 10^4$ cm^{-3} at which the primordial gas is found to fragment in cosmological simulations (e.g. Bromm and Larson 2004). With $A_{10} \simeq 5 \times 10^{-8}$ s^{-1} for this transition (e.g. Nakamura and Umemura 2002), we thus find the critical HD abundance to be approximately

$$X_{\text{HD,crit}} \sim 10^{-6} \,. \tag{19}$$

If the abundance of HD is lower than $X_{\text{HD,crit}}$, the gas will not have time to cool to $T_{\text{gas}} \simeq T_{\text{CMB}}$ during its collapse. As shown in Fig. 3, for the case of the primordial gas cooling in the first galaxies a large fraction of the gas at densities $n \geq 10$ cm^{-3} has an HD abundance greater than $X_{\text{HD,crit}}$, whereas in the case of cooling in the minihalos hosting the first stars, the HD fraction.

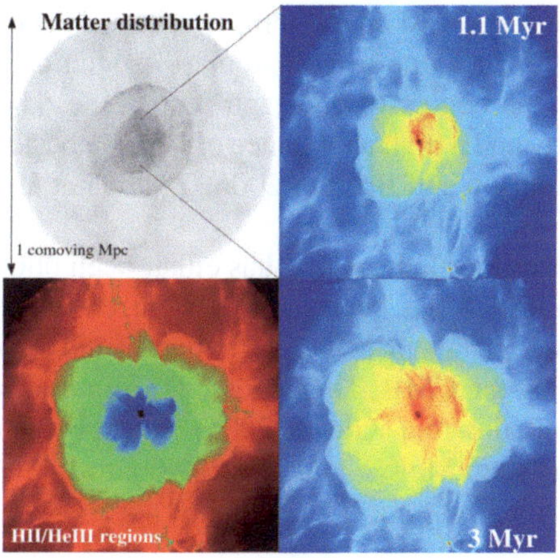

Fig. 5 Expansion of the H II region around a Pop III.1 star formed in a minihalo: the large-scale density distribution at $z = 26$ (*top left*), and the projected gas density at 1.1 (*top right*) and 3 Myr (*bottom right*) after the central star turns on. The bottom left panel shows the extent of the H II region (*green*) and that of the He III region (*blue*) at 3 Myr. When the central star turns off after this time, the strongly ionized primordial gas begins to cool and recombine, with H_2 and HD molecules forming in abundance. This, in turn, enhances the cooling properties of the gas, which may later collapse and form Pop III.2 stars in a more massive halo hosting a first galaxy. From Yoshida et al. (2007a)

A high abundance of HD can, in general, be formed whenever the primordial gas becomes ionized. This occurs through collisional ionization, as in the case of shock heating to temperatures above $\sim 10^4$ K in the virialization of atomic cooling halos, but also occurs when the first stars formed in minihalos emit high energy radiation which photoionizes the gas. As shown in Fig. 5, a massive Pop III star emits enough ionizing radiation to destroy almost all of the neutral hydrogen within a distance of a few physical kiloparsec[5] (e.g. Alvarez et al. 2006; Abel et al. 2007), via the reaction

$$\mathrm{H} + \gamma \rightarrow \mathrm{H}^+ + e^- . \tag{20}$$

Here the products are ionized hydrogen and a free electron, which has a kinetic energy equal to the energy of the ionizing photon minus the ionization potential of hydrogen, 13.6 eV. This free electron is ejected from the atom and shares it kinetic energy with other particles via collisions, thereby heating the gas to higher temperatures. Typically, an equilibrium temperature of $\sim 10^4$ K is established in H II regions, largely set by a balance between the rate at which the gas is photoheated via the above reaction and the rate at which it is cooled by the radiative recombination and resonance emission of hydrogen (e.g. Osterbrock and Ferland 2006). While in

[5]As H^+ is also referred to as H II, such photoionized regions formed around stars are called H II regions. Likewise, the radiation from massive stars, and especially massive Pop III stars, can doubly ionize helium within the so-called He III region (see Sect. 4).

the H II regions around active stars the temperature is thus too high for molecules to form in large abundances due to collisional dissociation, once the central star dies the hot ionized gas begins to cool and recombine. Under these conditions, molecules form rapidly and a high abundance of HD can be achieved (e.g. Nagakura and Omukai 2005; Johnson and Bromm 2006; Yoshida et al. 2007b; McGreer and Bryan 2008).

Therefore, overall, primordial gas that has either been photoionized by a Pop III star in a minihalo or which has been partially ionized during the virialization of an atomic cooling halo, may in principle collapse and cool all the way to the temperature floor set by the CMB. This is distinct from the case of the first Pop III star formation in minihalos (e.g. Glover 2005), and this distinction motivates the following terminology (e.g. McKee and Tan 2008; Greif et al. 2008; Bromm et al. 2009)[6]:

Pop III.1 The first generation of primordial stars formed in minihalos and not significantly affected by previous star formation.

Pop III.2 Primordial stars formed under the influence of a previous generation of stars, either by the ionizing or photodissociating radiation which they emit.

Based on the enhanced cooling of the gas due to high H_2 and HD fractions, it is expected that the typical mass scale of Pop III.2 stars is significantly lower than that of Pop III.1 stars (e.g. Uehara and Inutsuka 2000; Nakamura and Umemura 2002; Mackey et al. 2003; Machida et al. 2005; Nagakura and Omukai 2005; Johnson and Bromm 2006; Ripamonti 2007; Yoshida et al. 2007b). Following Eq. (4), the Jeans mass for gas that cools to the temperature of the CMB, which sets a rough upper limit for the mass of Pop III.2 stars formed from partially ionized primordial gas, is

$$M_{\rm J} \simeq 35 \left(\frac{1+z}{10}\right)^{\frac{3}{2}} \left(\frac{n}{10^4{\rm cm}^{-3}}\right)^{-\frac{1}{2}} {\rm M}_{\odot}\,, \tag{21}$$

where we have normalized to the same characteristic density at which the primordial gas fragments in Pop III.1 star formation (e.g. Bromm and Larson 2004). At this fixed density, the Jeans mass is roughly an order of magnitude lower than expected for the case of Pop III.1 star formation in minihalos. Also due to the lower temperature of the gas in the Pop III.2 case, the rate of accretion onto a protostar is similarly lower (Yoshida et al. 2007b):

$$\dot{M}_{\rm acc} \simeq 5 \times 10^{-5} \left(\frac{1+z}{10}\right)^{\frac{3}{2}} {\rm M}_{\odot}\ {\rm year}^{-1}\,, \tag{22}$$

compared to $\dot{M}_{\rm acc} \sim 10^{-3}\ {\rm M}_{\odot}\ {\rm year}^{-1}$ for the case of Pop III.1 stars (see Eq. (5)).

[6]Before the adoption of this terminology, Pop III.2 was formerly referred to as Pop II.5 in the literature (e.g. Mackey et al. 2003; Johnson and Bromm 2006).

While enhanced molecule abundances are likely to result in lower characteristic stellar masses, other mitigating effects also come into play in the formation of second generation primordial stars. One factor which likely becomes important for shaping the stellar IMF, particularly in atomic cooling halos (e.g. Wise and Abel 2007b; Greif et al. 2008), is the development of supersonic turbulence (see e.g. Mac Low and Klessen 2004; Clark et al. 2011b). Also, the degree to which the abundances of H_2 and HD can be raised in the first galaxies is dependent on the strength of the molecule-dissociating radiation field generated by the first generations of stars (e.g. Wolcott-Green and Haiman 2011). In the next Section, we shall see that an elevated radiation field may not only result in higher Pop III star masses, but may also result in the formation of the seeds of the first supermassive black holes.

2.2 *Suppression of Cooling by the Photodissociation of Molecules*

The assembly of the first galaxies becomes much more complex with the formation of the first stars, in part because they emit high energy radiation that alters the primordial gas in dramatic ways (e.g. Ciardi and Ferrara 2005). As in the case of primordial gas in the minihalos in which the first stars form, in the first galaxies one of the primary cooling processes is the emission of radiation from molecular hydrogen, which can be easily destroyed by the high energy radiation emitted by the first stars can easily destroy these molecules (e.g. Haiman et al. 1997; Omukai and Nishi 1999; Ciardi et al. 2000; Glover and Brand 2001; Ricotti et al. 2001). So called Lyman–Werner (LW) photons, with energies $11.2\ \mathrm{eV} \leq h\nu \leq 13.6\ \mathrm{eV}$ excite H_2, leading in turn to its dissociation into atomic hydrogen[7] (Stecher and Williams 1967):

$$H_2 + \gamma \rightarrow H_2^* \rightarrow 2H\,. \tag{23}$$

With the destruction of H_2 molecules, the primordial gas cools less rapidly and this signals a change in the rate at which gas can collapse into minihalos and form Pop III stars. An estimate of the minimum LW radiation field necessary to significantly delay star formation in a minihalo can be found by comparing the timescale $t_{\rm form}$ for the formation of H_2 to the timescale for its photodissociation. For a general radiation field the photodissociation time can be expressed as $t_{\rm diss} \simeq 3 \times 10^4\ {J_{21}}^{-1}$ year, where the specific intensity $J_{\rm LW}$ of the LW radiation field is defined as $J_{\rm LW} = J_{21} \times 10^{-21}\ \mathrm{erg\ s^{-1}\ cm^{-2}\ Hz^{-1}\ sr^{-1}}$ (e.g. Abel et al. 1997); here, J_{21} is a dimensionless parameter normalized to a typical level of the radiation field.

[7]While for simplicity we limit our discussion to the photodissociation of H_2, HD molecules are also destroyed via this mechanism.

To estimate the formation time we note that, as shown in the left panel of Fig. 3, primordial gas collapsing into a minihalo is roughly adiabatic until its density rises to roughly $n \simeq 1\ \mathrm{cm}^{-3}$, at which point its temperature is $T \simeq 10^3$ K. Therefore, it is only at this characteristic density and higher that H_2 is effective at cooling the gas, in turn leading to gravitational collapse and the formation of stars. It is the formation time of H_2 in these conditions, which is $t_{\mathrm{form}} \simeq 10^6$ year, that is to be compared to the photodissociation timescale t_{diss}. Equating these two timescales, we find a critical LW radiation field intensity of the order of $J_{21} \simeq 10^{-2}$, at which the suppression of H_2 formation and cooling slows the process of Pop III star formation in minihalos (see Kitayama et al. 2001; Yoshida et al. 2003; Mesinger et al. 2006; Wise and Abel 2007a; Johnson et al. 2008; Trenti and Stiavelli 2009).

Figure 6 shows the results of cosmological simulations of the collapse of primordial gas into minihalos, under the influence of different levels of a constant LW radiation field. As the panels in the figure show, for higher J_{LW} the primordial gas in a given minihalo collapses to form stars at lower redshift z_{coll}, when the halo has grown to a higher mass M_{vir} and has a higher virial temperature T_{vir}. The results of these simulations corroborate our estimate of the critical LW background, as T_{vir} and M_{vir} increase most dramatically at $J_{21} \simeq 0.04$.

Because the mean free path of LW photons is generally large, up to ~10 physical Mpc, a roughly uniform background field is quickly established when the first stars begin emitting radiation (e.g. Haiman et al. 1997). We can estimate the level of the H_2-dissociating background radiation, as a function of the cosmological average star formation rate $\dot{\rho_*}$ per unit comoving volume, by assuming that massive stars which live for a time t_* produce the LW flux and that η_{LW} LW photons are produced for each baryon in stars (see Greif and Bromm 2006). We then obtain for the number density n_γ of H_2-dissociating photons

$$n_\gamma \simeq \eta_{\mathrm{LW}} \frac{\dot{\rho_*} t_* X_{\mathrm{H}}}{m_{\mathrm{H}}} (1+z)^3 \ , \tag{24}$$

where m_{H} is the mass of the hydrogen atom, $X_{\mathrm{H}} \simeq 0.76$ is the fraction of baryonic mass in hydrogen, and the mass density in stars is $\simeq \dot{\rho_*} t_*$. Converting this to the photon energy density $u_\gamma = h\nu n_\gamma$, we obtain an estimate of J_{LW} as a function of the star formation rate per comoving volume:

$$J_{\mathrm{LW}} \simeq \frac{u_\gamma c}{4\pi\nu} = \frac{hc}{4\pi} \eta_{\mathrm{LW}} \frac{\dot{\rho_*} t_* X_{\mathrm{H}}}{m_{\mathrm{H}}} (1+z)^3 \ , \tag{25}$$

where c is the speed of light. In terms of J_{21}, this is

$$J_{21} \simeq 0.2 \left(\frac{\eta_{\mathrm{LW}}}{10^4}\right) \left(\frac{\dot{\rho_*}}{10^{-3} \mathrm{M_\odot year^{-1} Mpc^{-3}}}\right) \left(\frac{1+z}{10}\right)^3 \tag{26}$$

where we have assumed an average lifetime $t_* = 5 \times 10^6$ year for stars that produce the bulk of H_2-dissociating radiation (e.g. Leitherer et al. 1999; Schaerer 2002). For a population of metal-enriched stars formed with a Salpeter-like IMF, as is inferred for the Milky Way today, $\eta_{\mathrm{LW}} \simeq 4 \times 10^3$; however, for metal-free stellar population

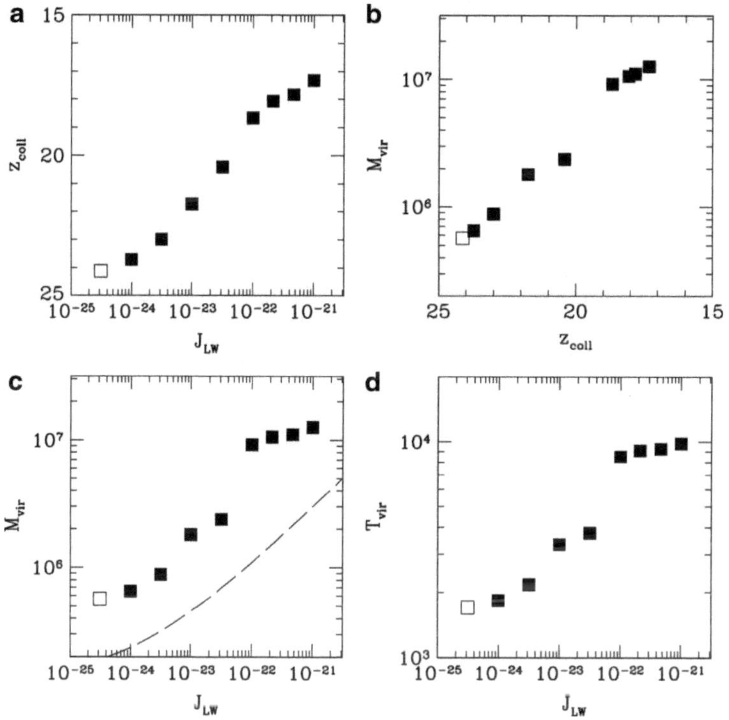

Fig. 6 The properties of simulated cosmological dark matter halos when the primordial gas first collapses to form Pop III stars, under the influence of various levels of the background Lyman–Werner photodissociating flux J_{LW} (shown here in units of erg s^{-1} cm^{-2} Hz^{-1} sr^{-1}): star formation redshift z_{coll} vs. J_{LW}, M_h vs. z_{coll}, T_{vir} vs. J_{LW}, and M_h vs. J_{LW} (*clockwise from top left panel*). As the intensity of the background H_2-dissociating flux increases, a given halo must grow to a larger mass and have a higher virial temperature before the primordial gas can cool, collapse and form stars. At a flux of $J_{LW} \simeq 4 \times 10^{-23}$ erg s^{-1} cm^{-2} Hz^{-1} sr^{-1} (i.e. $J_{21} \simeq 0.04$), there is a steep increase in the T_{vir} and M_h of star-forming halos due to the rate of H_2 photodissociation becoming comparable to the rate of H_2 formation. From O'Shea and Norman (2008)

with a top-heavy IMF, this can be as high as $\eta_{LW} \simeq 2 \times 10^4$ (see e.g. Greif and Bromm 2006). For simplicity, equation (26) is normalized to an intermediate value. Furthermore, while the star formation rate at very high redshift is not known, we have here normalized to a rough value expected in the standard ΛCDM picture of cosmological structure formation at $z \simeq 10$ (see e.g. Tornatore et al. 2007; Haiman 2009; Trenti and Stiavelli 2009).

A further estimate of the cosmological background J_{LW}, in particular that near the end of the epoch of reionization, can be found by assuming that the flux just above the Lyman limit (i.e. at $h\nu \geq 13.6$ eV) is sufficient to reionize the universe (see e.g. Bromm and Loeb 2003a; Shang et al. 2010) and that the sources producing the ionizing flux also produce a comparable flux in the LW energy range, 11.2–13.6 eV.

Relating the number density of hydrogen nuclei at redshift z to the number density n_γ of ionizing photons required to keep hydrogen photoionized in the IGM, we obtain

$$n_\gamma \simeq N_\gamma \frac{\Omega_b \rho_{crit} X_H}{m_H} (1+z)^3 , \tag{27}$$

where N_γ is the number of ionizing photons per hydrogen nucleus required to keep the universe reionized and $\Omega_b \rho_{crit}$ is the cosmological average mass density of baryons at $z = 0$, expressed as a fraction Ω_b of the critical density ρ_{crit} for a flat universe. Assuming that all LW photons which are emitted from sources within galaxies escape into the IGM, and taking it that only a fraction f_{esc} of ionizing photons are able to escape due to the higher optical depth to photoionization, we find an estimate of the background flux as

$$J_{LW} \simeq \frac{1}{f_{esc}} \frac{hc}{4\pi} \frac{N_\gamma \Omega_b \rho_{crit} X_H}{m_H} (1+z)^3 , \tag{28}$$

where again we have converted from photon energy density $u_\gamma = h\nu n_\gamma$ to units of specific intensity as in Eq. (25). Expressing this in terms of J_{21}, we have

$$J_{21} \simeq 400 \left(\frac{N_\gamma}{10}\right) \left(\frac{f_{esc}}{0.1}\right)^{-1} \left(\frac{1+z}{10}\right)^3 , \tag{29}$$

where we have normalized N_γ to the value estimated by Wyithe and Loeb (2003), and f_{esc} is normalized to a typical value found in cosmological radiative transfer simulations (e.g. Ricotti and Shull 2000; Ciardi and Ferrara 2005; Wise and Cen 2009; Razoumov and Sommer-Larsen 2010; Yajima et al. 2011).

This estimated level of the cosmological background radiation field during reionization is well above the critical level of $J_{21} \simeq 0.04$ required for suppressing the rate of Pop III star formation in minihalos, and this may have important implications for the nature of the stars that are formed. In particular, under the influence of such an elevated LW background, due to the destruction of the H_2 molecules which cool the gas, the temperature of the primordial gas when it finally collapses to form a star can be considerably higher than in the absence of a background H_2-dissociating radiation field (O'Shea and Norman 2008). This, in turn, results in a higher Jeans mass and protostellar accretion rate, likely leading to more massive Pop III stars forming in the presence of a high LW background flux.

While the LW radiation field is in general relatively uniform, near individual galaxies it can be locally higher than the cosmological average (see Dijkstra et al. 2008; Ahn et al. 2009), as shown in Fig. 7. In rare regions where the LW background radiation is exceptionally high, a different outcome besides Pop III star formation in dark matter halos may result: the formation of a black hole by direct collapse (e.g. Bromm and Loeb 2003a). For this to occur, the LW radiation field must be at a level high enough to destroy molecules not just in the outskirts of halos where the primordial gas begins to cool via emission from H_2 molecules, but also high enough to destroy H_2 even in the central dense regions of the halo (but see Begelman and Shlosman 2009; Mayer et al. 2010).

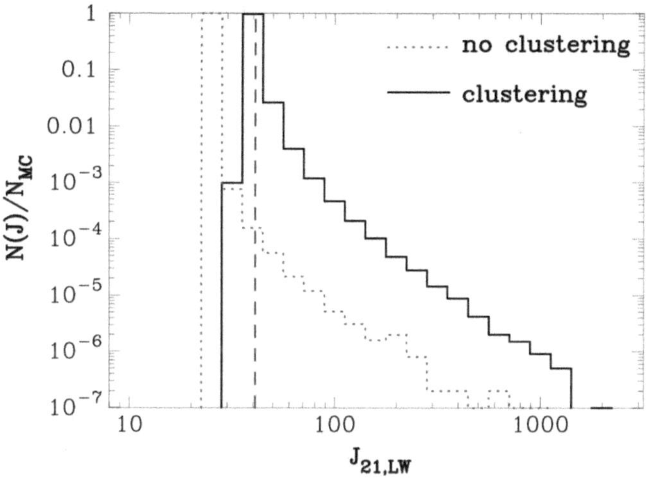

Fig. 7 The probability distribution of the Lyman–Werner flux J_{LW} to which atomic cooling halos are exposed at $z = 10$, for a model accounting for the clustering of the galaxies which emit LW radiation (*black solid histogram*) and for one in which galaxies are assumed to be distributed uniformly (*red dotted histogram*). The mean value of the LW background found in these models is $J_{21} \simeq 40$ (*dashed blue line*), which is well above the level needed to suppress the rate of Pop III star formation in minihalos (see Fig. 6). While rare, a small fraction of halos are exposed to a LW flux $J_{21} \geq 100$, high enough to completely suppress star formation in atomic cooling halos until a $> 10^4$ $M_\odot$ black hole forms by direct collapse instead (see Shang et al. 2010). From Dijkstra et al. (2008)

Figure 8 shows the results of cosmological simulations from which the minimum J_{21} required for the formation of a black hole by direct collapse can be estimated. As shown in the bottom-right panel, for $J_{21} \geq 100$ the H_2 fraction in the gas is kept to a low level at which H_2 cooling does not lower the temperature of the gas significantly below the virial temperature of $T_{vir} \simeq 10^4$ K of the halo (Shang et al. 2010).[8] Therefore, when the gas finally collapses, the accretion rate of primordial gas will be very high, of the order of $\simeq 0.1$ $M_\odot$ year^{-1}, as can be seen from Eq. (5). This is roughly two orders of magnitude higher than the accretion rate onto Pop III protostars formed in H_2-cooled gas at $T \simeq 200$ K, and the result is predicted to be an extremely massive "quasi-star" which quickly collapses to form a black hole with a mass $\geq 10^4$ $M_\odot$ (e.g. Bromm and Loeb 2003a; Koushiappas et al. 2004; Begelman et al. 2006; Spaans and Silk 2006; Lodato and Natarajan 2006; Regan and Haehnelt 2009). While this level of the background LW radiation field is expected to be higher

[8]It is important to note that the spectrum of the radiation producing the LW background must also be taken into account. While the results shown in Fig. 8 are derived under the assumption that the LW background is generated by stars with an effective surface temperature of 10^4 K, appropriate for Pop II stars, higher levels of the LW flux are required to suppress H_2 formation if, for instance, it is generated by massive Pop III stars with effective surface temperatures of $\simeq 10^5$ K (see e.g. Shang et al. 2010).

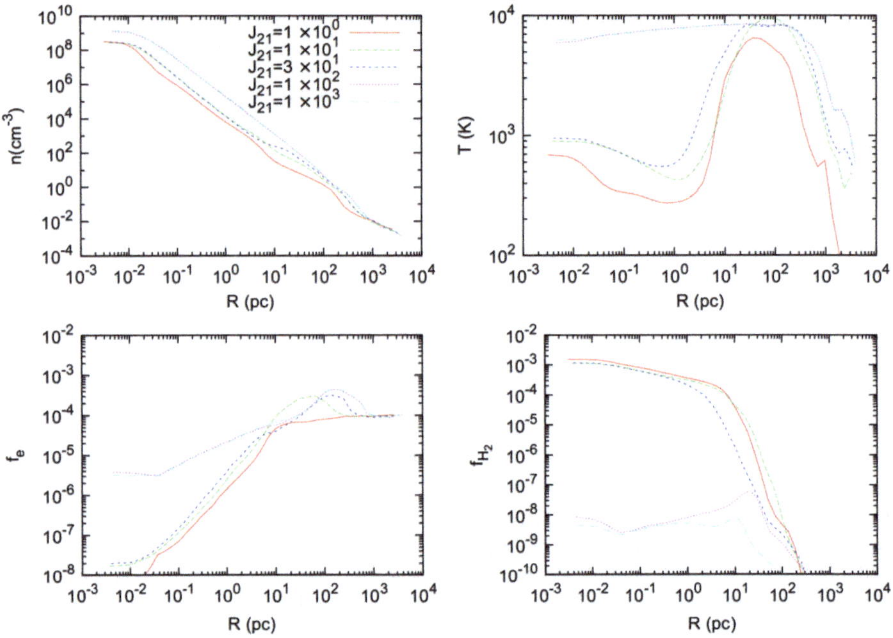

Fig. 8 Properties of the primordial gas in a cosmological simulation, for various elevated levels of the LW background radiation field $J_{\rm LW}$, here assumed to be produced by stars with an effective surface temperature 10^4 K, as is appropriate for metal-enriched Pop II stars. Clockwise from top left: number density, temperature, H_2 fraction, and free electron fraction, each shown as a function of the distance from the center of an atomic cooling halo. A LW flux with $J_{21} \geq 100$ can prevent the formation of a sufficient fraction of H_2 to cool the gas below $\simeq 6 \times 10^3$ K, even at very high densities. Due to the large Jeans mass (4) and accretion rate (5) this implies, instead of hosting Pop III star formation, atomic cooling halos which are exposed to such an intense LW flux are likely to host the formation of a $\geq 10^4$ $M_\odot$ black hole by direct collapse. From Shang et al. (2010)

than the average, as shown in Fig. 7, due to the clustering of the stars and galaxies producing LW radiation there may be a significant number density of black holes formed by direct collapse in the early universe. Indeed, some of these may be the seeds of the supermassive black holes observed at $z \leq 6$ (see e.g. Haiman 2009).

2.3 *The Impact of Radiation from Accreting Black Holes on the Primordial Gas*

In addition to the radiation emitted by the first generations of stars, black holes formed and assembled into the first galaxies can also produce radiation which dramatically impacts the primordial gas. In particular, the effects of the radiation emitted from black holes formed by direct collapse can be especially strong, as there is an initially large reservoir of gas that can be accreted onto the nascent black hole (see Johnson et al. 2011). To draw a comparison between the radiation emitted from stars in the first galaxies to that emitted during the accretion of gas onto black holes, we can calculate the temperature of the accretion disk and compare it to the typical effective temperature of a star. For a steady accretion flow, the temperature T of the accretion disk can be estimated by balancing the rate at which the disk is heated with the rate at which it cools. The heating is due to the gravitational potential energy of matter falling through the disk being dissipated by viscosity; for material falling through the disk and onto the black hole at a rate $\dot{M}_{\rm BH}$, the resultant heating rate per unit area of the disk, Λ, can be estimated on dimensional grounds as (e.g. Pringle 1981)

$$\Gamma \simeq \frac{GM_{\rm BH}\dot{M}_{\rm BH}}{r^3} \,. \tag{30}$$

Assuming the disk is optically thick, then the rate at which the disk cools per unit area, Λ, can be estimated using the Stefan–Boltzmann law: $\Lambda = \sigma_{\rm SB} T^4$, where $\sigma_{\rm SB}$ the Stefan–Boltzmann constant. Equating these rates yields a temperature profile for the disk. The profile thus obtained is very close to the following formal solution, but for a correction near the inner edge of the disk $r_{\rm in}$ where the viscous heating rate goes to zero:

$$\begin{aligned} T(r) &= \left(\frac{3}{8\pi}\frac{GM_{\rm BH}\dot{M}_{\rm BH}}{\sigma_{\rm SB} r^3}\right)^{\frac{1}{4}} \left[1-\left(\frac{r_{\rm in}}{r}\right)^{-\frac{1}{2}}\right] \\ &\simeq 10^6{\rm K}\left(\frac{M_{\rm BH}}{10^4{\rm M}_\odot}\right)^{-\frac{1}{4}}\left(\frac{r}{10r_{\rm s}}\right)^{-\frac{3}{4}} \,. \end{aligned} \tag{31}$$

In the second part of the equation we have normalized to a black hole mass of 10^4 $\rm M_\odot$, appropriate for the initial mass of a black hole formed by direct collapse. We have also normalized the radius to 10 Scharzschild radii $r_{\rm s} = 2GM_{\rm BH}/c^2$, which is well outside the inner edge of the accretion disk, $r_{\rm in} \leq 3r_{\rm s}$. Finally, we have assumed accretion to take place at the Eddington rate $\dot{M}_{\rm Edd}$, at which the outward force due to electron scattering of the emitted radiation balances the inward gravitational force acting on the accreting gas:

$$\dot{M}_{\rm Edd} = \frac{4\pi GM_{\rm BH}m_{\rm H}}{\epsilon c\sigma_{\rm T}} = 2\times10^{-5}\left(\frac{\epsilon}{0.1}\right)^{-1}\left(\frac{M_{\rm BH}}{10^4{\rm M}_\odot}\right){\rm M}_\odot\ {\rm year}^{-1} \,. \tag{32}$$

Here $\sigma_{\rm T} = 6.65\times10^{-25}$ cm^2 is the Thomson cross section for the scattering of photons off electrons, and ϵ is the ratio of the radiated energy to the rest mass energy of the accreting material, normalized to a value appropriate for a slowly rotating black hole.

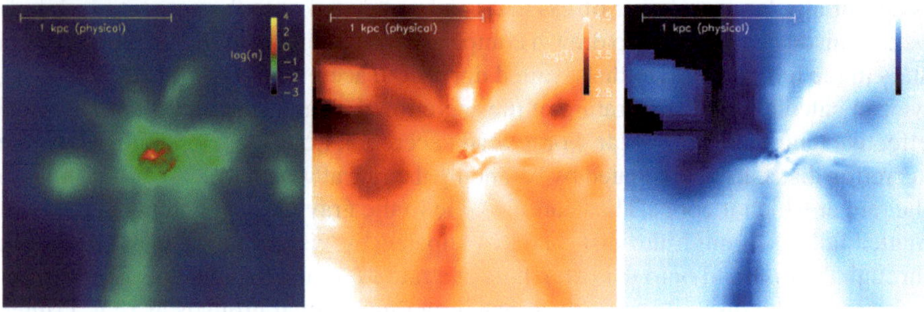

Fig. 9 The projected number density (*left panel*), temperature (*middle panel*), and H II fraction (*right panel*) of the gas in the vicinity of an accreting 2.5×10^4 $M_\odot$ black hole formed by direct collapse in an atomic cooling halo at $z \sim 12$. The ionizing radiation emitted from the accretion disk of the black hole has a strong impact on the gas in the halo, heating it to temperatures $\simeq 3 \times 10^4$ K and causing expansion and disruption of the dense gas in the center of the halo from which the black hole feeds. From Johnson et al. (2011)

As shown in Eq. (31), the temperature near the inner edge of the accretion disc of a rapidly accreting black hole can be as high as $T \simeq 10^7$ K. This is much higher than the effective temperature of even a very massive Pop III star, which is roughly two orders of magnitude lower. In turn, this implies that accreting black holes in the first galaxies emit both copious ionizing radiation and substantial LW radiation, as well as high energy X-rays (see e.g. Ricotti and Ostriker 2004; Kuhlen and Madau 2005). In the case of a black hole formed by direct collapse, the resultant photoheating of the gas in the host atomic cooling halo can drive its temperature to $\simeq 3 \times 10^4$ K, as shown in Fig. 9 (Johnson et al. 2011). Along with the associated high radiation pressure, this results in the expansion of the gas surrounding the black hole. The resultant drop in the density of the accreting gas translates into a decrease in the accretion rate of the black hole, which can be estimated by assuming gas which is gravitationally bound to the black hole falls towards it at the sound speed (see e.g. Bondi 1952). With the radius within which gas is bound to the black hole given by $r_B = 2GM_{BH}/(c_s^2 + v_{BH}^2)$, where v_{BH} is the velocity of the black hole relative to the gas, the accretion rate is estimated as the rate at which mass passes within a distance r_B of the black hole:

$$\dot{M}_{BH} \simeq \pi r_B^2 \mu m_H n \left(v_{BH}^2 + c_s^2\right)^{\frac{1}{2}} = \frac{4\pi G^2 M_{BH}^2 \mu m_H n}{(v_{BH}^2 + c_s^2)^{\frac{3}{2}}}$$
$$= 4 \times 10^{-6} \left(\frac{M_{BH}}{10^4 M_\odot}\right)^2 \left(\frac{\mu}{0.6}\right)^{\frac{5}{2}} \left(\frac{n}{10^2 \mathrm{cm}^{-3}}\right) \left(\frac{T}{10^4 \mathrm{K}}\right)^{-\frac{3}{2}} M_\odot \ \mathrm{year}^{-1}. \quad (33)$$

In the second part of the equation we have assumed a black hole at rest with respect to the gas ($v_{BH} = 0$) and we have again related the gas temperature to the sound speed using $3k_B T/2 = \mu m_H c_s^2/2$. As the accretion rate is directly proportional to the density of the accreting gas and inversely proportional to its temperature, that the high energy radiation emitted from the accretion disk acts to heat and rarify the gas means that the accretion rate itself is regulated by the radiation generated in the process. Indeed, the Eddington rate given by Eq. (32) provides an estimate of the maximum rate at which gas can be accreted in the face of the intense radiation that is emitted. However, hydrodynamics calculations of accretion onto black holes formed in the first galaxies suggest that $\dot{M}_{BH}$ is on average well below the Eddington rate because of both strong radiative feedback during accretion (see e.g. Pelupessy et al. 2007; Alvarez et al. 2009; Milosavljević et al. 2009; Park and Ricotti 2011; Johnson et al. 2011) and low gas densities (e.g. Yoshida 2006; Johnson and Bromm 2007). This poses a challenge for the rapid growth of black holes in the early universe.

A further challenge to the model of black hole formation by direct collapse is the enrichment of the primordial gas with the first heavy elements (e.g. Omukai et al. 2008; Safranek-Shrader et al. 2010), which can easily cool the gas more efficiently than either H_2 or HD molecules. We turn next to the broader question of how the first supernovae, which enrich the gas, transform the process of star formation in the first galaxies.

3 Metal Enrichment and the Onset of Population II Star Formation

We have seen that the characteristic mass of objects that form from the runaway gravitational collapse of gas, stars and in extreme cases black holes, depends critically on the temperature of the collapsing gas. The hotter the gas, the larger the Jeans mass and the higher the rate at which gas accretes onto the collapsed object. Therefore, the injection of heavy elements by the first supernovae represents a fundamental transition in star formation, in that new coolants are added to the primordial gas. As a result, the characteristic mass of stars formed from the first metal-enriched gas is likely to be lower than the characteristic mass of primordial stars. Here, we investigate the transition between these two modes of star formation.

3.1 The First Supernovae and Metal Enrichment

It is one of the hallmark predictions of modern cosmology that the first heavy elements, such as carbon, oxygen, and iron, are produced in the cores of stars and in supernovae, rather than in the Big Bang (e.g. Burbidge et al. 1957). Thus, when the first stars explode as supernovae, first metals forged in their cores are violently ejected into the primordial gas. In this, the first supernovae introduce not just new chemical elements, but also tremendous amounts of mechanical energy that disrupt their environments. Indeed, as given in Sect. 1, the definition that we have chosen for the first galaxies pertains to this: in the first galaxies, formed in haloes with virial temperatures $T_{vir} \simeq 10^4$ K, the gas can not be completely expelled by a single

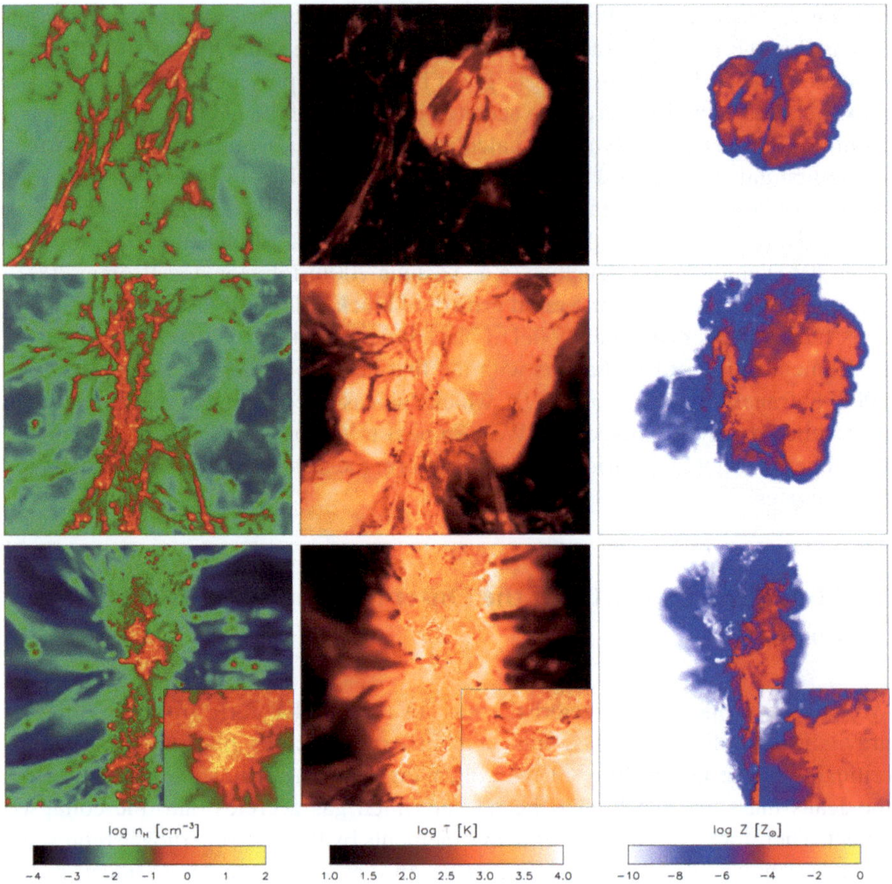

Fig. 10 The evolution of the remnant of a powerful Pop III.1 supernova, exploding in a minihalo at $z \sim 20$. *From left to right*: the number density, temperature, and metallicity of the gas along the line of sight. *From top to bottom*: a time series showing the simulation 15, 100, and 300 Myr after the SN explosion. Each panel is 100 comoving kpc on a side, while the inlays show the central 10 comoving kpc. The metals are initially distributed by the bulk motion of the SN remnant, and later by turbulent motions induced by photoheating from other stars and the re-collapse of the shocked gas. As can be seen in the inlays, the gas collapses to high densities once again in an atomic cooling halo in which a first galaxy later forms from the enriched gas. From Greif et al. (2010)

powerful supernova, as is the case in the minihalos hosting the first stars (see e.g. Bromm et al. 2003; Kitayama and Yoshida 2005; Greif et al. 2007; Whalen et al. 2008).

The effects of a powerful Pop III.1 supernova on the primordial gas are shown in Fig. 10, as gleaned from the cosmological simulation presented in Greif et al. (2010). Consistent with the results presented in Fig. 1, the gas within the minihalo hosting the progenitor Pop III star is completely blown out into the surrounding IGM. There the primordial gas is shock-heated to several thousand Kelvin and enriched to metallicities of up to $\sim 10^{-3}$ $Z_\odot$ within of the order of 10^7 year. The evolution of the supernova remnant can be well described analytically, as it passes through the four distinct phases of an explosion with energy $E_{SN} = 10^{52}$ erg in a medium with particle number density $n \leq 1$ cm^{-3}, as expected for a Pop III.1 progenitor star with a mass of the order of 100 $M_\odot$ (e.g. Fryer et al. 2001; Heger and Woosley 2002; Whalen et al. 2008). At first, the blast wave from the supernova propagates outwards at a roughly constant velocity v_{sh}; in this, the so-called free expansion phase, the distance r_{sh} which the shock has traveled from the site of the explosion by time t_{sh} is given simply by

$$r_{sh} \simeq v_{sh} t_{sh} \simeq \left(\frac{2E_{SN}}{M_{ej}}\right)^{\frac{1}{2}} t_{sh} \simeq 3 \left(\frac{E_{SN}}{10^{52}\mathrm{erg}}\right)^{\frac{1}{2}} \left(\frac{M_{ej}}{100\mathrm{M}_\odot}\right)^{-\frac{1}{2}} \left(\frac{t_{sh}}{10^3\mathrm{year}}\right) \mathrm{pc}\,. \tag{34}$$

At this stage all of the energy of the supernova is in the kinetic energy of the ejecta, which has an initial mass M_{ej}. When the shock has swept up an amount of mass comparable to the original ejecta mass, the shock enters the so-called Sedov-Taylor phase in which the energy of the blast wave is conserved while an increasing amount of mass M_{sw} is swept up by the shock. In this phase we therefore have $v_{sh} = dr_{sh}/dt_{sh} \simeq (2E_{SN}/M_{sw})^{1/2}$, which yields for the shock radius

$$r_{sh} \simeq 24 \left(\frac{E_{SN}}{10^{52}\mathrm{erg}}\right)^{\frac{1}{5}} \left(\frac{n}{1\mathrm{cm}^{-3}}\right)^{-\frac{1}{5}} \left(\frac{t_{sh}}{10^3\mathrm{year}}\right)^{\frac{2}{5}} \mathrm{pc}\,, \tag{35}$$

where we have used $M_{sw} = 4\pi/3{r_{sh}}^3 \mu m_H n$, with $\mu = 0.6$, which is appropriate for an ionized primordial gas. The transition between the free expansion and Sedov-Taylor phase is evident in Fig. 11, which charts the propagation of the blast wave of a powerful 10^{52} erg primordial supernova in a simulated cosmological minihalo, similar to that shown in Fig. 10.

The third phase, also shown in Fig. 11, sets in when a substantial fraction of the original energy in the blast wave has been radiated away, principally by recombination and resonance line cooling of the hydrogen and helium composing the primordial gas (Greif et al. 2007; Whalen et al. 2008), but also to some extent by bremsstrahlung and inverse Compton scattering of the CMB by free electrons, the latter being most important at high redshift due to the steep increase of the energy density of the CMB with redshift (e.g. Oh 2001). Known as the pressure-driven snowplow phase, at this stage the high pressure gas behind the blast wave powers its expansion, and the equation of motion thus becomes

$$\frac{d(M_{sw}v_{sh})}{dt_{sh}} = 4\pi r_{sh}^2 P_b\,, \tag{36}$$

where P_b is the pressure in the hot bubble interior to the blast wave. As discussed in Greif et al. (2007), at the radius where the transition to the snowplow phase begins, the density profile of the gas is close to that of an isothermal gas, $n \propto r_{sh}^{-2}$; within this radius, the density profile is much flatter due to the strong photoheating of the gas by the progenitor star (e.g. Kitayama and Yoshida 2005; and Whalen et al. 2008).

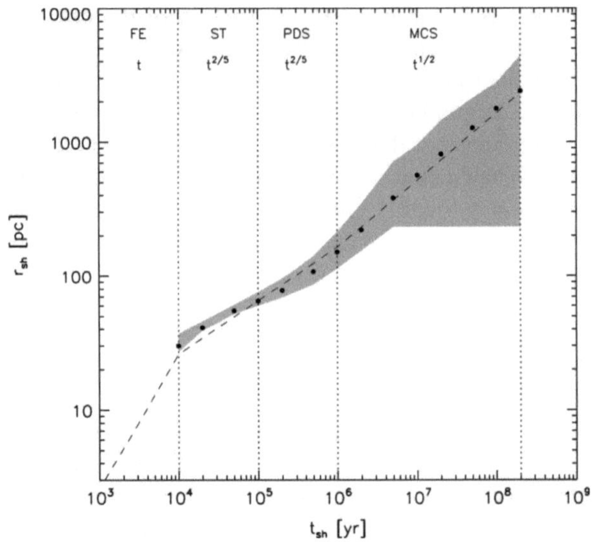

Fig. 11 The evolution of the remnant of a 10^{52} erg Pop III.1 supernova in a cosmological minihalo. The analytical solution (*dashed lines*) discussed in Sect. 3.1 accurately describes the expansion of the blast wave, the mass-weighted extent of which is shown by the *black dots*. As the results shown here are from a cosmological simulation, the inhomogeneous density field into which the shock propagates leads to some dispersion in the distance that it extends in different directions (*gray region*). As shown in Fig. 10 for a similar cosmological simulation, after of the order of 10^8 year the supernova shock stalls and the metal-enriched gas recollapses into the growing halo which hosted the progenitor star. From Greif et al. (2007)

Therefore, as the pressurized bubble expands adiabatically, in the snowplow phase we have $M_{sw} \propto r_{sh}$ and $P_b \propto r_{sh}^{-5}$. This, in turn, allows a solution to the equation of motion with $r_{sh} \propto t_{sh}^{2/5}$, just as in the previous Sedov-Taylor phase.

The final transition occurs when the bubble behind the blast wave has cooled and the pressure behind the shock no longer affects it dynamically. While by this time a large fraction of the energy of the supernova has been radiated away, the momentum that has accumulated in the dense shell of gas that forms behind the shock is conserved. As the density profile of the ambient gas is still $n \propto r^{-2}$ at this point, the conservation of momentum implies that the quantity $M_{sw}v_{sh} \propto r_{sh}\, dr_{sh}/dt_{sh}$ is a constant. Thus, in this final phase of the supernova remnant $r_{sh} \propto t_{sh}^{1/2}$, as shown in Fig. 11.

The explosion of a Pop III.1 star with a mass of $\sim$200 $M_\odot$ as is shown Figs. 10 and 11, is expected to release up to 10^{53} erg as well as all of the up to $\sim$100 $M_\odot$ in metals produced in the core of the star (e.g. Heger and Woosley 2002; Heger et al. 2003; Karlsson et al. 2012). The metal-enriched gas that is ejected into the IGM by the supernova explosion expands preferentially into low density regions, as shown in the middle row of panels in Fig. 10. This can be seen more explicitly in Fig. 12, which shows the metallicity distribution of the gas enriched by a similar primordial supernova as a function of density and temperature (Wise and Abel 2008).

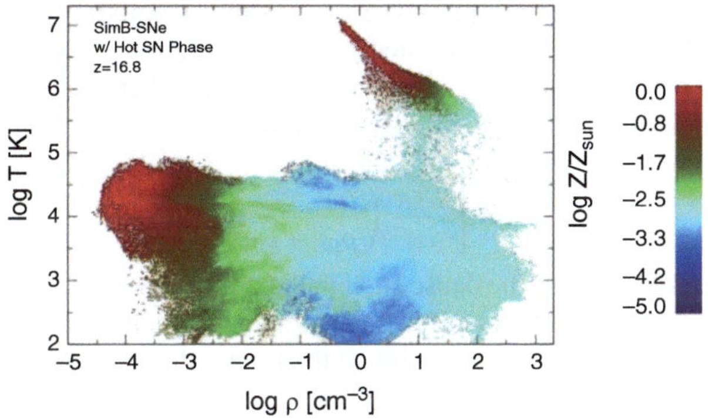

Fig. 12 The metallicity of the gas enriched by the violent expulsion of metals in a powerful Pop III.1 supernova, similar to that shown in Figs. 10 and 11, as a function of the density and temperature of the gas. The highest metallicity gas is in the low density regions into which the blast wave propagates most rapidly. As also shown in Fig. 10, the metallicity of the higher density gas re-collapsing into the halo hosting the progenitor star is $Z \sim 10^{-3}\ Z_{\odot}$. From Wise and Abel (2008)

The dark matter in the halo hosting the progenitor star is not nearly so violently disrupted as is the gas swept up in the blast wave, and in fact the host halo continues growing until its gravity is strong enough for the cooling, metal-enriched gas to collapse into it again. As shown in the bottom panels in Fig. 10, this occurs when the host halo has grown massive enough to host a first galaxy, as the gas is shock-heated to a temperature of $\sim 10^4$ K at the virial radius $r_{\rm vir} \sim 1$ physical kpc from the center of the $\sim 10^8\ M_{\odot}$ halo. As shown in both Figs. 10 and 12, the metallicity of the gas that re-collapses into the growing host halo is typically $\sim 10^{-3}\ Z_{\odot}$. Therefore, it is expected that stars formed in first galaxies enriched by powerful Pop III.1 supernovae are likely enriched to this level (e.g. Karlsson et al. 2008; Greif et al. 2010; Wise et al. 2010). Such stars would be the first Pop II stars, and as we shall see many of these stars may still be present today, 13 Gyr after the formation of the first galaxies.

3.2 *The Mixing of Metals with the Primordial Gas*

Here we consider two distinct situations in which the metal-enriched ejecta of primordial supernovae mix with the primordial gas, drawing on the results of the cosmological simulations of Pop III supernovae discussed in Sect. 3.1. Firstly, we shall estimate the timescale on which the primordial gas in the IGM that is swept up by the blast wave becomes mixed with the ejecta. Then, we will consider the likelihood that the primordial gas in minihalos that are overrun by the blast wave is mixed with the ejecta, thereby precluding Pop III star formation in those halos.

When the supernova shock finally stalls after $\sim 10^8$ years, the dense shell of swept-up gas is accelerated towards the growing halo embedded in the under-dense shocked gas. Such a configuration is Rayleigh-Taylor unstable and small perturbations of the shell can quickly grow, leading to mixing of the primordial gas in the shell with the metal-enriched gas in the interior. As a stability analysis shows, a small perturbation on a length scale $\epsilon << l_{\rm sh}$, where $l_{\rm sh}$ is the thickness of the dense shell, will grow exponentially, at a rate

$$\frac{d\epsilon}{dt} = \left[\frac{2\pi g}{l_{\rm sh}}\left(\frac{\rho_{\rm sh}-\rho_{\rm b}}{\rho_{\rm sh}+\rho_{\rm b}}\right)\right]^{\frac{1}{2}} \epsilon \,. \tag{37}$$

Here g is the acceleration of the dense shell in the direction of its interior, and $\rho_{\rm sh}$ and $\rho_{\rm b}$ are the densities of the shell and the interior metal-enriched bubble, respectively. Assuming that $\rho_{\rm sh} >> \rho_{\rm b}$, we can estimate the timescale on which the perturbation will grow as (e.g. Madau et al. 2001)

$$t_{\rm RT} \simeq \frac{\epsilon}{\frac{d\epsilon}{dt}} \simeq \left(\frac{2\pi g}{l_{\rm sh}}\right)^{-\frac{1}{2}} \simeq 6\left(\frac{l_{\rm sh}}{10{\rm pc}}\right)^{\frac{1}{2}} {\rm Myr}\,, \tag{38}$$

where in the last expression the gravitational acceleration towards the growing host halo is taken to be $g \simeq GM_{\rm h}/r_{\rm sh}^2$, with $M_{\rm h} = 10^8$ $M_{\odot}$, roughly the mass to which the host halo grows during the expansion of the blast wave. We have also used $r_{\rm sh} = 3$ kpc, which is roughly the spatial extent of the supernova shock when it finally stalls (e.g. Greif et al. 2007). Even for a shell as thick as $\sim$100 pc, the timescale on which the metal-enriched interior material mixes with the $\sim 10^5$ $M_{\odot}$ of primordial gas swept up by the blast wave is much shorter than the timescale on which the gas re-collapses into the host halo, which is $\sim 10^8$ year. Therefore, the gas which re-collapses into the host halo is expected to be well-mixed with the metals ejected in the supernova explosion.

While it is thus apparent that the low density gas swept up in the IGM can be efficiently mixed with the metal-enriched material ejected in Pop III supernovae, the blast waves from these powerful explosions can also impact the more dense primordial gas inside neighboring minihalos. A key question therefore is whether the metals are also mixed with this dense gas, resulting metal-rich pop II stars upon their collapse, instead of Pop III stars. In this case of a supernova blast wave overtaking a dense cloud of self-gravitating gas in a minihalo, there is the possibility of the dense gas becoming Kelvin–Helmholtz unstable, in which case vortices develop at the boundary with the fast-moving metal-enriched gas, and the two will mix with one another. However, for this to occur the dense gas cloud must not be too tightly bound by gravity. In particular, for a given relative velocity between the minihalo and the blast wave, $v_{\rm sh}$, the gas will be mixed due to the Kelvin–Helmholtz instability at the virial radius $r_{\rm vir}$ of the halo, if (e.g. Murray et al. 1993; Cen and Riquelme 2008)

$$v_{\rm sh} \geq \left(\frac{g r_{\rm vir}}{2\pi}\frac{\rho_{\rm vir}}{\rho_{\rm b}}\right)^{\frac{1}{2}} \simeq 10\left(\frac{M_{\rm h}}{10^6 h^{-1} {\rm M}_{\odot}}\right)^{\frac{1}{3}}\left(\frac{1+z}{20}\right)^{\frac{1}{2}} {\rm km\,s^{-1}} \simeq v_{\rm circ}\,, \tag{39}$$

where for the second expression we have used $g = GM_{\rm h}/{r_{\rm vir}}^2$ and we have assumed a density contrast between the gas at the virial radius and that of the shock

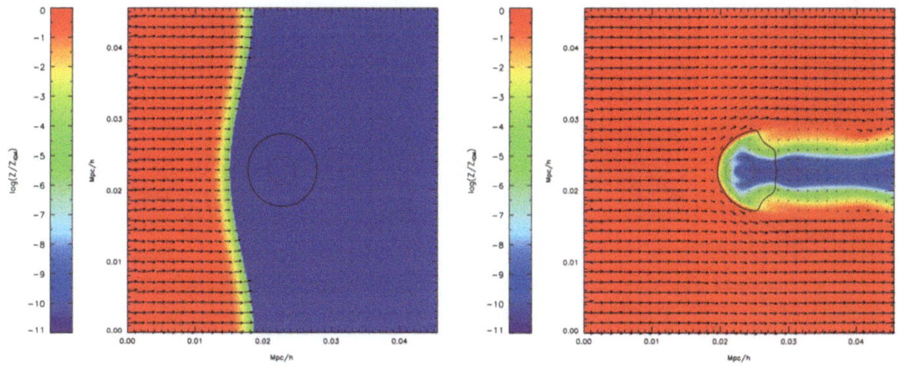

Fig. 13 The mixing of the primordial gas within a halo with mass $M_{\rm h} = 10^7\ {\rm M}_\odot$ with metal-enriched gas overtaking the halo at a velocity $v_{\rm sh} = 30\,{\rm km\,s^{-1}}$. Shown is the metallicity of the gas at $z = 9$ (*left panel*) and later at $z = 6$ (*right panel*). The velocity field, along with contours of $\rho = \rho_{\rm vir}$ (*solid lines*), is depicted in each of the panels. The primordial gas is well-mixed with the metal-enriched gas at radii $r \geq r_{\rm vir}$, and is mixed to a lesser degree at smaller radii, in basic agreement with Eq. (40). Note that the most dense gas in the center of the halo remains pristine; as it is from this gas that stars would most likely form, such a halo is likely to host Pop III star formation, despite being overtaken by high velocity metal-enriched gas. From Cen and Riquelme (2008)

$\rho_{\rm vir}/\rho_{\rm b} = 10$, consistent with the results of the cosmological simulations of Pop III supernvovae shown in Sect. 3.1. Thus, we see that the gas near the virial radius will be mixed if the speed of the shock exceeds the circular velocity $v_{\rm circ}$ of the halo, which is likely the case for a minihalo with mass $M_{\rm h} \sim 10^6\ {\rm M}_\odot$ at $z \lesssim 20$.

However, as the dense gas is often more deeply embedded in the halo from which stars form, the metal-enriched gas may have to be propagating at a significantly higher velocity in order to impact the nature of star formation in the halo. To estimate how fast the shock must be in order to mix the gas a distance r from the center of the halo. We conclude that the gas in the halo has a density profile that is roughly isothermal, with $\rho \simeq \rho_{\rm vir}\ (r/r_{\rm vir})^{-2}$. Using the same rough scaling also for the dark matter, we substitute ρ for $\rho_{\rm vir}$ and r for $r_{\rm vir}$ in Eq. (39) to estimate the shock speed required for mixing via the Kelvin–Helmholtz instability:

$$v_{\rm sh} \gtrsim 100 \left(\frac{M_{\rm h}}{10^6 h^{-1} {\rm M}_\odot}\right)^{\frac{1}{3}} \left(\frac{1+z}{20}\right)^{\frac{1}{2}} \left(\frac{r}{0.1 r_{\rm vir}}\right)^{-1} {\rm km\,s^{-1}}\,, \tag{40}$$

where we have implicitly assumed the same constant $\rho_{\rm b}$ as in Eq. (39). Therefore, we see that it is only relatively fast shocks that are able to efficiently mix the metal-enriched material with the dense, pristine gas in the interior of a primordial minihalo. The results of this analysis are in agreement with those of the simulations of high velocity shocks impacting minihalos, as shown in Fig. 13 (Cen and Riquelme 2008).

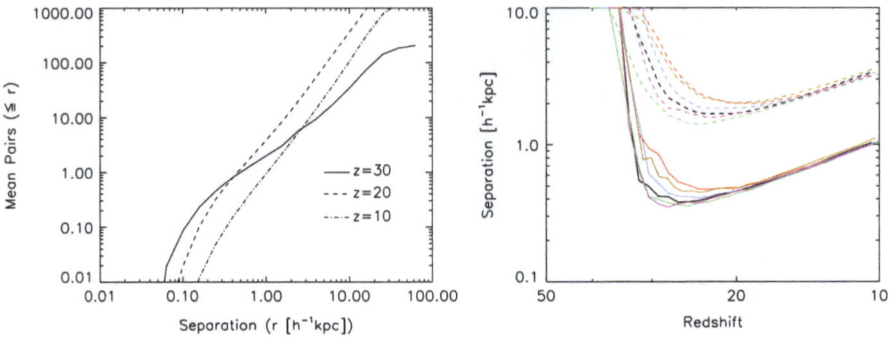

Fig. 14 Clustering of Pop III star-forming minihalos in a simulation of the formation of a Milky Way-like halo. *Left panel*: mean number of progenitor halo pairs, as a function of their physical separation at the three redshifts indicated in the panel. *Right panel*: mean separation of progenitor haloes in each of several similar simulations. Each halo has one neighbour within the distance given by the solid line, and 10 neighbours within the distance given by the *dashed lines*. While the radiation produced by one Pop III star in a given halo is likely to impact its neighbors and perhaps delay star formation, as discussed in Sect. 2, it is in general unlikely that the metal-enriched ejecta from Pop III supernovae will mix with the dense star-forming gas in neighboring halos. From Gao et al. (2010)

To gauge the likelihood that a cosmological minihalo is indeed impacted by a Pop III supernova shock that is sufficiently strong to enrich the material in its central regions, we turn to Fig. 14, which shows the average distance between Pop III star-forming minihalos in a cosmological simulation of the formation of a halo similar to that of the Milky Way (Gao et al. 2010). This figure shows that due to the clustering of such halos, the average distance between them is smaller than would be expected from a simple estimate drawn from their abundance assuming a homogeneous distribution. In particular, especially at high redshifts ($z \geq 20$), the halos are closely clustered, with an average separation of roughly $\sim$500 pc. From this we can estimate the average speed at which the blast wave from a Pop III supernova in one minihalo impacts its nearest neighboring minihalo, using the results presented in Fig. 11. As can be seen from that figure, the typical speeds with which the shock propagates at $r_{\rm sh} \sim 500$ pc from the explosion site are roughly 20–40 km s^{-1}. This is high enough to disrupt the gas near the virial radius of a neighboring halo, but not high enough to mix the metal-enriched ejecta with the dense star-forming gas in the interior of the halo at $r < 0.1 r_{\rm vir}$, as given by Eq. (40). Indeed, a similar result is found for the case of $v_{\rm sh} = 30\,{\rm km\,s^{-1}}$ in simulations in which the mixing of the gas is resolved, as shown in Fig. 13. Therefore, we conclude that the inefficiency of mixing poses a substantial challenge for the metals ejected in Pop III supernova explosions to enrich other star-forming halos and prevent Pop III star formation from occurring (see also Wyithe and Cen 2007; Wise and Abel 2008; Greif et al. 2010).

While here we have presented simple analytical estimates of the degree to which metals ejected in the first supernovae are mixed with the primordial gas via hydrodynamical instabilities, both in the IGM and in neighboring minihalos, other processes also contribute to mixing metals into the primordial gas (see e.g. Ferrara et al. 2000; Karlsson et al. 2012; Maio et al. 2011). Perhaps chief among these is the turbulence which develops as gas rapidly flows into the centers of the atomic cooling halos in which the first galaxies form (Wise and Abel 2007b; Greif et al. 2008) and acts to enhance the rate at which mixing takes place on small scales via diffusion (see e.g. Tenorio-Tagle 1996; Klessen and Lin 2003; Karlsson 2005; Pan and Scalo 2007). Once star formation begins in these halos, turbulent mixing is also facilitated by the energy injected by supernova explosions (e.g. Mori et al. 2002; Wada and Venkatesan 2003; Vasiliev et al. 2008), and the fraction of un-enriched primordial gas in the first galaxies is expected to continually drop with time (e.g. de Avillez and Mac Low 2002). We turn next to discuss the impact that the first metals, once mixed into the primordial gas, have on the cooling of the gas and alter the nature of star formation.

3.3 Metal Cooling in the First Galaxies

In Sect. 2.1 we discussed how cooling by the molecule HD, which may be formed in abundance in partially ionized primordial gas, can lower the temperature of the gas to the lowest temperature possible via radiative cooling, that of the CMB. Here we draw on the same formalism introduced there to show how just a small amount of metals mixed into the primordial gas can allow it to cool to low temperatures even more efficiently. While a number of heavy elements contribute to the cooling of low-metallicity gas, here we shall take a simplified approach and focus only on cooling by carbon, which is likely to have been released in abundance in the first supernova explosions (e.g. Heger and Woosley 2002, 2010; Tominaga et al. 2007).

To begin, we note that once the first generations of stars form and a background radiation field is established, as discussed in Sect. 2.2, besides dissociating H_2 this background radiation field can easily ionize neutral carbon (e.g. Bromm and Loeb 2003b). This makes available the potent coolant C II which, even in the Galaxy today, is important for cooling the gas to very low temperatures in dense star-forming clouds (e.g. Stahler and Palla 2004). To see how the presence of C II in the first galaxies affects the cooling of the gas, we first note that at low temperatures this ion is readily collisionally excited from its ground $^2P_{1/2}$ state to the first excited $^2P_{3/2}$ state. The energy difference between these two fine-structure states is just

$$\frac{h\nu_{10}}{k_B} \simeq 92\ \mathrm{K}\,, \tag{41}$$

where here ν_{10} denotes the frequency of the photon emitted in the radiative decay of the first excited state back to the ground state. As the energy difference is even smaller than that between the ground and first excited rotational states of HD, C II offers the potential to more efficiently cool the gas than HD, as even lower energy collisions are able to excite the ion. In addition, the Einstein coefficient for spontaneous radiative decay is $A_{10} = 2.4 \times 10^{-6}\ \mathrm{s}^{-1}$, almost two orders of magnitude higher than that for the $J = 1 \rightarrow 0$ transition of HD.

We can obtain a conservative lower limit for the cooling rate of C II via this transition by considering the cooling of gas in the low density regime, in which the rate of collisional excitations is balanced by the rate of radiative decays, that is at densities $n < n_{\rm crit}$, where the critical density $n_{\rm crit}$ is defined as that above which the rate of collisional deexcitations exceeds the rate of radiative deexcitations. For the transition of C II that we are considering, $n_{\rm crit} = 3 \times 10^3$ cm^{-3}. In this case, the cooling rate is given as (e.g. Stahler and Palla 2004)

$$\begin{aligned}\Lambda_{\rm CII}(n < n_{\rm crit}) &\simeq \frac{g_1}{g_0} n_0 n_{\rm H} \gamma_{10} h\nu_{10} e^{\frac{-h\nu_{10}}{k_{\rm B} T_{\rm gas}}} \\ &= 1.5 \times 10^{-23} \left(\frac{X_{\rm CII}}{10^{-6}}\right) \left(\frac{n_{\rm H}}{10^3 {\rm cm}^{-3}}\right)^2 e^{-\frac{92{\rm K}}{T_{\rm gas}}} \, {\rm erg\ s^{-1}\ cm^{-3}}, \qquad (42)\end{aligned}$$

where $\gamma_{10} n_{\rm H} = 6 \times 10^{-10}\, n_{\rm H}$ s^{-1} is the rate at which a given C II ion in the ground state is excited due to a collision with a neutral hydrogen atom, $g_{\rm i}$ is the statistical weight of the ith excited state, $X_{\rm CII}$ is the fractional abundance of C II relative to hydrogen, and $T_{\rm gas}$ is the temperature of the gas. For densities $n > n_{\rm crit}$, in turn, the cooling rate varies linearly with the density of the gas, since in this case the level populations are in LTE, as given by Eq. (10), and the rate of radiative decay is no longer balanced by the rate of collisional excitation. In this case, we have

$$\begin{aligned}\Lambda_{\rm CII}(n > n_{\rm crit}) &\simeq \frac{g_1}{g_0} n_0 A_{10} h\nu_{10} e^{\frac{-h\nu_{10}}{k_{\rm B} T_{\rm gas}}} \\ &= 6 \times 10^{-22} \left(\frac{X_{\rm CII}}{10^{-6}}\right) \left(\frac{n_{\rm H}}{10^4 {\rm cm}^{-3}}\right) e^{-\frac{92{\rm K}}{T_{\rm gas}}} \, {\rm erg\ s^{-1}\ cm^{-3}}. \qquad (43)\end{aligned}$$

The right panel of Fig. 15 shows, along with the cooling rates of a number of other metal species, the cooling rate due to C II given above, as a function of gas temperature, for $n < n_{\rm crit}$. The cooling rates of oxygen, iron, and silicon that are shown can be obtained following the first part of Eq. (42) using the atomic data corresponding to those elements (see e.g. Santoro and Shull 2006; Maio et al. 2007). We note, however, that the cooling rate per C II ion is higher than that of any of the other metal species shown, as well as being at least an order of magnitude higher than the cooling rate per molecule of any of the primordial species shown in the left panel, at temperatures $T_{\rm gas} \leq 100$ K. Therefore, we can focus on this chemical species as a means to derive a simple estimate of the minimum abundance of heavy elements required to significantly alter the cooling properties of the primordial gas, and perhaps thereby alter the nature of star formation.

A rough estimate of the minimum carbon abundance required for the characteristic fragmentation mass to change from the relatively large value expected for primordial gas in the case of Pop III.1 star formation can be found by considering the properties of the primordial gas when fragmentation takes place. At this stage,

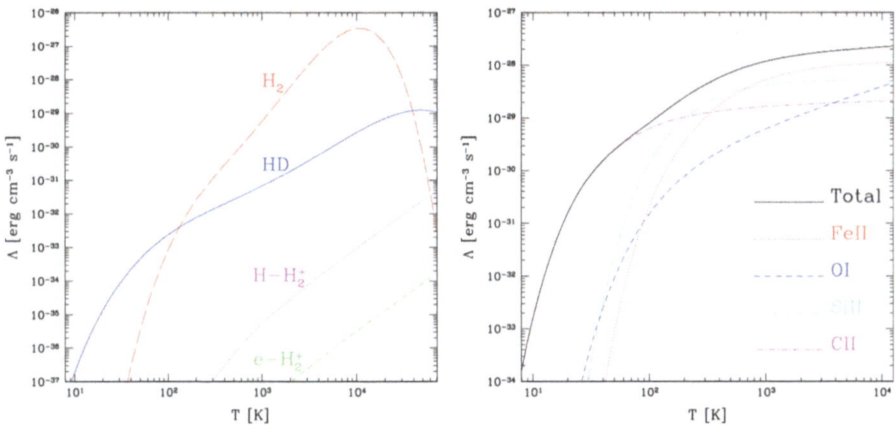

Fig. 15 *Left panel*: Cooling rates for primordial gas with a hydrogen number density $n_{\rm H} = 1\ {\rm cm}^{-3}$ and the following fractions for the different species: $X_{\rm HD} = 10^{-8}$, $X_{\rm H_2} = 10^{-5}$, $X_{\rm H_2^+} = 10^{-13}$, $X_{\rm e^-} = 10^{-4}$. The H_2 cooling rate (*long-dashed line*) is plotted together with those of HD (*solid*), H-impact $H_2{}^+$ (*dotted line*) and e-impact $H_2{}^+$ (*short-dashed line*). *Right panel*: Cooling rates of various metal species as a function of temperature, for a gas also with hydrogen number density $n_{\rm H} = 1\ {\rm cm}^{-3}$, for each metal species a fractional abundance of 10^{-6} is assumed. The cooling rate per C II ion is higher than the cooling rate per particle of any of the other species shown here, at $T_{\rm gas} \leq 100$ K. From Maio et al. (2007)

the so-called loitering phase in the collapse of the primordial gas in minihalos, $T_{\rm gas} \sim 200$ K and $n \sim 10^4\ {\rm cm}^{-3}$ (e.g. Abel et al. 2002; Bromm et al. 2002). Hence, the Jeans mass (4) is of the order of 100 $M_\odot$ and, if the gas does not cool efficiently then a massive Pop III star, or perhaps a binary or small multiple system, will likely form (e.g. Turk et al. 2009; Stacy et al. 2010; Clark et al. 2011a; Greif et al. 2011). However, if the gas cools to lower temperatures, then the Jeans mass becomes smaller and the gas is expected to fragment into smaller clumps; in turn, this is expected to translate into the formation of less massive stars.

Following the discussion in Sect. 2.1, we note that in order for the gas to cool efficiently at this stage, the cooling rate must exceed the rate at which the gas is heated adiabatically by compression during its collapse (Bromm and Loeb 2003b). Taking the adiabatic heating rate to be $\Gamma_{\rm ad} \sim 1.5 n k_{\rm B} T_{\rm gas}/t_{\rm ff}$, where $t_{\rm ff} \simeq (G\rho)^{-\frac{1}{2}}$ is the free-fall time and the cooling rate $\Lambda_{\rm CII}$ is given by Eq. (43), this condition is satisfied if $X_{\rm CII} > 7 \times 10^{-8}$. Assuming that all carbon is in the form of C II and taking it that the solar abundance of carbon is $\sim 3 \times 10^{-4}$ by number, this yields a critical carbon abundance of $[{\rm C/H}]_{\rm crit} \simeq -3.5$.[9]

[9]Here we use the common notation for abundance ratios relative to those of the sun given by $[{\rm X/Y}] = \log_{10}(N_{\rm X}/N_{\rm Y}) - \log_{10}(N_{\rm X}/N_{\rm Y})_\odot$, where $N_{\rm X}$ and $N_{\rm Y}$ are the numbers of nuclei of elements X and Y, respectively.

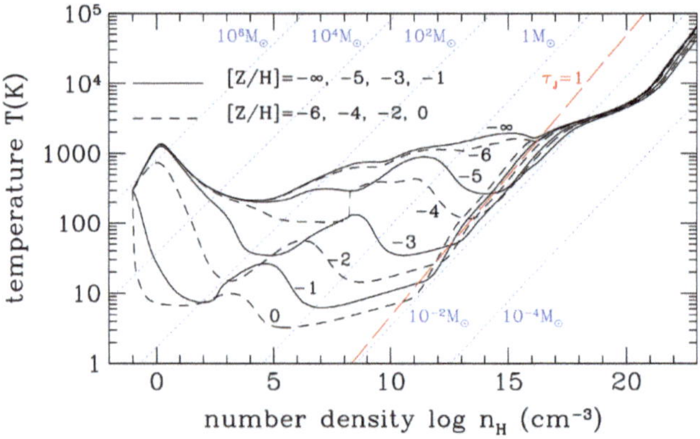

Fig. 16 Temperature evolution of collapsing prestellar clouds with different metallicities, as a function of density. The evolutionary tracks of clouds with metallicities [Z/H] $= -\infty$ ($Z=0$), -5, -3, and -1 (-6, -4, -2, and 0) are shown by *solid (dashed) lines*. The *dotted lines* denote various values of the Jeans mass M_J. Near the critical density of H_2, $n \sim 10^4$ cm^{-3}, at which the cooling of primordial gas becomes less efficient, the temperature continues to drop as the gas collapses to higher densities for metallicities $Z \geq 10^{-4}\ Z_\odot$; this is due to the cooling provided by C II and other atomic species, as discussed in the text. At higher densities, $n \geq 10^{10}$ cm^{-3}, the gas is able to cool efficiently even at a metallicity of $Z \sim 10^{-5}\ Z_\odot$, if dust is present, as is the case in the calculation shown here. From Omukai et al. (2005)

While other elements, such as oxygen, iron, and silicon, also contribute to the cooling of metal-enriched gas, this abundance of carbon relative to the solar value is similar to what is found for the overall critical metallicity Z_{crit} / $Z_\odot \sim 10^{-3.5}$ that is typically found in detailed calculations including atomic cooling[10] (e.g. Bromm et al. 2001; Omukai et al. 2005; Smith and Sigurdsson 2007; Smith et al. 2009; Aykutalp and Spaans 2011; but see Jappsen et al. 2009a,b). Figure 16 shows the results of one such calculation, in which the temperature evolution of the gas is modeled as it collapses to high densities, for various values of the metallicity of the gas. For the case of metal-free gas, the temperature of the gas increases after the loitering phase at $n \sim 10^4$ cm^{-3}; in this case, fragmentation at mass scales smaller than of the order of 100 $M_\odot$ is thus unlikely. However, when $Z \geq 10^{-4}$ $Z_\odot$, close the value we found above for the critical carbon abundance, the gas cools as it collapses to densities $n > 10^4$ cm^{-3} and consequently the fragmentation scale decreases appreciably compared to the primordial case. Hence, less massive stars are likely to form in gas enriched to this level.

We note also a second drop in the temperature of the gas at higher densities for even lower metallicities in Fig. 16; this decrease in temperature for $Z \geq 10^{-5}\ Z_\odot$

[10]As the cooling rates of the various atomic species each contribute separately to the total cooling rate, it is the combination of their individual abundances which determines whether the "critical metallicity" is achieved (see e.g. Frebel et al. 2007).

occurs because of dust cooling. While the dust fraction in extremely metal-poor gas is not known, dust formation in early supernovae (e.g. Nozawa et al. 2003; Schneider et al. 2004; Cherchneff and Dwek 2010) may yield it high enough for the thermal evolution of the gas to be affected at $n \geq 10^{10}$ cm^{-3}, as shown here, even for such extremely low metallicities. In this case, the critical metallicity for low-mass star formation may be smaller than we estimated above for the case of cooling by atomic species such as C II, perhaps as low as $Z_{\rm crit} \sim 10^{-5}\ Z_{\odot}$ (see e.g. Schneider et al. 2006; Clark et al. 2008).

In general, simulations of the evolution of low-metallicity star-forming gas give the same general result that the fragmentation scale, as well as the protostellar accretion rate, is higher for metal-free gas than for metal-enriched gas, and hence that the typical masses of Pop III stars are higher than those of Pop II stars (e.g. Bromm et al. 2001; Smith and Sigurdsson 2007; but see Jappsen et al. 2009a,b). However, as shown in Fig. 17, recent very high resolution cosmological simulations suggest that low-mass protostars formed in clusters may be ejected from the dense central regions of primordial minihalos due to dynamical interactions, in which case their growth may be limited due to the accretion of gas being dramatically slowed (Greif et al. 2011; see also Clark et al. 2011a). In this event, it is possible that low-mass stars may indeed form from primordial gas, although they may only constitute a small fraction of all Pop III stars (e.g. Tumlinson 2006; Madau et al. 2008). If their masses were less than $\simeq$0.8 M$_{\odot}$, then such low-mass stars could be detectable as un-enriched dwarfs or red giants in the Galaxy even today (Johnson and Khochfar 2011), although there is a strong possibility that their surfaces would be enriched due to accretion of metals from the interstellar medium (see e.g. Suda et al. 2004; Frebel et al. 2009; Komiya et al. 2010). To date, however, no low-mass stars with overall metallicity below of the order of $10^{-4}\ Z_{\odot}$ have been detected, which is consistent with the critical metallicity being set by cooling due to atomic species such as carbon and oxygen (see Frebel et al. 2007).

While the additional avenues for radiative cooling provided by even trace amounts of metals clearly alter the evolution of the gas and the process of star formation, other factors also play a role in dictating the thermal and dynamical state of the gas in the first galaxies. Magnetic fields may impede the large-scale collapse of the gas into dark matter halos (e.g. Schleicher et al. 2009; Rodrigues et al. 2010; de Souza et al. 2011) or alter the collapse of the gas at smaller scales during star formation (e.g. Kulsrud et al. 1997; Silk and Langer 2006; Xu et al. 2008; Schleicher et al. 2010a). Also, cosmic rays generated in the first supernova explosions are an additional source of ionization that can speed the formation of molecules and so enhance the cooling of the gas (see Vasiliev and Shchekinov 2006; Stacy and Bromm 2007; Jasche et al. 2007). Finally, the impact of the turbulence generated by both the accretion of gas from the IGM and supernovae in the first galaxies may dramatically impact the process of star formation, in general acting to decrease the mass scale at which the gas fragments and forms stars (e.g. Padoan et al. 2007; Clark et al. 2008, 2011b; Prieto et al. 2011).

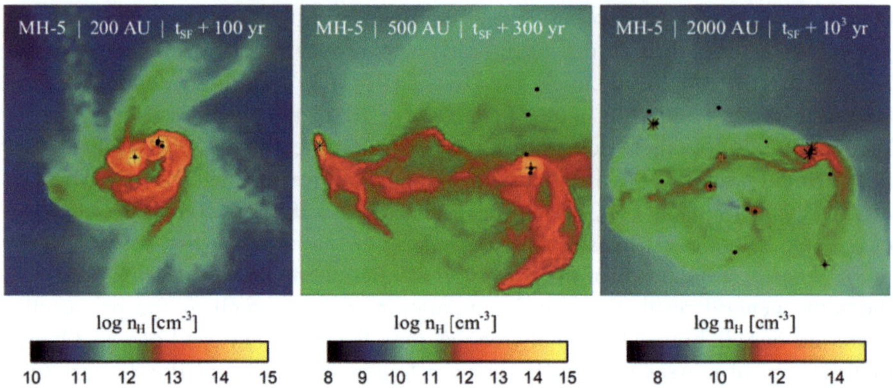

Fig. 17 The state of the central gas cloud in a primordial minihalo at 100 year (*left panel*), 300 year (*middle panel*), and 10^3 year (*right panel*) after the formation of the first protostar in a very high resolution cosmological simulation. Shown here is the hydrogen density projected along the line of sight. Blackdots, crosses and stars denote protostars with masses below $1M_\odot$, between $1M_\odot$ and $3M_\odot$, and above $3M_\odot$, respectively. The gas fragments into a relatively rich cluster of protostars with a range of masses. Dynamical interactions can lead to the ejection of low-mass protostars, while more massive protostars tend to remain at the center of the cloud and continue to accrete from the surrounding envelope of gas. Thus, while massive primordial stars are likely to form in such clusters, some fraction of Pop III stars with masses below $\sim$1 $M_\odot$ may also form. From Greif et al. (2011)

4 Observational Predictions and the Outlook for Identifying the First Galaxies

While the enrichment of the primordial gas by metals ejected in the first supernovae is likely to preclude primordial star formation in a large fraction of the first galaxies (Johnson et al. 2008; Wise and Abel 2008; Greif et al. 2010; Maio et al. 2010), it is also not likely that metal enrichment abruptly ends the epoch of Pop III star formation after the formation of the first stars. As discussed in Sect. 2.2, it is possible for the photodissociating background radiation established by early generations of stars to slow the collapse of the primordial gas, potentially delaying a large fraction of Pop III star formation and metal enrichment until later times. Also, as discussed in Sect. 3.2, the mixing of the first metals with the primordial gas, especially within minihalos, may not occur efficiently. Therefore, it is a distinct possibility that Pop III star formation continues well after the formation of the first stars (e.g. Scannapieco et al. 2003; Tornatore et al. 2007; Trenti et al. 2009; Maio et al. 2010), and that substantial primordial star formation may be detectable in the first galaxies.

It is therefore critical to predict observable signatures of Pop III star formation, in order that it can be identified in high redshift galaxies (e.g. Zackrisson 2011). Some distinctive signatures derive from the high surface temperatures of primordial stars, which arise due to a relatively low opacity in the stellar interior. This low opacity translates into a smaller radii R_* for primordial stars than for their metal-enriched counterparts. In turn, because stellar luminosity scales as $L_* \propto R_*^2 \, T_*^4$, for a given luminosity the surface temperature T_* of a primordial star will be higher than a metal-enriched star (e.g. Siess et al. 2002; Lawlor et al. 2008). For very massive primordial stars, the surface temperature is very high, roughly $\sim 10^5$ K (Bromm et al. 2001; Schaerer 2002). Owing to this high temperature, primordial stars emit copious high energy radiation, a relatively large fraction of which is able to ionize not only hydrogen (H I), but also helium (He I and He II).

As a substantial portion of the ionizing photons emitted from stars in early galaxies are absorbed by the relatively dense gas in the interstellar medium before escaping into the IGM (e.g. Wood and Loeb 2000; Gnedin et al. 2008; Wise and Cen 2009; Razoumov and Sommer-Larsen 2010; Paardekooper et al. 2011; Yajima et al. 2011), the energy in these photons is reprocessed into emission lines arising from the recombination of the ionized species (e.g. Osterbrock and Ferland 2006). For the case of primordial stars, because a relatively large fraction of the emitted radiation ionizes He II, the photons emitted during the recombination of He III to He II produce strong emission at characteristic wavelengths. The most prominent recombination line emitted from such He III regions, with a wavelength of 1,640 Å, emerges from the radiative decay of the lone electron in this ion from the $n = 3$ to the $n = 2$ state.[11] The most prominent emission lines from the recombination of ionized hydrogen in the H II regions surrounding primordial stars are the same as expected from metal-enriched stars, Lyα and Hα, which arise from the radiative decay from the $n = 2 \rightarrow 1$ and $n = 3 \rightarrow 2$ energy levels of hydrogen, respectively. The key observational signature of primordial star formation, as opposed to metal-enriched star formation, is a relatively large ratio of the luminosity emitted in the helium line, He II λ1,640, to that emitted in the hydrogen lines (see e.g. Tumlinson et al. 2001; Oh et al. 2001; Schaerer 2003; Raiter et al. 2010).

Figure 18 shows the luminosity emitted in each of the three recombination lines mentioned above from an instantaneous burst of Pop III star formation in a first galaxy formed in a halo of mass $M_{\rm h} \sim 10^8$ M$_\odot$ at $z \sim 12$, as gleaned from cosmological radiative transfer simulations (Johnson et al. 2009). Each of the panels shows the line luminosities for a different combination of the characteristic stellar mass of the stars (either 25 or 100 M$_\odot$) and of the total stellar mass (either 2,500 or 25,000 M$_\odot$). Even assuming such large characteristic masses for Pop III stars and that such a large fraction (either $\sim$1 or $\sim$10 %) of the gas in the first galaxies is converted into stars, the luminosities of the recombination lines are likely to be too dim to detect with telescopes in the near future. To see this, we can estimate the total flux F that would be in these lines at $z = 0$ as

[11] While photons are also emitted in transitions to the $n = 1$ state, the IGM is optically thick to these photons before reionization due to absorption by neutral hydrogen, and so they are not expected to be observable from the first galaxies.

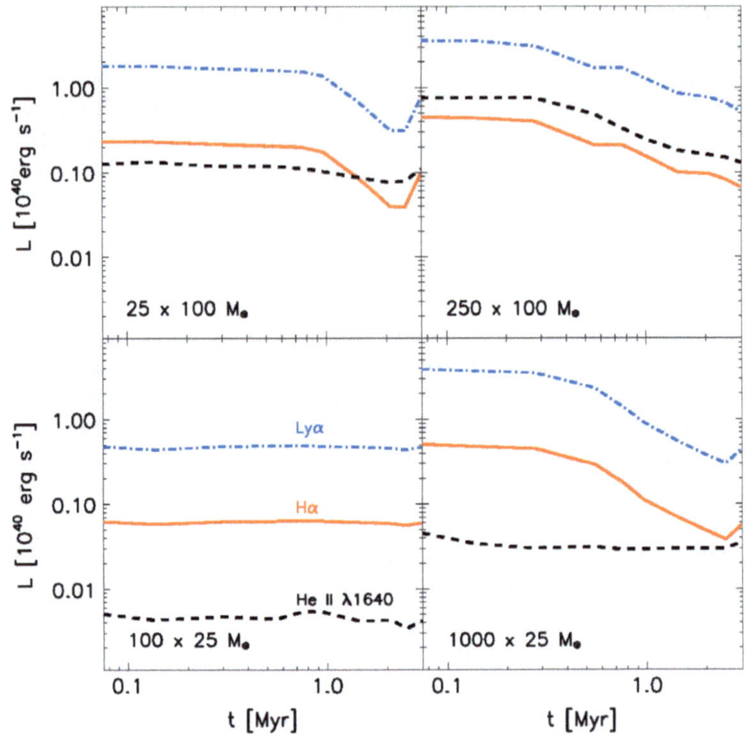

Fig. 18 The luminosity of Pop III star clusters in a first galaxy at $z \sim 12$, as a function of the time from their formation, in three recombination lines: Lyα (*dot-dashed blue*), Hα (*solid red*), and He II λ1,640 (*dashed black*). The four panels correspond to four different combinations of stellar IMF and total stellar mass; these are, clockwise from top-left: twenty-five 100 $M_\odot$ stars, two hundred fifty 100 $M_\odot$ stars, one thousand 25 $M_\odot$ stars, and one hundred 25 $M_\odot$ stars. The relative luminosities of detected He II and H I recombination lines can provide information about the stellar metallicity and IMF; more massive and more metal-poor stars emit more high energy radiation that can ionize He II, which leads to strong He II λ1,640 emission relative to Hα and Lyα. From Johnson et al. (2009)

$$F = \frac{L}{4\pi D_{\rm L}^2(z)} \simeq 10^{-20}\left(\frac{L}{10^{40}{\rm erg\,s^{-1}}}\right)\left(\frac{1+z}{10}\right)^{-2} {\rm erg\,s^{-1}\,cm^{-2}}\,, \qquad (44)$$

where L is the luminosity in a given line and $D_{\rm L}$ is the luminosity distance to redshift z. At $z \geq 10$, even the most luminous line, Lyα, would be seen at $z = 0$ with a flux of $\lesssim 4 \times 10^{-20}$ erg s^{-1} cm^{-2}, which is well below the flux limit of $\sim 2\times 10^{-19}$ erg s^{-1} cm^{-2} of surveys planned for the JWST (Gardner et al. 2006; Windhorst et al. 2006).

Instead of the first galaxies, hosted in halos with masses of $\sim 10^8$ $M_\odot$ at $z \geq 10$, it thus appears likely that observations in the next decade may reveal somewhat

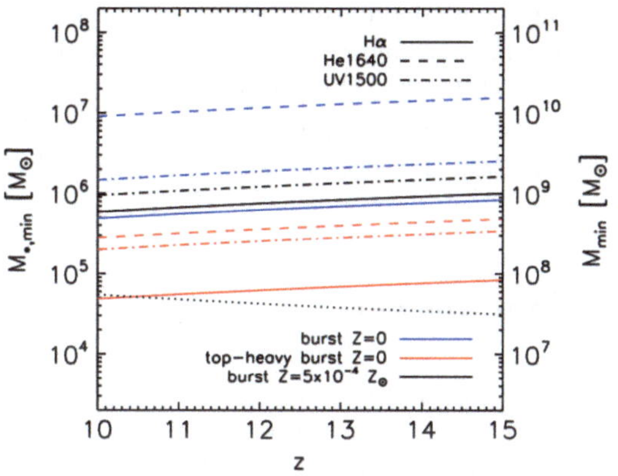

Fig. 19 Stellar masses $M_{*,\min}$ of the lowest mass starburst observable through the detection of the Hα line (*solid curves*), the He II λ1,640 line (*dashed curves*), or the continuum at 1,500 Å(*dash-dotted curves*) with JWST, assuming an exposure time of 10^6 s and a signal-to-noise ratio of S/N = 10. Stellar masses derived from the Schaerer (2003) zero-metallicity starbursts with a standard Salpeter-like IMF, zero-metallicity starbursts with a top-heavy IMF, and low-metallicity starbursts, are shown in *blue*, *red*, and *black*, respectively. The right axis shows the masses $M_{\min} = 10^3\ M_{*,\min}$ of halos expected to host a starburst with stellar mass $M_{*,\min}$. The *dotted curve* shows the mass of a first galaxy-sized halo, with virial temperature $T_{\rm vir} = 10^4$ K. From Pawlik et al. (2011)

more developed galaxies hosted in more massive halos (e.g. Barkana and Loeb 2000; Ricotti et al. 2008; Johnson et al. 2009; Pawlik et al. 2011), although there is the possibility of detecting less developed galaxies if their flux is magnified by gravitational lensing (see e.g. Zackrisson 2011). As shown in Fig. 19, the JWST is predicted to be capable of detecting both He II λ1,640 and Hα from metal-free starbursts in halos with masses $\geq 3 \times 10^8\ M_{\odot}$, if the IMF is very top-heavy. Though, if the typical mass of Pop III stars is $<50\ M_{\odot}$, it is likely that He II λ1,640 will only be detectable from significantly more massive stellar clusters, expected to form in similarly more massive halos. However, because more massive halos are formed from the mergers of smaller halos which themselves may have hosted star formation, it may be predominantly metal-enriched Pop II stars that form in the galaxies which will be detected by the JWST (e.g. Johnson et al. 2008).[12]

That said, there is the possibility that substantial Pop III star formation takes place even well after the epoch of the first galaxies (i.e. at $z < 10$), either due to inefficient mixing of primordial and metal-enriched gas (e.g. Jimenez and Haiman 2006; Pan and Scalo 2007; Wyithe and Cen 2007; Dijkstra and Wyithe 2007; Cen 2010) or to the collapse of primordial gas into late-forming atomic cooling halos (e.g. Tornatore et al. 2007; Trenti et al. 2009; Johnson 2010). Pop III star formation at such late times could be detected more easily, in large part because the emission

[12]It is also likely that other, complementary next generation facilities, such as the *Atacama Large Millimeter Array* (e.g. Combes 2010), will detect only metal-enriched star-forming galaxies.

line flux increases strongly with decreasing redshift, as shown in Eq. (44). However, at such low redshifts the background ionizing radiation field that builds up during reionization can strongly inhibit the infall of primordial gas into halos, limiting the amount of Pop III star formation that can occur even in metal-free galaxies (e.g. Efstathiou 1992; Gnedin 2000; Tassis et al. 2003; Dijkstra et al. 2004).

5 Summary and Conclusion

In this chapter we have discussed a wide range of the physical processes that must be accounted for in the theoretical modeling of the first galaxies. We have made important distinctions between the formation of the first stars in minihalos and star formation in the atomic cooling halos hosting the first galaxies, highlighting how the cooling properties of the gas assembled into the first galaxies are altered by high energy radiation and by the injection of heavy elements from the first supernovae. While this has not been a complete review of the theory of the formation of the first galaxies, it has hopefully served to illustrate, from basic principles where possible, much of the physics that comes into play in their study. The reader is referred to the many excellent articles in the bibliography below for more in-depth study on the topic.

In closing, it is critical to point out that without accounting for all of the effects we have discussed together, one is left with an incomplete understanding of the first galaxies. For instance, as we have seen, the radiation from the first stars can ionize the gas and trigger HD cooling, but it can also easily destroy H_2 and HD molecules. As well, while the first supernovae may enrich much of the gas from which the first galaxies form to a level above the critical metallicity needed for low-mass Pop II star formation, much of the dense gas in minihalos may not be efficiently mixed with the metal-enriched ejecta. Similarly, black holes may only form by direct collapse in rare regions in which the LW background radiation field is elevated, but the same stars which likely produce this radiation may also enrich the gas when they explode as supernovae, possibly precluding this mode of black hole formation. A complete and consistent picture of the formation of the first galaxies only emerges when accounting for star and black hole formation, metal enrichment, and radiative feedback all together in the full cosmological context.

Making this task especially daunting is the range of scales that must be taken into account. The gas clouds which collapse to form stars are on sub-parsec scales, metal-enrichment from the first supernovae occurs on parsec to kiloparsec scales, and the radiation emitted by the first stars can impact regions on kiloparsec or even megaparsec scales. Thus, simulations must ultimately resolve an enormous range of scales in order to capture all of the important physical processes that come into play. While we have introduced the results of numerous analytical calculations and simulations, none of them alone captures all of the processes we have discussed simultaneously. Indeed, this stands as one of the primary challenges to making detailed predictions of the nature of the first galaxies.

Acknowledgements The author is grateful to the editors for the invitation to contribute this Chapter, as well as to Bhaskar Agarwal, Volker Bromm, Umberto Maio, and Eyal Neistein for helpful comments on an earlier draft of this work.

References

Abel, T., Anninos, P., Zhang, Y., Norman, M.L.: Modeling primordial gas in numerical cosmology. New Astron. **2**, 181–207 (1997)
Abel, T., Bryan, G.L., Norman, M.L.: The formation of the first star in the Universe. Science **295**, 93–98 (2002)
Abel, T., Wise, J.H., Bryan, G.L.: The H II Region of a primordial star. Astrophys. J. **659**, L87–L90 (2007)
Ahn, K., Shapiro, P.R., Iliev, I.T., Mellema, G., Pen, U.-L.: The inhomogeneous background of H_2-dissociating radiation during cosmic reionization. Astrophys. J. **695**, 1430–1445 (2009)
Alvarez, M.A., Bromm, V., Shapiro, P.R.: The H II region of the first star. Astrophys. J. **639**, 621–632 (2006)
Alvarez, M.A., Wise, J.H., Abel, T.: Accretion onto the first stellar-mass black holes. Astrophys. J. **701**, L133–L137 (2009)
Aykutalp, A., Spaans, M.: The complexity that the first stars brought to the Universe: Fragility of metal riched gas in a radiation field. Astrophys. J. **737**, 63 (2011)
Barkana, R., Loeb, A.: High-redshift galaxies: their predicted size and surface brightness distributions and their graviational lensing probability. Astrophys. J. **531**, 613–623 (2000)
Barkana, R., Loeb, A.: In the beginning: the first sources of light and the reionization of the Universe. Phys. Rep. **349**, 125–238 (2001)
Begelman, M.C., Volonteri, M., Rees, M.J.: Formation of supermassive black holes by direct collapse in pre-galactic haloes. Mon. Not. R. Astron. Soc. **370**, 289–298 (2006)
Begelman, M.C., Shlosman, I.: Angular momentum transfer and lack of fragmentation in self-gravitating accretion flows. Astrophys. J. **702**, L5–L8 (2009)
Bondi, H.: On spherically symmetrical accretion. Mon. Not. R. Astron. Soc. **112**, 195–204 (1952)
Bromm, V., Coppi, P.S., Larson, R.B.: The formation of the first stars. I. The primordial star-forming cloud. Astrophys. J. **564**, 23–51 (2002)
Bromm, V., Ferrara, A., Coppi, P.S., Larson, R.B.: The fragmentation of pre-enriched primordial objects. Mon. Not. R. Astron. Soc. **328**, 969–976 (2001)
Bromm, V., Kudritzki, R.P., Loeb, A.: Generic spectrum and ionization efficiency of a heavy initial mass function for the first stars. Astrophys. J. **552**, 464–472 (2001)
Bromm, V., Larson, R.B.: The first stars. Annu. Rev. Astron. Astrophys. **42**, 79–118 (2004)
Bromm, V., Loeb, A.: Formation of the first supermassive black holes. Astrophys. J. **596**, 34–46 (2003a)
Bromm, V., Loeb, A.: The formation of the first low-mass stars from gas with low carbon and oxygen abundances. Nature **425**, 812–814 (2003b)
Bromm, V., Yoshida, N.: The first galaxies. Annu. Rev. Astron. Astrophys. **49**, 373–407 (2011)
Bromm, V., Yoshida, N., Hernquist, L.: The first supernova explosions in the Universe. Astrophys. J. **596**, L135–L138 (2003)
Bromm, V., Yoshida, N., Hernquist, L., McKee, C.F.: The formation of the first stars and galaxies. Nature **459**, 49–54 (2009)
Burbidge, E., Burbidge, G.R., Fowler, W.A., Hoyle, F.: Synthesis of the elements in stars. Rev. Mod. Phys. **1957**, 547–650 (1957)
Cen, R.: The state of star formation and the intergalactic medium at $z \sim 6$. Astrophys. J. **725**, 115–120 (2010)
Cen, R., Riquelme, M.A.: Lower metal enrichment of virialized gas in minihalos. Astrophys. J. **674**, 644–652 (2008)

Cherchneff, I., Dwek, E.: The chemistry of Population III Supernova Ejecta. II. The nucleation of molecular clusters as a diagnostic for dust in the early Universe. Astrophys. J. **713**, 1–24 (2010)
Ciardi, B., Ferrara, A.: The first cosmic structures and their effects. Space Sci. Rev. **116**, 625–705 (2005)
Ciardi, B., Ferrara, A., Abel, T.: Intergalactic H_2 photodissociation and the soft ultraviolet background produced by Population III objects. Astrophys. J. **533**, 594–600 (2000)
Clark, P.C., Glover, S.C.O., Smith, R.J., Greif, T.H., Klessen, R.S., Bromm, V.: The formation and fragmentation of disks around primordial protostars. Science **331**, 1040–1042 (2011a)
Clark, P.C., Glover, S.C.O., Klessen, R.S.: The first stellar cluster. Astrophys. J. **672**, 757–764 (2008)
Clark, P.C., Glover, S.C.O., Klessen, R.S., Bromm, V.: Gravitational fragmentation in turbulent primordial gas and the initial mass function of Population III stars. Astrophys. J. **727**, 110–127 (2011b)
Combes, F.: ALMA and the first galaxies. Am. Inst. Phys. Conf. Proc. **1294**, 9–16 (2010)
de Avillez, M.A., Mac Low, M.-M.: Mixing timescales in a Supernova-driven interstellar medium. Astrophys. J. **581**, 1047–1060 (2002)
de Souza, R.S., Rodrigues, L.F.S., Opher, R.: Random primordial magnetic fields and the gas content of dark matter haloes. Mon. Not. R. Astron. Soc. **410**, 2149–2155 (2011)
Dijkstra, M., Haiman, Z., Mesinger, A., Wyithe, J.S.B.: Fluctuations in the high-redshift Lyman-Werner background: close halo pairs as the origin of supermassive black holes. Mon. Not. R. Astron. Soc. **391**, 1961–1972 (2008)
Dijkstra, M., Haiman, Z., Rees, M.J., Weinberg, D.H.: Photoionization feedback in low-mass galaxies at high redshift. Astrophys. J. **601**, 666–675 (2004)
Dijkstra, M., Wyithe, J.S.B.: Very massive stars in high-redshift galaxies. Mon. Not. R. Astron. Soc. **379**, 1589–1598 (2007)
Efstathiou, G.: Suppressing the formation of dwarf galaxies via photoionization. Mon. Not. R. Astron. Soc. **256**, 43–47 (1992)
Ferrara, A., Pettini, M., Shchekinov, Y.: Mixing metals in the early Universe. Mon. Not. R. Astron. Soc. **319**, 539–548 (2000)
Flower, D.R., Le Bourlot, J., Pineau des Forêts, G., Roueff, E.: The cooling of astrophysical media by HD. Mon. Not. R. Astron. **314**, 753–758 (2000)
Frebel, A., Johnson, J.L., Bromm, V.: Probing the formation of the first low-mass stars with stellar archaeology. Mon. Not. R. Astron. Soc. **380**, L40–L44 (2007)
Frebel, A., Johnson, J.L., Bromm, V.: The minimum stellar metallicity observable in the Galaxy. Mon. Not. R. Astron. Soc. **392**, L50–L54 (2009)
Fryer, C.L., Woosley, S.E., Heger, A.: Pair-instability supernovae, gravity waves, and gamma-ray transients. Astrophys. J. **550**, 372–382 (2001)
Galli, D., Palla, F.: The chemistry of the early Universe. Astron. Astrophys. **335**, 403–420 (1998)
Gao, L., Theuns, T., Frenk, C.S., Jenkins, A., Helly, J.C., Navarro, J., Springel, V., White, S.D.M.: The earliest stars and their relics in the Milky Way. Mon. Not. R. Astron. Soc. **403**, 1238–1295 (2010)
Gardner, J.P., et al.: The James Webb Space Telescope. Space Sci. Rev. **123**, 485–606 (2006)
Glover, S.C.O.: The formation Of the first stars in the Universe. Space Sci. Rev. **117**, 445–508 (2005)
Glover, S.C.O., Brand, P.W.J.L.: On the photodissociation of H_2 by the first stars. Mon. Not. R. Astron. Soc. **321**, 385–397 (2001)
Gnedin, N.Y.: Effect of reionization on structure formation in the Universe. Astrophys. J. **542**, 535–541 (2000)
Gnedin, N.Y., Kravtsov, A.V., Chen, H.-W.: Escape of ionizing radiation from high-redshift galaxies. Astrophys. J. **672**, 765–775 (2008)
Greif, T.H., Bromm, V.: Two populations of metal-free stars in the early Universe. Mon. Not. R. Astron. Soc. **373**, 128–138 (2006)
Greif, T.H., Glover, S.C.O., Bromm, V., Klessen, R.S.: The first galaxies: chemical enrichment, mixing, and star formation. Astrophys. J. **716**, 510–520 (2010)

Greif, T.H., Johnson, J.L., Bromm, V., Klessen, R.S.: The first supernova explosions: energetics, feedback, and chemical enrichment. Astrophys. J. **670**, 1–14 (2007)

Greif, T.H., Johnson, J.L., Klessen, R.S., Bromm, V.: The first galaxies: assembly, cooling and the onset of turbulence. Mon. Not. R. Astron. Soc. **387**, 1021–1036 (2008)

Greif, T.: Simulations on a moving mesh: The clustered formation of population III protostars. Astrophys. J. **737**, 75 (2011)

Haiman, Z.: Observing the first stars and black holes. In: Thronson, H.A., Stiavelli, M., Tielens, A. (eds.) Astrophysics in the Next Decade, pp. 385–418. Springer, New York (2009)

Haiman, Z., Rees, M.J., Loeb, A.: Destruction of molecular hydrogen during cosmological reionization. Astrophys. J. **476**, 458–463 (1997)

Heger, A., Fryer, C.L., Woosley, S.E., Langer, N., Hartmann, D.H.: How massive single stars end their life. Astrophys. J. **591**, 288–300 (2003)

Heger, A., Woosley, S.E.: The nucleosynthetic signature of Population III. Astrophys. J. **567**, 532–543 (2002)

Heger, A., Woosley, S.E.: Nucleosynthesis and evolution of massive metal-free stars. Astrophys. J. **724**, 341–373 (2010)

Jappsen, A.-K., Klessen, R.S., Glover, S.C.O., Mac Low, M.-M.: Star formation at very low metallicity. IV. Fragmentation does not depend on metallicity for cold initial conditions. Astrophys. J. **694**, 1065–1074 (2009a)

Jappsen, A.-K., Mac Low, M.-M., Glover, S.C.O., Klessen, R.S., Kitsionas, S.: Star formation at very low metallicity. V. The greater importance of initial conditions compared to metallicity thresholds. Astrophys. J. **694**, 1161–1170 (2009b)

Jasche, J., Ciardi, B., Enßlin, T.A.: Cosmic rays and the primordial gas. Mon. Not. R. Astron. Soc. **380**, 417–429 (2007)

Jimenez, R., Haiman, Z.: Significant primordial star formation at redshifts z 3–4. Nature **440** 501–504 (2006)

Johnson, J.L.: Population III star clusters in the reionized Universe. Mon. Not. R. Astron. Soc. **404**, 1425–1436 (2010)

Johnson, J.L., Bromm, V.: The cooling of shock-compressed primordial gas. Mon. Not. R. Astron. Soc. **366**, 247–256 (2006)

Johnson, J.L., Bromm, V.: The aftermath of the first stars: massive black holes. Mon. Not. R. Astron. Soc. **374**, 1557–1568 (2007)

Johnson, J.L., Greif, T.H., Bromm, V.: Occurence of metal-free galaxies in the early Universe. Mon. Not. R. Astron. Soc. **388**, 26–38 (2008)

Johnson, J.L., Greif, T.H., Bromm, V., Klessen, R.S., Ippolito, J.: The first galaxies: signatures of the initial starburst. Mon. Not. R. Astron. Soc. **399**, 37–47 (2009)

Johnson, J.L., Khochfar, S.: Suppression of accretion onto low-mass Population III stars. Mon. Not. R. Astron. Soc. **413**, 1184–1191 (2011)

Johnson, J.L., Khochfar, S., Greif, T.H., Durier, F.: Accretion on to black holes formed by direct collapse. Mon. Not. R. Astron. Soc. **410**, 919–933 (2011)

Karlsson, T.: Stochastic chemical enrichment in metal-poor systems. I. theory. Astron. Astrophys. **439**, 93–106 (2005)

Karlsson, T., Bromm, V., Bland-Hawthorn, J.: Pre-galactic metal enrichment the chemical signature of the first stars. Rev. Mod. Phys., in press (arXiv:1101.4024) (2012)

Karlsson, T., Johnson, J.L., Bromm, V.: Uncovering the chemical signature of the first stars in the Universe. Astrophys. J. **679**, 6–16 (2008)

Kitayama, T., Ikeuchi, S.: Formation of subgalactic clouds under ultraviolet background radiation. Astrophys. J. **529**, 615–634 (2000)

Kitayama, T., Susa, H., Umemura, M., Ikeuchi, S.: Criteria for the formation of Population III objects in the ultraviolet background radiation. Mon. Not. R. Astron. Soc. **326**, 1353–1366 (2001)

Kitayama, T., Yoshida, N.: Supernova explosions in the early Universe: evolution of radiative remnants and the halo destruction efficiency. Astrophys. J. **630**, 675–688 (2005)

Klessen, R.S., Lin, D.N.: Diffusion in supersonic turbulent compressible flows. Phys. Rev. E **67** (2003)

Komiya, Y., Habe, A., Suda, T., Fujimoto, M.Y.: Formation history of metal-poor halo stars with the hierarchical model and the effect of interstellar matter accretion on the most metal-poor stars. Astrophys. J. **717**, 542–561 (2010)

Koushiappas, S.M., Bullock, J.S., Dekel, A.: Massive black hole seeds from low angular momentum material. Mon. Not. R. Astron. Soc. **354**, 292–304 (2004)

Kuhlen, M., Madau, P.: The first miniquasar. Mon. Not. R. Astron. Soc. **363**, 1069–1082 (2005)

Kulsrud, R.M., Cen, R., Ostriker, J.P., Ryu, D.: The protogalactic origin for cosmic magnetic fields. Astrophys. J. **480**, 481–491 (1997)

Larson, R.B.: Thermal physics, cloud geometry and the stellar initial mass function. Mon. Not. R. Astron. Soc. **359**, 211–222 (2005)

Lawlor, T.M., Young, T.R., Johnson, T.A., MacDonald, J.: Single and binary evolution of Population III stars and their supernova explosions. Mon. Not. R. Astron. Soc. **384**, 1533–1543 (2008)

Leitherer, C., Schaerer, D., Goldader, J.D., González Delgado, R.M., Robert, Carmelle, Kune, D.F., de Mello, D.F., Devost, D., Heckman, T.M.: Starburst99: synthesis models for galaxies with active star formation. Astrophys. J. Suppl. **123**, 3–40 (1999)

Lodato, G., Natarajan, P.: Supermassive black hole formation during the assembly of pre-galactic discs. Mon. Not. R. Astron. Soc. **371**, 1813–1823 (2006)

Mac Low, M.-M., Klessen, R.S.: Control of star formation by supersonic turbulence. Rev. Mod. Phys. **76**, 125–194 (2004)

Machacek, M.E., Bryan, G.L., Abel, T.: Simulations of pregalactic structure formation with radiative feedback. Astrophys. J. **548**, 509–521 (2001)

Machida, M.N., Tomisaka, K., Nakamura, F., Fujimoto, M.Y.: Low-mass star formation triggered by supernovae in primordial clouds. Astrophys. J. **622**, 39–57 (2005)

Mackey, J., Bromm, V., Hernquist, L.: Three epochs of star formation in the high-redshift Universe. Astrophys. J. **586**, 1–11 (2003)

Madau, P., Ferrara, A., Rees, M.J.: Early metal enrichment of the intergalactic medium by pregalactic outflows. Astrophys. J. **555**, 92–105 (2001)

Madau, P., Kuhlen, M., Diemand, J., Moore, B., Zemp, M., Potter, D., Stadel, J.: Fossil remnants of reionization in the halo of the Milky Way. Astrophys. J. **689**, L41–L44 (2008)

Maio, U., Dolag, K., Ciardi, B., Tornatore, L.: Metal and molecule cooling in simulations of structure formation. Mon. Not. R. Astron. Soc. **379**, 963–973 (2007)

Maio, U., Ciardi, B., Dolag, K., Tornatore, L., Khochfar, S.: The transition from Population III to Population II-I star formation. Mon. Not. R. Astron. Soc. **407**, 1003–1015 (2010)

Maio, U., et al.: The interplay between chemical and mechanical feedback from the first generation of stars. Mon. Not. R. Astron. Soc. **414**, 1145 (2011)

Mayer, L., Kazantzidis, S., Escala, A., Callegari, S.: Direct formation of supermassive black holes via multi-scale gas inflows in galaxy mergers. Nature **466**, 1082–1084 (2010)

McGreer, I.D., Bryan, G.L.: The impact of HD cooling on the formation of the first stars. Astrophys. J. **685**, 8–20 (2008)

McKee, C.F., Tan, J.C.: The formation of the first stars. II. Radiative feedback processes and implications for the initial mass function. Astrophys. J. **681**, 771–797 (2008)

Mesinger, A., Bryan, G.L., Haiman, Z.: Ultraviolet radiative feedback on high-redshift protogalaxies. Astrophys. J. **648**, 835–851 (2006)

Milosavljević, M., Bromm, V., Couch, S.M., Oh, S.P.: Accretion onto 'Seed' black holes in the first galaxies. Astrophys. J. **698**, 766–780 (2009)

Mori, M., Ferrara, A., Madau, P.: Early metal enrichment by pregalactic outflows. II. Three-dimensional simulations of blow-away. Astrophys. J. **571**, 40–55 (2002)

Murray, S.D., White, S.D.M., Blondin, J.M., Lin, D.N.C.: Dynamical instabilities in two-phase media and the minimum masses of stellar systems. Astrophys. J. **407**, 588–596 (1993)

Nagakura, T., Omukai, K.: Formation of population III stars in fossil HII regions: significance of HD. Mon. Not. R. Astron. Soc. **364**, 1378 (2005)

Nakamura, F., Umemura, M.: The stellar initial mass function in primordial galaxies. Astrophys. J. **569**, 549–557 (2002)
Nozawa, T., Kozasa, T., Umeda, H., Maeda, K., Nomoto, K.: Dust in the early Universe: Dust formation in the ejecta of Population III supernovae. Astrophys. J. **598**, 785–803 (2003)
Oh, S.P.: Reionization by hard photons. I. X-rays from the first star clusters. Astrophys. J. **553**, 499–512 (2001)
Oh, S.P., Haiman, Z.: Second-generation objects in the Universe: Radiative cooling and collapse of halos with virial temperatures above 10^4 K. Astrophys. J. **569**, 558–572 (2002)
Oh, S.P., Haiman, Z., Rees, M.J.: He II recombination lines from the first luminous objects. Astrophys. J. **553**, 73–77 (2001)
Omukai, K., Nishi, R.: Photodissociative regulation of star formation in metal-free pregalactic clouds. Astrophys. J. **518**, 64–68 (1999)
Omukai, K., Palla, F.: Formation of the first stars by accretion. Astrophys. J. **589**, 677–687 (2003)
Omukai, K., Schneider, R., Haiman, Z.: Can supermassive black holes form in metal-enriched high-redshift protogalaxies? Astrophys. J. **686**, 801–814 (2008)
Omukai, K., Tsuribe, T., Schneider, R., Ferrara, A.: Thermal and fragmentation properties of star-forming clouds in low-metallicity environments. Astrophys. J. **626**, 627–643 (2005)
O'Shea, B.W., Norman, M.L.: Population III star formation in a ΛCDM Universe. II. Effects of a photodissociating background. Astrophys. J. **673**, 14–33 (2008)
Osterbrock, D.O., Ferland, G.J.: Astrophysics of Gaseous Nebulae and Active Galactic Nuclei. University Science Books, Sausalito (2006)
Paardekooper, J.-P., et al.: The escape of ionizing radiation from high redshift dwarf galaxies. Astron. Astrophys. **530**, 87 (2011)
Padoan, P., Nordlund, Å., Kritsuk, A.G., Norman, M.L., Pak Shing, L.: Two regimes of turbulent fragmentation and the stellar initial mass function from primordial to present-day star formation. Astrophys. J. **661**, 972–981 (2007)
Pan, L., Scalo, J.: Mixing of primordial gas in lyman break galaxies. Astrophys. J. **654**, L29–L32 (2007)
Park, K., Ricotti, M.: Accretion into intermediate mass black holes regulated by radiation feedback I. Parametric study for spherically symmetric accretion. Astrophys. J. **739**, 2 (2011)
Pawlik, A.H., Milosavljević, M., Bromm, V.: The first galaxies: assembly of disks and prospects for direct detection. Astrophys. J. **731**, 54–70 (2011)
Pelupessy, F.I., Di Matteo, T., Ciardi, B.: How rapidly do supermassive black holes seeds grow at early times? Astrophys. J. **665**, 107–119 (2007)
Prieto, J., Padoan, P., Jimenez, R., Infante, L.: Pop. III stars from turbulent fragmentation at redshift $\sim$11. Astrophys. J. **731**, L38–L43 (2011)
Pringle, J.E.: Accretion discs in astropysics. Ann. Rev. Astron. Astrophys. **19**, 137–162 (1981)
Raiter, A., Schaerer, D., Fosbury, R.A.E.: Predicted UV properties of very metal-poor starburst galaxies. Astron. Astrophys. **523**, A64 (2010)
Razoumov, A.O., Sommer-Larsen, J.: Ionizing radiation from z = 4-10 galaxies. Astrophys. J. **710**, 1239–1246 (2010)
Read, J.I., Pontzen, A.P., Viel, M.: On the formation of dwarf galaxies and stellar haloes. Mon. Not. R. Astron. Soc. **371**, 885–897 (2006)
Regan, J.A., Haehnelt, M.G.: Pathways to massive black holes and compact star clusters in pre-galactic dark matter haloes with virial temperatures $\geq$ 10,000 K. Mon. Not. R. Astron. Soc. **396**, 343–353 (2009)
Ricotti, M., Gnedin, N.Y., Shull, J.M.: Feedback from galaxy formation: production and photodissociation of primordial H_2. Astrophys. J. **560**, 580–591 (2001)
Ricotti, M., Gnedin, N.Y., Shull, J.M.: The fate of the first galaxies. III. Properties of primordial dwarf galaxies and their impact on the intergalactic medium. Astrophys. J. **685**, 21–39 (2008)
Ricotti, M., Ostriker, J.P.: X-ray pre-ionization powered by accretion on the first black holes - I. A model for the WMAP polarization measurement. Mon. Not. R. Astron. Soc. **375**, 547–562 (2004)

Ricotti, M., Shull, J.M.: Feedback from galaxy formation: escaping ionizing radiation from galaxies at high redshift. Astrophys. J. **542**, 548–558 (2000)
Ripamonti, E.: The role of HD cooling in primordial star formation. Mon. Not. R. Astron. Soc. **376**, 709–718 (2007)
Rodrigues, L.F.S., de Souza, R.S., Opher, R.: Suppression of small baryonic structures due to a primordial magnetic field. Mon. Not. R. Astron. Soc. **406**, 482–485 (2010)
Safranek-Shrader, C., Bromm, V., Milosavljecić, M.: Fragmentation in the first galaxies. Astrophys. J. **723**, 1568–1582 (2010)
Santoro, F., Shull, J.M.: Critical metallicity and fine-structure emission of primordial gas enriched by the first stars. Astrophys. J. **643**, 26–37 (2006)
Scannapieco, E., Schneider, R., Ferrara, A.: The detectability of the first stars and their cluster enrichment signatures. Astrophys. J. **589**, 35–52 (2003)
Schaerer, D.: On the properties of massive Population III stars and metal-free stellar populations. Astron. Astrophys. **382**, 28–42 (2002)
Schaerer, D.: The transition from population III to normal galaxies: Lyalpha and HeII emission and the ionizing properties of high redshift starburst galaxies. Astron. Astrophys. **397**, 527 (2003)
Schleicher, D.R.G., Banerjee, R., Sur, S., Arshakian, T.G., Klessen, R.S., Beck, R., Spaans, M.: Small-scale dynamo action during the formation of the first stars and galaxies. I. The ideal MHD limit. Astron. Astrophys. **522**, A115–A124 (2010a)
Schleicher, D.R.G., Galli, D., Glover, S.C.O., Banerjee, R., Palla, F., Schneider, R., Klessen, R.S.: The influence of magnetic fields on the thermodynamics of primordial star formation. Astrophys. J. **703**, 1096–1106 (2009)
Schleicher, D.R.G., Spaans, M., Glover, S.C.O.: Black hole formation in primordial galaxies: chemical and radiative conditions. Astrophys. J. **712**, L69–L72 (2010b)
Schneider, R., Ferrara, A., Salvaterra, R.: Dust formation in very massive primordial supernovae. Mon. Not. R. Astron. Soc. **351**, 1379–1386 (2004)
Schneider, R., Omukai, K.: Metals, dust and the cosmic microwave background: fragmentation of high-redshift star-forming clouds. Mon. Not. R. Astron. Soc. **402**, 429–435 (2010)
Schneider, R., Omukai, K., Inoue, A.K., Ferrara, A.: Fragmentation of star-forming clouds enriched with the first dust. Mon. Not. R. Astron. Soc. **369**, 1437–1444 (2006)
Shang, C., Bryan, G.L., Haiman, Z.: Supermassive black hole formation by direct collapse: keeping protogalactic gas H_2 free in dark matter haloes with virial temperatures $T_{\rm vir} \geq 10^4$ K. Mon. Not. R. Astron. Soc. **402**, 1249–1262 (2010)
Shapiro, P.R., Kang, H.: Hydrogen molecules and the radiative cooling of pregalactic shocks. Astrophys. J. **318**, 32–65 (1987)
Siess, L., Livio, M., Lattanzio, J.: Structure, evolution, and nucleosynthesis of primordial stars. Astrophys. J. **570**, 329–343 (2002)
Silk, J., Langer, M.: On the first generation of stars. Mon. Not. R. Aston. Soc. **371**, 444–450 (2006)
Smith B.D., Sigurdsson, S.: The transition from the first stars to the second stars in the early Universe. Astrophys. J. **661**, L5–L8 (2007)
Smith, B.D., Turk, M.J., Sigurdsson, S., O'Shea, B.W., Norman, M.L.: Three modes of metal-enriched star formation in the early Universe. Astrophys. J. **691**, 441–451 (2009)
Spaans, M., Silk, J.: Pregalactic black hole formation with an atomic hydrogen equation of state. Astrophys. J. **652**, 902–906 (2006)
Stacy, A., Bromm, V.: Impact of cosmic rays on Population III star formation. Mon. Not. R. Astron. Soc. **382**, 229–238 (2007)
Stacy, A., Greif, T.H., Bromm, V.: The first stars: formation of binaries and small multiple systems. Mon. Not. R. Astron. Soc. **403**, 45–60 (2010)
Stahler, S.W., Palla, F.: The Formation of Stars. Wiley, Weinheim (2004)
Stahler, S.W., Shu, F.H., Taam, R.E.: The evolution of protostars. I - Global formulation and results. Astrophys. J. **241**, 637–654 (1980)
Stecher, T.P., Williams, D.A.: Photodestruction of hydrogen molecules in H I regions. Astrophys. J. **149**, L29–L30 (1967)

Suda, T., Aikawa, M., Machida, M.N., Fujimoto, M.Y., Iben, I.: Is HE 0107-5240 a primordial star? The characteristics of extremely metal-poor carbon-rich stars. Astrophys. J. **611**, 476–493 (2004)

Tan, J.C., McKee, C.F.: The formation of the first stars. I. Mass infall rates, accretion disk structure, and protostellar evolution. Astrophys. J. **603**, 383–400 (2004)

Tassis, K., Abel, T., Bryan, G.L., Norman, M.L.: Numerical simulations of high-redshift star formation in dwarf galaxies. Astrophys. J. **587**, 13–24 (2003)

Tenorio-Tagle, G.: Interstellar matter hydrodynamics and the dispersal and mixing of heavy elements. Astrophys. J. **111**, 1641–1650 (1996)

Tominaga, N., Umeda, H., Nomoto, K.: Supernova nucleosynthesis in population III 13-50 $M_\odot$ stars and abundance patterns of extremely metal-poor stars. Astrophys. J. **660**, 516–540 (2007)

Tornatore, L., Ferrara, A., Schneider, R.: Population III stars: hidden or disappeared? Mon. Not. R. Astron. Soc. **382**, 945–950 (2007)

Trenti, M., Stiavelli, M.: Formation rates of Population III stars and chemical enrichment of halos during the reionization era. Astrophys. J. **694**, 879–892 (2009)

Trenti, M., Stiavelli, M., Shull, J.M.: Metal-free gas supply at the edge of reionization: late-epoch Population III star formation. Astrophys. J. **700**, 1672–1679 (2009)

Tumlinson, J.: Chemical evolution in hierarchical models of cosmic structure I. Constraints on early stellar initial mass function. Astrophys. J. **641**, 1–20 (2006)

Tumlinson, J., Giroux, M.L., Shull, J.M.: Probing the first stars with hydrogen and helium recombination emission. Astrophys. J. **550**, L1–L5 (2001)

Turk, M.J., Abel, T., O'Shea, B.: The formation of Population III binaries from cosmological initial conditions. Science **325**, 601–605 (2009)

Uehara, H., Inutsuka, S.: Does deuterium enable the formation of primordial brown dwarfs? Astrophys. J. **531**, L91–L94 (2000)

Vasiliev, E.O., Shchekinov, Y.A.: The influence of ultrahigh-energy cosmic rays on star formation in the early Universe. Astron. Rep. **50** 778–784 (2006)

Vasiliev, E.O., Vorobyov, E.I., Shchekinov, Y.A.: First supernovae in dwarf protogalaxies. Astron. Astrophys. **489**, 505–515 (2008)

Wada, K., Venkatesan, A.: Feedback from the first supernovae in protogalaxies: the fate of the generated metals. Astrophys. J. **591**, 38–42 (2003)

Whalen, D., Hueckstadt, R.M., McConkie, T.O.: Radiative and kinetic feedback by low-mass primordial stars. Astrophys. J. **712**, 101–111 (2010)

Whalen, D., van Veelen, B., O'Shea, B.W., Norman, M.L.: The destruction of cosmological minihalos by primordial supernovae. Astrophys. J. **682**, 49–67 (2008)

Windhorst, R.A., Cohen, S.H., Jansen, R.A., Conselice, C., Yan, H.: How JWST can measure first light, reionization and galaxy assembly. New Astron. Rev. **50**, 113–120 (2006)

Wise, J.H., Abel, T.: Suppression of H_2 cooling in the Ultraviolet background. Astrophys. J. **671**, 1559–1567 (2007a)

Wise, J.H., Abel, T.: Resolving the formation of Protogalaxies. I. Virialization. Astrophys. J. **665**, 899–910 (2007b)

Wise, J.H., Abel, T.: Resolving the formation of Protogalaxies. III. Feedback from the first stars. Astrophys. J. **685**, 40–56 (2008)

Wise, J.H., Cen, R.: Ionizing photon escape fractions from high-redshift dwarf galaxies. Astrophys. J. **693**, 984–999 (2009)

Wise, J.H., Turk, M.J., Norman, M.L. Abel, T.: The birth of a galaxy: primordial metal enrichment and population II stellar populations. Astrophys. J. (arXiv:1011.2632)

Wolcott-Green, J., Haiman, Z.: Suppression of HD-cooling in protogalactic gas clouds by Lyman-Werner radiation. Mon. Not. R. Astron. Soc. **412**, 2603–2616 (2011)

Wood, K., Loeb, A.: Escape of ionizing radiation from high-redshift galaxies. Astrophys. J. **545**, 86–99 (2000)

Wyithe, J.S.B., Cen, R.: The extended star formation history of the first generation of stars and the reionization of cosmic hydrogen. Astrophys. J. **659**, 890–907 (2007)

Wyithe, J.S.B., Loeb, A.: Reionization of hydrogen and helium by early stars and quasars. Astrophys. J. **586**, 693–708 (2003)
Xu, H., O'Shea, B.W., Collins, D.C., Norman, M.L., Li, H., Li, S.: The biermann battery in cosmological MHD simulations of Population III star formation. Astrophys. J. **688**, L57–L60 (2008)
Yajima, H., Choi, J.-H., Nagamine, K.: Escape fraction of ionizing photons from high-redshift galaxies in cosmological SPH simulations. Mon. Not. R. Astron. Soc. **412**, 411–422 (2011)
Yoshida, N.: From the first stars to the first galaxies. New Astron. **50**, 19–23 (2006)
Yoshida, N., Abel, T., Hernquist, L., Sugiyama, N.: Simulations of early structure formation: primordial gas clouds. Astrophys. J. **592**, 645–663 (2003)
Yoshida, N., Oh, S.P., Kitayama, T., Hernquist, L.: Early cosmological H II/He III regions and their impact on second-generation star formation. Astrophys. J. **663**, 687–707 (2007a)
Yoshida, N., Omukai, K., Hernquist, L.: Formation of massive primordial stars in a reionized gas. Astrophys. J. **667**, L117–L120 (2007b)
Yoshida, N., Omukai, K., Hernquist, L.: Protostar formation in the early Universe. Science **321**, 669–671 (2008)
Zackrisson, E.: Pointing the James Webb Space Telescope through lensing clusters – can the first stars and galaxies be detected? (arXiv:1101.4033) (2011)
Zackrisson, E., et al.: The spectral evolution of the first galaxies. I. James Webb space telescope detection limits and color criteria for population III galaxies. Astrophys. J. **740**, 13 (2011)

Observing the First Galaxies

James S. Dunlop

Abstract I endeavour to provide a thorough overview of our current knowledge of galaxies and their evolution during the first billion years of cosmic time, corresponding to redshifts $z > 5$. After first summarizing progress with the seven different techniques which have been used to date in the discovery of objects at $z > 5$, I focus thereafter on the two selection methods which have yielded substantial samples of galaxies at early times, namely Lyman-break and Lyman-α selection. I discuss a decade of progress in galaxy sample selection at $z \simeq 5 - 8$, including issues of completeness and contamination, and address some of the confusion which has been created by erroneous reports of extreme-redshift objects. Next I provide an overview of our current knowledge of the evolving ultraviolet continuum and Lyman-α galaxy luminosity functions at $z \simeq 5 - 8$, and discuss what can be learned from exploring the relationship between the Lyman-break and Lyman-α selected populations. I then summarize what is known about the physical properties of these galaxies in the young universe, before considering the wider implications of this work for the cosmic history of star formation, and for the reionization of the universe. I conclude with a brief summary of the exciting prospects for further progress in this field in the next 5–10 years. Throughout, key concepts such as selection techniques and luminosity functions are explained assuming essentially no prior knowledge. The intention is that this chapter can be used as an introduction to the observational study of high-redshift galaxies, as well as providing a review of the latest results in this fast-moving research field up to the end of 2011.

J.S. Dunlop (✉)
Institute for Astronomy, University of Edinburgh, Royal Observatory, Edinburgh, EH9 3HJ, UK
e-mail: jsd@roe.ac.uk

T. Wiklind et al. (eds.), *The First Galaxies*, Astrophysics and Space Science Library 396,
DOI 10.1007/978-3-642-32362-1_5,

1 Introduction

One conclusion of this chapter will be that the very "first" galaxies have almost certainly not yet been observed. But in recent years we have undoubtedly witnessed an observational revolution in the study of early galaxies in the young Universe which, for reasons outlined briefly below, I have chosen to define as corresponding to redshifts $z > 5$ (a good, up-to-date overview of the physical properties of galaxies at $z = 2 - 4$ is provided by Shapley 2011).

The discovery and study of galaxies at redshifts $z > 5$ is really the preserve of the twenty-first century, and has been one of the most spectacular achievements of astronomy over the last decade. From the ages of stellar populations in galaxies at lower redshifts it was known that galaxies must exist at $z > 5$ (e.g. Dunlop et al. 1996), but observationally the $z = 5$ "barrier" wasn't breached until 1998, and then only by accident (Dey et al. 1998). Although this discovery of a Lyman-α emitting galaxy at $z = 5.34$ was serendipitous, it in effect represented the first successful application at $z > 5$ of the long-proposed (e.g. Patridge and Peebles 1967a,b) and oft-attempted (e.g. Koo and Kron 1980; Djorgovski et al. 1985; Pritchet and Hartwick 1990; Pritchet 1994) technique of searching for "primeval" galaxies in the young universe on the basis of bright Lyman-α emission. This discovery was important not just for chalking up the next integer value in redshift, but also because this was the first time that the redshift/distance record for any extra-galactic object was held by a "normal" galaxy which had *not* been discovered on the basis of powerful radio or optical emission from an active galactic nucleus (AGN). Later the same year, two more galaxies selected at $z > 5$ on the basis of their starlight (Fernandez-Soto et al. 1999) were spectroscopically confirmed at $z = 5.34$ by Spinrad et al. (1998), and the Lyman-α selection record was advanced to $z = 5.64$ (Hu et al. 1998).

In this chapter I will explain how these breakthroughs heralded a new era in the study of the high-redshift Universe, in which conceptually simple but technologically challenging techniques have now been successfully applied to discover thousands of galaxies at $z > 5$, and to extend the redshift record out to $z \simeq 9$. The key instrumental/observational advances which have facilitated this work are the last two successful refurbishments of the *Hubble Space Telescope* (*HST*; first with the ACS optical camera, and most recently with the near-infrared WFC3/IR imager), the provision of wide-field optical and near-infrared imaging on 4–8-m class ground-based telescopes (Suprime-Cam on 8.2-m Subaru telescope, WFCAM on the 3.8-m UK InfraRed Telescope (UKIRT), and ISAAC/Hawk-I on the 8.2-m Very Large Telescope (VLT)), the remarkable performance of the 85-cm *Spitzer Space Telescope* at mid-infrared wavelengths, and finally the advent of deep red-sensitive optical spectroscopy on the 10-m Keck telescope (with LRIS and DEIMOS), the VLT (with FORS2), and on Subaru (with FOCAS).

I will also endevour to summarize what we have learned about the *properties* of these early galaxies from this multi-frequency, multi-facility investigation and, as a result, what new information we have gleaned about the evolution of the universe during the first $\simeq$1 Gyr of cosmic time. I conclude with a very brief discussion of the prospects for further progress over the next decade; a more detailed description of future facilities is included elsewhere in this volume.

The cosmological parameters of relevance to this work are summarized briefly in the next section. Where required, all magnitudes are reported in the AB system, where $m_{AB} = 31.4 - 2.5\log(f_\nu/1\,\mathrm{nJy})$ (Oke and Gunn 1983).

2 Why Redshift z > 5

It is perhaps useful to first pause briefly to review what a redshift of $z = 5$ actually means, and why it matters. Redshift, z, is, of course, simply a straightforward way to quantify the ratio of the observed wavelength (λ_o) to the emitted wavelength (λ_e) of light:

$$1 + z = \frac{\lambda_o}{\lambda_e}. \tag{1}$$

Because, for $v \ll c$ the relativistic longitudinal Doppler effect can be approximated by $z = v/c$ it is sometimes stated that redshifts $z > 1$ cannot be interpreted in terms of recession velocities, but the relativistic Doppler effect is:

$$1 + z = \frac{\lambda_o}{\lambda_e} = \sqrt{\frac{1 + v/c}{1 - v/c}} \tag{2}$$

and so $z = 5$ corresponds to $v = 0.946c$.

However, in a Universe with matter, at least some of any observed redshift should be attributed to gravitational effects, and in any case the precise recessional velocity of a galaxy several billion years ago is of little real interest. What is more helpful is to recognise that, since light doesn't interact with itself, the stretching of the wavelength of light must simply reflect the overall expansion of the Universe, i.e.

$$1 + z = \frac{\lambda_o}{\lambda_e} = \frac{R(t_{now})}{R(t_e)} \tag{3}$$

where $R(t)$ is simply the scale factor which describes the time evolution of our apparently isotropic, homogeneous Universe.

Thus, when we observe a galaxy at $z = 5$ we are observing light which was emitted from that galaxy when the Universe was 1/6th of its present size (and at the highest redshifts currently probed, $z \simeq 9$, the Universe was 1/10th of its present size).

The precise age at which the Universe was 1/6th of its present size of course depends on the dynamics of the expansion. With our current "best-bet" concordance cosmology of a flat Universe with a matter density parameter of $\Omega_m = 0.27$, a vacuum energy (or dark energy) density parameter of $\Omega_\Lambda = 0.73$, and a Hubble Constant $H_0 = 71\,\mathrm{kms^{-1}Mpc^{-1}}$ (WMAP7; Komatsu et al. 2011; Larson et al. 2011), $z = 5$ corresponds to an age of 1.2 Gyr, equivalent to $\simeq$9 % of current cosmic time. Thus, to a very reasonable approximation, the study of the universe at $z > 5$ can be thought of as a direct window into the first Gyr, or first 10 % of the growth and evolution of cosmic structure.

Finally, at the risk of stating the obvious, it must always be remembered that different redshifts correspond not only to different times, but also to different places. Thus, when we presume to connect observations of galaxies at different redshifts to derive an overall picture of cosmic evolution, we are implicitly assuming homogeneity; i.e. that "back-then, over there" is basically the same as "back-then, over here". For this to be true it is crucial that surveys for high-redshift galaxies contain sufficient cosmological volume to be "representative" of the Universe at the epoch in question. As we shall see, at $z > 5$ this remains a key challenge with current observational facilities.

3 Finding Galaxies at z > 5: Selection Techniques

There are, in principal, several different ways to attempt to pinpoint extreme-redshift galaxies amid the overwhelming numbers of lower-redshift objects on the sky. The two methods that have proved most effective in recent years both involve optical to near-infrared observations of rest-frame ultraviolet light, and both rely on neutral Hydrogen. The first method, the so called Lyman-break technique, selects Lyman-break galaxies (LBGs) via the distinctive "step" introduced into their blue ultraviolet continuum emission by the blanketing effect of neutral hydrogen absorption (both within the galaxy itself, and by intervening clouds along the observer's line-of-sight; see Fig. 1). The second method selects galaxies which are Lyman-α emitters (LAEs), via their highly-redshifted Lyman-α emission lines, produced by hydrogen atoms in their interstellar media which have been excited by the ultraviolet light from young stars. Both of these techniques have now been used to discover large numbers of galaxies out to $z \geq 7$, and are therefore discussed in detail in the two subsections below.

The only real drawback of these two techniques is that they are only capable of selecting galaxies which are young enough to produce copious amounts of ultraviolet light, and are sufficiently dust free for a fair amount of this light to leak out in our direction. In an attempt to find galaxies at $z > 5$ which are at least slightly older (remembering there is only $\simeq$1 Gyr available) some authors (e.g. Wiklind et al. 2008) have endeavoured to select galaxies on the basis of the Balmer break, even though, at $\lambda_{rest} = 3{,}646\,\text{Å}$ this break is moved to $\lambda_{obs} > 2.4\,\mu\mathrm{m}$ at $z > 5$. Since this lies beyond the near-infrared wavelength range accessible from the ground, this work is only possible due to the power of the IRAC camera on board *Spitzer*, which can be used to observe from 3 to 8 μm. As discussed later

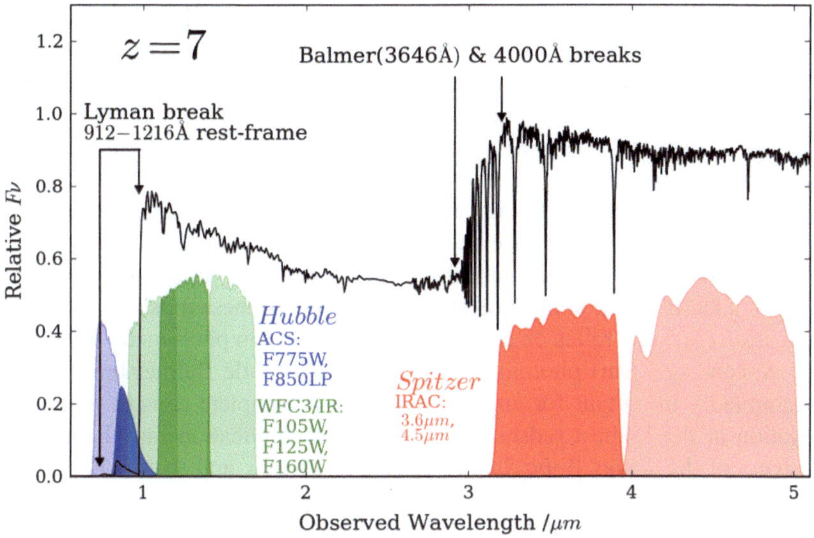

Fig. 1 An illustration of the redshifted form of the rest-frame ultraviolet spectral energy distribution (SED) anticipated from a young galaxy at $z \simeq 7$, showing how the ultraviolet light is sampled by the key red optical (i_{775}, z_{850}) and near-infrared (Y_{105}, J_{125}, H_{160}) filters on-board *HST* (in the ACS and WFC3/IR cameras respectively), while the longer-wavelength rest-frame optical light is probed by the 3.6 μm and 4.4 μm IRAC channels on-board *Spitzer*. Wavelength is plotted in the observed frame, with flux-density plotted as relative f_ν (i.e. per unit frequency). The spectrum shows the sharp drop at $\lambda_{rest} = 1{,}216$ Å due to the strong "Gunn–Peterson" absorption by intervening neutral hydrogen anticipated at this redshift (here predicted following Madau (1995); see also the observed spectrum of the most distant quasar shown in Fig. 2). Longward of this "Lyman-break" the spectrum shown is simply that of the intrinsic integrated galaxy starlight as predicted for a 0.5 Gyr-old galaxy by the evolutionary spectral synthesis models of Bruzual and Charlot (2003) (using Padova-1994 tracks, assuming constant star formation, zero dust extinction, and 1/5th solar metallicity). The characteristic sharp step in the galaxy continuum at $\lambda_{rest} = 1{,}216$Å (which at $z \simeq 7$ is predicted to result in a very red $z_{850} - Y_{105}$ colour) holds the key to the effective selection of Lyman-break galaxies at $z > 5$, as discussed in detail in Sect. 3.1. The theoretical spectrum shown here does not include the Lyman-α emission line which is produced by excitation/ionization of hydrogen atoms in the inter-stellar medium of the galaxy; this offers the main current alternative route for the selection of high-redshift galaxies (see Fig. 8 and Sect. 3.2), and the only realistic hope for spectroscopic confirmation of galaxy redshifts at $z > 5$ with available instrumentation. Also not shown are other nebular emission lines at rest-frame optical wavelengths, which can complicate the apparent strength of the key Balmer or 4,000Å break as measured by the IRAC photometry. In the absence of serious line contamination, the strength of this break offers a key estimate of the age of the stellar population, with consequent implications for a meaningful measurement of galaxy stellar mass. The gap between the WFC3/IR and IRAC filters can be filled for brighter objects with ground-based K-band imaging, but will not be covered from space until the advent of *JWST* (courtesy S. Rogers)

in Sect. 5.1, *Spitzer* has certainly proved very effective at measuring the strength of Balmer breaks in high-redshift galaxies which have already been discovered via their ultraviolet emission but, to date, Balmer-break *selection* has yet to uncover a galaxy at $z > 5$ which could not have been discovered via other techniques (i.e. the only spectroscopically-confirmed Balmer-break selected galaxy in the sample compiled by Wiklind et al. is also a Lyman-break galaxy). This lack of success may of course simply be telling us that there are not many (or indeed any) galaxies at these early epochs of the correct age and star-formation history ($\geq$0.5 Gyr-old and no longer forming stars) to be *better* selected via their Balmer break than their ultraviolet emission; in an era of essentially limitless gas fuel, and almost universal star-formation activity this would not be altogether surprising. However, it may also be the case that Balmer-break selection is simply premature with current facilities; the spectral feature itself is not nearly as strong (a drop in flux density of a factor of $\simeq$2 at most) as the Lyman-break at these redshifts, and at $z > 5$ its detection currently relies on combining *Spitzer* 3.6 μm photometry with ground-based K-band (2.2 μm) photometry (Fig. 1). Thus, while Balmer-break selection is undoubtedly important for ensuring we have a complete census of the galaxy population at the highest redshifts, its successful application may have to await the advent of the *James Webb Space Telescope* (*JWST*), and so it is not discussed further here.

A fourth technique, which is only now coming of age at $z > 5$, involves the selection of extreme-redshift galaxies via sub-mm/mm observations of their redshifted thermal dust emission. By definition this approach is incapable of detecting the *very* first primeval galaxies, devoid of any of the elements required for dust, but chemical enrichment appears to be a very rapid process, and both dust and molecular emission have certainly been detected in objects at $z > 6$ (Walter et al. 2003; Robson et al. 2004; Wang et al. 2010). Although this dust emission is powered by ultraviolet emission from young stars, it is already clear that at least some of the galaxies successfully discovered via sub-mm observations at more modest redshifts display such strong dust extinction that they could not have been selected by rest-frame ultraviolet observations. Thus, while we might expect dust to become less prevalent at extreme redshifts (and there is evidence in support of this presumption—e.g. Bouwens et al. 2009a; Zafar et al. 2010; Bouwens et al. 2010a; Dunlop et al. 2012; Finkelstein et al. 2012) it will be important to pursue sub-mm/mm selection at $z > 5$ over the coming decade to ensure a complete picture of early galaxy evolution. As with Balmer-break selection, this is a technique which (for technical reasons) is still in its infancy, although excitingly the first spectroscopic redshift at $z > 5$ for a mm-selected galaxy has recently been measured purely on the basis of redshifted CO line emission ($z \simeq 5.3$; Riechers et al. 2010). Over the next few years this whole field should be revolutionized by the advent of the Atacama Large Millimetre Array (ALMA).

A fifth approach, is that high-redshift "galaxies" continue to be located on the basis of both optical and radio emission powered by accretion onto their central super-massive black holes. Indeed, the quasar redshift record has recently crossed the $z = 7$ threshold ($z = 7.085$; Mortlock et al. 2011), and significant numbers of quasars are now known at $z > 6$ (e.g. Fan et al. 2003, 2006; Willott et al. 2010). High-redshift quasars are rare but, because of their brightness, have the potential to provide much useful information on the state of the inter-galactic medium (IGM) at early times (e.g. Carilli et al. 2010), as well as providing signposts towards regions of enhanced density in the young universe. However, the very strong active nuclear emission which facilitates the discovery of high-redshift quasars also makes it extremely difficult to detect, never mind study, the stellar populations in their host galaxies (e.g. Targett, Dunlop and McLure 2012), and so they are inevitably of limited use for the detailed investigation of early galaxy evolution. By contrast it is perfectly possible to study the stellar populations in high-redshift radio galaxies (e.g. McCarthy 1993; Dunlop et al. 1996; Seymour et al. 2007), and indeed for many years essentially all of our knowledge of galaxies at $z > 3$ was derived from the optical–infrared–sub-mm study of objects which were originally selected on the basis of radio-frequency synchrotron emission powered by super-massive black holes (e.g. Lilly 1988; Dunlop et al. 1994; Rawlings et al. 1996). However, in recent years the search for increasingly high-redshift radio galaxies has rather run out of steam; the $z = 5$ threshold was passed in 1999 ($z = 5.197$; van Breugel et al. 1999), but 12 years later the radio-galaxy redshift record remains unchanged. This difficulty in further progress is perhaps not unexpected, given the now well-established decline in the number density of powerful radio sources beyond $z \simeq 3$, and the unhelpfully strong k-correction provided by steeply-falling power-law synchrotron emission (Dunlop and Peacock 1990; Rigby et al. 2011). Nevertheless, searches for higher-redshift radio galaxies will continue, motivated at least in part by the desire to find even a few strong radio beacons against which to measure the 21-cm analogue of the Lyman-α forest as we approach the epoch of reionization (the "21-cm forest"; Carilli, Gnedin and Owen 2002; Furlanetto and Loeb 2002; Mack and Wyithe 2012). However, at least for now, radio-continuum selected objects offer little direct insight into galaxy evolution at the very highest redshifts. A thorough review of what is currently known about distant radio galaxies and their environments is provided by Miley and De Breuck (2008), who also include a compendium of known high-redshift radio galaxies.

A relatively new sixth, and remarkably effective route to pinpointing high-redshift objects has recently arrived with the discovery of Gamma-Ray Bursts (GRBs). These are now regularly detected via monitoring with gamma-ray satellites such as *Swift* (Gehrels et al. 2004), and then rapidly followed up with a range of ground-based observations (e.g. Fynbo et al. 2009). Long-duration GRBs are thought to arise from the death of very massive, possibly metal-poor stars (Woosley and Bloom 2006), and observationally have been associated with Type 1c supernovae (e.g. Hjorth et al. 2003). Regardless of their precise physical origin, they have proved to be very luminous events which are visible out to the highest redshifts, $z > 8$ (the gamma-ray positions are poor, but rapid follow-up can pinpoint the fading optical/near-infrared afterglow unambiguously and, if quick enough, can also yield robust redshift information). GRBs broke the $z = 5$ "barrier" very quickly after their discovery, with a redshift of $z = 6.295$ measured for GRB 050904 by Haislip et al. (2006) and Kawai et al. (2006). Another GRB at $z > 5$ followed the next year with the discovery of GRB 060927 at $z = 5.467$ (Ruiz-Veasco et al. 2007). Two years later, GRBs wrested the redshift record from quasars and LAEs, with Greiner et al. (2009) reporting a redshift of $z = 6.7$ for GRB 080913. Then, most spectacularly, a GRB became the first spectroscopically confirmed object at $z > 8$, with GRB 090423 being convincingly shown to lie at $z = 8.23$ (Salvaterra et al.

2009; Tanvir et al. 2009). Most recently, it has been argued that GRB 090429B lies at $z \sim 9.4$ (Cucchiara et al. 2011) but the robustness of this (photometric) redshift is currently a matter of debate. Given this impressive success in redshift record breaking, the reader may be surprised to learn that I have chosen not to consider GRBs further in this chapter. The reason is that, to date, while the hosts of many lower-redshift GRBs have been uncovered (e.g. Perley et al. 2009) follow-up observations targeted on the positions of faded GRB remnants at $z > 5$ have yet to yield useful information on their host galaxies. This is, of course, an interesting result in its own right. It indicates that, as arguably expected, GRBs largely occur in faint dwarf galaxies which lie below the sensitivity limits of even our very best current instrumentation. Specifically, the follow-up *HST* WFC3/IR imaging of the $z = 8.23$ GRB 090423 reaching $J_{125} \simeq 28.5$ has failed to detect the host galaxy (Tanvir et al. in prep), while the host of GRB 090429B is apparently undetected to $Y_{105} \simeq 28$ (Cucchiara et al. 2011). Thus, while as discussed by Robertson and Ellis (2012), high-redshift GRBs can already provide important insights into global cosmic star-formation history, their usefulness as transient signposts towards extreme-redshift galaxies is unlikely to be properly exploited until the advent of *JWST*.

Finally, over the next decade we are likely to see the emergence of a seventh technique for finding extreme-redshift galaxies via radio-wavelength spectroscopy. Specifically, following the first successful mm-to-radio CO-line redshift determinations, in addition to the above-mentioned targeted CO line follow-up of pre-selected mm/sub-mm sources with ALMA, we can expect to see "blind" spectroscopic surveys for CO and for highly-redshifted 21-cm atomic Hydrogen emission with the new generation of radio facilities (e.g. Carilli 2011).

3.1 Lyman-Break Selection

In the absence of dust obscuration, young star-forming galaxies are expected to be copious emitters of UV continuum light, with a star-formation rate $SFR = 1\,M_{\odot}$ year^{-1} predicted to produce a UV luminosity at $\lambda_{rest} \simeq 1{,}500$ Å of $f_{\nu} \simeq 8 \times 10^{27}$ erg s^{-1} Hz^{-1} for a Salpeter (1955) initial mass function (Madau et al. 1998). For reference, this corresponds to an absolute magnitude of $M_{1{,}500} \simeq -18$ which, at $z \simeq 7$, translates to an observed near-infrared J-band magnitude of $J \simeq 28.5$. As we shall see, this is very comparable to the detection limit of the deepest *HST* WFC3/IR imaging currently available.

The basic idea of selecting distant objects (galaxies or quasars) via the signature introduced by hydrogen absorption of this ultraviolet light goes back several decades (e.g. Meier 1976a,b). As first successfully implemented in the modern era by Guhathakurta et al. (1990) and Steidel and Hamilton (1992), the aim was to select galaxies at $z \sim 3$ by searching for sources in which the Lyman-limit at $\lambda_{rest} = 912$ Å had been redshifted to lie between the U and B_j filters at $\lambda_{obs} \simeq 3{,}600$ Å. All ultraviolet-bright astrophysical objects display an intrinsic drop in their spectra

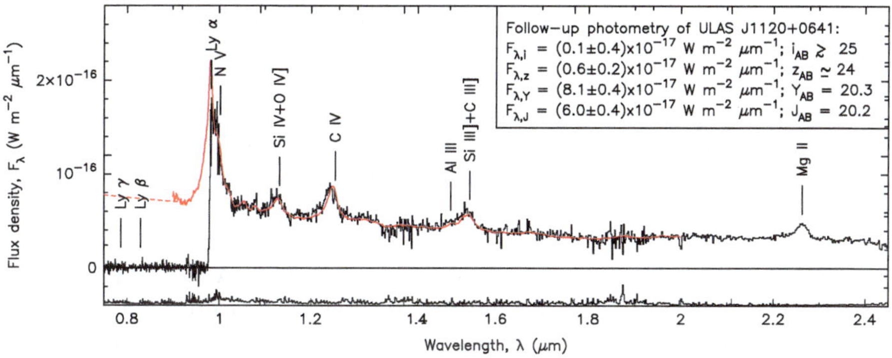

Fig. 2 The near-infrared spectrum of the most distant known quasar ULASJ112001.48 + 064124.3, the first (and, to date, only) quasar discovered at redshifts $z > 7$ ($z = 7.085$; Mortlock et al. 2011). The data are shown in *black*, with the 1-σ error spectrum shown at the base of the plot. Despite being observed only 0.77 billion years after the Big Bang, this quasar has an intrinsic spectrum essentially identical to that displayed by lower-redshift quasars with, for example, strong Carbon lines indicating approximately solar metallicity (the *red curve* shows the average spectrum of 169 quasars in the redshift range $2.3 < z < 2.6$). However, shortward of $\lambda_{rest} \simeq 1{,}216\,\text{Å}$ the spectrum provides an excellent demonstration of the Gunn–Peterson effect, whereby the increased fraction of neutral hydrogen along the line-of-sight has completely obliterated the UV continuum emission from the quasar. This sudden drop in flux-density shortward of Lyman-α is the key spectral feature which facilitates not just the selection of rare extreme-redshift quasars such as this, but also the selection of fainter but much more numerous "Lyman-break galaxies" (LBGs) at redshifts $z > 5$ (see Fig. 1) (courtesy D. Mortlock)

at $\lambda_{rest} = 912\,\text{Å}$ (which corresponds to the ionization energy of the hydrogen atom in the ground state), and the expectation was that, in young galaxies, this drop would be very strong (roughly an order-of-magnitude in flux density) due to a combination of the hydrogen edge in stellar photospheres, and photo-electric absorption by the interstellar neutral hydrogen gas (expected to be abundant in young galaxies). At the highest redshifts, the ever denser intervening neutral hydrogen clouds also produce increasing Lyman-α absorption (between energy levels 1 and 2 in the hydrogen atom) resulting in an ever-thickening Lyman-α forest which impacts on the continuum of the target galaxy between $\lambda_{rest} = 1{,}216\,\text{Å}$ and $\lambda_{rest} = 912\,\text{Å}$. At moderate redshifts the average blanketing effect of this forest simply produces an additional (and useful) signature in the galaxy spectrum in the form of an apparent step in the continuum below Lyman-α (a factor of $\sim$2 drop in flux density at $\lambda_{rest} = 1{,}216\,\text{Å}$ at $z \sim 3$; Madau 1995). However, as discussed further below (and illustrated in Figs. 1 and 2) ultimately the forest becomes so optically thick that it kills virtually all of the galaxy light at $\lambda_{rest} < 1{,}216\,\text{Å}$, rendering the original 912 Å break irrelevant, and Lyman-break selection in effect becomes the selection of objects with a sharp break at $\lambda_{rest} = 1{,}216\,\text{Å}$.

The beauty of the Lyman-break selection technique is that it can be applied using imaging with broad-band filters, allowing potentially large samples of high-redshift galaxies to be selected for spectroscopic follow-up and confirmation. When selecting galaxies in this way, what one is looking for are objects which are repeatedly visible (and fairly blue) in the longer wavelength images, but then effectively disappear in the bluest image under consideration. For this reason such objects are often called "dropout" galaxies. Thus,"U-dropouts" (or simply "U-drops") are galaxies which disappear in the U-band filter, and are therefore expected to have their Lyman limit moved to $\lambda_{obs} \simeq 3{,}500$ Å implying a redshift $z \sim 3$ (in practice $2.5 \leq z \leq 3.5$). Similarly, "B-drops" (or "G-drops") are expected to be galaxies at $z \sim 4$, while "V-drops" should have $z \sim 5$. Thus, deep broad-band optical imaging can be used to select samples of galaxies in bands of increasing redshift.

The simple act of colour selection yields redshifts accurate to $\delta z \simeq 0.1 - 0.2$. Consequently, with the aid of simulations to estimate the effective redshift distribution and cosmological volume probed by each specific drop-out criterion, luminosity functions (LFs) can be derived in broad redshift bands without recourse to optical spectroscopy. However, for proper assessment of completeness/contamination, and the determination of redshifts with sufficient accuracy to allow robust clustering measurements, spectroscopic follow-up is essential.

The huge break-throughs enabled by the successful application of the "dropout" technique in the 1990s are perhaps best exemplified by the work of Steidel and collaborators (who were able to use the 10-m Keck telescope to spectroscopically confirm large samples of LBGs, enabling LF and clustering measurements—e.g. Steidel et al. 1996, 2000) and by the study of Madau et al. (1996) who applied the technique to the deep *HST* WFPC2 U_{300}, B_{450}, V_{606}, I_{814} imaging in the Hubble Deep Field (HDF; Williams et al. 1996; Ferguson et al. 2000) to produce the first measurement of the average cosmic star-formation density out to $z \sim 4$. A full overview of this "low-redshift" work is beyond the scope of this Chapter, but a thorough review of the success of the Lyman-break technique in enabling the discovery and study of galaxies in the redshift range $2 < z < 5$ can be found in Giavalisco (2002).

The ensuing decade has seen rapid progress from $z \simeq 5$ to $z \simeq 8$, in part because this selection technique is, in principle, even more straightforward at $z \geq 5$ than at lower redshifts. This is because by $z = 5$ the Lyman-α forest produced by intervening clouds of neutral Hydrogen is expected to be so dense that the anticipated break in the continuum level at $\lambda_e \simeq 1{,}216$ Å is $\simeq$1.8 mag., or a factor $\simeq$5 in average flux density (Madau 1995). This is more than twice as strong as any of the other intrinsically strong breaks displayed by the starlight from galaxies (e.g. the $\lambda = 4{,}000$ Å break in an old stellar population, produced by an accumulation of absorption lines from ionized metals, (especially Ca II H and K lines at 3,933 and 3,968 Å), or the $\lambda = 3{,}646$ Å Balmer break in a $\sim$0.5 Gyr-old post-starburst galaxy, most prominent in A stars, with $T \sim 10{,}000$ K). By $z > 6.5$, observations of the highest-redshift quasars indicate that essentially all flux shortward of Lyman-α

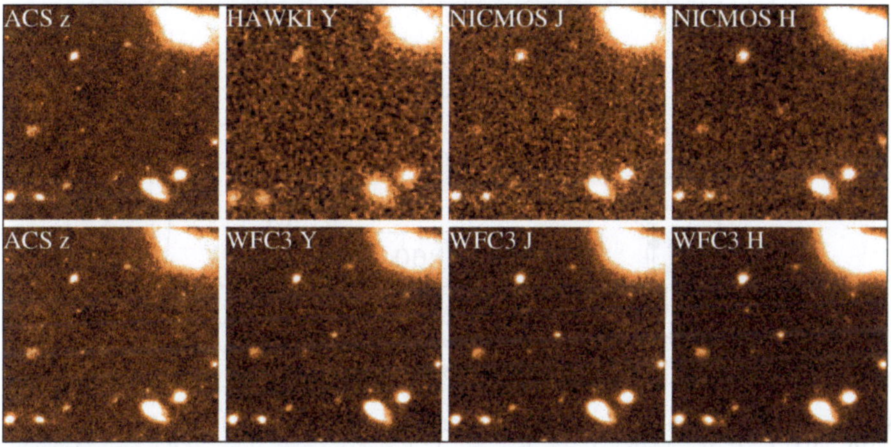

Fig. 3 The Lyman-break selection of a $z \simeq 7$ galaxy uncovered in the Hubble Ultra-Deep Field (HUDF). The upper row of plots shows postage stamps of the available data at z_{850}, Y, J_{110}, H_{160} prior to the advent of the new WFC3/IR near-infrared camera on *HST* in 2009. The lower row of plots shows the hugely-improved near-infrared imaging provided by WFC3/IR for the same object; it can be clearly seen that this galaxy is strongly detected in the three longest-wavelength passbands (H_{160}, J_{125} and Y_{105}) but drops out of the z_{850} image altogether, due to the presence of the Lyman-break redshifted to $\lambda_{obs} \simeq 1\,\mu$m, as was illustrated in Fig. 1 (courtesy R. McLure)

is extinguished (Fig. 2), and LBG selection effectively becomes the selection of galaxies with a complete "Gunn–Peterson Trough" (Gunn and Peterson 1965).

Thus, given sufficiently good signal:noise, and appropriate broad-band filters, the selection of Lyman-break galaxies at $z > 5$ should be easy and reasonably clean, and indeed has proved to be so once detector and telescope developments were successfully combined to deliver the necessary deep, red-sensitive imaging.

3.1.1 Lyman-Break Galaxies at z > 5

The main reason for a delay in progress in LBG selection beyond $z \simeq 5$ was the need for sufficiently deep imaging in at least two wavebands longer than the putative Lyman break; as illustrated in Figs. 1, 3, 4 and 5, at least two colours (hence three wavebands) are needed to confirm both the existence of a strong spectral break, and a blue colour longward of the break (as anticipated for a young, ultraviolet-bright galaxy; see Sect. 3.1.3 on potential contaminants). This need was finally met with the refurbishment of the *HST* in March 2002 with a new red-sensitive optical camera, the Advanced Camera for Surveys (ACS), and a new cooling system for the Near Infrared Camera and Multi-Object Spectrometer (NICMOS). Crucially, the ACS was quickly used to produce and release the deepest ever optical image of the sky, the 4-band (B_{435}, V_{606}, i_{775}, z_{850}) Hubble Ultra Deep Field (HUDF; Beckwith et al. 2006), covering an area of $\simeq$11 arcmin^2 to typical depths of

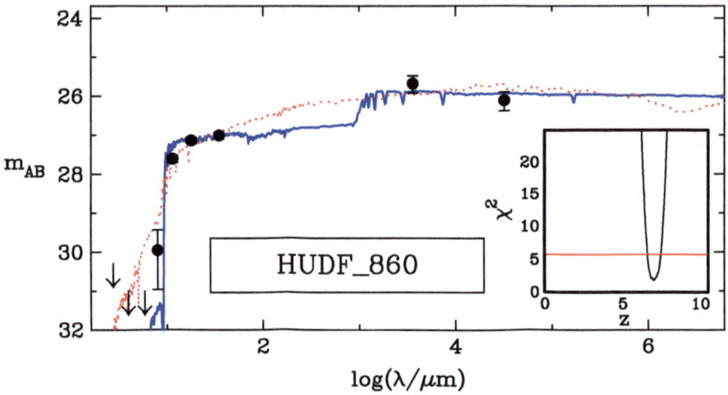

Fig. 4 An example of the galaxy-template SED-fitting analysis employed by McLure et al. (2009, 2010, 2011) for high-redshift galaxy selection, which makes optimum use of the available multi-wavelength photometry (including de-confused *Spitzer* IRAC fluxes; McLure et al. 2011). Based on the evolutionary synthesis models of Bruzual and Charlot (2003), not only redshift, but also age, star-formation history, dust extinction/reddening, mass and metallicity are all varied in search of the best-fitting solution. This also enables robust errors to be placed on the range of acceptable photometric redshifts, after marginalising over all other parameters. In this case the photometry provides more than adequate accuracy and dynamic range to exclude all redshift solutions other than that indicated by the *blue line*, which yields $z \simeq 6.96 \pm 0.25$. The *thin dotted red line* shows the best-fitting alternative solution at low redshift, albeit in this case this alternative is completely unacceptable

$m_{AB} \simeq 29$ for point sources. This field (or at least 5.7 arcmin^2 of it) was also imaged with NICMOS, in the J_{110} and H_{160} bands by Thompson et al. (2005, 2006) to depths of $m_{AB} \simeq 27.5$. Around the same time the ACS was also used as part of the Great Observatories Deep Survey (GOODS) program to image two 150 arcmin^2 fields (again in B_{435}, V_{606}, i_{775}, z_{850}) to more moderate depths, $m_{AB} \simeq 27.5 - 26.5$ (GOODS-North, containing the HDF, and GOODS-South, containing the HUDF; Giavalisco et al. 2004). Deep *Spitzer* IRAC imaging (at 3.6, 4.5, 5.6, 8 μm) was also obtained over both GOODS fields, and a co-ordinated effort was made to obtain deep K_s-band imaging for GOODS-South from the ground with ISAAC on the 8.2-m VLT (Retzlaff et al. 2010).

The result was a flood of papers reporting the discovery of "i-drop" galaxies at $z \simeq 6$ (Bouwens et al. 2003, 2004a, 2006; Bunker et al. 2003, 2004; Dickinson et al. 2004; Stanway et al. 2003, 2004, 2005; Yan and Windhorst 2004; Malhotra et al. 2005; Beckwith et al. 2006; Grazian et al. 2006), and even an (arguably premature, but partially successful) attempt to uncover "z_{850}-drop" galaxies at $z \simeq 7$ (Bouwens et al. 2004c) and set limits at even higher redshifts (Bouwens et al. 2005).

Spectrosopic follow-up was rapidly achieved for several of the brighter "i-drops" yielding the first spectroscopically-confirmed LBGs at $z \simeq 6$ (Bunker et al. 2003; Lehnert and Bremer 2003; Vanzella et al. 2006; Stanway et al. 2007), and some of these were even successfully detected with *Spitzer* at 3.6 μm and 4.5 μm, yielding

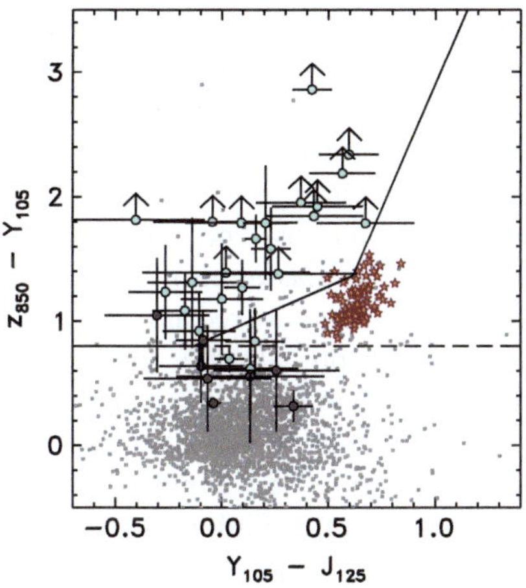

Fig. 5 An illustration provided by Finkelstein et al. (2010) of some of the limitations of using a simple, strict, colour–colour criterion to select high-redshift LBGs. *Filled circles* indicate high-redshift galaxies selected by SED fitting, with the *dark-grey circles* indicating galaxies with $6.0 < z_{phot} < 6.3$ and the *blue circles* highlighting galaxies at $6.3 < z_{phot} < 7.5$. *Arrows* are 1-σ limits. The *solid lines* show the selection cuts adopted by Oesch et al. (2010a), and the *dashed line* is from Yan et al. (2010) (both designed to select LBGs at $z \simeq 7$). The *small grey squares* are low-redshift galaxy contaminants with $z_{phot} < 6.0$, and the *red stars* indicate the colours of galactic brown-dwarf stars. The colour cuts result in the inclusion of many contaminants as well as the exclusion of genuine high-redshift candidates that are identified via full SED fitting (which makes more optimal use of all the available data, including marginal detections at optical wavelengths) (courtesy S. Finkelstein)

some first estimates of their stellar masses and star-formation histories (e.g. Labbé et al. 2006; Yan et al. 2006; Eyles et al. 2007)

Further spectroscopic follow-up of $z \geq 5$ LBGs in the GOODS fields has been steadily pursued with Keck and the VLT over the last few years, (e.g. Stark et al. 2009, 2010, 2011; Vanzella et al. 2009) yielding interesting results on mass density, evolution, and Lyman-α emission from LBGs which are discussed further in Sects. 5.4 and 5.5.

From 2005, progress in wide-area red optical and near-infrared imaging with Suprime-Cam on the Subaru telescope, and WFCAM (Casali et al. 2007) on UKIRT (via the UKIDSS survey; Lawrence et al. 2007) led to the first significant samples of brighter $z \simeq 6$ galaxies being selected from ground-based surveys covering areas approaching $\simeq$1 deg^2 (Kashikawa et al. 2004; Shimasaku et al. 2005; Ota et al. 2005; McLure et al. 2006, 2009; Poznanski et al. 2007; Richmond et al. 2009). As discussed further below in Sect. 4.1, this work complements the deeper but much smaller-area *HST* surveys by providing better sampling of the bright end of the LF.

Motivated by the availability of 12-band CFHT + Subaru + UKIRT + *Spitzer*-IRAC photometry in the UKIDSS Ultra Deep Survey (UDS) field (coincident with the Subaru/XMM-Newton Deep Survey (SXDS); Furusawa et al. 2008), McLure et al. (2006) also introduced a new approach to selecting galaxies at $z > 5$, replacing simple two-colour "dropout" criteria with multi-band redshift estimation via model spectral energy distribution (SED) fitting (a technique commonly adopted at lower redshifts—e.g. Mobasher et al. 2004, 2007; Cirasuolo et al. 2007, 2010). This approach has the advantage of using all of the data in a consistent way (including multiple non-detections) and captures the uncertainty in redshift (and resulting uncertainty in stellar masses etc) for each individual object (Fig. 4). In addition, it provides better access to redshift ranges where the simple two-colour dropout technique is sub-optimal (due, for example, to the Lyman-break lying within rather than at the edge of a filter bandpass; see Fig. 5). It can also yield a redshift probability distribution for each source (e.g. Finkelstein et al. 2010; although there is a debate to be had about appropriate priors), and explicitly exposes alternative acceptable redshift solutions (e.g. Dunlop et al. 2007), enabling targeted spectroscopic follow-up to reject these if desired. Nevertheless, careful simulation work is still required to estimate incompleteness and contamination, and the SED-fitting approach can arguably be harder for others to replicate than simple two-colour selection.

One disadvantage of ground-based imaging is the potential for $z > 5$ LBG sample contamination by cool dwarf stars (see Sect. 3.1.3). On the other hand, because the LBG candidates are relatively bright, spectroscopic follow-up has proved very productive, and has now yielded Lyman-α redshifts in the range $z \simeq 6 - 6.5$ for $\simeq$30 LBGs selected from ground-based surveys (Nagao et al. 2004, 2005, 2007; Jiang et al. 2011; Curtis-Lake et al. 2012). These spectroscopic programs provide not only more accurate redshifts, but also enable measurement of the prevalence and strength of Lyman-α emission from LBGs as a function of redshift and continuum luminosity. Such measurements have the potential to shed light on the connection between LBGs and LAEs, the cosmic evolution of dust, and the process of reionization (see Sects. 4.3 and 6.2)

At $z > 6.5$ ground-based selection of LBGs becomes extremely difficult due to the difficulty in reaching the necessary near-infrared depths, although quite how difficult depends of course on the shape of the LF at $z \simeq 7$. Such progress as has been made with Subaru and the VLT between $z \simeq 6.5$ and $z \simeq 7.3$ is discussed in the next subsection.

3.1.2 Lyman-Break Galaxies at z = 7 – 10

By redshifts $z \simeq 7$, the Lyman break moves to $\lambda_{obs} \simeq 1\,\mu$m, beyond the sensitivity regime of even red-sensitive CCD detectors. As a result, efforts to uncover LBGs at $z > 6.5$ were largely hamstrung by the lack of sufficiently-deep near-infrared imaging, until the installation of the long-awaited new camera, WFC3, in the *HST* in May 2009. Due to its exquisite sensitivity and (by space standards) wide field-of-view (4.8 arcmin^2), the infrared channel of this camera, WFC3/IR, offered a $\sim$40-fold improvement in mapping speed over NICMOS for deep near-infrared surveys. This, coupled with the availability of an improved near-infrared filter set, immediately rendered obsolete the few heroic early attempts to uncover LBGs at $z > 6.5$ with NICMOS (e.g. Bouwens et al. 2004; Bouwens et al. 2010c), even those assisted by gravitational lensing (e.g. Richard et al. 2008; Bradley et al. 2008; Bouwens et al. 2009a; Zheng et al. 2009).

The remarkable improvement offered by WFC3/IR at near-infrared wavelengths is illustrated in Fig. 3, which shows the imaging data available before and after Sept 2009 for (arguably) the only moderately-convincing "z_{850}-drop" $z \simeq 7$ galaxy uncovered with NICMOS + ACS (Bouwens et al. 2004; Oesch et al. 2009; McLure et al. 2010). These images are extracted from the first (Sept. 2009) release of the WFC3/IR $Y_{105}, J_{125}, H_{160}$ imaging of the HUDF, taken as part of the HUDF-09 treasury program (PI: Illingworth). This reached previously unheard-of depths $m_{AB} \simeq 28.5$, and immediately transformed our knowledge of galaxies at $z > 6.5$, with four independent groups reporting the first substantial samples of galaxies with $6.5 < z < 8.5$ (Oesch et al. 2010a; Bouwens et al. 2010b; McLure et al. 2010; Bunker et al. 2010; Finkelstein et al. 2010). Both the above-mentioned alternative approaches to LBG selection were applied to these new data, with McLure et al. (2010) and Finkelstein et al. (2010) undertaking full SED fitting (Fig. 4), while the other groups applied standard two-colour "drop-out" criteria (Fig. 5). Three independent reductions of the raw data were also undertaken prior to LBG selection. Given this, the level of agreement between the $6.5 < z < 8.5$ source lists was (and remains) undeniably impressive. The era of galaxy study at $z > 7$ has now truly arrived.

The initial HUDF WFC3/IR data release was rapidly followed by the release of the WFC3 Early Release Science (ERS) data (Windhorst et al. 2011). The infrared component of this dataset comprised 2-orbit depth WFC3/IR imaging in $Y_{098}, J_{125}, H_{160}$ over 10 pointings in the northern part of the GOODS-South field, and thus complemented the HUDF imaging by delivering imaging of $\simeq$40 arcmin^2 to $m_{AB} \simeq 27.5$. The intervening two years have seen the completion of the HUDF-09 program, involving deeper imaging of the HUDF itself to $m_{AB} \simeq 29$, and Y_{105}, J_{125}, H_{160} imaging of two parallel fields to $m_{AB} \simeq 28.5$. The combined HUDF-09 and ERS dataset has now been analysed in detail for LBGs at $z > 6.5$ (again by several independent groups; Wilkins 2010; Wilkins et al. 2011a; Bouwens et al. 2011b; Lorenzoni et al. 2011; McLure et al. 2011), and has yielded samples of $\sim$70 candidate LBGs at $z \sim 7$, $\sim$50 at $z \sim 8$, and possibly one galaxy at $z \sim 10$ (Bouwens et al. 2011a; Oesch et al. 2012). As with the original data release, despite disagreement over certain individual sources (see, for example, the careful cross-checking performed by McLure et al. 2011) there is generally good agreement over the $z \sim 7$ and $z \sim 8$ galaxy samples, especially if attention is restricted to the brighter objects. However, where the data have been pushed to the limit, potential contamination by low-redshift interlopers becomes more of an issue (see below), and in particular there is some debate over the robustness of the $z \sim 10$ galaxy. This discovery relies on detection in a single band (H_{160}) because the proposed Lyman-break lies at the long-wavelength edge of the J_{125} filter. Therefore, while there is little doubt that this is a real object, there is currently no direct observational evidence that it displays a blue slope longward of the break. As illustrated in

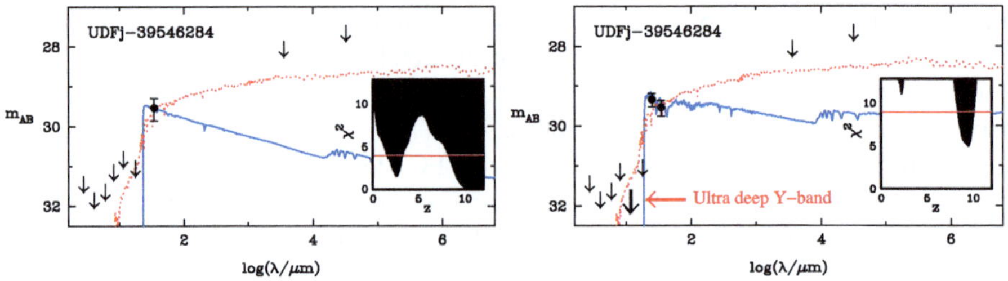

Fig. 6 Recovering a reliable galaxy population at $z \simeq 9-10$ requires *ultra-deep* exposures across the Lyman-break and a strategically-chosen deployment of at least *two* WFC3/IR filters for source detection. (*Left*): The marginal nature of the $z \simeq 10$ candidate claimed by Bouwens et al. (2011a) based on a sole H_{160} detection. In addition to a possible high z SED (*blue line*), an acceptable solution also exists at $z \simeq 2$ (*red dotted line*). The inset shows χ^2 as a function of redshift. Possible 'flux-boosting' at H_{160} is an additional concern. (*Right*): Deeper Y_{105} imaging (as planned in 2012), coupled with the security of two detections (J_{140} & H_{160}) above the Lyman break, should allow secure identification of this source and the elimination the low-z solution if this galaxy really does lie at extreme redshifts $z > 9$ (simulated J_{140} photometry was here inserted assuming $z_{true} \simeq 9.5$). The planned ultra-deep WFC3/IR UDF12 imaging of the HUDF in *HST* Cycle 19 may detect up to $\simeq$20 sources beyond $z \simeq 8.5$ to $H_{160} = 29.5$

Fig. 6, to push LBG selection beyond $z \simeq 8$ really requires still deeper imaging, and the additional use of the J_{140} filter (to provide two detections of the galaxy ultraviolet continuum above the Lyman-break at $z \sim 9$). Such a program has now been approved in the HUDF, and is planned with *HST* WFC3/IR in summer 2012.

In addition to this ground-breaking ultra-deep near-infrared imaging, wider field surveys with WFC3/IR are now underway. In particular Trenti et al. (2011, 2012) and Yan et al. (2011) have recently used parallel WFC3/IR imaging to search for "brighter" Y-drop $z \sim 8$ LBGs, yielding several candidates which are potentially bright enough to be amenable to spectroscopic follow-up with ground-based near-infrared spectrographs. The 3-year, 902-orbit, Cosmic Assembly Near-infrared Deep Extragalactic Survey (CANDELS) Treasury Program has also recently commenced (Grogin et al. 2011; Koekemoer et al. 2011).[1] This will ultimately deliver WFC3/IR imaging (with parallel ACS optical imaging) to $m_{AB} \sim 27$ over $\simeq$0.25 deg^2 spread over 5 different well-studied fields, including deeper survey regions reaching $m_{AB} \sim 28$ over $\simeq$0.04 deg^2 (split between GOODS-North and GOODS-South). This survey is expected to provide the area and depth required to enormously clarify our understanding of the prevalence and properties of moderate luminosity (L^*) galaxies at $z \simeq 6.5 - 8.5$.

Finally, progress is also expected from WFC3/IR imaging of lensing clusters. Imaging of the Bullet Cluster has already yielded several $z \simeq 7$ LBG candidates (Hall et al. 2012) and a second major (524 orbit) *HST* multi-cycle Treasury Program,

[1] http://candels.ucolick.org

the Cluster Lensing and Supernoa Survey with Hubble (CLASH)[2], will deliver multi-band WFC3 imaging of 25 clusters over the next 3 years (Postman et al. 2012).

These rapidly-growing samples of WFC3/IR-selected LBGs at $z > 6.5$ are providing a wealth of new information on galaxies and their evolution in the first billion years, not least because many of the brighter ones have also proved to be detectable at 3.6 μm with *Spitzer* IRAC. As a result, even without spectroscopic redshifts, it has already been possible not only to obtain the first meaningful measurements of the galaxy luminosity function at $z \sim 7$ and $z \sim 8$ (see Sect. 5.4) but also to explore the physical properties of these young galaxies (i.e. masses, stellar populations, sizes; see Sect. 5.5).

Nevertheless, spectroscopic follow-up is being vigorously pursued (e.g. Schenker et al. 2012). It is to be hoped that the new, wider area WFC3/IR surveys yield brighter $z \sim 7-8$ LBGs which are amenable to spectroscopic follow-up, as effective ground-based near-infrared spectroscopy of the most distant galaxies revealed via the HUDF09 imaging at $m_{AB} \simeq 28.5$ has, unsurprisingly, proved extremely challenging. In particular, while Lehnert et al. (2010) reported a spectroscopic redshift $z \simeq 8.55$ for the most distant credible HUDF LBG discovered by McLure et al. (2010) and Bouwens et al. (2010b), this observation took 15 h of integration with the near-infrared spectrograph SINFONI on the VLT, and the claimed marginal detection of Lyman-α has *not* been confirmed by independent follow-up spectroscopy (Bunker et al., in prep). Indeed, as discussed further below, follow-up spectroscopy of even the brighter $z \simeq 7$ LBG candidates selected from ground-based surveys has, to date, not been particularly productive, for reasons that are still a matter of some debate (see Sect. 4.3). But it must be noted that the current lack of spectroscopic redshifts should *not* be taken as implying that most of the $z \simeq 7$ and $z \simeq 8$ are not robust, as given sufficiently deep photometry all potential contaminants can be excluded, and a redshift estimated accurate to $\delta z \simeq \pm 0.1$. In fact, it may well be the case that, by $z \simeq 7$, many galaxies do not produce measurable Lyman-α emission, and much of the current ongoing spectroscopic effort is really directed at trying to better quantify the evolution of Lyman-α emission from LBGs, a measurement which has the potential to shed light on the physics of reionization (see Sects. 4.3 and 6.2).

From the ground, LBG selection has now been pursued with some success right up to (but not significantly beyond) $z \simeq 7$, due to advent of deep Y-band imaging on both Subaru/Suprime-Cam (Ouchi et al. 2009b) and Hawk-I on the VLT (Castellano et al. 2010a,b). From the deep Y-band and z-band imaging of both the Subaru Deep Field (SDF) and GOODS-North, Ouchi et al. (2009b) reported 22 z'-drops to a depth of $y = 25.5 - 26$ over a combined area of $\simeq 0.4\,\mathrm{deg}^2$, but the lack of comparably-deep near-infrared data at longer wavelengths forced them to make major corrections (by about a factor $\simeq 2$) for contamination. Nevertheless, three of these LBGs now have spectroscopically-confirmed redshifts at $z \sim 7$ based on Lyman-α emission-line detections with the DEIMOS spectrograph on the Keck

[2]http://www-int.stsci.edu/~postman/CLASH

telescope (Ono et al. 2012). The VLT Hawk-I imaging undertaken by Castellano et al. (2010a,b) covered a smaller area ($\simeq$200 arcmin^2), but to somewhat deeper depths, and has yielded $\simeq$20 z-drops to $Y \simeq 26.5$. Spectroscopic follow-up of this sample with FORS2 on the VLT has now provided five Lyman-α spectroscopic redshifts in the range $6.7 < z < 7.1$ (Fontana et al. 2010; Vanzella et al. 2011; Pentericci et al. 2011).

Given the current concerns over the validity of the Lehnert et al. (2010) redshift, at the time of writing the robust *spectroscopic* redshift record for an LBG (or indeed any galaxy or quasar) stands at $z = 7.213$ (Ono et al. 2012). Further spectroscopic follow-up at $z \simeq 7$ is, of course, in progress, but wide-area ground-based exploration of the bright end of the LBG luminosity function at even higher redshifts must await deeper Y, J, H, K-band imaging (now underway with UltraVISTA; see Sect. 7).

3.1.3 Contaminants and Controversies

Spectroscopic follow-up (or improved multi-frequency photometry) of LBG samples has revealed, not unexpectedly, that three different types of interloper can contaminate samples of LBGs at $z > 5$.

The first class of contaminant comprises very red dusty galaxies, or AGN, at lower redshifts. Such objects can produce a rapid drop in flux density over a relatively short wavelength range, which can sometimes be so severe as to be mistaken for a Lyman-break, especially if the two filters designed to straddle the break are actually not immediately adjacent in wavelength (e.g. z' and J). Because such red dusty objects do not rapidly turn over to produce very blue colours at longer wavelengths, LBG sample contamination by such objects is not too serious an issue provided **(i)** a sufficiently strong Lyman-break criterion is enforced (unfortunately not always the case), **(ii)** sufficiently deep multi-band imaging is available at longer wavelengths to properly establish the longer wavelength SED slope, and **(iii)** LBG selection is confined to young, unreddened, reasonably-blue galaxies. However, if, as attempted by Mobasher et al. (2005), one seeks to select more evolved objects without very blue slopes longward of the proposed Lyman-break, then things can become difficult. As shown by Dunlop et al. (2007), very dusty objects at $z \simeq 1.5-2.5$ can easily be mistaken for evolved, high-mass LBGs at $z \simeq 5-6$, and templates with reddening as extreme as $A_V > 6$ sometimes need to be considered to reveal the alternative low-redshift solution. Such reddening is extreme, but the point is that "dropout" selection specifically designed to find LBGs at $z > 5$ transpires to also be an excellent method for selecting the rare, most extremely-reddened objects in the field at the appropriate lower redshifts. As discussed by Dunlop et al. (2007), often *Spitzer* MIPS 24 μm detections can help to reveal low-redshift dust-enshrouded interlopers, but this experience illustrates how difficult it will be to robustly uncover any significantly-evolved or reddened galaxies at $z > 5$.

This confusion lies behind several dubious/eroneous claims of extreme-redshift galaxies in the literature. Examples include not only the supposed $z \simeq 6$ ultra-massive galaxy uncovered by Mobasher et al. (2005) in the HUDF, but also the claimed discovery of a bright $z \simeq 9$ galaxy reported by Henry et al. (2008) (subsequently retracted when deeper optical imaging revealed a significant detection in the i'-band; Henry et al. 2009). It is also the likely reason that most of the $z > 6.5$ galaxies tentatively uncovered by Hickey et al. (2010) from the VLT Hawk-I Y-band imaging of GOODS-South have proved to be false (in the light of the subsequent ERS + CANDELS WFC3/IR imaging of the field) and, as illustrated in Fig. 7, is part of the explanation (in combination with eroneous photometric error analysis) for recent claims of very bright $z \simeq 7$ galaxies in the COSMOS field (despite supposed "tentative" spectroscopic confirmation at $z \simeq 7$ for two objects; Capak et al. 2011b). Fortunately, continuity arguments indicate that this may become less of a problem when attempting to select LBGs at the highest redshifts and faintest magnitude limits, as the reddened lower-redshift interloper population seems to become (relatively) much less prevalent in this region of parameter space.

The second class of contaminant comprises cool galactic stars, specifically M, L and (in the case of $z \simeq 7$ LBG selection) T dwarfs. This is a long-established problem in the colour-selection of high-redshift radio-quiet quasars which are unresolved in all but the very deepest images (Hewett et al. 2006). However, as discussed by many authors (e.g. McLure et al. 2006; Stanway et al. 2008a,b; Vanzella et al. 2009; Hickey et al. 2010) the compactness of high-redshift galaxies (see Sect. 5.4) means that contamination by cool dwarf stars has also become an important issue in the search for high-redshift LBGs. The problem is most acute for ground-based surveys both because most $z > 5$ LBGs are unresolved with even good ground-based seeing, and because the brighter LBGs are so much rarer on the sky (McLure et al. 2009; Capak et al. 2011b) than the fainter more numerous population revealed by the deeper *HST* imaging.

The particular problem of T-dwarf contamination of $z \simeq 7$ LBG searches has arguably been under-estimated until very recently, in part because our knowledge of T dwarfs has evolved in tandem with LBG searches over the last decade (Knapp et al. 2004; Chiu et al. 2008; Burningham et al. 2010). Specifically, early $z \simeq 7$ LBG "dropout" criteria appear to have assumed that T dwarfs did not display colours redder than $z - J \simeq 1.8$ (e.g. Bouwens et al. 2004), but cooler dwarfs have since been found with $z - J > 2.5$ (e.g. Burningham et al. 2008, 2010; Delorme et al. 2008; Leggett et al. 2009; Lucas et al. 2010; Liu et al. 2011). For ground-based $z \simeq 7$ LBG searches, the key to excluding dwarf-star contamination lies in having sufficiently-accurate multi-band infrared photometry since, for example, T-dwarfs have redder $Y - J$ colours (by $\simeq$1 mag) than genuine $z \simeq 7$ LBGs (and different IRAC colours; Stanway et al. 2008a). This is a further argument in favour of multi-band SED fitting which, given Y, J, H, K and IRAC photometry can often reveal a stellar contaminant on the basis of failure to achieve an acceptable fit with any galaxy template (as shown in the lower panel of Fig. 7). Given the above-mentioned high level of spectroscopic completeness achieved by Curtis-Lake et al. (2012) and Jiang et al. (2011) (and the results of stacking analyses; McLure et al. 2006, 2009) it seems unlikely that the published ground-based $z \simeq 6$ LBG samples are seriously contaminated by dwarf stars, but the situation remains more confused for bright surveys at $z \simeq 7$.

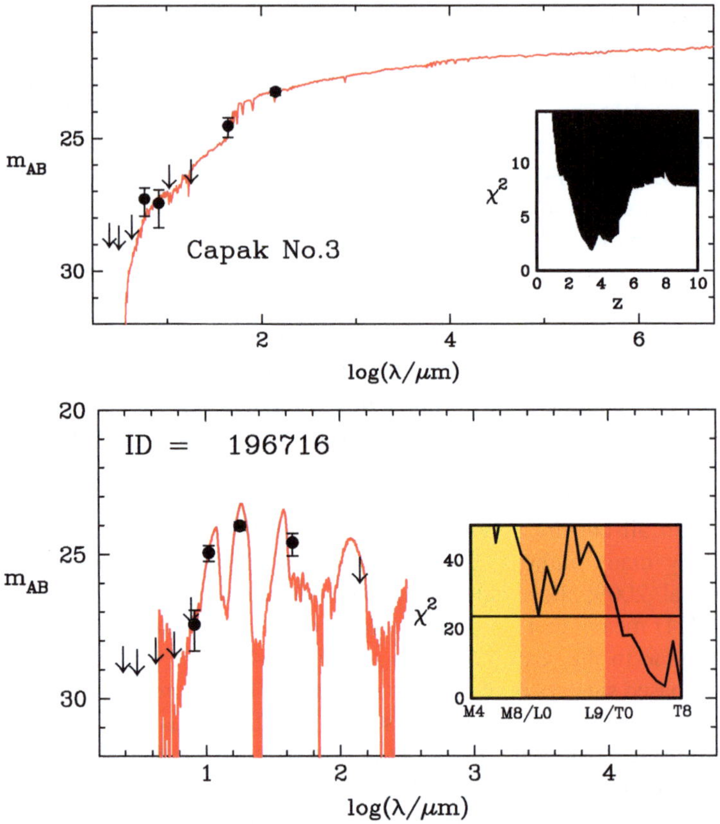

Fig. 7 Examples of two different types of interlopers which can contaminate LBG samples, especially those selected at brighter magnitudes from ground-based imaging. The *upper panel* shows the best SED fit and χ^2 versus z for a COSMOS galaxy claimed by Capak et al. (2011b) to meet the standard LBG selection criterion at $z \simeq 7$, and to be tentatively confirmed by near-infrared spectroscopy at $z = 7.69$. In fact, with the improved near-infrared photometry provided by the UltraVISTA survey it is clear that no acceptable redshift solution for this galaxy exists at $z > 5$ and that the original object selection was based on inadequate photometric error analysis (Bowler et al. 2012). The true best-fitting model solution corresponds to a moderately dusty galaxy at $z \simeq 3.5$. The *lower panel* shows another object selected from the UltraVISTA imaging in the COSMOS field which really does meet the standard $z' - J{:}J - H$ colour criterion for a $z = 7$ galaxy, but which as shown here is in fact a T-dwarf galactic star. In this case the photometry (especially in the crucial Y-band) is of high enough quality that no acceptable solution could be found with a galaxy SED at any redshift, but this is not always the case. Fortunately, both these types of contaminant become (at least statistically) less of a problem for $z \simeq 7$ LBG surveys at fainter magnitudes, because the most dusty galaxies at $z \simeq 2-4$ tend to be high-mass objects, and because the number counts of cool dwarf stars fall (or at least certainly plateau) beyond $J = 24$ (Ryan et al. 2011) due to the scale-height of the galactic plane

Fortunately, due to the combination of image depth, small field-of-view, and high angular resolution, T-dwarf contamination of the $z \simeq 7$ LBG samples revealed by the new deep WFC3/IR imaging is expected to be extremely small. This is confirmed by considering that the typical absolute J-band (AB) magnitude of T-dwarf stars is $J \simeq 19$ (Leggett et al. 2009). At the depths probed by the WFC3/IR imaging of the HUDF, a T-dwarf contaminant would thus have to be located at a distance of $0.5 - 1.0\,\mathrm{kpc}$. Given this distance is $2 \rightarrow 3$ times the estimated galaxy thin disk scale-height of $\simeq$300 pc (e.g. Reid and Majewski 1993; Pirzkal et al. 2009), it is clear that significant contamination is unlikely. This is not to suggest that dwarf stars cannot be found at such distances as, for example, Stanway et al. (2008a) report the discovery of M dwarfs out to distances of $\simeq$10 kpc. However, the surface density is low, with the integrated surface density over all M-dwarf types contained within $\simeq$1 kpc amounting to $\simeq 0.07\,\mathrm{arcmin}^{-2}$. Extrapolating these results to T dwarfs is somewhat uncertain, but a comparable surface density for L and T-dwarf stars is supported by the search for such stars in deep fields undertaken by Ryan et al. (2005, 2011). The results of this work suggest that the 4.5-arcmin^2 field-of-view of WFC3/IR data should contain $\leq$0.5 T-dwarf stars down to a magnitude limit of $z_{850} = 29$.

The final class of contaminant, as revealed by the lower-redshift secondary solutions in SED-based redshift estimation (McLure et al. 2010; Finkelstein et al. 2010) consists of fairly blue, $\simeq$0.5-Gyr old post-starburst galaxies which display a strong Balmer break. Given sufficient signal-to-noise there is really no room for confusion, as the Balmer break can never approach the strength of the anticipated Lyman-break at $z > 5$ (e.g. before it faded the $z = 8.2$ GRB displayed $Y - J > 4$). However, the SED-fits shown by McLure et al. (2010) demonstrate that, with inadequate photometric dynamic range, a Balmer break at $z \simeq 2$ can be mistaken for a Lyman-break at $z \simeq 8$. Fortunately the potential contaminants occupy a rather specific regime of parameter space (i.e. they must lie in a narrow redshift range, a narrow age range, be virtually dust-free, and have very low stellar masses to be confused with $z \simeq 7 - 8$ LBGs selected at the faintest magnitudes) and continuity arguments can be advanced that they are likely rare (e.g. Bouwens et al. 2011b), but the real lesson here is the importance of ensuring that any imaging shortward of any putative Lyman-break is sufficiently deep to exclude lower-redshift interlopers (not necessarily easy with the deepest WFC3/IR imaging, given the depth of the available complementary ACS optical imaging).

3.2 *Lyman-α Selection*

The intrinsic Lyman-α emission from young galaxies is expected to be strong, reaching large rest-frame equivalent widths $EW_{rest} \simeq 200\,\mathrm{Å}$ if driven by star formation (Charlot and Fall 1993). A star-formation rate of $SFR = 1\,\mathrm{M_{\odot}\,year^{-1}}$ corresponds to a Lyman-α luminosity of $\simeq 1 \times 10^{42}\,\mathrm{erg\,s^{-1}}$ (Kennicutt 1998).

However, for many years, blank-field searches for Lyman-α emitters (LAEs) at even moderate redshifts were disappointingly unsuccessful (e.g. Koo and Kron 1980; Pritchet and Hartwick 1990), raising fears that observable Lyman-α in high-redshift galaxies might be severely compromised by dust, because of the potentially long path lengths traversed by Lyman-α photons through the interstellar medium due to resonant scattering (Charlot and Fall 1991, 1993). However, as mentioned at the beginning of this chapter, by the end of the twentieth century a few $z > 5$ LAEs had been uncovered through the complementary techniques of long-slit spectroscopy (covering small areas but a broad redshift range) and narrow-band imaging (covering larger areas but a narrower redshift range). For a while these two techniques were competitive (Stern et al. 2000) but, with the advent of genuinely wide-field CCD imaging cameras on 8-m class telescopes, narrow-band searches for LAEs have surged ahead, and have proved spectacularly successful in uncovering large samples of galaxies at $z > 5$ (e.g. Ouchi et al. 2005, 2008).

Modern narrow-band imaging searches are sensitive to Lyman-α rest-frame equivalent widths down to $EW_{rest} \simeq 15\,\text{Å}$ (helped at high redshift by the fact that $EW_{obs} = (1+z)EW_{rest}$) and limiting line flux-densities $f \simeq 5 \times 10^{-18}\,\mathrm{erg\,s^{-1}\,cm^{-2}}$. At $z \simeq 7$ this corresponds to a Lyman-α luminosity $L \simeq 2.5 \times 10^{42}\,\mathrm{erg\,s^{-1}}$ which, in the absence of obscuration, is equivalent to a star-formation rate $SFR \simeq 2\,\mathrm{M_{\odot}\,year^{-1}}$. Thus, like the most sensitive LBG surveys at high redshifts, LAE selection can now detect galaxies at $z \simeq 7$ with a star-formation rate comparable to that of the Milky Way (e.g. Chomiuk and Povich 2011).

The basic technique involves comparing images taken through a narrow-band (100 – 200 Å wide) filter with a broad-band (or nearby narrow-band) image at comparable wavelengths. At very high redshifts, the efficiency of this approach is sensibly optimized by designing filters to image in low-background regions between the OH atmospheric emission lines which begin to plague substantial wavelength ranges beyond $\lambda_{obs} \simeq 7{,}000\,\text{Å}$. For this reason, samples of $z > 5$ LAEs are generally confined to pragmatically-selected redshift bands (Fig. 8).

Narrow-band searches for LAEs complement broad-band surveys for LBGs by probing a largely distinct region of parameter space. The weaknesses of narrow-band searches are that they probe smaller redshift ranges and hence smaller cosmological volumes (for a given survey area), and obviously can only uncover that fraction of the galaxy population which actually displays relatively bright Lyman-α emission. They are also subject to severe contamination by emission-line galaxies at lower-redshifts, which can only be sorted out via follow-up spectroscopy, or additional broad-band (or further tuned narrow-band) imaging (see Sect. 3.2.2). On the other hand, narrow-band imaging is sensitive to objects with much fainter continua than can be detected in LBG surveys, delivers targets for follow-up spectroscopy which are at least already known to contain an emission line, and is extremely effective at uncovering large-scale structures where many objects lie within a relatively narrow redshift band (e.g. Capak et al. 2011a).

As with LBG selection, a detailed overview of LAE studies at $z < 5$ is beyond the scope of this chapter, but a helpful overview of this "lower-redshift" work is provided by Ouchi et al. (2003), who first used narrow-band imaging through

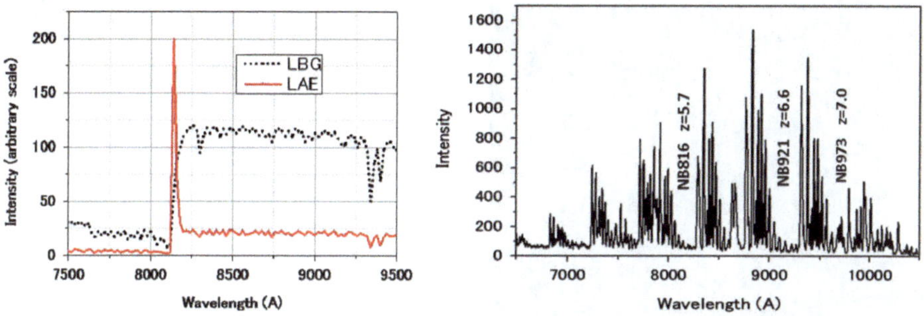

Fig. 8 The selection of high-redshift galaxies via Lyman-α emission. The *left-hand panel* illustrates the typical spectrum of a Lyman-α emitter (*solid line*) compared with a Lyman-break galaxy (*broken line*) at an assumed redshift $z = 5.7$, showing the Lyman-α emission-line redshifted from $\lambda_e = 1{,}216$Å to $\lambda_{obs} = 8150$Å and the stellar continuum long-ward of the Lyman-α emission line. The *right-hand plot* shows the OH night sky emission bands, highlighting the few gaps within which narrow-band filters can be most effectively targeted. The Subaru narrow band filters whose transmission profiles are matched to these dark windows are used to detect LAEs at $z = 5.7$ (NB816), $z = 6.6$ (NB921) and $z = 7.0$ (NB973), as discussed in Sect. 3.2.1 (courtesy M. Iye)

the NB711 filter on Subaru to uncover substantial numbers of LAEs at $z \simeq 4.8$. The successful use of Lyman-α selection at $z > 5$ is now described in detail below.

3.2.1 Lyman-α Galaxies at z > 5

After passing the $z = 5$ threshold in 1998, the redshift record for LAEs rapidly advanced beyond $z = 6.5$ (Hu et al. 2002; Rhoads et al. 2003, 2004), and indeed LAEs were to provide the most distant known objects for the rest of the decade.

Since 2004, the discovery of LAEs at $z > 5$ has been largely driven by narrow-band imaging with the wide-field optical camera Suprime-Cam on the Subaru telescope, coupled with follow-up spectroscopy with the FOCAS spectrograph on Subaru, and the LRIS and DEIMOS spectrographs on Keck. A consortium of Subaru astronomers developed the required series of narrow-band filters at ever increasing wavelengths. As shown in Fig. 8, the band-passes of these filters are designed to fit within the most prominent dark gaps between the bands of strong telluric OH emission which come to increasingly-dominate the night-sky spectrum at $\lambda_{obs} > 7{,}000$ Å.

A filter at 8,160Å (NB816) is able to target Lyman-α emission at $z \simeq 5.7$. This was used by Ouchi et al. (2005) to produce a very large sample of $\simeq$500 $z \simeq 5.7$ LAEs from imaging of the Subaru XMM-Newton Deep Survey field (SXDS; Furusawa et al. 2008) and by Shimasaku et al. (2006) to produce another large and independent sample of $z \simeq 5.7$ LAEs from imaging of the Subaru Deep Field (SDF; Kashikawa et al. 2004). The NB816 filter was also used by Ajiki et al.

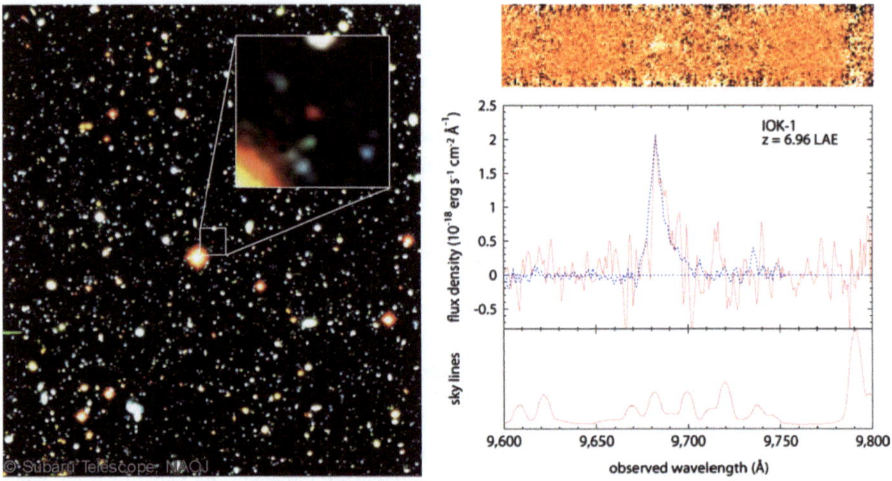

Fig. 9 The most distant spectroscopically-confirmed LAE selected via narrow-band imaging, the galaxy IOK-1, is shown as a *red blob* in the colour postage-stamp insert image which covers 8 × 8 arcsec. The entire field of view shown in the larger image covers 254 × 284 arcsec (North is up and East to the left). The 2-dimensional and 1-dimensional Subaru FOCAS spectrum of IOK-1 is shown in the *right-hand panels* (Iye et al. 2006). The spectrum clearly shows an asymmetric Lyman-α emission line at a wavelength corresponding to a redshift $z \simeq 6.96$ (courtesy M. Iye)

(2006) to image both GOODS fields, and a fourth sample of NB816-selected LAEs was uncovered in the COSMOS field by Murayama et al. (2007).

Imaging of these survey fields through another, redder filter (NB921) led to the first substantial samples of potential LAEs at $z \simeq 6.6$ (Taniguchi et al. 2005; Kashikawa et al. 2006; Ouchi et al. 2010), and imaging of the SDF through the even redder NB973 filter yielded what remains to this day the most distant narrow-band selected galaxy. This LAE, IOK-1, was spectroscopically confirmed at $z = 6.96$ by Iye et al. (2006) and was, for four years, the most distant object known. The discovery image and spectrum of IOK-1 is shown in Fig. 9; the spectrum clearly shows the asymmetric emission-line profile which is characteristic of Lyman-α emission at extreme redshift (produced by neutral Hydrogen absorption of the blue wing of the emission line; Hu et al. 2010) and helps to enable single-line spectroscopic confirmation of narrow-band selected LAE candidates at these high redshifts (see below for potential contaminants).

The NB973 filter has now been used on Subaru to provide a few more candidate LAEs at $z \simeq 7$ (Ota et al. 2008, 2010a). Most recently, following refurbishment of Suprime-Cam with new red-sensitive CCDs, the NB1006 filter has been installed to allow searches for LAEs at $z \simeq 7.3$ (Iye et al. 2008).

Complementary deeper (but smaller-area) narrow-band searches for LAEs at $z > 7$ have recently been conducted on the VLT, but have not yet yielded any spectroscopically-confirmed candidates (Cuby et al. 2007; Clément et al. 2012). As discussed above in the context of the spectroscopic follow-up of the highest redshift LBGs, there of course exists the interesting possibility that Lyman-α emission may not be so easily produced by many galaxies as we enter the epoch of reionization (see Sect. 4.3.6). This issue may soon be clarified by further deeper narrow-band imaging searches in the near-infrared. Finally, it is probably fair to say that existing attempts to uncover extreme redshift LAEs up to $z \simeq 10$ via long-slit infrared spectroscopy targeted on the critical lines in strong-lensing clusters remain controversial (Stark et al. 2007b).

3.2.2 Potential Contaminants

It must be emphasized that narrow-band selected LAE candidates at $z > 5$ need to be confirmed with spectroscopy because the vast majority of objects with a narrow-band excess will be contaminants. Many of these are genuine emission-line objects (galaxies or AGN) at lower redshifts, with the narrow-band excess being produced by, for example, CIV emission at 1,549 Å, MgII at 2,798 Å, [OII] at 3,727 Å, [OIII] at 5,007 Å, or H-α at 6,563 Å. Isolation of genuine extreme-redshift LAEs is of course helped by the fact that, like LBGs, they should show essentially no emission at wavelengths shortward of $\lambda_{rest} = 1{,}216$ Å. Thus, broad-band imaging at bluer wavelengths can be used to reject many low-redshift objects without recourse to spectroscopy. A second alternative to spectroscopy as a means to rule out at least some sub-samples of lower-redshift emission-line objects is observation through a second narrow-band filter at a wavelength specifically designed to pick up a second emission line (e.g. Sobral et al. 2012). However, this is rarely practical, and at least multi-object spectroscopy is reasonably efficient when targeting a subset of objects which are already known to likely display detectable emission lines.

Another potential source of LAE sample contamination is transient objects (e.g. variable AGN or supernovae) because often the narrow-band image is compared with a broad-band image which was taken one or two years earlier. Finally, the sheer size of the images means that rare, apparently significant (5-σ) noise peaks can occur in a single narrow-band image, and these need to be excluded by either repeated imaging or spectroscopy (this is the same single-band statistical detection problem which can afflict searches for extreme-redshift LBGs in the longest-wavelength broad-band filter; Bouwens et al. 2011a).

4 Luminosity Functions

The evolving luminosity function is generally regarded as the best way to summarize the changing demographics of high-redshift galaxies. It is defined as the number of objects per unit comoving volume per unit luminosity, and the data are most often fitted to a Schechter function (Schechter 1976):

$$\frac{dn}{dL} = \phi(L) = \left(\frac{\phi^*}{L^*}\right)\left(\frac{L}{L^*}\right)^{\alpha} e^{-(L/L^*)} \quad (4)$$

where ϕ^* is the normalization density, L^* is a characteristic luminosity, and α is the power-law slope at low luminosity, L. The faint-end slope, α, is usually negative ($\alpha \simeq -1.3$ in the local Universe; e.g. Hammer et al. 2012) implying large numbers of faint galaxies.

In the high-redshift galaxy literature, the UV continuum luminosity function is usually presented in units of per absolute magnitude, M, rather than luminosity L, in which case, making the substitutions $\phi(M)dM = \phi(L)d(-L)$ and $M - M^* = -2.5\log(L/L^*)$, the Schechter function becomes

$$\phi(M) = \frac{\ln 10}{2.5}\phi^* 10^{0.4(\alpha+1)(M-M^*)} \exp\left[-10^{0.4(M-M^*)}\right] \tag{5}$$

and this function is usually plotted in log space (i.e. $\log[\phi(M)]$ vs. M).

The Schechter function can be regarded as simply one way of describing the basic shape of any luminosity function which displays a steepening above a characteristic luminosity L^* (or below a characteristic absolute magnitude M^*). Alternative functions, such as a double power-law can often also be fitted, and traditionally have been used in studies of the luminosity function of radio galaxies and quasars (e.g. Dunlop and Peacock 1990). Given good enough data, especially extending to the very faintest luminosities, such simple parameterizations of the luminosity function are expected to fail, but the Schechter function is more than adequate to describe the data currently available for galaxies at $z \geq 5$. A recent and thorough overview of the range of approaches to determining and fitting luminosity functions, and the issues involved, can be found in Johnston (2011).

At the redshifts of interest here, the luminosity functions derived from optical to near-infrared observations are rest-frame ultraviolet luminosity functions. Continuum luminosity functions for LBGs are generally defined at $\lambda_{rest} \simeq 1{,}500$ Å or $\lambda_{rest} \simeq 1{,}600$ Å, while the luminosity functions derived for LAEs involve the integrated luminosity of the Lyman-α line. Because of the sparcity of the data at the highest redshifts, and the typical redshift accuracy of LBG selection, the evolution of the luminosity function is usually described in unit redshift intervals, although careful simulation work is required to calculate the volumes actually sampled by the filter-dependent selection techniques used to select LBGs and LAEs. Detailed simulations (involving input luminosities and sizes) are also required to estimate incompleteness corrections when the survey data are pushed towards the detection limit, and the form of these simulations can have a significant effect on the shape of the derived luminosity functions, especially at the faint end (as discussed by, for example, Grazian et al. 2011).

Different reported Schechter-function fits can sometimes exaggerate the discrepancies between the basic data gathered by different research groups. In particular, without good statistics and dynamic range, there can be severe degeneracies between ϕ^*, L^* and α, and very different values can be deduced for these parameters even when the basic statistics (e.g. integrated number of galaxies above the flux-density limit) are not very different (e.g. Robertson 2010).

Table 1 Example comoving cosmological volumes sampled by different types and scales of high-redshift galaxy surveys at $z \simeq 6$ and $z \simeq 7$

Survey type	Redshift range	Area	Volume/Mpc^3	Example reference
LBG WFCAM/VISTA	$z = 5.5 - 6.5$	1 deg^2	10,000,000	(McLure et al. 2009)
LAE Suprime-Cam(x4)	$z = 5.7 \pm 0.05$	1 deg^2	1,000,000	Ouchi et al. (2008)
LBG Suprime-Cam	$z = 6.5 - 7.1$	0.25 deg^2	1,000,000	Ouchi et al. (2009)
LAE Suprime-Cam	$z = 6.6 \pm 0.05$	0.25 deg^2	200,000	Kashikawa et al. (2011)
LBG HUDF/WFC3	$z = 6.5 - 7.5$	4.5 arcmin^2	10,000	Oesch et al. (2010a)
LBG CANDELS/WFC3	$z = 6.5 - 7.5$	0.2 deg^2	1,500,000	Grogin et al. (2011)

This is an important point, because the luminosity-integral of the evolving luminosity function

$$j(L) = \int_L^\infty L\phi(L)dL = \phi^* L^* \Gamma(2 + \alpha, L/L^*) \tag{6}$$

(where Γ is the incomplete gamma function) is often used to estimate the evolution of average *luminosity density* as a function of redshift (from which the cosmic history of star-formation density and ionizing photons can be inferred; e.g. Robertson et al. 2010). For this reason care must be taken not to over-interpret the implications of extrapolating the fitted function (e.g. Su et al 2011), especially when, as appears to be the case at very high redshift (see below), the faint-end slope is very steep. Formally, luminosity density diverges for $\alpha < -2$ if the luminosity function is integrated to zero, but in practice the integral needs to be terminated at some appropriate faint luminosity (see Sect. 6.1). The key point is that, for any steep faint-end slope even approaching $\alpha \simeq -2$, the value of the integral depends critically on α and the adopted value of the faint-end luminosity cutoff (which, for obvious reasons, is still a matter of debate and could be a function of environment; Hammer et al. 2012).

At the bright end of the luminosity function the problem is generally not completeness but small-number statistics, and authors are often tempted to push their survey to produce a derived value for a brightest luminosity bin which depends on only a handful of objects. Given the small numbers, contamination by even rare populations (such as the brown dwarf stars discussed in Sect. 3.1.3) can often be a problem at the bright end. An additional issue for a steeply-falling luminosity function is correcting for "Eddington bias", which tends to boost apparent average luminosity at the bright end. This again requires careful simulation to achieve a consistent solution. Finally it must be remembered that all luminosity functions are afflicted to some extent by cosmic variance (Sommerville et al. 2004), and ultimately high-redshift surveys need to cover sufficient area (helped by covering independent lines of sight) to offer a representative picture of the galaxy population at each epoch.

The comoving cosmological volumes sampled by various example LBG and LAE surveys at $z \simeq 6$ and $z \simeq 7$ are given for convenient comparison in Table 1.

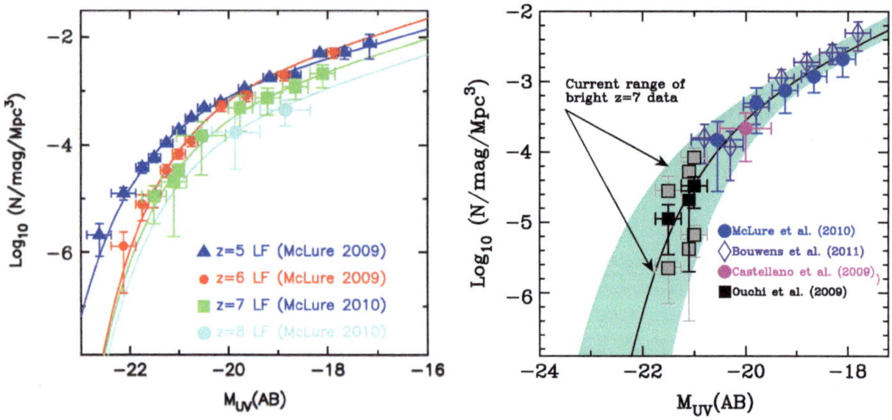

Fig. 10 The UV continuum LF of LBGs, and its high-redshift evolution. The left-hand panel shows the $z \simeq 5$ and $z \simeq 6$ LFs determined by McLure et al. (2009), along with the LFs at $z \simeq 7$ and $z \simeq 8$ determined by McLure et al. (2010) from the recent *HST* WFC3 imaging (the results obtained from a similar analysis by Bouwens et al. 2011b are summarized in Fig. 11). The right-hand panel demonstrates the extent of general agreement over the form and normalization of the UV LF at $z \simeq 7$ derived by different groups using both ground-based (Subaru & VLT) and *HST* WFC3/IR data. While the overall level of agreement is impressive, this plot also shows current tension over the true value of the faint-end slope, and the lack of information at the very bright end of the LF (courtesy R. McLure)

4.1 High-Redshift Evolution of the LBG Luminosity Function

The last $\simeq$5 years have seen a rolling series of papers on the LBG UV luminosity function at $z > 5$, based purely on the ever-improving *HST* ACS, NICMOS, and now WFC3/IR deep imaging datasets (Bouwens et al. 2006, 2007, 2008, 2011b; Oesch et al. 2007, 2010a; Trenti et al. 2010)

In a complementary effort, McLure et al. (2009) focussed on determining the bright-end of the LBG luminosity function at $z \simeq 5$ and $z \simeq 6$ from ground-based data, before extending this work to $z \simeq 7$ and $z \simeq 8$ with WFC3/IR (McLure et al. 2010). In addition, the ground-based determination of the LBG luminosity function has recently been pushed out to $z \simeq 7$ by Ouchi et al. (2009) and Castellano et al. (2010a,b).

In general, the results of these various studies are in very good agreement. Specifically, McLure et al. (2009) combined their ground-based data on bright LBGs with the Bouwens et al. (2007) data on fainter *HST*-selected LBGs to determine the form of the UV luminosity function at $z \simeq 5$ and $z \simeq 6$, and derived Schechter-function parameter values in excellent agreement with Bouwens et al. (2007). The form and evolution of the LBG luminosity function deduced from this work is shown in the left-hand panel of Fig. 10, including the McLure et al. (2010) extension to $z \simeq 7$ and $z \simeq 8$. The simplest way to summarize these results is that the available

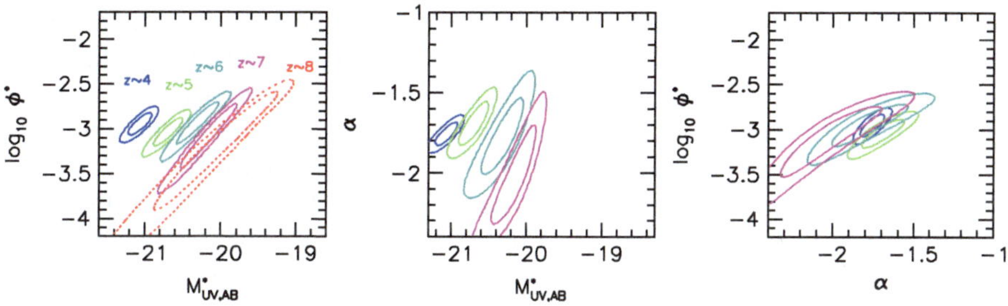

Fig. 11 Sixty-eight percent and ninety-five percent likelihood contours on the model Schechter-function parameters derived by Bouwens et al. (2011b) from their determination of the UV (rest-frame ~1,700Å) continuum LF at $z \sim 7$ (*magenta lines*) and $z \sim 8$ (*dotted red lines*). Also shown for comparison are the LF determinations at $z \sim 4$ (*blue lines*), $z \sim 5$ (*green lines*), and $z \sim 6$ (*cyan lines*) from Bouwens et al. (2007). No $z \sim 8$ contours are shown in the center and right panels given the large uncertainties on the $z \sim 8$ Schechter parameters. Fairly uniform evolution in the UV LF (*left and middle panels*) is seen as a function of redshift, although there remains significant degeneracy between ϕ^* and M^*. Most of the evolution in the LF appears to be in M^* (particularly from $z \sim 7$ to $z \sim 4$). Within the current uncertainties, there is no evidence for evolution in ϕ^* or α (rightmost panel)(courtesy R. Bouwens)

data are consistent with $\alpha = -1.7$ over the full redshift range $z \simeq 5-7$, and that the characteristic luminosity declines by a factor of two from $z \simeq 5$ ($M^* \simeq -20.7$) to $z \simeq 6$ ($M^* \simeq -20.0$) (as always one must caution this does not necessarily imply pure luminosity evolution of individual objects; see, for example, Stark et al. 2009).

From $z \simeq 6$ to $z \simeq 8$ there is good agreement that the number density of LBGs continues to decline but uncertainties and degeneracies in the fitted Schechter-function parameters means that it is currently hard to establish whether this evolution is better described as density or luminosity evolution. For example, McLure et al. (2010) concluded that the $z \simeq 7$ and $z \simeq 8$ luminosity functions are consistent with having the same overall shape at $z \simeq 6$, but with ϕ^* a factor of $\simeq$2.5 and 5 lower, respectively. Ouchi et al. (2009) also concluded in favour of a drop in ϕ^* between $z \simeq 6$ and $z \simeq 7$. Meanwhile, as shown in Fig. 11, the results of Bouwens et al. (2011b) appear to favour some level of continued luminosity evolution (perhaps also combined with a decline in ϕ^* beyond $z \simeq 6$), but their best-fitting values for ϕ^*, M^* and α as a function of redshift are still consistent with the results of McLure et al. (2009, 2010) within current uncertainties (note that at $z \simeq 8$ current data do not really allow a meaningful Schechter-function fit).

There are, however, emerging (and potentially important) areas of tension. The right-hand panel of Fig. 10 shows the generally good level of agreement over the basic form of the LBG LF at $z \simeq 7$ (i.e. $\phi^* \simeq 0.8 \times 10^{-3}$ Mpc^{-3} and $M^* \simeq -20.1$; Ouchi et al. 2009; McLure et al. 2010; Bouwens et al. 2011b), but also reveals issues at both the faint and bright ends (issues which we can hope will be resolved as the dataset on LBGs at $z \simeq 7-8$ continues to improve and grow).

At the faint end there is growing debate over the slope of the luminosity function. As summarized above, essentially all workers are in agreement that the faint-end slope, α, is steeper by $z \simeq 5$ than in the low-redshift Universe, where $\alpha \simeq -1.3$. But recently, Bouwens et al. (2011b), pushing the new WFC3/IR to the limit with very small aperture photometry, have provided tentative evidence that the faint-end slope α may have steepened to $\alpha \simeq -2.0$ by $z \simeq 7$. This "result" is illustrated in Fig. 10, which shows the confidence intervals on the Schechter parameter values deduced by Bouwens et al. (2011b) from $z \simeq 4$ to $z \simeq 7$. Clearly the data are still consistent with $\alpha = -1.7$ over this entire redshift range, but given the luminosity function has definitely steepened between $z \simeq 0$ and $z \simeq 5$, further steepening by $z \simeq 7$ is certainly not implausible, and (as discussed above and in Sect. 6) would have important implications for the integrated luminosity density, and hence for reionization. Figure 11 also nicely illustrates the problems of degeneracies between the Schechter parameters; clearly it will be hard to pin down α without better constraints on ϕ^* and M^* which can only be provided by the larger area surveys such as CANDELS and UltraVISTA (Robertson 2010). Another key issue is surface brightness bias. As discussed in detail by Grazian et al. (2011), because, for a given total luminosity, *HST* is better able to detect the most compact objects (especially in the very small ($\simeq$0.3-arcsec diameter) apertures adopted by Bouwens et al. 2011b), the estimated completeness of the WFC3/IR surveys at the faintest flux densities is strongly dependent on the assumed size distribution of the galaxy population. Thus, it appears that potentially all of any current disagreement over the faint-end slope at $z \simeq 7$ can be traced to different assumptions over galaxy sizes and hence different completeness corrections. Finally, there are of course the usual issues over cosmic variance, with the faintest points on the luminosity function being determined from the WFC3/IR survey of the HUDF which covers only $\simeq$4 arcmin^2. However, as discussed in detail by Bouwens et al. (2011b), it appears that large-scale structure uncertainties do not pose a very big problem for luminosity function determinations in the luminosity range $-21 < M^* < -18$.

At the bright end, Fig. 10 illustrates that the problem is mainly lack of data, which in turn can be traced to a lack of large-area near-infrared surveys of sufficient depth and multi-frequency coverage. As discussed above, current ground-based surveys for LBGs at $z \simeq 7$ are limited to those undertaken by Ouchi et al. (2009) and Castellano et al. (2010a,b) and suffer from somewhat uncertain contamination due to lack of sufficiently deeper longer-wavelength data. Nevertheless, both Ouchi et al. (2009) and Castellano et al. (2010b) conclude that a decline in the number density of brighter LBGs between $z \simeq 6$ and $z \simeq 7$ is now established with better than 95 % confidence, even allowing for cosmic variance (the contrary results of Capak et al. 2011b can be discounted for the reasons discussed in Sect. 3.1.3). Significant further improvement in our knowledge of the bright end of the LBG luminosity function at $z \simeq 7$ and $z \simeq 8$ can be expected over the next $\simeq$3 years, from CANDELS (Grogin et al. 2011), the WFC3/IR parallel programs (Trenti et al. 2011, 2012; Yan et al. 2011) and from UltraVISTA (McCracken et al. 2012; Bowler et al. 2012).

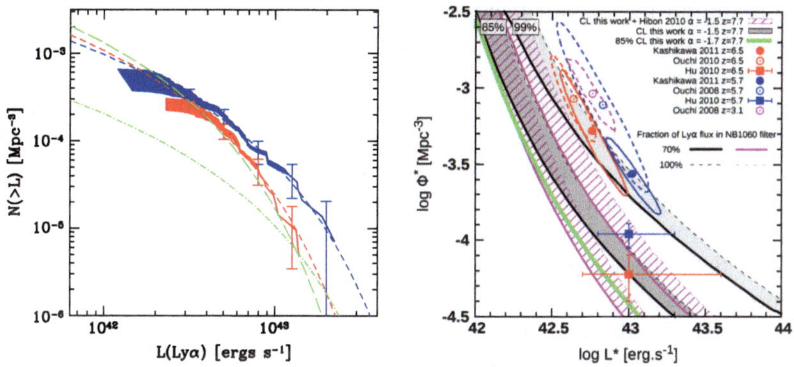

Fig. 12 Current constraints on the Lyman-α LF at high redshift. The left-hand plot, taken from Kashikawa et al. (2011), shows a comparison of the *cumulative* Lyman-α LFs of LAEs at $z = 5.7$ (*blue-shaded region*) and at $z = 6.5$ (*red-shaded region*). The upper-edge of each shaded region is based on the assumption that all photometrically-selected candidates in the two SDF samples are indeed LAEs, while the lower-edge is derived purely on the spectroscopically-confirmed sample at each redshift. The *short-dashed lines* (*red* for $z = 6.5$ and *blue* for $z = 5.7$) show the fitted Schechter LFs assuming $\alpha = -1.5$. For comparison, the *green long-dashed line* shows the Lyman-α LF at $z = 6.5$ determined from the larger area SXDS survey by Ouchi et al. (2010), and the *green dot-dashed line* shows the $z = 6.5$ Lyman-α LF determined by Hu et al. (2010) (courtesy N. Kashikawa). The right-hand plot, taken from Clément et al. (2012) summarizes our current knowledge (including some controversial disagreement) of ϕ^* and L^* for Schechter-function fits to the Lyman-α LF at $z = 5.7$ and $z = 6.6$ (again assuming $\alpha = -1.5$), as well as attempting to set joint limits on these two parameters at $z = 7.7$ (see text in Sect. 4.2 for details; courtesy B. Clément)

4.2 *High-Redshift Evolution of the Lyman-α Luminosity Function*

In contrast to the steady decline seen in the LBG ultraviolet continuum luminosity function at high-redshift, there is little sign of any significant change in the Lyman-α luminosity function displayed by LAEs from $z \simeq 3$ to $z \simeq 5.5$. Indeed, as shown in Fig. 12, Ouchi et al. (2008) and Kashikawa et al. (2011) have presented evidence that the Lyman-α luminosity function displayed by LAEs selected via narrow-band imaging (and extensive spectroscopic follow-up) at $z \simeq 5.7$ is, within the uncertainties, essentially identical to that seen at $z \simeq 3$. Both studies were unable to constrain the faint-end slope of the Lyman-α luminosity function but, assuming $\alpha = -1.5$, reported fiducial values for the other Schechter parameters at $z = 5.7$ of $\phi^* \simeq 8 \times 10^{-4}\,\mathrm{Mpc}^{-3}$ and $L^*_{\mathrm{Lyff}} = 7 \times 10^{42}\,\mathrm{erg\,s}^{-1}$.

Why the Lyman-α luminosity function should display different evolution to the LBG continuum luminosity function is a subject of considerable current interest. The relationship between LBGs and LAEs is discussed at more length in the next subsection (including the evolution of the ultraviolet *continuum* luminosity function of LAEs), but the key point to bear in mind here is that the evolution of the

Lyman-α luminosity function inevitably reflects not just the evolution in the number density and luminosity of star-forming galaxies, but also cosmic evolution in the escape fraction of Lyman-α emission. This latter could, for example, be expected to increase with increasing redshift due to a decrease in average dust content, and/or at some point decrease with increasing redshift due to an increasingly neutral IGM.

Staying with the direct observations for now, at still higher redshifts the situation is somewhat controversial. Following Kashikawa et al. (2006), Ouchi et al. (2010) have extended the Subaru surveys of narrow-band selected LAEs at $z \simeq 6.6$ to the SXDS field. They conclude that there is a modest ($\simeq$20–30 %) decline in the Lyman-α LF over the redshift interval $z \simeq 5.7$–6.6 (also shown in the left-hand panel of Fig. 12), and that this decline is best described as luminosity evolution, with $L^*_{\rm Lyff}$ falling from $\simeq 7$ to $\simeq 4.5 \times 10^{42}$ erg s^{-1}, while ϕ^* remains essentially unchanged at $\simeq 8 \times 10^{-4}$ Mpc^{-3} (see right-hand panel of Fig. 12, again assuming $\alpha = -1.5$).

The results from Ouchi et al. (2008, 2010) are based on large LAE samples but with only moderate levels of spectroscopic confirmation. These results have recently been contested by Hu et al. (2010). As summarized in the right-hand panel of Fig. 12, Hu et al. (2010) report a comparable value of L^* at $z \simeq 5.7$, but ϕ^* an order of magnitude lower. They then also report a modest drop in LAE number density by $z \simeq 6.6$, but conclude this is better described as density evolution (with ϕ^* dropping by a factor of $\simeq$2).

Also included in Fig. 12 are the latest results from Kashikawa et al. (2011), who used further spectroscopic follow-up to increase the percentage of spectroscopically confirmed LAEs in the SDF narrow-band selected samples of Taniguchi et al. (2005) and Shimasaku et al. (2006) to 70 % at $z \simeq 5.7$ and 81 % at $z \simeq 6.6$. The outcome of the resulting luminosity function reanalysis appears, at least at $z \simeq 5.7$, to offer some hope of resolving the situation, with Kashikawa et al. (2011) reporting a value for ϕ^* somewhat lower than (but consistent with) the value derived by Ouchi et al. (2008), and at least closer to the ϕ^* value reported by Hu et al. (2010). But at $z \simeq 6.6$ the results from Kashikawa et al. (2011) remain at odds with Hu et al. (2010), with ϕ^* still an order of magnitude higher, and modest luminosity evolution since $z \simeq 5.7$ (if anything offset by slight positive evolution of ϕ^*, resulting in any significant decline in number density being confined to the more luminous LAEs). Given the high spectroscopic confirmation rates in the new Kashikawa et al. (2011) samples, the claim advanced by Hu et al. (2010) that the previous Ouchi et al. (2008, 2010) and Kashikawa et al. (2006) studies were severely affected by high contamination rates in the narrow-band selected samples now seems untenable. Rather, it appears much more probable that the Hu et al. (2010) samples are either affected by incompleteness (and hence they have seriously under-estimated ϕ^* for LAEs at high redshift), or that our knowledge of the Lyman-α luminosity function at $z \simeq 6.6$ is still severely confused by the affects of cosmic variance and/or patchy reionization (see, for example, Nakamura et al. 2011), an issue which is discussed further in Sect. 5.5.

The recent work of Cassata et al. (2011), based on a pure spectroscopic sample of (mostly) serendipitous Lyman-α emitters found in deep VIMOS spectroscopic surveys with the VLT, also yields results consistent with an unchanging Lyman-α luminosity function from $z \simeq 2$ to $z \simeq 6$. In addition, their estimated values of ϕ^* and $L^*_{\rm Ly\alpha}$ at $z = 5-6$ are in excellent agreement with those reported by Ouchi et al. (2008) at $z \simeq 5.7$. Interestingly, because the VIMOS spectroscopic surveys can probe to somewhat deeper Lyman-α luminosities than the narrow-band imaging surveys, this work has also provided useful constraints on the evolution of the faint-end slope, α, at least at moderate redshifts. Specifically, Cassata et al. (2011) conclude that α steepens from $\simeq -1.6$ at $z \simeq 2.5$ to $\alpha = -1.8$ at $z \simeq 4$. Direct constraints at the highest redshifts remain somewhat unclear, but the clear implication is that, as for the LBG luminosity function, the faint-end slope is significantly steeper than $\alpha = -1.5$ at $z > 5$ (and hence it is probably more appropriate to consider ϕ^* and $L^*_{\rm Ly\alpha}$ values reported by authors assuming $\alpha = -1.7$ or even $\alpha \simeq -2$).

Finally, also shown in Fig. 12 are limits on the luminosity function parameter values at $z \simeq 7.7$ (albeit assuming $\alpha = -1.5$), imposed by the failure of Clément et al. (2012) to detect any LAEs from deep HAWK-I VLT 1.06 μm narrow-band imaging of three 7.5 × 7.5 arcmin fields (probing a volume $\sim 2.5 \times 10^4$ Mpc^3). The ability of Clément et al. (2012) to draw crisp conclusions from this work is hampered by the confusion at $z \simeq 6.6$, with the above-mentioned different LFs of Ouchi et al. (2010), Hu et al. (2010) and Kashikawa et al. (2011) predicting 11.6, 2.5 and 13.7 objects respectively in the Hawk-I imaging (if the luminosity function remains unchanged at higher redshifts). Clément et al. (2012) conclude that an unchanged Lyman-α luminosity function can be excluded at $\simeq$85 % confidence, but that this confidence-level could rise towards $\sim$99 % if one factors in significant quenching of IGM Lyman-α transmission due to a strong increase in the neutral Hydrogen fraction as we enter the epoch of reionization (an issue discussed further in the next subsection). However, the issue of whether or not the Lyman-α luminosity function really declines beyond $z \simeq 7$ undoubtedly remains controversial (e.g. Ota et al. 2010a; Tilvi et al. 2010; Hibon et al. 2010, 2011, 2012; Krug et al. 2012) and further planned surveys for LAEs at $z \geq 7$ are needed to address this question (e.g. Nilsson et al. 2007).

4.3 The LBG-LAE Connection

The recent research literature in this field is littered with extensive and sometimes confusing discussions over the differences and similarities between the properties of LBGs and LAEs. In the end, however, the LAE population must be a subset of the LBG population, and the reported differences must be due to the biases (sometimes helpful) which are introduced by the different selection techniques. One key area of much current interest is to establish whether/how the fraction of LBGs which emit observable Lyman-α varies with cosmic epoch, because this has the potential to provide key information on the evolution of dust and gas in galaxies,

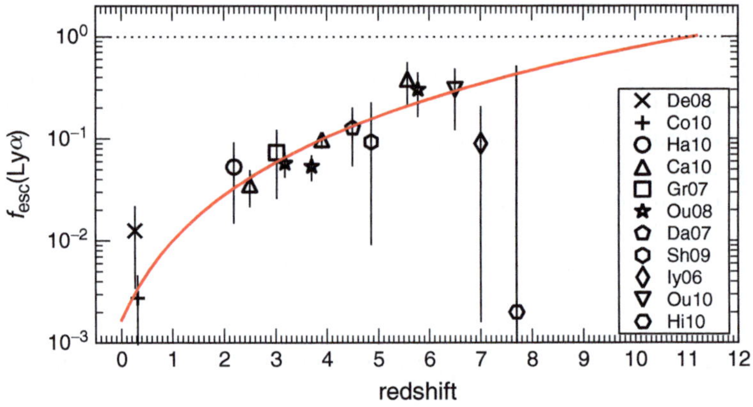

Fig. 13 The redshift evolution of volume-averaged Lyman-α escape fraction, $f_{\rm esc}^{\rm Ly\alpha}$, as deduced by Hayes et al. (2011), normalized to $\simeq$5 % at $z \simeq 2$ via comparison of the Lyman-α and H-α integrated luminosity functions, and deduced at higher redshifts by comparison of the Lyman-α and UV continuum luminosity functions discussed in Sects. 4.1 and 4.2. The solid red line shows the best-fitting power-law to points between redshift 0 and 6, which takes the form $(1 + z)^{2.6}$, and appears to be a good representation of the observed points over this redshift range. It intersects with the $f_{\rm esc}^{\rm Ly\alpha} = 1$ line (*dotted*) at redshift $z = 11.1$ (courtesy M. Hayes)

and on the neutral hydrogen fraction in the IGM. There are a number of lines of attack being vigorously pursued, and I start by considering how we might reconcile the apparently very different high-redshift evolution of the Lyman-α and LBG ultraviolet luminosity functions (as summarized in the previous two subsections).

4.3.1 Comparison of the LBG UV and LAE Lyman-α Luminosity Functions

First, given the steady negative evolution of the LBG luminosity function from $z \simeq 3$ to $z \simeq 6$, and the apparently unchanging form and normalization of the Lyman-α luminosity function over this period, it seems reasonable to deduce that *on average* the fraction of Lyman-α photons emerging from star-forming galaxies *relative to the observed continuum emission* increases with increasing redshift out to at least $z \simeq 6$. Indeed Hayes et al. (2011) have used this comparison to deduce that the *volume averaged* Lyman-α escape fraction, $f_{\rm esc}^{\rm Ly\alpha}$, grows according to $f_{\rm esc}^{\rm Ly\alpha} \propto (1 + z)^{2.5}$ (normalized at $\simeq$5 % at $z \simeq 2$ through a comparison of the Lyman-α and H-α luminosity functions, currently feasible only at $z \simeq 2$). This result is shown in Fig. 13, which also shows that extrapolation of the fit to higher redshifts would imply $f_{\rm esc}^{\rm Ly\alpha} = 1$ at $z \simeq 11$ in the absence of any new source of Lyman-α opacity, providing clear motivation for continuing the comparison of the Lyman-α and LBG continuum luminosity functions to higher redshift if at all possible (see below).

4.3.2 The LAE UV Continuum Luminosity Function

The *italics* in the preceding paragraph have been chosen with care, because we must proceed carefully. This is because the situation is confused by the fact that, *for those galaxies selected as LAEs* (via, for example, narrow-band imaging as discussed in detail above), the ratio of average Lyman-α emission to ultraviolet continuum emission apparently stays unchanged or even *decreases* with increasing redshift. We know this from studies of the ultraviolet continuum luminosity function of LAEs, which we have deliberately avoided discussing until now because there are complications in interpreting the UV continuum luminosity function of objects which have been selected primarily on the basis of the contrast between Lyman-α emission and UV continuum emission. Nevertheless, Ouchi et al. (2008) have convincingly shown that while the Lyman-α luminosity function of LAEs holds steady between $z \simeq 3$ and $z \simeq 5.7$, the UV continuum luminosity function of the same objects actual grows with redshift, more than bucking the negative trend displayed by LBGs. Then, from $z = 5.7$ to $z \simeq 6.6$, as the Lyman-α luminosity function shows the first signs of gentle decline, Kashikawa et al. (2011) find that the UV continuum luminosity function of the LAEs stops increasing, but seems to hold steady. Another way of saying this is that the average equivalent width of Lyman-α emission in LAEs is constant or if anything slightly falling with increasing redshift. Indeed, Kashikawa et al. (2011) report that the median value of Lyman-α equivalent width falls from $EW_{rest} \simeq 90$ Å at $z \simeq 5.7$ to $EW_{rest} \simeq 75$ Å at $z \simeq 6.6$ although, interestingly, there is a more pronounced extreme EW_{rest} tail in their highest-redshift sample.

Possible physical reasons for why this happens are discussed further below but, whatever the explanation, it is clear that the UV continuum luminosity function of LAEs cannot keep rising indefinitely, or it will at some point exceed the UV luminosity function of LBGs, which is impossible. Indeed, the two luminosity functions appear to virtually match at $z \simeq 6$. Specifically, as shown in Fig. 14, and as first demonstrated by Shimasaku et al. (2006), by $z \simeq 6$, LAE selection down to $EW_{rest} \simeq 20$ Å recovers essentially all LBGs with $M_{1,500} < -20$, to within a factor $\simeq 2$. It is thus no surprise that the UV luminosity function of LAEs must freeze or commence negative evolution somewhere between $z \simeq 6$ and $z \simeq 7$, as by then it must start to track (or fall faster than) the evolution of the (parent) LBG population.

4.3.3 The Prevalence of Lyman-α Emission from LBGs

If this is true, then it must also follow that the fraction of LBGs which display Lyman-α emission with $EW_{rest} > 20$ Å in follow-up spectroscopy must also rise from lower redshifts to near unity at $z \simeq 6$. There has been some controversy over this issue, but recent observations appear to confirm that this is indeed the case. First, Stark et al. (2011) have reported that, with increasing redshift, an increasing fraction of LBGs display strong Lyman-α emission such that, by $z \simeq 6$, over 50 % of faint LBGs display Lyman-α with $EW_{rest} > 25$ Å. Similarly high Lyman-α "success

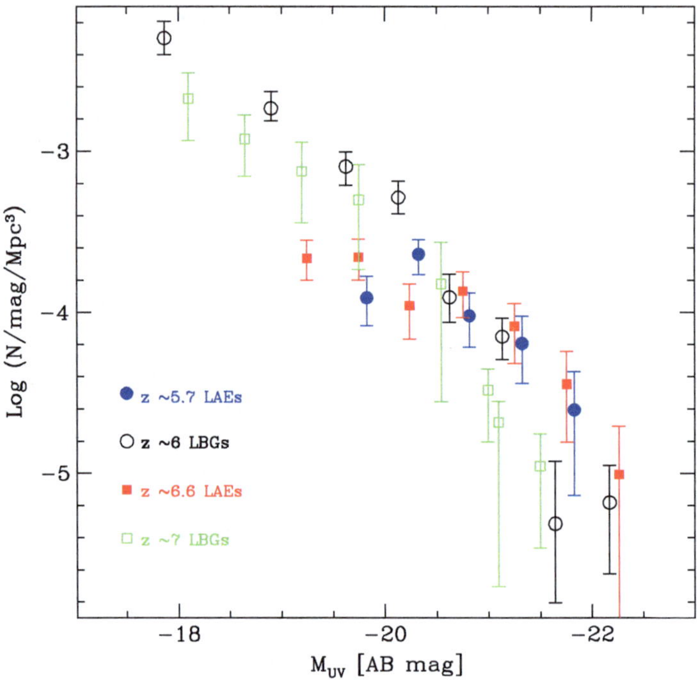

Fig. 14 A comparison of the high-redshift UV continuum LFs of galaxies selected as LBGs and galaxies selected as LAEs. Shown here are the UV continuum LFs for LBGs at $z \simeq 6$ as determined by Bouwens et al. (2007), for LBGs at $z \simeq 7$ as determined by McLure et al. (2010), for LAEs at $z \simeq 5.7$ as determined by Shimasaku et al. (2006), and for LAEs at $z \simeq 6.6$ as as determined by Shimasaku et al. (2006). The LAE UV LFs become incomplete at $M_{UV} > -21$ because of the limited depth of the ground-based broad-band imaging in the large Subaru survey fields (compared to the deeper *HST* data used to derive the LBG LFs). However, at brighter magnitudes the agreement between the $z \simeq 6$ LBG LF and the two LAE-derived LFs at $z \simeq 5.7$ and 6.6 is very good (courtesy P. Dayal)

rates" have now been reported for more luminous ($\simeq 2L^*$) LBGs at $z \simeq 6$ by Curtis-Lake et al. (2012) (Fig. 15), and by Jiang et al. (2011).

4.3.4 Reconciliation to $z \simeq 6$

It thus appears that the average volumetric increase in Lyman-α emission relative to ultraviolet continuum emission as summarized by Hayes et al. (2011) to produce Fig. 13 is due to an increase with redshift in the *fraction* of star-forming galaxies which emit at least some detectable Lyman-α emission rather than a systematic increase in the Lyman-α to continuum ratio of objects which are selected as LAEs at all epochs.

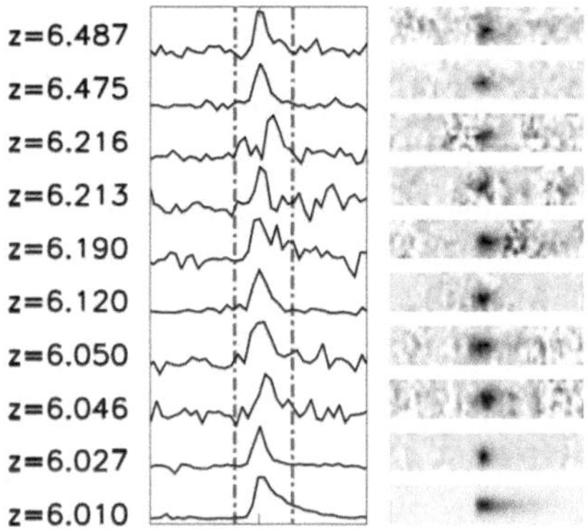

Fig. 15 A high success rate in the detection of Lyman-α emission from bright LBGs at $z = 6 - 6.5$. VLT FORS2 spectra of 10 $z > 6$ LBGs selected from the UDS/SXDS field are shown; this represents 70 % of the targeted high-redshift sample (Curtis-Lake et al. 2012). The extracted one-dimensional (1D) spectra are shown on the left, with the corresponding two-dimensional (2D) spectra on the right; Lyman-α emission (often obviously asymmetric) is clearly detected from all of these objects (courtesy E. Curtis-Lake)

As already discussed above, one thing which is clear, and perhaps surprising, is that by $z \simeq 6$, narrow-band selection of LAEs from the wide-area Subaru surveys seems to be an excellent way of determining the bright end of the complete LBG UV luminosity function, indicating that the fraction of LBGs which display $EW_{rest} >$ 10 Å is approaching unity by this redshift. We stress that this is not the same as seeing *all* of the Lyman-α emission from the star-forming population; it is perfectly possible for virtually all the LBGs to emit enough Lyman-α to be detected in deep LAE surveys, while still having some way to go before the overall volume-average Lyman-α escape fraction could be regarded as approaching unity. Put another way, it is not unreasonable to conclude that the detection rate of bright LBGs in LAE surveys reaches 100 % at $z \simeq 6$, while volume-averaged $f_{\rm esc}^{\rm Ly\alpha}$ as plotted in Fig. 13 has only reached $\simeq$40–50 %. As discussed below, there may be good astrophysical reasons why the volume-averaged $f_{\rm esc}^{\rm Ly\alpha}$ never reaches 100 %.

Given that, at least for $M_{1,500} < -20$, current LAE and LBG surveys appear to be seeing basically the same objects at $z \simeq 6$, it makes sense to consider what, if anything, can be deduced about the fainter end of the LBG UV luminosity function from those LAEs which are *not* detected in the continuum. As can be seen from Fig. 14, the ground-based imaging from which the LAE samples are selected, not unexpectedly runs out of steam at fluxes which correspond to $M_{1,500} \simeq -21$ at $z \simeq 6.6$. But there still of course remain many (in fact the vast majority of) LAEs which have no significant continuum detections, and these objects are not only useful for the Lyman-α luminosity function, but potentially also carry information on the faint end of the LBG UV luminosity function. The question is of course how to extract this information. One could follow-up all the LAEs which are undetected in the ground-based broad imaging with *HST* WFC3/IR to determine their UV luminosities (i.e. $M_{1,500}$); this would undoubtedly yield many more detections allowing further extension of the UV continuum luminosity function of LAEs to fainter luminosities. However, this would still not overcome another incompleteness problem which is that, as narrow-band searches are limited not just by equivalent width, but also by basic Lyman-α luminosity, the subset of LBGs detectable in the LAE surveys becomes confined to those objects with increasingly extreme values of EW_{rest} as we sample down to increasingly faint UV luminosities. Then, even with deep WFC3/IR follow-up of detectable LAEs, and even assuming all LBGs emit some Lyman-α, we would still be forced to infer the total number of faint LBGs by extrapolating from the observable extreme equivalent-width tail of the LAE/LBG population assuming an equivalent-width distribution appropriate for the luminosity and redshift in question.

This is difficult, and presents an especially severe problem at high-redshift, where our knowledge of Lyman-α equivalent-width distributions is confined to the highest luminosities. Nevertheless, Kashikawa et al. (2011) have attempted it, and discuss in detail how they tried to arrive at an appropriate equivalent-width distribution as a function of UV luminosity at $z \simeq 6.6$. A key issue is that it is difficult, if not impossible, to determine the UV continuum luminosity dependence of Lyman-α EW_{rest} in the underlying LBG population from the equivalent width distribution displayed by the narrow-band selected LAEs themselves, as this is in general completely dominated by the joint selection effects of Lyman-α luminosity and equivalent width. There has indeed been much controversy over this issue, with Nilsson et al. (2009) claiming that, at $z \simeq 2-3$ where LAE surveys display good dynamic range, there is no evidence for any UV luminosity dependence of the EW_{rest} distribution, contradicting previous claims that there was a significant anti-correlation between EW_{rest} and UV luminosity. Of course what is really required is complete spectroscopic follow-up of LBGs over a wide UV luminosity range, to determine the distribution of EW_{rest} as a function of $M_{1,500}$, free from the biases introduced by LAE selection. This has been attempted by Stark et al. (2010, 2011), and it is the results of this work that Kashikawa et al. (2011) have employed to try to estimate the faint end of the LBG UV luminosity function at $z \simeq 6$ from the number counts of faint (but still extreme equivalent width) LAEs extracted from the narrow-band surveys. The problem with this is that even the state-of-the-art work of Stark et al. (2011) only really provides EW_{rest} distributions in two luminosity bins at $z \simeq 6$, and the apparent luminosity dependence inferred from this work is called into some question by the success in Lyman-α detection in bright LBGs at $z \simeq 6$ by Curtis-Lake et al. (2012) and Jiang et al. (2011). Thus, at present, our understanding of the luminosity dependence of the Lyman-α EW_{rest} distribution displayed by LBGs remains poor, and is virtually non existent for LBGs with $M_{1,500} > -19$ at $z > 6$.

Nevertheless, the experiment is of interest, and the resulting UV LF derived from the Lyman-α LF by Kashikawa et al. (2011) is more like a power-law than a Schechter function. Moreover, the implied faint end slope is extremely steep; if a Schechter function fit is enforced, $\alpha = -2.4$ results, even though the UV LF was inferred from a Lyman-α Schechter function with $\alpha = -1.5$, which is probably too

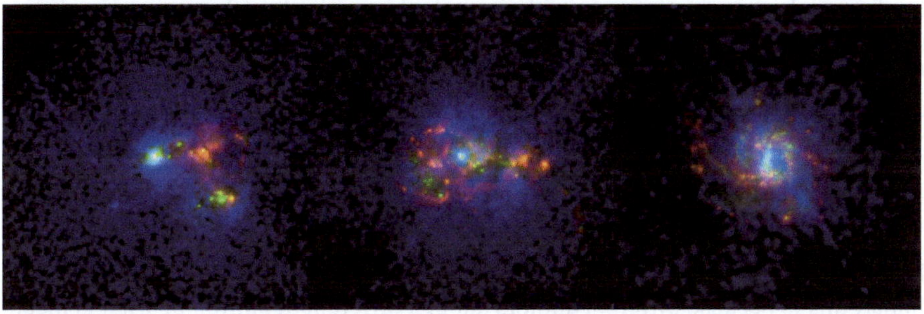

Fig. 16 Three nearby star-forming galaxies imaged as part of the *HST* Lyman-α imaging program LARS. Green shows the UV continuum and traces the massive stars, with the ionized nebulae they produce shown in Red (tracing $H\alpha$). The Lyman-α photons must also be produced in these nebulae, but the Lyman-α image (shown in Blue) reveals all these galaxies to be morphologically very different in $H\alpha$ and Lyman-α due to the resonant scattering of the Lyman-α photons. This is at least qualitatively similar to what is found for high-redshift LAEs, in which the Lyman-α emission is in general more extended and diffuse than the UV continuum light (courtesy of Matt Hayes)

flat. It is not yet clear what to make of this result, but it would appear that either the equivalent width distribution of Lyman-α from faint LBGs is biased to even higher values than assumed (so that even extreme equivalent-width LAEs sample a larger fraction of the LBG population at faint $M_{1,500}$ than anticipated by Kashikawa et al. 2011), or the incompleteness in the faint LBG surveys has been under-estimated. This latter explanation seems unlikely given the already substantial incompleteness corrections made by Bouwens et al. (2011b), but is not entirely impossible if LAEs pick up not just the compact LBGs seen in the HST surveys, but also a more extended population which is missed with HST but is uncovered by ground-based imaging (which is less prone to surface-brightness bias). This, however, also seems unlikely; while recent work has certainly demonstrated that the Lyman-α emission from high-redshift galaxies is often quite extended (e.g. Finkelstein et al. 2011; Steidel et al. 2011), consistent with theoretical predictions (e.g. Zheng et al. 2011), all evidence suggests that the *UV continuum emission* from these same objects is at least as compact as LBGs at comparable redshifts (i.e. typically $r_h \leq 1$ kpc at $z > 5$; Cowie et al. 2011; Malhotra et al. 2012; Gronwall et al. 2011). The fact that, due to the complex radiative transfer of Lyman-α photons, the Lyman-α morphologies of young galaxies are expected to be complex and in general more extended than their continuum morphologies is supported by new observational studies of low-redshift Lyman-α emitting galaxies as illustrated in Fig. 16 (from the Lyman Alpha Reference Sample—LARS; *HST* Program GO12310). However, from the point of view of luminosity-function comparison, the key point is that while extended low-surface brightness Lyman-α emission might be hard to detect with *HST*, such LAEs will still not be missed by deep *HST* broad-band LBG surveys, if virtually all of them display compact continuum emission. The quest to better constrain the true form of the faint-end slope of the UV LF will continue, not least because it is of crucial importance for considering whether and when these young galaxies could have reionized the Universe (see Sect. 6.2). Deeper and more extensive *HST* WFC3/IR imaging over the next few years has the potential to clarify this still currently controversial issue.

4.3.5 The Nature of LAEs

It is easy to become confused by the (extensive) literature on the properties of LAEs. In part this is because different authors adopt a different definition of what is meant by the term LAE. For some, an LAE is any galaxy which displays detectable Lyman-α emission, including objects originally selected as LBGs and then followed up with spectroscopy. All spectroscopically-confirmed LBGs at $z > 5$ must of course be emitters of Lyman-α radiation, and so in an astrophysical sense they are indeed LAEs. However, in practice most LAE studies are really confined to objects which have been *selected* on the basis of Lyman-α emission. Furthermore, many of these studies then proceed to deliberately confine attention to those LAEs which could *not* also have been selected as LBGs from the data in hand. Of course there are often good reasons for doing this. For example, Ono et al. (2010) in their study of the typical UV properties of LAEs at $z \simeq 5.7$ and $z \simeq 6.6$ first excised 39 of the 295 LAEs from their sample because they were individually detected at IRAC wavelengths, before proceeding to stack the data for the remaining LAEs to explore their average continuum colours. This makes sense given the objective of this work was to explore the properties of those objects which were not detected with IRAC, but such deliberate focus on the extreme equivalent-width subset of the LAE population does sometimes run the danger of exaggerating the differences between the LBG and LAE populations.

To put it another way, in many respects the properties of LAEs, selected on the basis of large EW_{rest}, are largely as would be anticipated from the extreme Lyman-α equivalent-width tail of the LBG population. I have already argued above that the observational evidence on luminosity functions suggests LAEs are just a subset of LBGs, and that by $z \simeq 6$ the increased escape of Lyman-α means that the two populations are one and the same. At least some existing *HST*-based comparisons of LAEs and LBGs support this viewpoint (e.g. Yuma et al. 2010), as do at least some theoretical predictions (e.g. Dayal and Ferrara 2012). It is then simply to be expected that the subset of galaxies selected on the basis of extreme Lyman-α EW_{rest} (and hence also typically faint UV continuum emission) will, on average, have lower stellar masses, younger ages, and lower-metallicities than typical LBGs discoverable by current continuum surveys (e.g. Ono et al. 2012).

At present, therefore, there is really no convincing evidence that LAEs are anything other than a subset of LBGs. This is not meant to denigrate the importance of LAE studies; faint narrow-band selection provides access to a special *subset* of the UV-faint galaxy population over much larger areas/volumes than current deep *HST* continuum surveys. But the really interesting questions are whether this extreme equivalent-width subset represents an increasingly important fraction of

LBGs with decreasing UV continuum luminosity and, conversely, whether some subset of this extreme equivalent-width population *cannot* be detected in current deep *HST* continuum imaging. To answer the first question really requires ultra-deep spectroscopic follow-up at $z \simeq 6-7$ of objects *selected as LBGs* spanning a wide range of continuum luminosity $M_{1,500}$. To answer the second question, following Cowie et al. (2011), further deep *HST* WFC3/IR imaging of objects *selected as LAEs* is desirable to establish what subset (if any) of the LAE population lacks sufficiently compact UV continuum emission to be selected as a faint LBG given the surface brightness biases inherent in the high-resolution deep *HST* imaging.

4.3.6 Beyond z ≃ 6.5: A Decline in Lyman-α?

Both the follow-up spectroscopy of LBGs, and the discovery of LAEs via narrow-band imaging become increasingly more difficult as we approach $z \simeq 7$, due to the declining sensitivity of silicon-based detectors at $\lambda \simeq 1\,\mu$m, the increasing brightness of night sky emission and, of course, the reduced number density of potential targets (as indicated by the evolution of the LBG luminosity function discussed above). Nevertheless, even allowing for these difficulties, there is now growing (albeit still tentative) evidence that Lyman-α emission from galaxies at $z \simeq 7$ is significantly less prevalent than at $z \simeq 6$. Specifically, while spectroscopic follow-up of LBGs with $z_{phot} > 6.5$ has indeed yielded several Lyman-α emission-line redshifts up to $z \simeq 7$ (see Sect. 3.1), these same studies all report a lower success rate ($\simeq$15–25 %) than encountered at $z \simeq 6$ (Pentericci et al. 2011; Schenker et al. 2012; Ono et al. 2012). In addition, such Lyman-α emission as is detected seems typically not very intense, with an especially significant lack of intermediate Lyman-α equivalent widths, $EW_{rest} \simeq 20-55$ Å. The significance of the inferred reduction in detectable Lyman-α obviously becomes enhanced if judged against extrapolation of the rising trend in Lyman-α emission out to $z \simeq 6$, as discussed above, and plotted in Fig. 13 (Stark et al. 2011; Hayes et al. 2011; Curtis-Lake et al. 2012).

These results may be viewed as confirming a trend perhaps already hinted at by the reported modest decline in the Lyman-α luminosity function between $z \simeq 5.7$ and $z \simeq 6.6$ (as outlined above in Sect. 4.2), and the tentative (albeit controversial) indications of further decline at $z \geq 7$. In summary, there is growing evidence of a relatively sudden reduction in the transmission of Lyman-α photons between $z \simeq 6$ and $z \simeq 7$. Given that the galaxy population itself appears, on average, to become increasingly better at releasing Lyman-α photons to the observer out to $z \simeq 6$ (perhaps due to a global decline in average dust content), the most natural and popular interpretation of this decline at $z \simeq 7$ is a significant and fairly rapid increase in the neutral hydrogen fraction in the IGM.

This has several implications. First, it suggests that further comparison of LAEs and LBGs over the redshift range $z \simeq 6-7$ may well have something interesting to tell us about reionization (at least its final stages; see Sect. 6.2). Second, it implies that spectroscopic redshift determination/confirmation of LBGs at $z > 7$ is likely to be extremely difficult, and that, for astrophysical reasons, we may be forced to rely on photometric redshifts at least until the advent of genuinely deep near-to-mid infrared spectroscopy with *JWST* (capable of detecting longer-wavelength emission lines including Hα). Third it suggests that the spectacular success of LAE selection via narrow-band imaging out to $z \simeq 6.6$ could be hard to replicate at higher redshifts, and hence that the future study of galaxies and their evolution at $z \simeq 7-10$ may well be driven almost entirely by Lyman-break selection.

5 Galaxy Properties

5.1 Stellar Masses

Stellar mass is one of the most important and useful quantities that can be estimated for a high-redshift galaxy. There are two reasons for this. First, since it represents the time-integral of past star-formation activity, it can be compared directly with observed star-formation rates in even higher-redshift galaxies to set model-independent constraints on plausible modes of galaxy evolution (e.g. Stark et al. 2009). Second, it enables fairly direct and unambiguous comparison with the predictions of different theoretical/computational models of galaxy formation, most of which deliver stellar mass functions as one of their basic outputs (e.g. Bower et al. 2006; De Lucia and Blaizot 2007; Choi and Nagamine 2011; Finlator et al. 2011).

Unfortunately, however, accurate galaxy stellar masses are very hard to derive from data which only sample the rest-frame UV continuum. There are two well-known reasons for this. First, for any reasonable stellar initial mass function (IMF) the UV continuum in a galaxy is dominated by light from a relatively small number of short-lived massive stars, and thus depends critically on recent star-formation activity. Second, the UV continuum is much more strongly affected by dust extinction than is light at longer wavelengths, with 1 mag. of extinction in the rest-frame V-band producing $\simeq$3–4 mag. of extinction at $\lambda_{rest} \simeq 1{,}500$ Å (e.g. Calzetti et al. 2000).

Ideally, then, stellar masses should be estimated from the rest-frame near-infrared emission, at $\lambda_{rest} \simeq 1.6\,\mu$m. But, beyond $z \simeq 5$, this is redshifted to $\lambda_{obs} \geq 10\,\mu$m, and so this is not really practical until *JWST*. Nevertheless, photometry at any wavelength longer that $\lambda_{rest} \simeq 4{,}000$ Å is enormously helpful in reducing the uncertainty in stellar masses, and so the now-proven ability of *Spitzer* to detect LBGs and LAEs at $z \simeq 5-7$ in the two shortest-wavelength IRAC bands (at $\lambda_{obs} = 3.6\,\mu$m and $\lambda_{obs} = 4.5\,\mu$m) has been crucial in enabling meaningful estimates of their stellar masses (e.g. Yan et al. 2006; Stark et al. 2007a; Eyles et al. 2007; Labbé et al. 2006, 2010a,b; Ouchi et al. 2009a; González et al. 2010, 2011; McLure et al. 2011).

However, because even these *Spitzer* IRAC bands still sample rest-frame *optical* emission at $z > 5$ (see Fig. 1), the derived stellar-masses remain significantly affected by star-formation history, which thus needs to either be assumed (often a constant star-formation rate is simply adopted for a galaxy's entire history—e.g. González et al. 2010) or inferred from full SED fitting to as much multi-waveband photometry as is available (e.g. Labbé et al. 2010b). This is, of course, a natural bi-product of the SED fitting approach to deriving photometric redshifts, but it presents a number of challenges. First, it requires the careful combination of *Spitzer* and *HST* data which differ by an order of magnitude in angular resolution. Second, there are areas of disagreement between different evolutionary synthesis models of galaxy evolution (e.g. Jimenez et al. 2000; Bruzual and Charlot 2003; Maraston 2005; Conroy and Gunn 2010), although in practice these are not very serious when it comes to modelling the rest-frame ultraviolet-to-optical SEDS of galaxies which must be less than 1 billion years old (e.g. uncertainty and controversy over the strength of the asymptotic red giant branch is not really an issue when modelling the UV-to-optical SEDS of young galaxies; Maraston 2005; Conroy and Gunn 2010; Labbé et al. 2010a). Third, and probably most serious, there are often significant degeneracies between age, dust-extinction, and metallicity, which can be hard or impossible to remove given only moderate signal-to-noise photometry in only a few wavebands. Fourth, as has recently become more apparent (e.g. Ono et al. 2010; Labbé et al. 2010b; McLure et al. 2011; González et al. 2012), very different stellar masses can be produced depending on what is assumed about the strength of the nebular emission-lines and continuum from a galaxy's inter-stellar medium relatively to the continuum emission from its stars (an issue discussed further in Sect. 6.1). Fifth, one cannot escape the need to assume a stellar IMF to deduce a total stellar mass, as most of the mass is locked up in low-mass stars which are not detected!

It would be a mistake to over-emphasize the issue of the IMF, because it is an assumption which can also be changed in the model predictions (e.g. Davé 2008). Thus, for example, the systematic factor of $\simeq$1.8 reduction in inferred stellar mass that results from changing from the Salpeter (1955) IMF to that of Chabrier (2003) (see also Weidner et al. 2011) need not necessarily prevent useful comparison with theory. In addition, another key quantity which follows on from the derivation of stellar mass is relatively immune to the assumed IMF. This is the specific star-formation rate (*sSFR*), defined here as the ratio of star-formation rate to stellar mass already in place. The extrapolation to smaller masses invoked by the IMF assumption applies to both the numerator and denominator when calculating this quantity, making it reasonably robust (albeit still highly vulnerable to any uncertainties in dust extinction). This, combined with the attraction that *sSFR* encapsulates a basic measure of "current" to past star-formation activity, has made *sSFR* a key focus of many recent studies of high-redshift galaxies. Indeed, one of the most interesting, and controversial results to emerge from this work in recent years is that star-forming galaxies lie on a "main sequence" which can be characterised by a single value of *sSFR* which is a function of epoch. Moreover, as shown in Fig. 17, the current observational evidence suggests that this characteristic *sSFR*, after rising by a factor of $\simeq$40 from $z = 0$ to $z \simeq 2$ (Noeske et al. 2007; Daddi et al. 2007) plateaus at $2 - 3\,\mathrm{Gyr}^{-1}$ at all higher redshifts (e.g. Stark et al. 2009; González et al. 2010).

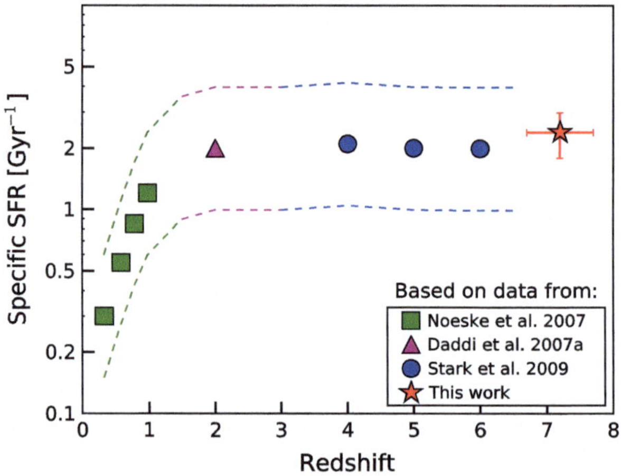

Fig. 17 Average $\langle sSFR \rangle$ determined at $z \sim 7$ by González et al. (2010) for a median stellar mass of $5 \times 10^9\,M_\odot$ compared to the average values determined by other authors at lower redshifts, but comparable stellar masses (Noeske et al. 2007; Daddi et al. 2007; Stark et al. 2009). This plot implies that $\langle sSFR \rangle$ stays remarkably constant, at $\simeq 2\,\mathrm{Gyr}^{-1}$ over the redshift range $2 < z < 7$, suggesting that the star-formation—mass relation does not evolve strongly during the first $\simeq 3$ billion years of galaxy evolution (courtesy V. González)

Our current best estimate of the evolving stellar mass function of LBGs at $z = 4$, 5, 6 & 7 is shown in Fig. 18. This summarizes the work of González et al. (2011), who used the WFC3/IR ERS data in tandem with the GOODS-South IRAC and *HST* ACS data to establish a mass to luminosity relation $M_{star} \propto L_{1,500}^{1.7}$ at $z \sim 4$, and then applied this to convert the UV LFs of Bouwens et al. (2007, 2010c) into stellar mass functions at $z \simeq 4$, 5, 6 & 7.

This plot also usefully illustrates both the level of agreement *and* current tension with current theoretical models, several of which predict extremely steep faint-end mass-function slopes, $\alpha_{mass} \simeq -2$ to -3 (e.g. Choi and Nagamine 2010; Finlator et al. 2011). The observationally-inferred mass-functions in Fig. 17 appear to display significantly flatter low-mass slopes than this, with $\alpha_{mass} \simeq -1.4$ to -1.6. However, it must be emphasized that Fig. 18 is based on the assumption that all galaxies at these redshifts follow the same mass-to-light relation as inferred at $z \simeq 4$ (and even this relation includes many objects without individual IRAC detections, and once again involves the simplifying assumption of constant star-formation rate). Thus the fact that the low-mass end of the mass function has a flatter slope than the faint-end slope of the UV LF is a simple consequence of applying $M_{star} \propto L_{1,500}^{1.7}$. In detail this is clearly wrong, but the question is how wrong?

This issue has recently been explored by McLure et al. (2011) who, for 21 galaxies at $z > 6.5$ which *do* have IRAC detections, explored the extent to which M_{star} for individual galaxies changes if the assumption of a universal $M_{star} - L_{1,500}$ relation is relaxed, and the full parameter space of age, star-formation history, dust-extinction and metallicity is explored in search of the best model fit. The results

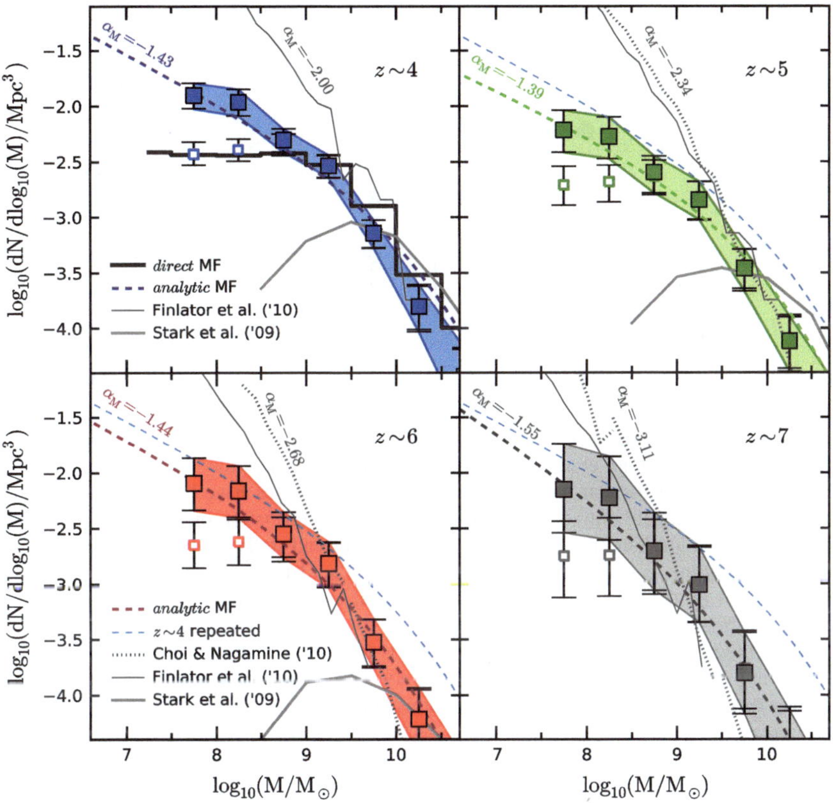

Fig. 18 The stellar mass functions of LBGs at $z \sim 4$, 5, 6, and 7 as produced by González et al. (2011) by applying the $M_{star} - L_{1,500}$ distribution found for $z \sim 4$ B-dropouts (within the WFC3/IR ERS field) to the UV LFs of Bouwens et al. (2007), Bouwens et al. (2010) at $z \sim 4-7$. For masses $M_{star} > 10^{9.5}\,M_{\odot}$, the $z < 7$ mass functions are in reasonable agreement with the earlier determinations by Stark et al. (2009) and McLure et al. (2009). The thick dashed curve in each panel represents the *analytic* mass functions derived from an idealized $M_{star} - L_{1,500}$ relation which, given the adopted form of $M_{star} \propto L_{1,500}^{1.7}$, inevitably display somewhat flatter low-mass slopes $\alpha_{mass} \simeq -1.4$ to -1.6, than the faint-end slopes in the LBG UV LFs ($\alpha = -1.7$ to -2.0; see Sect. 4.1). The $z \simeq 4$ *analytical* mass function is repeated in the other panels for comparison (thin dashed curve). The dotted and thin solid lines show the simulated mass functions from Choi and Nagamine (2010) and Finlator et al. (2011) (courtesy V. González)

of this analysis are shown in Fig. 19. For individual objects, M_{star} can vary considerably depending on the adopted model, and it appears that stellar mass can span a factor of up to $\simeq$50 at a given UV luminosity. However, for most objects it is found that the size of the mass uncertainty is generally limited to typically a factor of $\simeq$2–3, partly by the lack of available cosmological time (see also Labbé et al. 2010b); this is one (the only?) benefit of working at $z > 5$, namely that the impact on M_{star} of a plausible range of star-formation histories is damped somewhat by the fact that less than 1 billion years is available. Another interesting outcome of this

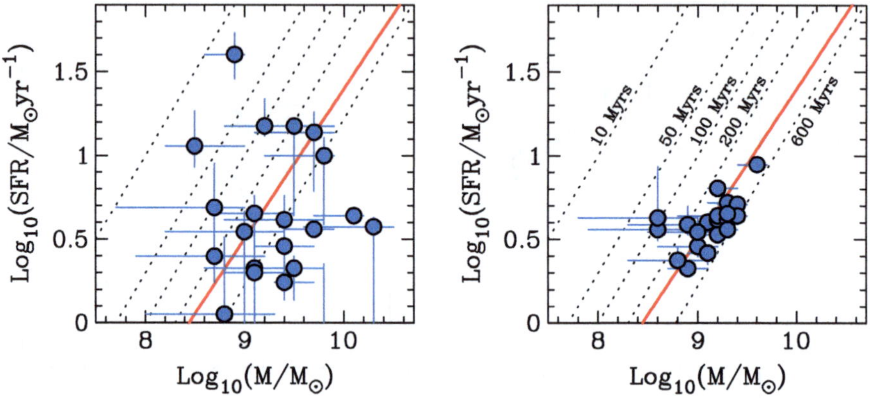

Fig. 19 Star-formation rate versus stellar mass for the twenty-one objects in the final robust $z > 6.5$ LBG sample of McLure et al. (2011) with IRAC detections at either 3.6 μm or 3.6 + 4.5 μm. In the left-hand panel the star-formation rates and stellar masses have been measured from the best-fitting SED template drawn from a wide range of star-formation histories, metallicities and reddening. In the right-hand panel the star-formation rates and stellar masses have been estimated from the best-fitting constant star-formation model. The 1σ errors on both parameters have been calculated by determining the $\Delta\chi^2 = 1$ interval, after marginalization over all other free parameters. The solid line in both panels is the SFR$-M_{star}$ relation derived by Daddi et al. (2007) for star-forming galaxies at $z \simeq 2$ and corresponds to a sSFR of $\simeq$2.5 Gyr^{-1}. The dotted lines illustrate how the SFR$-M_{star}$ relation for a galaxy with a constant star-formation rate and zero reddening varies as a function of stellar population age, as marked in the right-hand panel (courtesy R. McLure)

analysis is that, despite the increase in scatter in M_{star}, the *average* value was indeed still found to be consistent with $\langle sSFR \rangle \simeq 2 - 3\,\mathrm{Gyr}^{-1}$, and McLure et al. (2011) also confirmed that the mass-luminosity relation at $z \simeq 7$ is at least consistent with the $z \simeq 4$ relation adopted by González et al. (2011).

From these plots it can be seen that the typical stellar mass of an L^* LBG detected at $z \simeq 7$ in the current deep WFC3/IR imaging surveys is $M_{star} \simeq 2 \times 10^9\,\mathrm{M}_{\odot}$, and the faintest LBGs uncovered at these redshifts have masses as low as $M_{star} \simeq 1 \times 10^8\,\mathrm{M}_{\odot}$ (see also Finkelstein et al. 2010). This is impressive, as is the effort to establish the typical masses of faint LAEs from stacking of the available (somewhat shallower) IRAC imaging over the wider-area narrow-band Subaru surveys. This work indicates even lower typical stellar masses for LAEs selected at $z \simeq 5.7$ and $z \simeq 6.6$, with $M_{star} \simeq 1 - 10 \times 10^7\,\mathrm{M}_{\odot}$ (Ono et al. 2010).

This is not to say that a few significantly more massive LAEs have not been uncovered. For example, Ouchi et al. (2009a) discovered "Himiko", a giant LAE at $z = 6.595$, whose relatively straightforward IRAC detection implies a stellar mass in the range $M_{star} = 0.9{-}5 \times 10^{11}\,\mathrm{M}_{\odot}$. The example of Himiko shows that reasonably massive galaxies can be uncovered at $z \simeq 7$ given sufficiently large survey areas (in this case the parent $z \simeq 6.6$ narrow-band survey covered $\simeq$1 deg^2, sampling a comoving volume of $\simeq$800,000 Mpc^3; Ouchi et al. 2010). Such discoveries are "expected", and are entirely consistent with the mass functions shown in Fig. 18.

However, one should obviously be sceptical about claims of enormously massive (e.g. $M_{star} > 5 \times 10^{11}\,M_{\odot}$) galaxies at $z > 5$, especially at very high redshift and/or if discovered from very small surveys (e.g. Mobasher et al. 2005). In fact, Dunlop et al. (2007) found no convincing evidence of any galaxies with $M_{star} > 3 \times 10^{11}\,M_{\odot}$ at $z > 4$ in 125 arcmin^2 of the GOODS-South field, a result which is again consistent with the high-mass end of the mass functions shown in Fig. 18.

5.2 *Star-Formation Histories*

Given the growing evidence that $\langle sSFR \rangle$ is approximately constant at early times, it is tempting to conclude that the star-formation rates of high-redshift galaxies are *exponentially increasing* with time (rather than staying constant or exponentially decaying, as generally previously assumed—e.g. Eyles et al. 2007; Stark et al. 2009). While the analysis of McLure et al. (2011) indicates that it may be a mistake to assume all individual galaxies grow in this self-similar manner, and simple exponential growth is not really supported by the González et al. (2011) mass-luminosity relationship, on average something at least close to this scenario does indeed appear to be broadly consistent with much of the available data from $z \simeq 8$ down to $z \simeq 3$ (González et al. 2010; Papovich et al. 2011). Interestingly, the latest results from hydrodynamical simulations strongly predict that the star-formation rates of high-redshift galaxies should be increasing approximately exponentially (e.g. Finlator et al. 2011). However, the same simulations make the generic prediction that $\langle sSFR \rangle$ should continue to rise with increasing redshift beyond $z \simeq 2$, with $\langle sSFR \rangle \propto (1+z)^{2.5}$, tracking the halo mass accretion rate (e.g. Dekel et al. 2009). Thus, if the improving data continue to support an unevolving value for $\langle sSFR \rangle$ at high redshift, it will likely be necessary to invoke an additional feedback mechanism to suppress star-formation in the theoretical models at high redshift.

It is probably premature to attempt to say anything much more detailed about the evolution of the stellar populations in LBGs and LAEs at $z \simeq 5-8$. Without high-quality spectroscopy, the determination of detailed star-formation histories is inevitably confused by the complications and degeneracies arising from uncertain dust extinction, nebular emission, and metallicity, issues which are discussed a bit further below in the context of UV slopes and ionizing photon escape fractions. Certainly, the best-fitting star-formation histories for $z \simeq 7$ LBGs deduced by McLure et al. (2011) range from 10 Myr-old "Bursts" to models involving constant star-formation over 700 Gyr, and for each individual object a wide range of alternative star-formation histories is generally statistically permitted by the broad-band photometric data and the uncertainty in metallicity. Meanwhile, for LAEs, ages as young as 1–3 Myr have been inferred for the fainter, bluer objects (e.g. Ono et al. 2010).

Of course, as argued by Stark et al. (2009), in reality the star formation in these young galaxies may be highly intermittent or episodic. Interestingly, it is possible to try to estimate the "duty cycles" of LBGs and LAEs by reconciling their clustering properties with their number density. As discussed by Ouchi et al. (2010), the clustering of LAEs at $z \simeq 6.6$ can be used to infer a typical halo mass of $M_{halo} \simeq 10^8 - 10^9\,M_\odot$, and then comparison of the predicted number-density of such halos with the observed number density of LAEs implies that these galaxies/halos are observable as LAEs for 1–10 % of the time. These estimates are clearly still highly uncertain, not least because the clustering properties of LAEs and LBGs are still not very well determined (see Sect. 5.5 below), and indeed, based on the luminosity function comparison discussed above, it could be argued that the duty cycles of LAE and LBG activity must be virtually the same at $z \simeq 6 - 7$. Nevertheless, such calculations have the potential to provide genuinely useful constraints on duty cycles, as future surveys for $z \simeq 5 - 8$ LAEs and LBGs increase in both area and depth.

5.3 *Ultraviolet Slopes*

The first galaxies, by definition, are expected to contain very young stellar populations of very low metallicity. However, the possibility of detecting unambiguous observable signatures of such primordial stellar populations with current or indeed planned future instrumentation is currently a matter of considerable debate.

For example, one long-sought distinctive spectral signature of the first generation of galaxies is relatively strong HeII emission at $\lambda_{rest} = 1{,}640$Å (e.g. Shapley et al. 2003; Nagao et al. 2008; di Serego Alighieri et al. 2008). However, near-infrared spectroscopy of the sensitivity required to detect this line at $z > 7$ will certainly not be available until the *JWST*, and even then some theoretical predictions indicate that it is unlikely to be found in detectable objects (Salvaterra, Ferrara and Dayal 2011, but see also Pawlik, Milosavljevic and Bromm 2011).

By necessity, therefore, recent attention has focussed on whether the broad-band near-infrared photometry which has now been successfully used to discover galaxies at $z \simeq 6.5 - 8.5$ can actually be used to establish the rest-frame continuum slopes of the highest redshift galaxies. Specifically, very young, metal-poor stellar populations are arguably expected to result in substantially bluer continuum slopes around $\lambda_{rest} \simeq 1{,}500$Å than have been detected to date in galaxies discovered at any lower redshift $z < 6.5$ (e.g. Steidel et al. 1999; Meurer et al. 1999; Adelberger and Steidel 2000; Ouchi et al. 2004; Stanway et al. 2005; Bouwens et al. 2006; Hathi et al. 2008; Bouwens et al. 2009b; Erb et al. 2010).

It has become the normal convention to parameterise the ultra-violet continuum slopes of galaxies in terms of a power-law index, β, where $f_\lambda \propto \lambda^\beta$ (e.g. Meurer et al. 1999; thus, $\beta = -2$ corresponds to a source which has a flat spectrum in terms of f_ν, and hence has zero colour in the AB magnitude system). As discussed by several authors, while the bluest galaxies observed at $z \simeq 3 - 4$ have $\beta \simeq -2$, values as low (i.e. blue) as $\beta = -3$ can in principle be produced by a young, low-metallicity stellar population (e.g. Bouwens et al. 2010b; Schaerer 2002). However, as illustrated in Fig. 20, for this idealized prediction to actually be realized

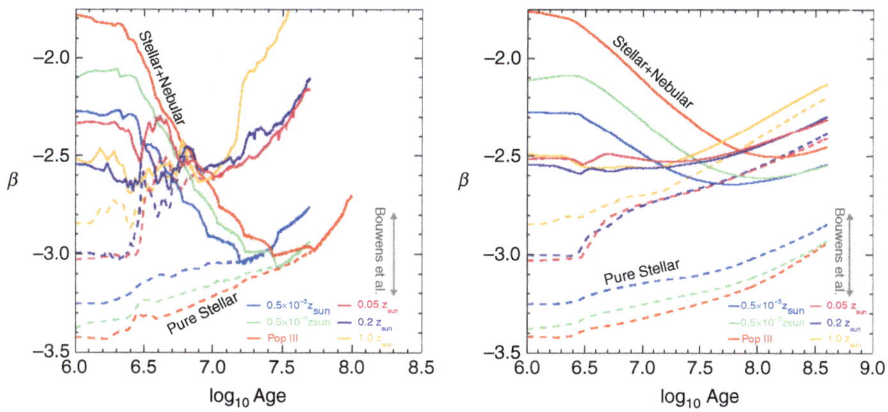

Fig. 20 Theoretical predictions of galaxy UV-slope β showing the values that are expected for stellar populations of different age and metallicity, and the reason for the interest generated by initial claims that faint galaxies at $z \simeq 7$ display $\langle\beta\rangle \simeq -3$ (Bouwens et al. 2010a). The left-hand panel shows the predicted evolution of β for instantaneous starbursts of differing metallicity, with and without predicted nebular continuum (Schaerer 2003). The right-hand panel shows equivalent information for the (arguably more realistic) case of continuous star-formation. More recent measurements of $\langle\beta\rangle$ have failed to find evidence for values $\langle\beta\rangle < -2.5$ at $z \simeq 7$, and have converged on $\langle\beta\rangle = -2.1 \pm 0.1$ for brighter (L^*) galaxies at $z \simeq 7$. These plots show that while extreme values of $\langle\beta\rangle < -2.8$ would imply both very low metallicity, *and* a high UV-photon escape fraction, the interpretation of the more moderate values actually observed (i.e. $\langle\beta\rangle = -2.0 \rightarrow -2.5$), is much less straightforward

in practice, several conditions have to be satisfied simultaneously, namely (i) the stellar population has to be very young (e.g. $t < 30\,\mathrm{Myr}$ for $Z \simeq 10^{-3}\,Z_\odot$, or $t < 3\,\mathrm{Myr}$ for $Z \simeq 10^{-2}\,Z_\odot$), (ii) the starlight must obviously be completely free from any significant dust extinction, and (iii) the starlight must also *not* be significantly contaminated by (redder) nebular continuum (a condition which has important implications for UV photon escape fraction, and hence reionization—see, for example, Robertson et al. 2010).

For this reason, the report by Bouwens et al. (2010a) (supported to some extent by Finkelstein et al. 2010) that the faintest galaxies detected at $z > 6.5$ apparently display an average value of $\langle\beta\rangle = -3.0 \pm 0.2$ was both exciting and arguably surprising, and was immediately subjected to detailed theoretical interpretation (e.g. Taniguchi et al. 2010).

However, a series of further observational studies of UV slopes over the last year have failed to confirm this result, revealing that the original measurement was biased towards excessively blue slopes in a subtle but significant way (e.g. Dunlop et al. 2012; Wilkins et al. 2011b; Finkelstein et al. 2012; Bouwens et al. 2012). It now seems likely that, for the faintest galaxies uncovered so far at $z \simeq 7$, the average UV power-law index lies somewhere in the range $\langle\beta\rangle = -2.5 \rightarrow -2.0$. As can be seen from Fig. 20, the correct interpretation of such slopes is unclear, as they can be produced by different mixes of age, metallicity, and nebular contributions; it is only in the extreme case of $\langle\beta\rangle \simeq -3.0$ that the interpretation in favour of exotic stellar populations and large escape fractions becomes relatively clean.

Although these more moderate UV slopes are arguably in better accord with theoretical expectations (e.g. Dayal and Ferrara 2012) it may still be the case that some individual galaxies at $z \simeq 7$ with $\beta = -3$ have already been discovered among the faint LBG or LAE samples (e.g. Ono et al. 2010). Unfortunately, however, this is impossible to check with current data because β is such a sensitive function of observed colour. At $z \simeq 7$, the estimate of β for an *HST*-selected LBG has currently to be based on a single colour, with

$$\beta = 4.43(J_{125} - H_{160}) - 2. \tag{7}$$

Thus, a perfectly "reasonable" photometric uncertainty of $\simeq$15 % in J_{125} and H_{160} translates to a $\simeq$20 % uncertainty in colour and hence to an 1-σ uncertainty of ± 0.9 in β. Better photometric accuracy, ideally combined with additional near-infrared wavebands to allow improved power-law or SED fitting (as is already possible at lower redshifts; Finkelstein et al. 2012; Castellano et al. 2012) is required to enable a proper investigation of the UV slopes of the faintest galaxies at $z \simeq 7-8$.

In contrast to this confusion at the faintest luminosities, there does at least seem to be general agreement that the brighter galaxies found at $z \simeq 7$ (with $M_{1,500} \simeq -21$) have $\langle\beta\rangle = -2.1 \pm 0.1$. This is basically as expected for a few hundred Myr-old star-forming galaxy, with solar metallicity and virtually no dust extinction, although other interpretations are possible (see Fig. 20). Certainly, for galaxies at these brighter luminosities, any evolution in $\langle\beta\rangle$ with redshift has generally been interpreted as arising predominantly from a change in the level of dust obscuration, as already discussed above in the context of Lyman-α emission. In particular, Bouwens et al. (2009b) have reported that the average value of β exhibited by brighter LBGs declines from $\langle\beta\rangle \simeq -1.5$ at $z \simeq 4$ to $\langle\beta\rangle \simeq -2$ at $z \simeq 6$, and have interpreted this as reflecting a progressive reduction in average extinction with increasing redshift. This is an important result, because it leads to the conclusion that the correction to be applied to observed UV luminosity density to account for dust-obscured star-formation steadily decreases with increasing redshift. This has obvious implications for the inferred history of cosmic star-formation density as discussed further below in Sect. 6.1. However, the precise redshift dependence of average extinction, and indeed whether there exists a clear relationship between UV luminosity and spectral slope at high redshift is still the subject of some controversy (Dunlop et al. 2012; Bouwens et al. 2012; Finkelstein et al. 2012; Castellano et al. 2012).

5.4 Galaxy Sizes and Morphologies

Only the most basic information is known about the morphologies of LBGs and LAEs at $z > 5$, for the simple reason that the objects are faint, and are thus generally detected at only modest signal:noise ratios. Oesch et al. 2010b investigated

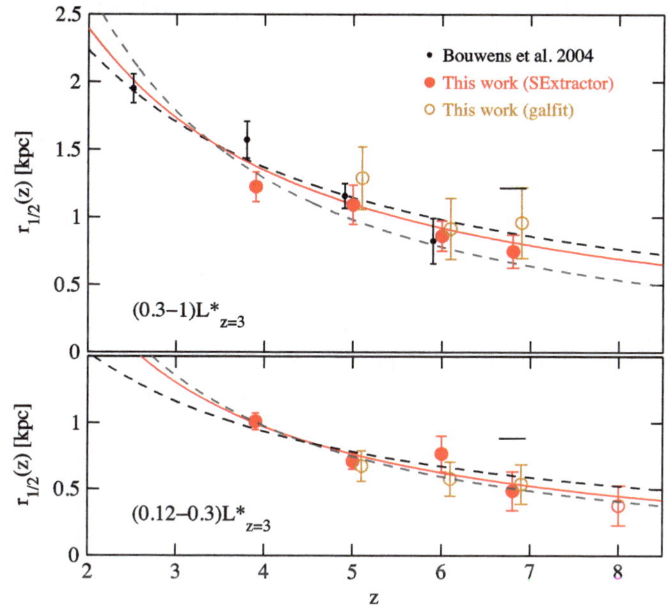

Fig. 21 The observed evolution of the mean size of Lyman-break galaxies reported by Oesch et al. (2010b), over the redshift range $z \sim 2-8$ in two different luminosity ranges $(0.3\text{-}1)L^*_{z=3}$ (*top*) and $(0.12\text{-}0.3)L^*_{z=3}$ (*bottom*) (where $L^*_{z=3}$ is the characteristic luminosity of a LBG at $z \simeq 3$). Different symbol styles correspond to different ways of analysing the data to extract size estimates. The dashed lines indicate the scaling expected for a fixed dark matter halo mass ($\propto (1+z)^{-1} \equiv\propto H(z)^{-2/3}$; *black*) or at fixed halo circular velocity ($\propto (1+z)^{-3/2} \propto H(z)^{-1}$; *gray*). The solid red lines indicate the best fit to the observed evolution, which is described as proportional to $(1+z)^{-m}$, with $m = 1.12 \pm 0.17$ for the brighter luminosity bin, and $m = 1.32 \pm 0.52$ at fainter luminosities (but both are formally identical, and consistent with $m \simeq 1$). The extent to which this apparent relationship is influenced by the surface-brightness bias inherent in deep *HST* imaging is still a matter of some debate (courtesy P. Oesch)

the first WFC3/IR imaging of LBGs at $z \simeq 7-8$, and were able to show that almost all of these galaxies are marginally resolved, with an average intrinsic size of $\simeq 0.7 \pm 0.3$ kpc. Thus, known extreme-redshift LBGs are clearly very compact (certainly too compact to be resolved with typical ground-based imaging), and the detection of extended features is, to date, rare. Comparison with the sizes of LBGs at lower redshift implies that average size decreases gently from $z \simeq 4$ to $z \simeq 7$, following approximately a relationship of the form $(1+z)^{-1}$ (Fig. 21). A slow decrease in average size at a fixed luminosity with increasing look-back time is anticipated from semi-analytic models of galaxy formation (e.g. Mo et al. 1998, 1999; Sommerville et al. 2008; Firmani et al. 2009), and is consistent with earlier observations of lower-redshift LBGs (e.g. Ferguson et al. 2004; Bouwens et al. 2004b) and disc galaxies (e.g. Buitrago et al. 2008). However, it is still unclear whether the apparently observed relationship at high redshift is, at least in part, a consequence of the fact that galaxy detection with *HST* is biased towards objects which have high surface brightness (e.g. Grazian et al. 2011). Oesch et al. (2010b) claim that this is not a significant problem, but *HST* imaging covering a greater dynamic range, and providing larger samples for stacking, should certainly help to clarify this issue in the near future.

At somewhat lower redshifts, Taniguchi et al. (2009) used the *HST* ACS single-orbit I_{814} imaging in the COSMOS field to attempt to investigate the morphologies of 85 LAEs at $z \simeq 5.7$, selected via Subaru narrow-band imaging. The results of this study are somewhat inconclusive, with only 47 LAEs being detected, and approximately half of these being apparently unresolved. Nevertheless, the result is a typical half-light radius of $\simeq$0.8 kpc, clearly not inconsistent with that displayed by LBGs at comparable redshifts. Taniguchi et al. (2009) also report that fits to the light profile of their stacked images favour a Sérsic index $n \sim 0.7$, more consistent with disc-like or irregular galaxies than with a de Vaucouleurs spheroid.

5.5 *Clustering*

A measurement of the clustering of high-redshift galaxies is of interest primarily for estimating the characteristic mass of the dark matter halos in which they reside. If this can be established with meaningful accuracy, then a comparison of the observed galaxy number density with that of the relevant halos (as predicted within the concordance Λ-CDM cosmological model) can yield an estimate of the halo occupation fraction, or, equivalently, the duty cycle of a given class of high-redshift galaxy.

The measurement of galaxy clustering at $z > 5$ is still in its infancy, due primarily to the small sample sizes delivered by current facilities. To date it has been most profitably pursued using the samples of several hundred LAEs selected over degree-scale fields via the narrow-band Subaru imaging targeted at $z \simeq 5.7$ (Ouchi et al. 2005) and $z \simeq 6.6$ (Ouchi et al. 2010). Figure 22 shows the distribution on the sky of the $\simeq$200 LAEs at $z \simeq 6.6$ uncovered in the SXDS field by Ouchi et al. (2010).

To quantify the significance and strength of any clustering present in such images, the standard technique is to calculate the Angular Correlation Function, $\omega(\theta)$, which represents the excess (or deficit) of objects at a given angular radius from a galaxy relative to that expected from a purely random distribution of galaxies with the observed number counts. This is usually calculated following the prescription of Landy and Szalay (1993), which gives

$$\omega_{\rm obs}(\theta) = [DD(\theta) - 2DR(\theta) + RR(\theta)]/RR(\theta), \tag{8}$$

where $DD(\theta)$, $DR(\theta)$, and $RR(\theta)$ are the numbers of galaxy–galaxy, galaxy–random, and random–random pairs normalized by the total number of pairs in each of the three samples.

Considerable care and simulation work is required to calculate $\omega_{\rm obs}(\theta)$ especially when, as shown by the grey regions in Fig. 22, several areas of the image have

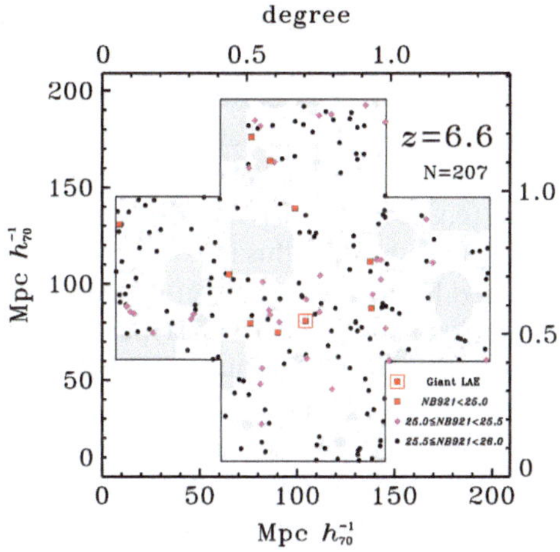

Fig. 22 The distribution on the sky of the 207 LAEs at $z = 6.565 \pm 0.054$ detected by Ouchi et al. (2010) via narrow-band imaging with Subaru in the SXDS field. *Red squares, magenta diamonds, and black circles* indicate the positions of bright ($NB921 < 25.0$), medium-bright ($25.0 \leq NB921 \leq 25.5$), and faint ($25.5 \leq NB921 \leq 26.0$) LAEs, respectively. The *red square surrounded by a red open square* indicates the giant LAE, *Himiko*, at $z = 6.595$ reported by Ouchi et al. (2009a), and already briefly discussed in Sect. 5.1 (courtesy M. Ouchi)

to be masked out due to bright stars or image artefacts. As described in detail by Ouchi et al. (2005, 2010), $\omega_{\rm obs}(\theta)$ is converted to a best estimate of $\omega(\theta)$, then used to derive a clustering amplitude A_ω assuming a power-law correlation function $\omega(\theta) = A_\omega \theta^{-\beta}$, in which β has generally to be fixed rather than fitted given the limited sample size (usually $\beta = 0.8$ is adopted on the basis of clustering analyses at lower redshift). Finally A_ω is converted into a physical correlation length r_0 using knowledge of the redshift distribution (which is relatively straightforward to establish for narrow-band selected LAEs).

The result of these studies is that significant clustering has now been detected in the LAE population at $z > 5$, and that the best estimate of the correlation length for LAEs at $z \simeq 5-7$ is $r_0 = 3-7$ Mpc (for $H_0 = 70\,{\rm km\,s^{-1}\,Mpc^{-1}}$). This in turn can be used to infer an average mass for the dark-matter halos hosting these LAEs of $10^{10}-10^{11}\,{\rm M_\odot}$. A similar analysis for LBGs at $5 < z < 6$ has been performed by McLure et al. (2009), who report a correlation length of $r_0 = 8^{+2}_{-1.5}$ Mpc (for $H_0 = 70\,{\rm km\,s^{-1}\,Mpc^{-1}}$) and a resulting characteristic dark-matter halo mass of $10^{11.5}-10^{12}\,{\rm M_\odot}$. This significantly larger halo mass for the LBGs is not unexpected, given that they are considerably rarer, more massive objects (than typical narrow-band selected LAEs), having been selected from a substantially larger cosmological volume, down to much brighter continuum flux limits. These results are consistent with those derived by Overzier et al. (2006) for LBGs at a mean redshift of $z \simeq 5.9$. Interestingly, the $z > 5$ LBGs studied by McLure et al. (2009) are bright enough to allow a reasonable estimate of average stellar mass, $\log_{10}(M/M_\odot) = 10.0^{+0.2}_{-0.4}$, which is consistent with the results of the clustering analysis for plausible values of the ratio of stellar to dark matter.

Unfortunately the large uncertainty in current estimates of characteristic halo mass, combined with the steepness of the halo mass function, means that such clustering measurements cannot yet be used to yield meaningfully-accurate duty cycles for LAEs and LBGs (e.g. Ouchi et al. (2010) report an inferred duty cycle of $\simeq 1\,\%$ for $z \simeq 6.6$ LAEs, but acknowledge this is extremely uncertain). Nonetheless, these pioneering studies provide hope for meaningful measurements with the much larger LAE and LBG samples anticipated from Hyper-Suprime Cam on Subaru (Takada 2010) and from the *EUCLID* Deep survey over the next decade (Laurejis et al. 2011).

As discussed in Sect. 6.2, such future large-scale surveys also have the potential to search for one of the long anticipated signatures of reionization, namely an enhancement in the clustering of LAEs relative to LBGs due to patchy reionization. At present, Ouchi et al. (2010) report no evidence for such a clustering amplitude boost at $z \simeq 6.6$.

6 Global Perspective

6.1 A Consistent Picture of Galaxy Evolution?

6.1.1 Cosmic Star-Formation History

To gain a broader of view of the time evolution of action in the Universe the evolving galaxy UV luminosity function discussed in Sect. 4.1 can be integrated (over luminosity) to yield the evolving *comoving UV luminosity density*. This might be of academic interest were it not for the fact that this quantity can, in principle, be converted into *star-formation density*, to gain a global view of the evolution of star-formation activity per unit comoving volume over cosmic time.

This calculation has been performed by many authors over a wide range of redshifts since being first promoted by Lilly et al. 1996 and Madau et al. 1996. It is, however, a calculation fraught with danger as it involves (and can depend critically upon) several extrapolations; the galaxy luminosity function has to be correctly extrapolated to the lowest luminosities, the stellar mass function has to be correctly extrapolated to the lowest masses, and any mass and/or time dependence of the obscuring effects of cosmic dust has to be correctly accounted for and removed. In addition, care has to be taken to account for highly-obscured populations which may be entirely missed in UV-selected galaxy samples.

A full review of the many, and continually improving studies of cosmic star-formation history is obviously beyond the scope of this $z > 5$ review. However, to place the high-redshift results in context, it is fair to say that there is now broad agreement that star-formation density rises by an order of magnitude as we look

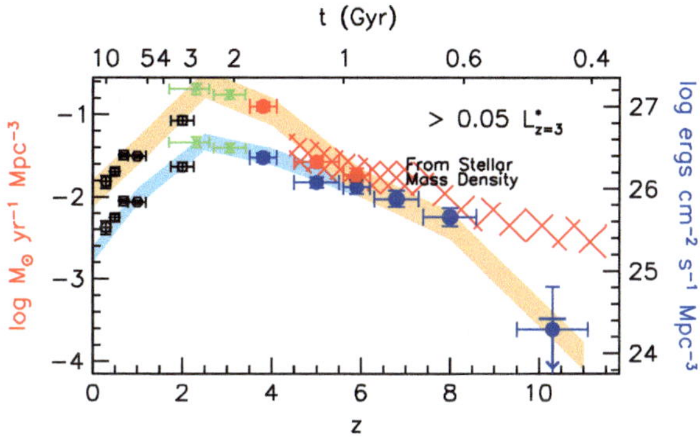

Fig. 23 A compilation of UV luminosity density measurements, and hence inferred evolution of cosmic star-formation rate density as presented by Bouwens et al. (2011b). The data-points at $z = 4 - 8$ are basically based on integrating the luminosity functions described in Fig. 11 down $M_{UV} \simeq -18$ AB mag, while data points at lower redshift from Reddy and Steidel (2009), Bouwens et al. (2007) and Schiminovich et al. (2005) are provided for context. The *blue data points and lower shaded regions* indicate UV luminosity density prior to any correction for dust obscuration. The *upper locus* indicates the effects of correcting for redshift dependent dust obscuration which, as apparent, is assumed here to decline to zero by $z \simeq 7$ following Bouwens et al. (2009b). The *red hatched region* is intended to indicate the SFR density derived from differentiating the growth in stellar mass density delineated below in Fig. 24. Results at $z > 8$ are probably not meaningful, as the stellar mass density is based on an assumed extrapolation, while the $z \simeq 10$ datapoint is based on the single $z \simeq 10$ galaxy in the HUDF claimed by Bouwens et al. (2011a) and additional tentative upper limits based on non detections (Oesch et al. 2012). (courtesy R. Bouwens)

back from $z = 0$ to $z = 1$, increases by a further factor of 2 or 3 by $z \simeq 2$, and then appears to plateau out to $z \simeq 3$ before declining at still higher redshifts (see e.g. Hopkins and Beacom 2006; Dunlop 2011). This evolution is shown in Fig. 23 (taken from Bouwens et al. 2011b), but this particular figure has been deliberately designed to focus on the apparently smooth and steady decline of star-formation density from $z \simeq 3$ to $z > 8$.

Crucial to the precise form of this plot is the assumed strength and redshift dependence of typical dust obscuration as a function of redshift, which in this case is assumed to decline from a factor of $\simeq$7 at $z \simeq 2$ to essentially zero at $z \simeq 7$. The true redshift dependence of dust extinction in LBGs of course remains a matter of debate, but several independent pieces of evidence point towards a high-redshift decline (e.g. Fig. 13), and even with dust corrections it is hard to escape the conclusion that SFR density is significantly lower at $z \simeq 7$ than at the peak epoch corresponding to $z \simeq 2 - 3$.

Finally, it is worth noting that while the datapoint at $z \simeq 10$ should probably be taken with a pinch of salt (but see also Oesch et al. 2012), the decline in SFR density from $z = 4$ to $z = 8$ is in fact fairly precipitous when viewed in terms of the $\simeq$1 Gyr of elapsed cosmic time. Current observations of the high-redshift UV LF thus support the view that we are witnessing the rapid emergence of the star-forming galaxy population.

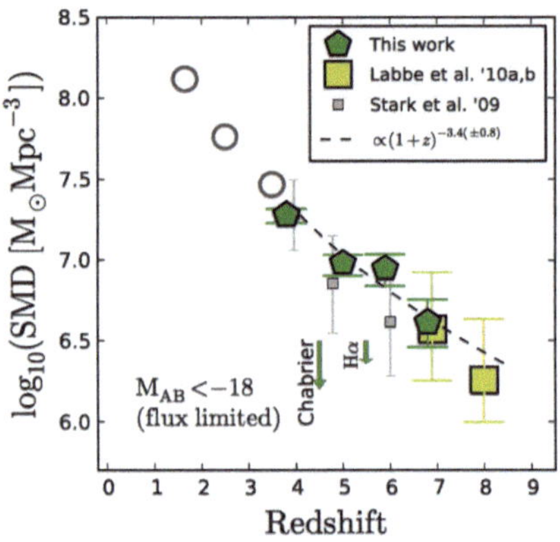

Fig. 24 Stellar Mass Density as a function of redshift for sources brighter than $M_{1,500} = -18$ as derived by González et al. (2011). These stellar-mass density values were produced by integrating the mass functions shown in Fig. 18 to $M_{1,500} = -18$ at $z = 4$, 5, 6, and 7. Shown for comparison are the stellar-mass density determinations from from Stark et al. (2009) corrected to the same $M_{1,500} = -18$ limit. The *low-redshift open circles* were derived by integrating the Marchesini et al. (2009) mass functions between $8.3 < \log_{10}\mathrm{M_{star}/M_\odot}) < 13$ and multiplying by 1.6 to match the Salpeter IMF. A constant SFH and $0.2\,\mathrm{Z_\odot}$ metallicity was assumed to derive all stellar masses at $z \geq 4$. The effect of a possible 20 % correction due to contamination by Hα is shown, as is the effect of using a different IMF. The integrated mass growth shown here is well described by $\log_{10}(\mathrm{SMD}) \propto (1 + \mathrm{z})^{-3.4\pm0.8}$ (courtesy V. González)

6.1.2 The Growth of Cosmic Stellar Mass Density

An alternative route to determining the cosmic history of star-formation is provided by integrating the evolving galaxy stellar mass functions (over stellar mass) to map out the build-up of *comoving stellar mass density* with cosmic time. In principle this can be used as a check on the validity of the assumptions (concerning, for example, dust) used to estimate cosmic star-formation history as described above, because the stellar mass density in place at any epoch should (modulo some stellar mass loss) equate to the time integral of all preceding star-formation activity. At $z > 5$ these calculations are arguably still premature. However, given data of sufficient quality, they may in fact be more straightforward and less uncertain than at more modest redshifts, principally because serious dust obscuration may be less of a problem in the young, relatively metal-poor Universe.

Figure 24 shows a compilation of recent determinations of stellar mass density at high redshift taken from González et al. (2011), with the dark-green pentagons in effect being based on integration of the stellar mass functions shown in Fig. 18 (again down to $M_{UV} \simeq -18$, for ease of comparison with Fig. 23). Despite the well-documented uncertainties in current measurements of the stellar mass function at early times, this figure indicates that there is now reasonably good evidence for a smooth, monotonic rise in the integrated stellar mass density with decreasing redshift. Moreover, while there have in the past been some problems in reconciling the growth in stellar mass density with the directly observed SFR density (e.g. Wilkins, Trenti and Hopkins 2008), this situation has improved (e.g. Reddy and Steidel 2009), and the reasonable agreement seen here between the time differential of Fig. 24 with the SFR density plotted in Fig. 23 (as indicated by the red hatched region) arguably provides some confidence that neither measurement is too far off. It also suggests that fears the IRAC fluxes from many $z \simeq 7$ galaxies are dominated by extreme nebular emission lines (rather than starlight) may have been somewhat exaggerated (Schaerer and Barros 2010; see also McLure et al. 2011).

It is however, probably premature to conclude that the agreement is sufficiently good to support the assumption of zero dust obscuration at high redshift, especially given the current limitations in constraining the low-mass end of the stellar mass function at $z > 4$. Deeper *HST* data will undoubtedly help further progress in this area, as of course will the higher resolution mid-infrared imaging to be delivered by *JWST* (see Sect. 7).

6.2 *Cosmic Reionization*

6.2.1 Current Constraints on Reionization

A second arena in which the integrated UV luminosity density of the evolving high-redshift galaxy population is of interest is in the study of cosmic reionization. The reionization of the hydrogen gas that permeates the Universe was a landmark event in cosmic history. It marked the end of the so-called cosmic "dark ages", when the first stars and galaxies formed, and when the intergalactic gas was heated to tens of thousands of degrees Kelvin from much smaller temperatures. This global transition had far-reaching effects on the formation of the early cosmological structures and left deep impressions on subsequent galaxy and star formation, some of which almost certainly persist to the present day.

The study of this epoch is thus arguably the key frontier in completing our understanding of cosmic history, and naturally the focus of much current research. Nevertheless, despite the considerable recent progress in both theory and observations (for recent reviews see Robertson 2010; McQuinn et al. 2010) all that is really established about this crucial era is that it was completed by redshift $z \simeq 6$ (as evidenced by the Gunn–Peterson troughs in the spectra of the most distant quasars; Fan et al. 2006) and probably commenced around $z \sim 15$ (as suggested by the latest WMAP microwave polarisation measurements, which favour a mean redshift of reionization of 10.4 ± 1.4; Komatsu et al. 2009. However, as discussed by Dunkley et al. (2009), within these bounds the reionization history is essentially unknown, and with current data we cannot even distinguish whether it was "sharp" or extended.

Unsurprisingly, therefore, understanding reionization is one of the key science goals for a number of current and near-future large observational projects. In particular, it is a key science driver for the new generation of major low-frequency radio projects (e.g. LOFAR, MWA and SKA) which aim to map out the cosmic evolution of the *neutral atomic* Hydrogen via 21-cm emission and absorption. However, radio observations at these high redshifts are overwhelmingly difficult, due to the faintness of the emission and the very strong foregrounds, and in any case such radio surveys cannot tell us about the sources of the ionizing flux.

A key and interesting question, then, is *whether and when* the apparently rapidly evolving UV-selected galaxy population is capable of delivering enough ionizing photons per unit time per unit volume which can escape from their host galaxies to reionize the inter galactic medium.

6.2.2 The Galaxy Population at z $\simeq$ 7, and the Supply of Reionizing Photons

Clearly the complete ionization of hydrogen in the intergalactic medium requires sustained sources of Lyman continuum photons with wavelengths $\lambda < 912\,\text{Å}$ (corresponding to the ionization energy of ground-state hydrogen, $E > 13.6\,\text{eV}$). If the emerging population of young faint galaxies revealed in the HST surveys is responsible for reionizing the Universe then, as discussed by many authors, the process of reionization should, at least in broad terms, follow their time-dependent density (e.g. Robertson et al. 2010; Trenti et al. 2010). However, it is not straightforward to establish the number density of ionizing photons delivered by galaxies at $z \simeq 6 - 10$, because they are essentially unobservable due to the fact they are absorbed by neutral hydrogen (as they must be if they are doing their job of reionizing the hydrogen gas). We are therefore forced to infer the number density of ionizing photons from the observable evolving UV luminosity density at $\lambda_{rest} \simeq 1{,}500\,\text{Å}$, coupled with estimates of the rate of ionizing photons produced per unit solar mass of star-formation activity, and an estimate of what fraction of the ionizing Lyman-continuum photons produced by young stars can actually escape their host galaxies to help with reionization of the surrounding inter-galactic medium (f_{esc}).

Finally, we also require some knowledge of the "clumpiness" (C) of the inter-galactic medium in the young Universe. For the IGM to be ionized simply requires that recombinations are balanced by ionizations. The recombination rate depends on the IGM temperature and the physical hydrogen density which declines with time according to the universal expansion factor $R(t)^{-3} \propto (1+z)^3$. However, it is enhanced in locally overdense regions by the clumping factor $C = \langle n_H^2 \rangle / \langle n_H \rangle^2$ (i.e. $C = 1$ corresponds to a uniform IGM). Early cosmological simulations indicated that the IGM clumping factor at $z \simeq 6$ could be as high as $C \simeq 30$ (e.g. Gnedin and Ostriker 1997) making reionization difficult due to self-shielding within dense clumps. However, more recent simulations suggest that the IGM clumping factor lies in the range $1 < C < 6$, making reionization easier to achieve (e.g. Bolton and Haehnelt 2007; Pawlik, Schaye and van Scherpenzeel 2009).

An illustration of current constraints on the ability of galaxies to reionize the Universe at $z \simeq 7$ is shown in Fig. 25 (adapted from Robertson et al. 2010). This is

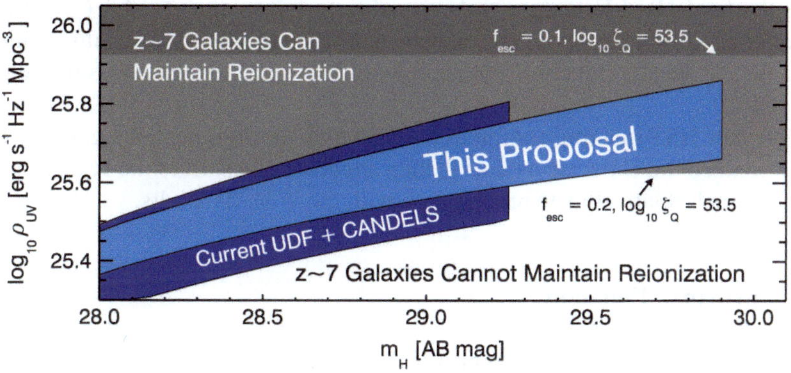

Fig. 25 Did Galaxies Reionize the Universe? Expected constraints on the rate of ionizing photons dn_{ion}/dt, from the combination of the HUDF and upcoming CANDELS *HST* surveys. The *shaded regions* show a 3σ uncertainty in the UV luminosity density ρ_{UV} and the improvement made possible via the planned deeper 29.9(AB) (stacked $J_{140} + H_{160}$) UDF12 imaging program. The *light and dark grey regions* show the ρ_{UV} ranges necessary to maintain reionization for escape fractions, $f_{esc} = 0.2$ and 0.1 respectively (for other assumptions see Robertson et al. 2010) (courtesy B. Robertson)

based on our current knowledge of the galaxy LF at this epoch, an assumed IGM clumping factor $C = 2$, a production rate of ionizing photons per unit star formation of $10^{53.5}\,\mathrm{s}^{-1}\mathrm{M}_{\odot}^{-1}$year, and alternative escape fraction assumptions of $f_{esc} = 0.1$ and $f_{esc} = 0.2$ (values for f_{esc} which at least have some tentative support from observations at $z \simeq 3$; Shapley et al. 2006; Iwata et al. 2009). All of these numbers remain highly uncertain, but it can be seen that for these, arguably not unreasonable assumptions, confirmation that the $z \simeq 7$ LF of McLure et al. (2010) continues to rise steeply down to $H \simeq 30$ AB mag may be sufficient to prove that the emerging population of young galaxies could have reionized the Universe by $z \simeq 7$. Of course this conclusion would become even more secure if the even steeper faint-end slope in the LF favoured by Bouwens et al. (2011) (see Sect. 4.1) is confirmed, and especially if more accurate determinations of the UV-slope parameter β favour high escape fractions (see Sect. 5.3). This provides strong motivation for the even deeper HST WFC3/IR observations in the HUDF planned in the UDF12 project in summer 2012 (*HST* programme GO-12498).

In addition, Fig. 25 conservatively assumes that we only count up the photons from the *observable* galaxies which, for the $H_{160} \simeq 30$ limit of the planned HST imaging, corresponds to an absolute magnitude limit of $M_{1,500} \simeq -17$. Extrapolation to still fainter luminosities will provide yet more ionizing photons, especially if the LF remains steep, and even more so if escape fraction rises with decreasing luminosity. However, it is currently unclear how far down in luminosity one can safely integrate. At low redshifts, *GALEX* imaging in the Coma and Virgo clusters suggests a turnover in the LF around $M_{UV} \simeq -14$ (Hammer et al. 2012; Boselli et al. 2011) corresponding to a deficit of dwarfs below masses of $10^8\,\mathrm{M}_{\odot}$, while the field UV LFs appear to keep rising to fainter magnitudes ($\simeq -11$; e.g. Treyer et al. 2005). However, the relevance of these low-redshift results is unclear, given that Schechter function fits at low-redshift yield a faint-end slope of only $\alpha = -1.4$.

If the new HST imaging continues to strengthen the argument that galaxies could have reionized the Universe by $z \simeq 7$, then attention will shift to the issue of how to reconcile such relatively late reionization with the WMAP results.

6.2.3 Lyman-α Emission

Additional information on the progress of reionization can be gleaned from observations of Lyman α emission from high-redshift galaxies, which have the potential to inform us about the ionization state of the IGM as a function of redshift.

This work complements detailed analyses of Lyman-α emission from high-redshift quasars, where studies of the size of proximity zone (i.e. the ionized region) around, for example, the $z = 7.085$ quasar LAS J1120 + 0641 (whose spectrum was shown in Fig. 2), have been used to argue that the IGM is significantly more neutral at $z \simeq 7$ than at $z \simeq 6.5$ (Mortlock et al. 2011; Bolton et al. 2011). The problem with such studies of very bright, but hence very rare objects is that it is hard to know whether the sightline is typical, and it is also hard to decide whether a small proximity zone may simply reflect the fact a given quasar has only recently turned on. Thus, while observations of Lyman-α emission from the much fainter galaxy population are obviously much more challenging, they offer the prospect of statistically representative results based on multiple sightlines.

As already discussed in detail in Sect. 4.3, and as concisely described by Finlator (2012), a partially neutral IGM has the effect of scattering the Lyman-α emission from a galaxy into a low surface brightness halo, and this has several measurable consequences which, for convenience, I summarize again briefly here.

First, the luminosity function of LAEs may evolve. However, as discussed in Sects. 4.2 and 4.3, the interpretation of current measurements of the evolving LAE LF is complicated by the fact any evolution seen reflects a mix in the underlying evolution of the galaxy mass function, evolution in the intrinsic ISM of the galaxies, and the desired signature of the evolving IGM. This complication, coupled with the very poor constraints currently available on the form of the LAE LF at $z \simeq 7$, mean that it appears too early to attempt to draw any definitive conclusions on the progress of reionization from this work.

Second, the Lyman-α escape fraction should evolve. As explained in Sect. 4.3.3, follow-up spectroscopy of objects selected as LBGs has, in principle, the ability to cleanly separate the evolution of the underlying galaxy population from the evolution of Lyman-α escape fraction. These observations are being keenly pursued, and current indications are that average Lyman-α escape fraction increases out to $z \simeq 6$, but that this trend shows signs of reversal at $z \simeq 7$ (Pentericci et al. 2011; Schenker et al. 2012; Ono et al. 2012). In addition, tentative claims have been advanced that this drop in escape fraction is more dramatic for faint objects than

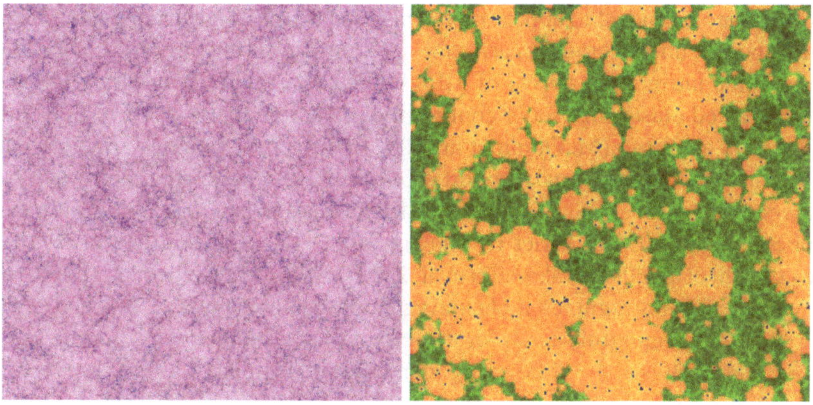

Fig. 26 *Left:* Early structure formation in ΛCDM (at $z = 6$) from an N-body simulation with $5{,}488^3$ (165 billion) particles and $(425\,\mathrm{h}^{-1}Mpc)^3$ volume. Shown are the dark-matter density (*pink*) and halos (*blue*). This synthetic image corresponds to 3.5×3.5 degrees on the sky. *Right:* The geometry of the epoch of reionization, as illustrated by a slice through a $(165\,\mathrm{Mpc})^3$ simulation volume at $z = 9$. Shown are the density (*green/yellow*), ionized fraction (*red/orange*), and ionizing sources (*dark dots*) (courtesy I. Iliev)

for bright ones, and that this may indicate reionization proceeds from high- to low-density environments, as suggested by an inside-out reionization model (Ono et al. 2012). However, the spectroscopy at $z \simeq 7$ is challenging, and current results are somewhat controversial and based on very small samples. Moreover, the $z \simeq 6$ baseline against which it can be judged is still in the process of being properly pinned down (Curtis-Lake et al. 2012). Nevertheless, continued work in this area has the potential to yield relatively clear-cut results, and should be enormously aided by the advent of the new generation of multi-object near-infrared spectrographs, as summarized below in Sect. 7.

Third, the mean shape of the Lyman-α line emission from galaxies should evolve, as any increasing neutral; fraction impacts on the blue side of the Lyman-α line more than the red. However, current constraints on this are necessarily confined to stacking measurements, and to date have proved inconclusive (e.g. Ouchi et al. 2010).

Finally, the clustering of LAEs should increase with redshift as we look back into the epoch of reionization. If galaxies produced the photons that reionized the IGM, then their clustering should have influenced the history of reionization, and the first galaxies are certainly expected to have been highly clustered. Specifically, by the time the neutral fraction of hydrogen has dropped to $\simeq$50 %, the average ionized region is expected to have a radius of $\simeq$10 Mpc (comoving), created and maintained by many hundreds of small galaxies working in concert. Because a Lyman-α photon redshifts out of resonance after travelling $\simeq$1 Mpc, most Lyman-α emission produced by the galaxies which together have created this bubble should emerge unscathed (i.e. unscattered) by the IGM. Consequently, as illustrated by the state-of-the-art simulations shown in Fig. 26, the clumpy nature of reionization means that LAEs are predicted to appear more clustered than LBGs during the reionization epoch. Indeed, under some scenarios the apparent clustering of LAEs can be well in excess of the intrinsic clustering of halos in the concordance cosmology. Observing such enhanced clustering would confirm the prediction that the *HII* regions during reionization are large (McQuinn et al. 2007). This prediction has arguably gained some tentative observational support from the latest large-area surveys for Lyα emitters at $z \simeq 6.5$, where it has been found that, depending on luminosity, the number density of LAEs varies by a factor of $2 - 10$ between different $\simeq 1/4\,\mathrm{deg}^2$ fields (Ouchi et al. 2010; Nakamura et al. 2011). However, Ouchi et al. (2010) report no evidence for any significant clustering amplitude boost at $z \simeq 6.6$, and it seems clear that the meaningful search for this effect must await surveys of large samples of LAEs and LBGs at $z > 7$ covering many square degrees.

7 Conclusion and Future Prospects

Over the last decade we have witnessed a revolution in our knowledge of galaxies in the first billion years of cosmic time. Arguably the next 10 years should be even more exciting.

The accurate measurement of the bright end of the evolving galaxy UV luminosity function should soon be improved by combining the data over $\simeq 0.2\,\mathrm{deg}^2$ from the *HST* CANDELS project (Grogin et al. 2011; Finkelstein et al. 2012; Oesch et al. 2012) with the brighter but even larger-area multi-colour imaging being produced by the new generation of ground-based near-infrared surveys, such as UltraVISTA (Fig. 27; Bowler et al. 2012; McCracken et al. 2012). Beyond this, the *EUCLID* satellite, as part of the "deep" component of its mission, is expected to survey several tens of square degrees down to $J \simeq 26$ mag. (Laurejis et al. 2011). Near-infrared narrow-band imaging surveys can also be expected to continue to expand in scope (e.g. ELVIS, at $z \simeq 8.8$; Nilsson et al. 2007). Crucially, these wide-area near-infrared imaging surveys will be complemented by well-matched wide-area optical imaging provided, for example, by Hyper-Suprime CAM on Subaru (Takada 2010).

At the faint end, attempts will continue to exploit the power of WFC3/IR on *HST* to the full, with further ultra-deep near-infrared imaging planned over several square arcmin of sky, and at the end of the decade NIRCam on the *JWST* should extend this work out to $z > 10$. At the same time the angular resolution limitations of IRAC on *Spitzer* will be overcome with MIRI on *JWST*, which should deliver *rest-frame optical* imaging of the highest-redshift galaxies of a quality comparable to that currently achieved with WFC3 on *HST* (Fig. 28). This should enormously improve our knowledge of the rest-frame optical morphologies, and the stellar masses of the highest-redshift galaxies. It should also enable much more accurate measurement of the UV-optical SEDs of faint galaxies at $z \simeq 5 - 10$, including accurate determination of their UV slopes (with resulting implications for age, metallicity and escape fraction as discussed above).

Fig. 27 A small sub-region of the first-year UltraVISTA near-infrared image of the COSMOS field, presented as a $Y + J + K_s$ colour image (McCracken et al. 2012). This 1.5 deg^2 image reaches $Y + J \simeq 25$ AB mag, and has already revealed $\simeq$200,00 galaxies, of which $\simeq 5$ appear to be massive galaxies at $z \simeq 7$ (Bowler et al. 2012). The final UltraVISTA imaging should reach 5–10 times deeper, and enable the search for rare massive galaxies out to $z \simeq 10$ (courtesy UltraVISTA/Terapix/CNRS/CASU)

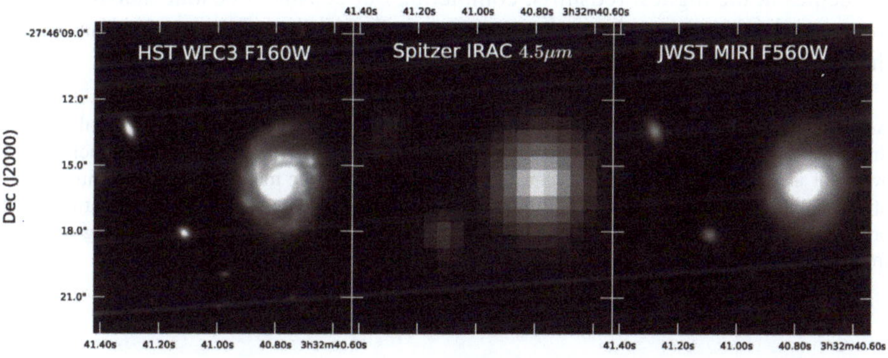

Fig. 28 *HST* WFC3 and *Spitzer* IRAC observations of a small sub-region of the HUDF field, alongside a simulated image of the same region as expected from MIRI on the *JWST*. The *HST* image is the product of 22 h of on-source integration with WFC3 in the H_{160} band. The simulated 5.6 μm *JWST* image was produced using the H-band morphology and the IRAC fluxes as input, and assumes 28 h of integration with MIRI. The unconfused mid-infrared imaging which will be delivered by MIRI at IRAC wavelengths will enable the study of the rest-frame optical morphologies out beyond $z \simeq 10$, and will also allow much more robust determinations of the stellar masses of the most distant galaxies (courtesy A. Rogers)

Spectroscopic follow-up of the brighter $z > 7$ galaxies (perhaps down to $J \simeq 28$) is set to be transformed by the new generation of ground-based multi-object near-infrared spectrographs including FMOS on Subaru (Kimura et al. 2010), KMOS on the VLT (Sharples et al. 2006), and MOSFIRE on Keck (McLean et al. 2008), before this work should be extended to even fainter magnitudes with NIRSpec on *JWST* (Birkmann et al. 2011). This should clarify the currently confused picture of Lyman-α emission from LBGs, with important implications for our understanding of the progress of reionization. Wide-field near-infrared GRISM spectroscopy with *EUCLID* may enable the first meaningful study of the clustering of LAEs relative to LBGs at $z > 7$, where an enhancement of the apparent clustering of Lyman-α emitters is a prediction of some models for reionization (e.g. McQuinn et al. 2007). The next generation of giant ground-based near-infrared telescopes equipped with sophisticated adaptive-optics systems (TMT, E-ELT, GMT) will also enable detailed near-infrared high-resolution spectroscopy of the most distant galaxies.

Finally, in the rather near future, we can expect a revolution in the search for and study of galaxies at $z > 5$ at sub-millimetre wavelengths. We already know that the most distant quasars are detectable in the sub-mm, so we can anticipate that a significant population of rare high-mass dusty galaxies should be uncovered by combining existing *Herschel* SPIRE imaging with longer-wavelength data from the SCUBA-2 Cosmology Legacy Survey now underway at the JCMT. Crucially, detailed millimetre spectroscopy of such objects will be relatively straightforward with ALMA. Over the next few years ALMA can also be exploited to undertake the first sub-mm surveys of sufficient depth and angular-resolution to complement the Ultra Deep Field studies previously only possible at shorter wavelengths with HST. This work should enormously clarify our understanding of the role of dust and molecules at the highest redshifts, completing our census of cosmic star-formation history at early times, and transforming our understanding of the production-rate of the first metals in the Universe.

Acknowledgements James Dunlop gratefully acknowledges the support of the Royal Society through a Wolfson Research Merit award and the support of the European Research Council via the award of an Advanced Grant. He also wishes to acknowledge the many substantial contributions of his collaborators in the study of high-redshift galaxies, and the (mostly) good-natured and productive rivalry engendered by the various competing groups working at this exciting research frontier.

References

Adelberger, K.L., Steidel, C.C.: ApJ **544**, 218 (2000)
Ajiki, M., et al.: ApJ **638**, 596 (2006)
Beckwith, S., et al.: AJ **132**, 1729 (2006)
Birkmann, S.M., et al.: SPIE **8150**, 10 (2011)
Bolton, Haehnelt: Mon. Not. R. Astron. Soc. **382**, 325 (2007)
Bolton, et al.: Mon. Not. R. Astron. Soc. **416**, 70 (2011)
Boselli, et al.: Astron. Astrophys. **528**, 107 (2011)

Bouwens, R.J., et al.: ApJ **595**, 589 (2003)
Bouwens, R.J., et al.: ApJ **606**, L25 (2004a)
Bouwens, R.J., et al.: ApJ **611**, L1 (2004b)
Bouwens, R.J., et al.: ApJ **616**, L79 (2004c)
Bouwens, R.J., et al.: ApJ **624**, L5 (2005)
Bouwens, R.J., et al.: ApJ **653**, 53 (2006)
Bouwens, R.J., et al.: ApJ **670**, 928 (2007)
Bouwens, R.J., et al.: ApJ **686**, 230 (2008)
Bouwens, R.J., et al.: ApJ **690**, 1764 (2009a)
Bouwens, R.J., et al.: ApJ **705**, 936 (2009b)
Bouwens, R.J., et al.: ApJ **708**, L69 (2010a)
Bouwens, R.J., et al.: ApJ **709**, L133 (2010b)
Bouwens, R.J., et al.: ApJ **725**, 1587 (2010c)
Bouwens, R.J., et al.: Nature **469**, 504 (2011a)
Bouwens, R.J., et al.: ApJ **737**, 90 (2011b)
Bouwens, et al.: Astrophys. J. **754**, 83 (2012)
Bower, R.J., et al.: MNRAS **370**, 645 (2006)
Bowler, et al.: Mon. Not. R. Astron. Soc. (arXive:1205.4270) (2012)
Bradley, L.D., et al.: ApJ **678**, 647 (2008)
Bremer, M.N., et al.: MNRAS **347**, L7 (2004)
Bruzual, G., Charlot, S.: MNRAS **344**, 1000 (2003)
Buitrago, F., et al.: ApJ **687**, L61 (2008)
Bunker, A.J., et al.: MNRAS **342**, L47 (2003)
Bunker, A.J., et al.: MNRAS **355**, 374 (2004)
Bunker, A.J., et al.: MNRAS **409**, 855 (2010)
Burningham, B., et al.: MNRAS **391**, 320 (2008)
Burningham, B., et al.: MNRAS **406**, 1885 (2010)
Calzetti, D., et al.: ApJ **533**, 682 (2000)
Capak, P.L., et al.: Nature **470**, 233 (2011a)
Capak, P., et al.: ApJ **730**, 68 (2011b)
Carilli, C.L., Gnedin, N.Y., Owen, F.: ApJ **577**, 22 (2002)
Carilli, C.L.: ApJ **730**, L30 (2011)
Carilli, C.L., et al.: ApJ **714**, 834 (2010)
Casali, M., et al.: A&A **467**, 777 (2007)
Cassata, P., et al.: A&A **525**, 143 (2011)
Castellano, M., et al.: A&A **511**, 20 (2010a)
Castellano, M., et al.: A&A **524**, 28 (2010b)
Catellano, et al.: Astron. Astrophys. **540**, 39 (2012)
Chabrier, G.: PASP **115**, 763 (2003)
Charlot, S., Fall, S.M.: ApJ **378**, 471 (1991)
Charlot, S., Fall, S.M.: ApJ **415**, 580 (1993)
Chiu, K., et al.: MNRAS **385**, L53 (2008)
Choi, J., Nagamine, K.: MNRAS **407**, 1464 (2010)
Choi and Nagamine: MNRAS **410**, 2579 (2011)
Chomiuk, L., Povich, M.S.: AJ **142**, 197 (2011)
Cirasuolo, M., et al.: MNRAS **380**, 585 (2007)
Cirasuolo, M., et al.: MNRAS **401**, 1166 (2010)
Clément, B., et al.: A&A **538**, 66 (2012)
Conroy, C., Gunn, J.E.: ApJ **712**, 833 (2010)
Cowie, L.L., Hu, E.M., Songaila, A.: **735**, L38 (2011)
Cuby, J.-G., et al.: A&A **461**, 911 (2007)
Cucchiara, A., et al.: ApJ **736**, 7 (2011)
Curtis-Lake, et al.: Mon. Not. R. Astron. Soc. **422**, 1425 (2012)
Daddi, E., et al.: ApJ **670**, 156 (2007)

Davé, R.: MNRAS **385**, 147 (2008)
Dayal, S., Ferrera, A.: Mon. Not. R. Astron. Soc. **421**, 2568 (2012)
De Lucia, G., Blaizot, J.: MNRAS **376**, 1054 (2007)
Dekel, A., et al.: Nature **457**, 451 (2009)
Delorme, P., et al.: A&A **482**, 961 (2008)
Dey, A., et al.: ApJ **498**, L93 (1998)
di Serego Alighieri, S., et al.: IAUS **255**, 75 (2008)
Dickinson, M., et al.: ApJ **600**, L99 (2004)
Djorgovski, S.G., Spinrad, H., McCarthy, P.: ApJ **299**, 1 (1985)
Douglas, L.S., et al.: MNRAS **400**, 561 (2009)
Dunkley, et al.: Astrophys. J. Suppl. **180**, 306 (2009)
Dunlop, J.S.: Science **333** (6039), 178 (2011)
Dunlop, J.S., Peacock, J.A.: MNRAS **247**, 19 (1990)
Dunlop, J.S., et al.: Nature **370**, 347 (1994)
Dunlop, J.S., et al.: Nature **381**, 581 (1996)
Dunlop, J.S., et al.: MNRAS **376**, 1054 (2007)
Dunlop, J.S., et al.: MNRAS **420**, 901 (2012)
Erb, D.K., et al.: ApJ **719**, 1168 (2010)
Eyles, L.P., et al.: MNRAS **374**, 910 (2007)
Fan, X., et al.: AJ **125**, 1649 (2003)
Fan, X., Carilli, C.L., Keating, B.: ARA&A **44**, 415 (2006)
Ferguson, H.C., Dickinson, M., Williams, R.: ARA&A **38**, 667 (2000)
Ferguson, H.C., et al.: ApJ **600**, 107 (2004)
Fernandez-Soto, A., Lanzetta, K.M., Yahil, A.: ApJ **513**, 34 (1999)
Finkelstein, S.L., et al.: **719**, 1250 (2010)
Finkelstein, S.L., et al.: ApJ **735**, 5 (2011)
Finkelstein, et al.: Astrophys. J. **758**, 93 (2012)
Finlator, K., et al.: MNRAS **410** 1703 (2011)
Finlator, K.: arXiv:1203.4862 (2012)
Firmani, C., et al.: MNRAS **396**, 1675 (2009)
Fontana, A., et al.: ApJ **725**, L205 (2010)
Fynbo, J.P.U., et al.: ApJS **185**, 526 (2009)
Furlanetto, S.R., Loeb, A.: ApJ **579**, 1 (2002)
Furusawa, H., et al.: ApJS **176**, 1 (2008)
Giavalisco, M.: ARA&A **40**, 579 (2002)
Giavalisco, M., et al.: ApJ **600**, L94 (2004)
Gehrels, N., et al.: ApJ **611**, 1005 (2004)
Gnedin, Ostriker: Astrophys. J. **486**, 581 (1997)
González, V., et al.: ApJ **713**, 115 (2010)
González, V., et al.: ApJ **735**, L34 (2011)
Gonzalez, et al.: Astrophys. J. **755**, 148 (2012)
Grazian, A., et al.: A&A **449**, 951 (2006)
Grazian, A., et al.: A&A **532**, 33 (2011)
Greiner, J., et al.: ApJ **693**, 1610 (2009)
Grogin, N.A., et al.: ApJS **197**, 35 (2011)
Gronwall, C., et al.: ApJ **743**, 9 (2011)
Guhathakurta et al.: ApJ **357**, 9 (1990)
Gunn, J.E., Peterson, B.A.: ApJ **142**, 1633 (1965)
Haislip, J.B., et al.: Nature **440**, 181 (2006)
Hall, N., et al.: ApJ **745**, 155 (2012)
Hammer, D.M., et al.: ApJ **745**, 177 (2012)
Hathi, N.P., Malhotra, S., Rhoads, J.E.: ApJ **673**, 686 (2008)
Hayes, M., et al.: ApJ **730**, 8 (2011)
Henry, A.L., et al.: ApJ **680**, L97 (2008)

Henry, A.L., et al.: ApJ **697**, 1128 (2009)
Hewett, P.C., et al.: MNRAS **367**, 454 (2006)
Hibon, P., et al.: A&A **515**, 97 (2010)
Hibon, P., et al.: ApJ **741**, 101 (2011)
Hibon, P., et al.: ApJ **744**, 89 (2012)
Hickey, S., et al.: MNRAS **404**, 212 (2010)
Hjorth, J., et al.: Nature **423**, 847 (2003)
Hopkins, A.M., Beacom, J.F.: ApJ **651** (2006)
Hu, E.M., Cowie, L.L., McMahon, R.G.: ApJ **502**, L99 (1998)
Hu, E.M., Cowie, L.L., McMahon, R.G.: ApJ **568**, L75 (2002)
Hu, E.M., et al.: AJ **127**, 563 (2004)
Hu, E.M., et al.: ApJ **725**, 394 (2010)
Iwata, et al.: Astrophys. J. **692**, 1287 (2009)
Iye, M., et al.: Nature **443**, 186 (2006)
Iye, M., et al.: SPIE **7016**, 2 (2008)
Jiang, L., et al.: ApJ **743**, 65 (2011)
Jimenez, R., et al.: ApJ **532**, 152 (2000)
Johnston, R.: A&ARv **19**, 41 (2011)
Kashikawa, N., et al.: PASJ **56**, 1101 (2004)
Kashikawa, N., et al.: ApJ **637**, 631 (2006)
Kashikawa, N., et al.: ApJ **648**, 7 (2006)
Kashikawa, N., et al.: ApJ **734**, 119 (2011)
Kawai, N., et al.: Nature **440**, 184 (2006)
Kennicutt, R.C. Jr: ARA&A **36**, 189 (1998)
Kimura, M., et al.: PASJ **62**, 1135 (2010)
Knapp, G.R., et al.: AJ **127**, 3553 (2004)
Kodaira, K., et al.: PASJ **55**, L17 (2003)
Koekemoer, A.M., et al.: ApJS **197**, 36 (2011)
Komatsu, et al.: Astrophys. J. Suppl. **180**, 330 (2009)
Komatsu, E., et al.: ApJS **192**, 18 (2011)
Koo, D.C., Kron, R.G.: PASP **545**, 537 (1980)
Krug, H.B., et al.: ApJ **745**, 122 (2012)
Labbé, I., et al.: **649**, L67 (2006)
Labbé, I., et al.: ApJ **708**, L26 (2010a)
Labbé, I., et al.: ApJ **716**, L103 (2010b)
Landy, S.D., Szalay A.S.: ApJ **412**, 64 (1993)
Larson D., et al.: ApJS **192**, 16 (2011)
Laurejis, R., et al.: (arXiv:1110.3193) (2011)
Lawrence, A., et al.: MNRAS **379**, 1599 (2007)
Leggett, S.K., et al.: ApJ **695**, 1517 (2009)
Lehnert, M.D., Bremer, M.: ApJ 593, L630 (2003)
Lehnert, M.D., et al.: Nature **467**, 940 (2010)
Lilly, S.J.: ApJ **333**, 161 (1988)
Lilly, S.J., et al.: ApJ **460**, L1 (1996)
Liu, M.C., et al.: ApJ **740**, 108 (2011)
Lorenzoni, S., et al.: MNRAS **414**, 1455 (2011)
Lucas, P.W., et al.: MNRAS **408**, L56 (2010)
Mack, K.J., Wyithe, J.S.B.: Mon. Not. R. Astron. Soc. (arXive:1101.5431) (2011)
Mack, K.J., Wyithe, J.S.B.: MNRAS submitted (arXiv:1101.5431) (2012)
Madau, P.: ApJ **441**, 18 (1995)
Madau, P., et al.: MNRAS **283**, 1388 (1996)
Madau, P., Pozzetti, L., Dickinson, M.: ApJ **498**, 106 (1998)
Malhotra, S., et al.: ApJ **626**, 666 (2005)
Malhotra, et al.: Astrophys. J. **750**, 36 (2012)

Marchesini, D., et al.: ApJ **701**, 1765 (2009)
Maraston, C.: MNRAS **362**, 799 (2005)
McCarthy, P.J.: ARA&A **31**, 639 (1993)
McCracken, et al.: Astron. Astrophys. **544**, 156 (2012)
McLean, I.S.: SPIE **7014**, 99 (2008)
McLure, R.J., et al.: MNRAS **372**, 357 (2006)
McLure, R.J., et al.: MNRAS **395**, 2196 (2009)
McLure, R.J., et al.: MNRAS **403**, 960 (2010)
McLure, R.J., et al.: MNRAS **418**, 2074 (2011)
McQuinn, M., et al.: MNRAS **381**, 75 (2007)
McQuinn, et al.: Mon. Not. R. Astron. Soc. **408**, 1945 (2010)
Meier, D.L.: ApJ **203**, 103 (1976a)
Meier, D.L.: ApJ **207**, 343 (1976b)
Meurer, G.R., Heckman, T.M., Calzetti, D.: ApJ **521**, 64 (1999)
Miley, G., De Breuck, C.: A&ARv **15**, 67 (2008)
Mo, H.J., et al.: MNRAS **295**, 319 (1998)
Mo, H.J., et al.: MNRAS **304**, 175 (1999)
Mobasher, B., et al.: ApJ **600**, L167 (2004)
Mobasher, B., et al.: ApJ **635**, 832 (2005)
Mobasher, B., et al.: ApJS **172**, 117 (2007)
Mortlock, D.J., et al.: Nature **474**, 616 (2011)
Murayama, T., et al.: ApJS **172**, 523 (2007)
Nagamine, K., et al.: PASJ **62**, 1455 (2010)
Nagao, T., et al.: ApJ **613**, L9 (2004)
Nagao, T., et al.: ApJ **634**, 142 (2005)
Nagao, T., et al.: A&A **468**, 877 (2007)
Nagao, T., et al.: ApJ **680**, 100 (2008)
Nakamura, E., et al.: MNRAS **412**, 2579 (2011)
Nilsson, K.K., et al.: A&A **471**, 71 (2007)
Nilsson, K.K., et al.: MNRAS **400**, 232 (2009)
Noeske, K.G., et al.: ApJ **660**, L43 (2007)
Oesch, P.A., et al.: ApJ **671**, 1212 (2007)
Oesch, P.A., et al.: ApJ **690**, 1350 (2009)
Oesch, P.A., et al.: ApJ **709**, L16 (2010a)
Oesch, P.A., et al.: ApJ **709**, L21 (2010b)
Oesch, P.A., et al.: ApJ **745**, 110 (2012)
Oke, J.B., Gunn, J.E.: ApJ **266**, 713 (1983)
Ono, Y., et al.: ApJ **724**, 1524 (2010)
Ono, Y., et al.: ApJ **744**, 83 (2012)
Ota, K., et al.: JKAS **38**, 179 (2005)
Ota, K., et al.: ApJ **677**, 12 (2008)
Ota, K., et al.: ApJ **722**, 803 (2010a)
Ota, K., et al.: PASJ **62**, 1167 (2010b)
Ouchi, M., et al.: ApJ **582**, 60 (2003)
Ouchi, M., et al.: ApJ **611**, 660 (2004)
Ouchi, M., et al.: ApJ **620**, L1 (2005)
Ouchi, M., et al.: ApJS **176**, 301 (2008)
Ouchi, M., et al.: ApJ **696**, 1164 (2009a)
Ouchi, M., et al.: ApJ **706**, 1136 (2009b)
Ouchi, M., et al.: ApJ **723**, 869 (2010)
Overzier, R.A., et al.: ApJ **648**, L5 (2006)
Papovich, C., et al.: MNRAS **412**, 1123 (2011)
Patridge, R.B., Peebles, P.J.E.: ApJ **147**, 868 (1967a)
Patridge, R.B., Peebles, P.J.E.: ApJ **148**, 377 (1967b)

Pawlik, Schaye, van Scherpenzeel: Mon. Not. R. Astron. Soc. **394**, 1812 (2009)
Pawlik, A.H., Milosavljevic, M., Bromm, V.: ApJ **731**, 54 (2011)
Pentericci, L., et al.: ApJ **743**, 132 (2011)
Perley, D.A., et al.: AJ **138**, 1690 (2009)
Pirzkal, N., et al.: ApJ **695**, 1591 (2009)
Postman, M., et al.: ApJS **199**, 25 (2012)
Poznanski, D., et al.: MNRAS **382**, 1169 (2007)
Pritchet, C.J., Hartwick, F.D.A.: ApJ **355**, 11 (1990)
Pritchet, C.J.: PASP **106**, 1052 (1994)
Rawlings, S., et al.: Nature **383**, 502 (1996)
Reddy, N.A., Steidel, C.C.: ApJ **692**, 778 (2009)
Reid, N., Majewski, S.R.: ApJ **409**, 635 (1993)
Retzlaff, J., et al.: A&A **511**, 50 (2010)
Richard, J., et al.: ApJ **685**, 705 (2008)
Richmond, M.W., et al.: PASJ **61**, 97 (2009)
Riechers, D., et al.: ApJ **720**, L131 (2010)
Rigby, E.E., et al.: MNRAS **416**, 1900 (2011)
Rhoads, J.E., et al.: AJ **125**, 1006 (2003)
Rhoads, J.E., et al.: ApJ **611**, 59 (2004)
Robertson, B.E.: ApJ **713**, 1266 (2010)
Robertson, B.E., et al.: Nature **468**, 55 (2010)
Robertson, B.E., Ellis, R.S.: ApJ **744**, 95 (2012)
Robson, I., et al.: MNRAS **351**, L29 (2004)
Ruiz-Veasco, A.E., et al.: ApJ **669**, 1 (2007)
Ryan, R.E., et al.: ApJ **631**, L159 (2005)
Ryan, R.E., et al.: ApJ **739**, 83 (2011)
Salpeter, E.E.: ApJ **121**, 161 (1955)
Salvaterra, R., et al.: Nature **461**, 1258 (2009)
Salvaterra, R., Ferrara, A., Dayal, P.: MNRAS **414**, 847 (2011)
Schaerer, D.: A&A **382**, 28 (2002)
Schaerer, D.: A&A **397**, 527 (2003)
Schaerer, D., Barros, S.: A&A **515**, 73 (2010)
Sharples, R., et al.: NewAR **50**, 370 (2006)
Schechter, P.: ApJ **203**, 297 (1976)
Schenker, M.A., et al.: ApJ **744**, 179 (2012)
Schiminovich, D., et al.: ApJ **619**, L47 (2005)
Seymour, N., et al.: ApJS **171**, 353 (2007)
Shapley, A.E., et al.: ApJ **588**, 65 (2003)
Shapley, et al.: Astrophys. J. **651**, 688 (2006)
Shapley, A.E.: 1998, ARA&A **49**, 525 (2011)
Shimasaku, N., et al.: PASJ **57**, 447 (2005)
Shimasaku, K., et al.: PASJ **58**, 313 (2006)
Sobral, D., et al.: MNRAS **420**, 1926 (2012)
Sommerville, R.S., et al.: ApJ **600**, L171 (2004)
Sommerville, R.S., et al.: ApJ **672**, 776 (2008)
Stiavelli, M., et al.: ApJ **610**, L1 (2004)
Spinrad, H., et al.: AJ **116**, 2617 (1998)
Stanway, E.R., Bunker, A.J., McMahon, R.G.: MNRAS **342**, 439 (2003)
Stanway, E.R., et al.: ApJ **607**, 704 (2004)
Stanway, E.R., McMahon, R.G., Bunker, A.J.: MNRAS **359**, 1184 (2005)
Stanway, E.R., et al.: MNRAS **376**, 727 (2007)
Stanway, E.R., et al.: MNRAS **384**, 348 (2008a)
Stanway, E.R., et al.: MNRAS **385**, 493 (2008b)
Stanway, E.R., et al.: MNRAS **386**, 370 (2008c)

Stark, D.P., et al.: ApJ **659**, 84 (2007a)
Stark, D.P., et al.: ApJ **633**, 10 (2007b)
Stark, D.P., et al.: ApJ **697**, 1493 (2009)
Stark, D.P., et al.: MNRAS **408**, 1628 (2010)
Stark, D.P., et al.: ApJ **728**, L2 (2011)
Steidel, D.P., Hamilton, D.: AJ **104**, 941 (1992)
Steidel, C.C., et al.: ApJ **462**, 17 (1996)
Steidel, C.C., et al.: ApJ **519**, 1 (1999)
Steidel, C.C., et al.: ApJ **532**, 170 (2000)
Steidel, C.C., et al.: ApJ **736**, 160 (2011)
Stern, D., Bunker, A., Spinrad, H., Dey, A.: ApJ **537**, 73 (2000)
Su, et al.: ApJ **738**, 123 (2011)
Takada, M.: AIPC **1279**, 120 (2010)
Taniguchi, Y., et al.: PASJ **57**, 165 (2005)
Taniguchi, Y., et al.: ApJ **701**, 915 (2009)
Taniguchi, Y., Shioya, Y., Trump, J.R.: ApJ **724**, 1480 (2010)
Tanvir, N., et al.: Nature **461**, 1254 (2009)
Targett, T., Dunlop, J.S., McLure, R.J.: MNRAS **420**, 3621 (2012)
Thompson, R.I., et al.: AJ **130**, 1 (2005)
Thompson, R.I., et al.: ApJ **647**, 787 (2006)
Tilvi, V., et al.: ApJ **721**, 1853 (2010)
Trenti, M., et al.: ApJ **714**, L202 (2010)
Trenti, M., et al.: ApJ **727**, L39 (2011)
Trenti, M., et al.: ApJ **746**, 55 (2012)
Treyer, M., et al.: ApJ **619**, L19 (2005)
van Breugel, W., et al.: ApJ **518**, L61 (1999)
Vanzella, E., et al.: A&A **454**, 423 (2006)
Vanzella, E., et al.: **695**, 1163 (2009)
Vanzella, E., et al.: ApJ **730**, L35 (2011)
Walter, F., et al.: Nature **424**, 406 (2003)
Wang, R., et al.: ApJ **714**, 699 (2010)
Weidner, C., et al.: MNRAS **412**, 295 (2011)
Wiklind, T., et al.: ApJ **676**, 781 (2008)
Wilkins, S.M., Trenti, N., Hopkins, A.: MNRAS **385**, 687 (2008)
Wilkins, S.M.: MNRAS **403**, 938 (2010)
Wilkins, S.M., et al.: MNRAS **411**, 23 (2011a)
Wilkins, S.M., et al.: MNRAS **417**, 717 (2011b)
Williams, R.E., et al.: AJ **112**, 1335 (1996)
Willott, C.J., et al.: AJ **139**, 906 (2010)
Windhorst, R.A.: ApJS **193**, 27 (2011)
Woosley, S.E., Bloom, J.S.: ARA&A **44**, 507 (2006)
Yan, H., Windhorst, R.A.: ApJ **612**, L93 (2004)
Yan, H., et al.: ApJ **634**, 109 (2005)
Yan, H., et al.: ApJ **651**, 24 (2006)
Yan, H., et al.: RAA **10**, 867 (2010)
Yan, H., et al.: ApJ **728**, L22 (2011)
Yuma, S., et al.: ApJ **720**, 1016 (2010)
Zafar, T., et al.: A&A **515**, 94 (2010)
Zheng, W., et al.: ApJ **697**, 1907 (2009)
Zheng, Z., et al.: ApJ **739**, 62 (2011)

The Formation of the First Massive Black Holes

Zoltán Haiman

Abstract Supermassive black holes (SMBHs) are common in local galactic nuclei, and SMBHs as massive as several billion solar masses already exist at redshift $z = 6$. These earliest SMBHs may grow by the combination of radiation-pressure-limited accretion and mergers of stellar-mass seed BHs, left behind by the first generation of metal-free stars, or may be formed by more rapid direct collapse of gas in rare special environments where dense gas can accumulate without first fragmenting into stars. This chapter offers a review of these two competing scenarios, as well as some more exotic alternative ideas. It also briefly discusses how the different models may be distinguished in the future by observations with *JWST, LISA* and other instruments.

1 Introduction

The discovery of about two dozen bright quasars with luminosities $\gtrsim 10^{47}$ erg s^{-1} at redshift $z \simeq 6$ suggests that some supermassive black (SMBHs) as massive as a few $\times 10^9$ $M_\odot$ have been already assembled when the universe was less than 1 Gyr old (see, e.g., Fan 2006 for a review). These high-redshift quasars are exceedingly rare, with a space density of order ~ 1Gpc^{-3}, and can only be found in large surveys of the sky, such as the Sloan Digital Sky Survey (SDSS), or the smaller-area but deeper CFHQS Willott et al. (2010) and UKIDSS Lawrence et al. (2007) surveys. These quasars overall appear to be "fully developed", with spectra and metallicity patterns

Z. Haiman (✉)
Department of Astronomy, Columbia University, 550 West 120th Street, New York, NY 10027, USA
e-mail: zoltan@astro.columbia.edu

T. Wiklind et al. (eds.), *The First Galaxies*, Astrophysics and Space Science Library 396, DOI 10.1007/978-3-642-32362-1_6, © Springer-Verlag Berlin Heidelberg 2013 that appear remarkably similar to their counterparts at moderate redshifts (Fan et al. 2003). Indeed, if one selects individual quasars with the same luminosity, their properties show little evolution with cosmic epoch.[1] This implies that the behavior of individual quasars is probably determined by local physics near the SMBH and is not directly coupled to the cosmological context in which the SMBH is embedded. However, it is clear that the quasar population as a whole does evolve over cosmic timescales. Observations from $0 \lesssim z \lesssim 6$ in the optical (e.g., the Anglo-Australian Telescope's Two Degree Field, or 2dF, and the Sloan Digital Sky Survey, or SDSS) and radio bands Shaver et al. (1996) show a pronounced peak in the abundance of bright quasars at $z \approx 2.5$. A similar behavior has been confirmed in X-ray observations (Silverman et al. 2008).

The cosmic evolution of quasar black holes between $0 \lesssim z \lesssim 6$ is likely driven by a mechanism other than local physics near the hole. This is reinforced by the fact that the timescale of activity of individual quasars is significantly shorter than cosmic timescales at $z \lesssim 6$, both on theoretical grounds ($\sim 4 \times 10^7$ year, the e-folding time for the growth of mass in a SMBH, whose accretion converts mass to radiation with an efficiency of $\epsilon = \dot{M}c^2/L_{\rm Edd} \sim 10\,\%$) and is limited by its own [Eddington] luminosity), and using the duty cycle of quasar activity inferred from various observations (also $\sim 10^7$ year but with large uncertainties, e.g. Martini (2004) and references therein; see also Haiman et al. (2004); Shankar (2009); Shankar et al. (2009, 2010)).

In the cosmological context, it is tempting to link the evolution of massive quasar black holes with that of dark matter (DM) halos condensing in a Cold Dark Matter (CDM) dominated universe, as the halo population naturally evolves on cosmic timescales (Efstathiou and Rees 1988). Indeed, this connection has proven enormously fruitful and has resulted in the following broad picture: the first massive astrophysical black holes appear at high redshifts ($z \gtrsim 10$) in the shallow potential wells of low mass ($\lesssim 10^8$ M$_\odot$) dark matter halos. These black holes grow by mergers and gas accretion, evolve into the population of bright quasars observed at lower redshifts, and eventually leave the SMBH remnants that are ubiquitous at the centers of galaxies in the nearby universe.

In this picture, the presence of few$\times 10^9$ M$_\odot$ SMBHs at $z > 6$ presents a puzzle (Haiman and Loeb 2001).[2] Metal-free stars, with masses ~ 100 M$_\odot$, are expected to form at redshifts as high as $z \gtrsim 25$ Abel et al. (2002); Bromm et al. (2002); Yoshida et al. (2008), and leave behind remnant BHs with similar masses (Carr et al. 1984; Heger et al. 2003). However, the natural time-scale, i.e. the Eddington time, for growing these seed BHs by $\gtrsim 7$ orders of magnitude in mass is comparable to the age of the universe (e.g. Haiman and Loeb 2001).

[1] A few possibly important exceptions to this are discussed in Sect. 4.1.

[2] More generally, the non-trivial cosmological implications of the existence of massive BHs at early times was noted already when quasars were first found at redshifts $z > 4$ (Turner 1991).

This makes it difficult to reach 10^9 $M_\odot$ without a phase of rapid (at least modestly super-Eddington) accretion, unless a list of optimistic assumptions are made in hierarchical merger models, in which multiple seed BHs are allowed to grow without interruption, and to combine into a single SMBH (Haiman 2004; Yoo and Miralda-Escudé 2004; Bromley et al. 2004; Shapiro 2005; Volonteri and Rees 2006; Pelupessy et al. 2007; Li et al. 2007; Sijacki et al. 2009; Tanaka and Haiman 2009).

An alternative class of explanations involves yet more rapid gas accretion or collapse (Oh and Haiman 2002; Bromm and Loeb 2003; Koushiappas et al. 2004; Lodato and Natarajan 2006; Spaans and Silk 2006; Begelman et al. 2006; Volonteri et al. 2008; Wise and Abel 2008; Regan and Haehnelt 2009; Schleicher et al. 2010; Shang et al. 2010). In this family of models, primordial gas collapses rapidly into a SMBH as massive as 10^4–10^6 $M_\odot$, either directly, or possibly by accreting onto a pre-existing smaller seed BH Volonteri and Rees (2005), or going through the intermediate state of a very massive star Bromm and Loeb (2003), a "quasistar" Begelman et al. (2008), or a dense stellar cluster (Omukai et al. 2008; Devecchi and Volonteri 2009). These so-called "direct collapse" models involve metal-free gas in relatively massive ($\gtrsim 10^8$ $M_\odot$) dark matter halos at redshift $z \gtrsim 10$, with virial temperatures $T_{\rm vir} \gtrsim 10^4$K. The gas that cools and collapses in these halos must avoid fragmentation, shed angular momentum efficiently, and collapse rapidly.

Many uncertainties about each of the above scenarios remain, and the astrophysical process(es) responsible for the formation of the earliest massive black holes (and indeed for the presence of SMBHs at all redshifts) remain poorly understood. In this review, we focus on the emergence of the first generation of black holes, though many of the important questions are quite general and apply equally to subsequent generations of black holes. This review is organized as follows. In Sect. 2, we describe theoretical expectations for the formation and growth of these black holes within the paradigm of hierarchical CDM cosmologies. In Sect. 3, we "zoom in" and consider the local physics of black hole formation, and various pathways which could lead to the early presence of supermassive black holes. In Sect. 4, we summarize several relevant recent observations that have implications for early black holes, and speculate on the power of future observations to probe the physics of the first BHs. We offer our conclusions in Sect. 5.

This chapter is an expanded an updated version of an earlier review (Haiman and Quataert 2004). Another recent, complimentary review on SMBH formation at high redshift can be found in Volonteri (2010).

2 First Structure Formation

In this section, we sketch some basic theoretical arguments relevant to the formation of structure in the early universe. We then discuss formation mechanisms for SMBHs.

2.1 *Cosmological Perturbations as the Sites of the First Black Holes*

Measurements of the Cosmic Microwave Background (CMB) anisotropies by the *Wilkinson Microwave Anisotropy Probe (WMAP)*, determinations of the luminosity distance to distant type Ia Supernovae, and other observations have led to the emergence of a robust "best fit" cosmological model with energy densities in CDM and "dark energy" of $(\Omega_{\rm M}, \Omega_{\Lambda}) \approx (0.3, 0.7)$ (see, e.g. Komatsu et al. (2011), for the seven-year *WMAP* results, and its combination with other datasets).

The growth of density fluctuations and their evolution into nonlinear dark matter structures can be followed in this cosmological model from first principles by semi-analytic methods (Press and Schechter 1974; Sheth et al. 2001). More recently, it has become possible to derive accurate dark matter halo mass functions directly in large cosmological N-body simulations Jenkins et al. (2001), with different codes agreeing at the 10 % level, and mass functions measured down masses as low as $\sim 10^6\,{\rm M}_{\odot}$ and high redshifts as high as $z \approx 30$ (e.g. Lukić et al. 2007; Reed et al. 2007).

Within the ΛCDM model, with a scale-invariant primordial power spectrum, robust predictions can therefore be made for the dark matter halos. Structure formation in such a universe is "bottom-up", with low-mass halos condensing first. Halos with the masses of globular clusters ($10^{5-6}\,{\rm M}_{\odot}$) are predicted to have condensed from $\sim 3\sigma$ peaks of the initial primordial density field as early as ~ 1 % of the current age of the universe, or at redshifts of $z \sim 25$. These predictions are limited mainly by the 5–10 % uncertainty in the normalization of the primordial power spectrum, σ_8, and by the need to extrapolate the power-spectrum 2–3 orders of magnitude below the scales on which it has been directly constrained. In warm dark matter models, with particle masses of order $\sim$1keV or less, free-streaming would result in a sharp exponential suppression of the fluctuation power on the relevant scales (masses below $10^8\,{\rm M}_{\odot}$), and could significantly reduce the number of DM halos at the earliest redshifts (Barkana et al. 2001; Yoshida et al. 2003).

It is natural to identify the first collapsed DM halos as the sites where the first astrophysical objects, including the first black holes, were born. The nature of the objects that form in these early dark matter halos is currently one of the most rapidly evolving research frontiers in astronomy.

2.2 *Chemistry and Gas Cooling at High Redshifts*

Baryonic gas that falls into the earliest nonlinear dark matter halos is unable to cool efficiently, and is shock heated to the characteristic virial temperatures less than a few hundred Kelvin. It has long been pointed out Binney (1977); Rees and Ostriker (1977); White and Rees (1978) that such gas needs to lose its thermal energy efficiently (within about a dynamical time) in order to continue contracting, or in order to fragment. In the absence of any dissipation, it would simply reach hydrostatic equilibrium and would eventually be incorporated into a more massive halo further down the halo merger hierarchy. While the formation of nonlinear dark matter halos can be followed from first principles, the cooling and contraction of the baryons, and the ultimate formation of stars or black holes in these halos, is much more difficult to model ab initio.

The gas content of a cosmological perturbation can contract together with the dark matter only in dark halos above the cosmological Jeans mass, $M_J \approx 10^4\ M_\odot[(1+z)/11]^{3/2}$, in which the gravity of dark matter can overwhelm thermal gas pressure. Recent work Tseliakhovich and Hirata (2010) has shown that immediately following recombination (at redshift $z \sim 1,000$), the baryons develop coherent streaming motions relative to the dark matter, with relative speeds of order 30 km s^{-1}, on scales of a few Mpc. These relative velocities decay as $\propto (1+z)$, and reduce to $\sim$1 km s^{-1} by $z \sim 30$, comparable to the velocity dispersions in the smallest dark matter halos at this epoch. In the somewhat more massive halos in which the baryons can typically cool efficiently (with velocity dispersions of several km s^{-1} at $z \lesssim 30$; see below), the streaming motions are expected to have at most a modest effect, reducing the gas fraction within the virial radius by a factor of $\sim$ two (Tseliakhovich et al. 2010; Maio et al. 2011; Stacy et al. 2011; Greif et al. 2011b; Dalal et al. 2010). However, as we will see below, the "stellar-seed" model for SMBH growth relies on one (or a few) "special" rare seed BHs that form in few $\times \sim 10^5\ M_\odot$ halos at redshifts as high as $z \sim 30$. This scenario appears vulnerable to the streaming motions.

In the earliest, chemically pristine clouds, radiative cooling is dominated by H_2 molecules. As a result, gas phase H_2 "astro-chemistry" is likely to determine the epoch when the first stars and black holes appear (primordial molecular chemistry, focusing on the role of H_2 early structure formation was reviewed by Abel and Haiman (2001)). Several papers have constructed complete gas-phase reaction networks and identified the two possible ways of gas-phase formation of H_2 via the H^- or H_2^+ channels. These were applied to derive the H_2 abundance under densities and temperatures expected in collapsing high redshift objects (Hirasawa 1969; Matsuda et al. 1969; Palla et al. 1983; Lepp and Shull 1984; Shapiro and Kang 1987; Kang et al. 1990; Kang and Shapiro 1992; Shapiro et al. 1994). Studies that incorporate H_2 chemistry into cosmological models and that address issues such as non-equilibrium chemistry, dynamics, or radiative transfer have appeared relatively more recently. Ref Haiman et al. (1996) used spherically symmetric simulations to study the masses and redshifts of the earliest objects that can collapse and cool via H_2; their findings were confirmed by a semi-analytic treatment (Tegmark et al. 1997). The first fully three dimensional cosmological simulations that incorporate H_2 chemistry and cooling date back to Ostriker and Gnedin (1996); Gnedin and Ostriker (1997) and Abel et al. (1997).

The basic picture that emerged from these papers is as follows. The H_2 fraction after recombination in the smooth "protogalactic" gas is small ($x_{H2} = n_{H2}/n_H \sim 10^{-6}$). At high redshifts ($z \gtrsim 100$), H_2 formation is inhibited, even in overdense regions, because the required intermediaries H_2^+ and H^- are dissociated by cosmic "microwave" background (CMB, but with the typical wavelength then in the infrared) photons. However, at lower redshifts, when the CMB photons redshift to lower energies, the intermediaries survive, and a sufficiently large H_2 abundance builds up inside collapsed clouds ($x_{H2} \sim 10^{-3}$) at redshifts $z \lesssim 100$ to cause cooling on a timescale shorter than the dynamical time. Sufficient H_2 formation and cooling is, however, possible only if the gas reaches temperatures in excess of $\sim$200 K or masses of a few $\times$ 10^5 $M_\odot[(1+z)/11]^{-3/2}$ (note that while the cosmological Jeans mass increases with redshift, the mass corresponding to the cooling threshold, which is well approximated by a fixed virial temperature, has the opposite behavior and decreases at high redshift). The efficient gas cooling in these halos suggests that the first nonlinear objects in the universe were born inside $\sim 10^5$ $M_\odot$ dark matter halos at redshifts of $z \sim$ 20–30, corresponding to an $\sim$3–4σ peaks of the primordial density peak (of course, yet rarer low-mass halos exist even earlier—the first one within our Hubble volume collapsing as early as $z \approx 60$ (Naoz et al. 2006)).

The behavior of metal-free gas in such a cosmological "minihalo" is a well posed problem that has been addressed in three dimensional numerical simulations. The first series of such simulations Abel et al. (2000, 2002); Bromm et al. (1999, 2002); Yoshida et al. (2003) were able to follow the contraction of gas to much higher densities than preceding studies. They have shown convergence toward a temperature/density regime of T $\sim$ 200 K, n $\sim$ 10^4 cm^{-3}, dictated by the critical density at which the excited states of H_2 reach equilibrium and cooling becomes less efficient (Galli and Palla 1998). These simulations suggested that the gas does not fragment further into clumps below sizes of $10^2 - 10^3$ $M_\odot$, but rather it forms unusually massive stars. Very recent simulations reached higher resolution than the earlier ones, and, in some cases, using sink particles, were able to continue their runs beyond the point at which the first ultra-dense clump develops (Turk et al. 2009; Stacy et al. 2010; Greif et al. 2011a; Prieto et al. 2011). These simulations suggest that at least in some cases, the gas in the central regions does, eventually, fragment into two or more distinct clumps, raising the possibility that the first stars formed in pairs, or even in higher-multiple systems.

The masses of the first stars would then presumably be reduced. The initial mass function (IMF) of the first stars is crucial, and is indeed one of the most important uncertainties for early BH formation. This is because massive stars would naturally leave behind black hole seeds, which can subsequently grow by mergers and accretion into the SMBHs. Interestingly, massive stars appear to have an "either/or" behavior. Non-rotating stars with masses between $\sim$40–140 $M_\odot$ and above $\sim$260 $M_\odot$ collapse directly into a black hole without an explosion, and hence without ejecting their metal yields into the surrounding medium, whereas stars in the range $\sim$140–260 $M_\odot$ explode as pair-instability supernovae without leaving a remnant (Heger et al. 2003). In contrast, stars with initial masses $\sim$25–40 $M_\odot$ still leave BH remnants but also eject metals, whereas those with masses $M \lesssim 25$ $M_\odot$ do not leave any BH remnants. This dichotomy is especially interesting because early massive stars are attractive candidates for polluting the IGM with metals at high redshifts (Madaua et al. 2001; Wasserburg and Qian 2000). It is likely that the first stars had a range of masses, in which case they could contribute to both metal enrichment and to the seed black hole population, with a relative fraction that depends sensitively on their initial mass function (IMF).

3 Massive Black Hole Formation

Having reviewed the general problem of structure formation at high redshifts, we now focus on the question of how the first SMBHs were assembled. It is worth emphasizing that this is an unsolved problem—indeed, it is not entirely clear even whether the first nonlinear objects in the universe were stars or black holes, and whether galaxies or their central black holes formed first (Haiman 2004). The leading ideas related to the formation of SMBHs at high redshifts can be broadly divided into three areas: (1) formation of seed black holes from "normal" stellar evolution and subsequent Eddington-limited accretion, (2) rapid direct collapse of gas to a SMBH, usually via a supermassive star/disk, and (3) formation of a SMBH (or an IMBH seed) by stellar dynamical processes in dense stellar systems, such as star clusters or galactic nuclei. It is, of course, possible that all of these processes could be relevant (Begelman and Rees 1978; Rees 1984).

3.1 Growth from Stellar-Mass Seeds

3.1.1 Basic Ingredients and Uncertainties

Perhaps the most natural possibility is that early SMBHs grow by gas accretion out of stellar-mass seed black holes, left behind by early generation of massive stars. If the subsequent gas accretion obeys the Eddington limit and the hole shines with a radiative efficiency of 10 %, then the time it takes for a SMBH to grow to the size of 3×10^9 $M_\odot$ from a stellar seed of $\sim$100 $M_\odot$ is $3 \times 10^7 \ln(3 \times 10^9/100)$ year $\sim$ 7×10^8 year. This is comparable to the age of the universe at $z = 6$ ($\sim 9 \times 10^8$ year for a flat ΛCDM universe with $H_0 = 70$ km s^{-1} Mpc^{-1} and $\Omega_M = 0.3$). Therefore, the presence of these SMBHs is consistent with the simplest model for black hole growth, provided that *(i) the seeds are present early on (at $z \gtrsim 15$; see below), (ii) and the near-Eddington growth is uninterrupted.* As the $\sim 10^{5-6}$ $M_\odot$ host halo of the initial seed BH gets incorporated into the $\sim 10^{12-13}$ $M_\odot$ host halo of the $z \approx 6$ SMBH, it grows by $\sim$7 orders in magnitude, and experiences a large number of mergers with other, comparable-sized, halos. These merger partners may (or may not) have a growing BH at their centers. Therefore, these mergers need to be taken into account, and the "stellar seed" model most likely can not be viewed as that of a single seed BH, growing in isolation.

Several authors have worked out the growth of SMBHs from stellar-mass seeds, by following the build-up of dark DM halos, and using simple prescriptions to track the formation of seed BHs, their subsequent growth by accretion, and their mergers. This can be done either semi-analytically Haiman and Loeb (2001); Wyithe and Loeb (2003b); Haiman (2004); Shapiro (2005), using Monte-Carlo realizations of the DM merger trees Yoo and Miralda-Escudé (2004); Bromley et al. (2004); Volonteri and Rees (2006); Tanaka and Haiman (2009), or based on cosmological hydrodynamics simulations (Li et al. 2007; Pelupessy et al. 2007; Sijacki et al. 2009). As noted in the Introduction, the uncertainties about the statistics of the DM halo merger trees are essentially negligible,[3] since DM halo formation has been directly resolved in numerical simulations at the relevant low masses (down to $\sim 10^6\ M_\odot$) and high redshifts (out to $z \approx 30$). The most important—and still highly uncertain—ingredients of this 'stellar seed' scenario can be summarized as follows.

(i) *What is the smallest possible mass (or virial temperature, $T_{\rm seed}$) for early DM halos in which PopIII stars can form?* A reasonable answer is $T_{\rm seed}$ = few × 100 K, which allows molecular H_2-cooling (Haiman et al. 1996; Tegmark et al. 1997).

(ii) *In what fraction ($f_{\rm seed}$) of these halos do seed BHs actually form?* This is a much more difficult question, since various feedback processes (due to radiation, metal pollution, or mechanical energy deposition) could suppress PopIII star formation in the vast majority of early low-mass halos (Haiman et al. (2000); Haiman and Holder (2003); see also a recent review Bromm et al. (2009)). Interestingly, the *WMAP* measurement of the electron scattering optical depth provides empirical evidence that such negative feedback took place early on and shaped the reionization history (Haiman and Bryan 2006). The answer also depends on the IMF of PopIII stars, since, as noted above, whether the stars leave a BH remnant or explode as pair instability SNe depends on their masses. The dividing mass, $\approx 25\ M_\odot$, was evaluated in non-rotating stellar evolution models Heger et al. (2003), whereas recent simulations indicate that the first stars in minihalos have significant rotation (Stacy et al. 2011). Rotation can help drive winds and prevent BH formation entirely, or can produce a hypernova and reduce the mass of the remnant BH. Finally the velocity dispersions of the lowest-mass minihalos are only a few km s^{-1}, only a factor of few higher than the residual bulk streaming motions between the gas and the DM halo at $z \gtrsim 20$. These streaming motions can therefore reduce the gas fractions in the earliest minihalos and also lower the stellar masses by driving turbulence (Greif et al. 2011b).

(iii) *What is the time-averaged accretion rate of the seed BHs?* This is conveniently parameterized by a duty cycle $f_{\rm duty}$, defined as the fraction of the mass accretion rate that would produce the Eddington luminosity, if $\epsilon \approx 10\,\%$ of the rest mass was converted to radiation (so that $f_{\rm duty} = 1$ is the fiducial Eddington rate). Radiative feedback is usually expected to lead to sub-Eddington rates

[3] At least in principle, since halo mass functions in large N-body simulations agree at the few percent level. In practice, however, there can be significant disagreements between Monte-Carlo halo merger trees made with different algorithms (Zhang et al. 2008).

(e.g. Alvarez et al. 2009), and in spherical symmetry, the accretion was recently to shown to be episodic, with $f_{\rm duty} \approx 0.3$ (Milosavljević et al. 2009). The expectation is therefore that $f_{\rm duty}$ is less than unity. In practice, if the accretion is radiatively inefficient, or if the radiation is trapped or is beamed and "leaks out", then $f_{\rm duty}$ could exceed unity (see more on this below).

(iv) *Finally, what happens when DM halos merge?* The simplest and most optimistic assumption is that the BHs promptly coalesce, as well. However, even if dynamical friction on the DM (and on any stars present in later stages of the merger hierarchy) is efficient, it is possible that, due to the radiation of its parent star, the remnant BHs are no longer embedded in dense enough gas to allow this. Furthermore, even if the BHs coalesce, the merged binary BH can suffer strong gravitational recoil at the time of the merger, due to the linear momentum carried by the anisotropic emission of gravitational waves (e.g. Baker et al. 2008 and references therein). Such a "kick" can eject the BH from the shallow potential wells (∼1km/s) of the early halos, and the BH will be effectively lost. While kicks for comparable-mass BHs with random spins are of the order of ∼100 km/s, the kick speed depends strongly on the mass ratio and on the spin vectors of the two BHs. In particular, kicks become very small ($\lesssim$1km/s) for mass ratios $q \equiv M_1/M_2 \lesssim 10^{-2}$, irrespective of BH spins (e.g. Baker et al. 2008). This may be key to avoid loosing growing seed BHs by ejection, and thus for the buildup of SMBHs early on.

3.1.2 Worked Illustrative Examples

In Fig. 1, we show SMBH mass functions at $z = 6$, illustrating the impact of three of the most uncertain basic assumptions above, taken from a recent example of the Monte Carlo merger tree approach (Tanaka and Haiman 2009). The mass functions were constructed from the merger histories of $\approx 10^5$ DM halos with masses $M > 10^8 M_\odot$ at redshift $z = 6$. The upper right region in each panel, demarcated by the red dashed rectangle, shows the observational constraint on the SMBH space density. Each galaxy was modeled with a spherically symmetric mass distribution consisting of a DM halo with a Navarro–Frenk–White (NFW) profile Navarro et al. (1997), and a more cuspy baryonic component (with $\rho \propto r^{-2.2}$, suggested by 3D simulations). At the time of a merger, the trajectories of kicked BH—ejections, or oscillations damped by dynamical friction—were followed explicitly by one-dimensional orbital calculations.

Figure 1 shows that a robust conclusion for a model to produce enough $z = 6$ SMBHs is that $f_{\rm duty} \gtrsim 0.6$—namely the $\approx$100 $M_\odot$ stellar seed BHs must accrete near the Eddington rate nearly all the time. [Note that this value is excluded in the spherically symmetric case (Milosavljević et al. 2009).] The initial BH occupation fraction also has to be $f_{\rm seed} \gtrsim 10^{-3}$. Another interesting, and less intuitive conclusion, is that if the initial seeds are rare ($f_{\rm seed} = 10^{-3}$–10^{-2}), then gravitational kicks do *not* have a big impact, and it makes little difference to the SMBH mass function whether spins are aligned or randomly oriented (this can be

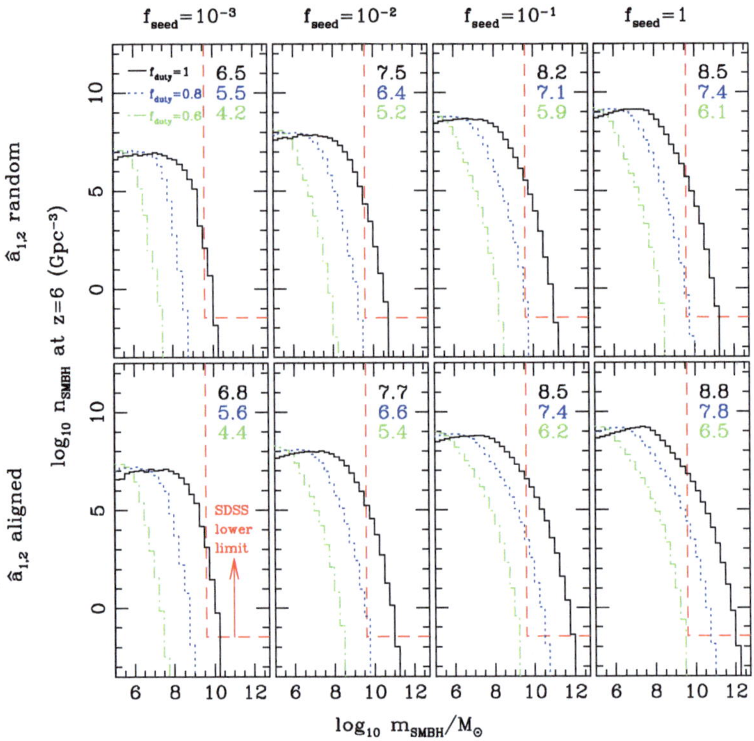

Fig. 1 The comoving number densities of SMBHs in different mass bins at redshift $z = 6$. The 24 different models shown in the figure assume different parameter combinations as follows. The columns, from left to right, adopt $f_{\rm seed} = 10^{-3}$, 10^{-2}, 10^{-1}, 1. The top row assumes a random binary spin orientation, and the bottom row assumes that BH spins are aligned with the binary's orbital angular momentum. In each panel, the time-averaged mass-accretion rates, in Eddington units, are assumed to be either $f_{\rm duty} = 1$ (*black solid curves*), $f_{\rm duty} = 0.8$ (*blue dotted*), and $f_{\rm duty} = 0.6$ (*green dash-dotted*). The numbers in the upper-right corners represent the total mass density in SMBHs $\log_{10}[\rho_\bullet/(M_\odot\ {\rm Mpc}^{-3})]$ for each model. The red dashed line demarcates the abundance of $z \approx 6$ SMBHs with $m \gtrsim 10^{9.6} M_\odot$ already observed in the SDSS (adapted from Tanaka and Haiman (2009))

seen by comparing the bottom and top panels in Fig. 1). This is because the few "lucky" seeds that form earliest (at $z \gtrsim 25$) have a chance to grow by $\gtrsim$ two orders of magnitude in mass before encountering their first merger. The masses of the two BHs at this first merger are then very unequal ($q = M_1/M_2 \lesssim 0.01$), making kick velocities too low to lead to ejection. It is important to emphasize, however, that the model trajectories for the kicked BHs assume spherical symmetry (Madau and Quataert 2004; Blecha and Loeb 2008; Gualandris and Merritt 2008; Tanaka and Haiman 2009). In gas-rich galaxies, most of the dynamical friction occurs due to the dense baryons at the center of the potential (Tanaka and Haiman 2009; Guedes et al. 2011). In asymmetric potentials, the kicked BH does not return to the central region—its oscillations are damped less quickly Guedes et al. (2009) and the accretion rate onto the oscillating hole is also suppressed.

An important additional issue is that in those models that satisfy the observational constraint on the SMBH abundance, the massive end of the SMBH mass function is extremely steep. This prediction is not surprising, as the most massive SMBHs reside in few $\times 10^{12} M_\odot$ halos, which probe the 5σ tail of the halo mass function at $z = 6$ (and there are indeed $\approx 10^8$ (!) times as many few $\times 10^9 M_\odot$ halos, which host $\sim 10^6 M_\odot$ BHs). As a result, the total mass density in SMBHs with masses above $\gtrsim 10^5 M_\odot$ BHs (shown by the numbers in the upper right corners in Fig. 1) are overpredicted by a factor of 10^2–10^3. Note that these numbers indicate the mass in SMBHs that avoided ejection due to kicks, and remained in galactic centers (in some of the models, a significant fraction of the BHs are ejected and form intergalactic BHs; there is no obvious means to detect these Tanaka and Haiman (2009)). The mass density of such nuclear SMBHs at $z \approx 0$ can be inferred from the observed correlations between BH masses and host galaxy properties (such as the masses or velocity dispersions of the host halos; e.g. Ferrarese (2002)). The result is several $\times 10^5 M_\odot \mathrm{Mpc}^{-3}$; furthermore, the expectation is that most ($\gtrsim 90$ %) of this mass was accreted well after $z = 6$ (Shankar 2009). Some strong feedback is therefore needed to eliminate this significant overprediction. Possible candidates for this are radiative feedback internal to halos, which maintains the "$M - \sigma$ relation" in ultra-high redshift, low-mass halos, or the termination of PopIII star formation, at redshifts as high as $z \sim 20$, due to Lyman Werner radiation Haiman and Bryan (2006) or metal pollution (Bromm et al. 2009).

Finally, it is worth emphasizing that the mass accretion rate corresponding to the Eddington limit—for the fiducial radiative efficiency of $\epsilon \equiv L/\dot{m}c^2 = 0.1$ for converting mass to radiation—would need to be exceeded only by a factor of a $\sim$few to make the growth from stellar seeds much easier. Modestly exceeding the Eddington rate is theoretically certainly plausible (see below): density inhomogeneities can allow radiation to leak out of low density regions while most of the accreting matter can be contained in high density regions. For example, magnetized radiation dominated accretion disks are subject to a "photon bubble" instability that nonlinearly appears to lead to strong density inhomogeneities (e.g. Begelman 2002). Nevertheless, observations have so far not revealed systems that sustain super-Eddington accretion for extended periods; it would then still have to be explained why the $z \approx 6$ quasar BHs have this uniquely different behaviour.

3.1.3 Accretion Versus Mergers

Mergers between halos can help build up the mass of individual black holes (without significantly changing the total mass of the population), provided that the central black holes in the halos coalesce rapidly. The mean accretion efficiency of ~ 10 % inferred from comparing the local black hole mass density with the integrated quasar light suggests that accretion dominates at least the last e-folding of the black hole mass (Yu and Tremaine 2002; Shankar 2009). Mergers may, however, be significant earlier on (Haiman et al. 2004). In addition, uncertainties in the *expected* radiative efficiency of black hole accretion limit how accurately one can constrain the growth of black hole mass by mergers. For example, if the typical efficiency was $\approx$40 %, as for a maximally rotating Kerr black hole, then the Eddington-limited mass accretion rate would be decreased correspondingly, and mergers could dominate black hole growth (on the other hand, note that multiple mergers would have a tendency to cancel the black hole spin; Hughes and Blandford (2003)). In order for mergers to contribute significantly to the growth of individual black hole masses, stellar seeds must be present in large numbers, in the most of the numerous minihalos that form at $z \gtrsim 15$, down to small halo masses.

The balance between growth through BH mergers and growth through gas accretion is indeed a key characteristic of any SMBH assembly scenario. For concreteness, consider possible merger histories for the $z = 5.82$ SDSS quasar SDSS 1044-0125 (Haiman and Loeb (2001), the following arguments would be stronger for more luminous quasars at higher redshift). One can estimate the mass of the dark matter halo harboring the quasar by its abundance. SDSS searched a comoving volume of $\sim$1 Gpc^3 to find each quasar. Assuming a duty cycle of a few times 10^7 years, one estimates that the dark matter halos corresponding to this space density have masses of 10^{13} $M_\odot$ (using the halo mass function in Jenkins et al. (2001), the original Press-Schechter formula Press and Schechter (1974) would give a similar answer). A 10^{13} $M_\odot$ halo at $z = 6$ typically has only $\sim$10 progenitors with circular velocities of $v > 50$ km s^{-1} (the other progenitors being smaller). This implies that mergers can only help build up the black hole mass if seed black holes are present in progenitor halos with much smaller masses. A cutoff in the black hole mass function in halos with circular velocities below $v = 50$ km s^{-1} would be justified if the cosmic ultraviolet background could suppress gas infall into smaller halos (Efstathiou 1992; Thoul and Weinberg 1996; Navarro and Steinmetz 1997; Kitayama and Ikeuchi 2000). However, one-dimensional gas collapse models with radiative feedback Dijkstra et al. (2004) have shown that this suppression is ineffective at redshifts beyond $z \gtrsim 6$. Thus, there is no obvious obstacle to forming seed black holes in halos down to $v \sim 10$ km s^{-1} (below this threshold, atomic H cooling becomes inefficient, and H_2-photodissociation can be a limitation).

In the illustrative models discussed in the previous section, which successfully reproduce the abundance of the $\times 10^9 M_\odot$ SDSS quasar BHs, gas accretion accounts for the vast majority of the growth (in the sense that if the seed BHs were simply added together without any further accretion, the resulting total BH mass at $z = 6$ would be reduced by many orders of magnitude; see Table 3 in Tanaka and Haiman (2009)). However, in versions of these models in which BH growth is assumed to be self-regulating, accretion is much less important. Such models essentially describe the most heavily merger-driven scenarios possible, requiring accretion-driven growth of as little as a factor of a few. This is not surprising: placing a seed black hole in each arbitrarily low mass progenitor halo, with the same black hole mass to halo mass ratio as inferred for the SDSS quasars ($M_\bullet / M_{\rm halo} \sim 10^{-4}$), could account for the observed black hole masses in quasars by $z = 6$, even without *any* gas accretion (Haiman et al. 2004).

A further important unsolved question is whether halo mergers necessarily lead to black hole mergers at all (see, e.g., Merritt and Milosavljević 2005 for a review). During a galaxy merger, the black holes sink via dynamical friction to the center of the galaxy and form a tight black hole binary in the nucleus. In normal galaxies with a stellar component in the nucleus, the black hole binary can continue to shrink by ejecting low-angular momentum stars that pass close to the binary (those in the "loss cone"). This process, however, clearly does not operate in the earliest stages of structure formation, when there are at most a few stars (if any) present in the merging mini-galaxies. Even at the later states, this process is inefficient, at least in spherical stellar systems, because the loss cone must be replenished by two-body relaxation. The black hole binary thus appears to stall and cannot coalesce even during a Hubble time (Begelman et al. 1980).

Several ideas for circumventing this difficulty have been proposed. At later stages, in triaxial stellar populations, low-angular momentum orbits are populated much more efficiently because the stellar orbits can be chaotic; the resulting binary decay times are in many cases significantly less than a Hubble time, even if only a few percent of the stellar mass is on chaotic orbits (e.g., Yu 2002; Merritt and Poon 2004; Preto et al. 2011; Mahmood Khan et al. 2011). In the earliest galaxies without large stellar populations, the coalescence of BHs must be facilitated by gas physics. If circumbinary gas is present and forms a thin accretion disk, then BH-disk interactions can drag the binary together, in a manner similar to Type II migration in planetary systems. The main difference from the planet case is that at least for nearly equal mass binaries, the secondary BH's mass will far exceed the mass of the disk, which slows down the migration (see Haiman et al. (2009) for a comprehensive discussion, and Gould and Rix (2000); Armitage and Natarajan (2002); Escala et al. (2005) for examples of earlier work). A binary BH embedded in a spherical gas cloud is also facilitated by gaseous torques Escala et al. (2005); this case is much less well explored, but is likely to be more relevant to the earliest stages of the growth in the stellar-seed models, in halos without stars and with potential wells too shallow to support a thin disk. Finally, if SMBHs are brought together by successive halo mergers at a rate higher than the rate at which they can coalesce, then one or more of the BHs can be ejected out of the nucleus of the merger remnant by the slingshot mechanism (Saslaw et al. 1974). This could have implications for SMBH mass build-up in principle; in practice, more recent work on the dynamics of triple BHs indicate that ejections are relatively rare, and in the majority of cases, at least two of the BHs coalesce (Hoffman and Loeb 2007; Amaro-Seoane et al. 2010).

3.1.4 Super-Eddington Mass Accretion

If mass is supplied to a black hole at $\dot{m} \equiv \dot{M}/\dot{M}_{Edd} \gg 1$, the photons are trapped in the inflowing gas because the photon diffusion time out of the flow becomes longer than the time it takes the gas to accrete into the black hole (e.g. Begelman 1978; Begelman and Meier 1982). The resulting accretion is thus not via the usual thin disk Shakura and Sunyaev (1973), but rather via a radiatively inefficient flow (RIAF). The luminosity is still set by the Eddington limit, but most of the gravitational binding energy released by the accretion process is not radiated away (being trapped in the flow).

It is attractive to assume that the growth of SMBHs at high redshifts proceeds via such an optically thick, photon trapped accretion flow with $\dot{m} \gg 1$. Indeed, it would be a remarkable coincidence if the mass supply rate were precisely $\sim\dot{M}_{Edd}$ (required for a thin accretion disk) during the entire growth of massive black holes. It is more likely that the mass supply rate is initially much larger in the dense environments of high redshift galaxies ($\dot{m} \gg 1$) and then slowly decreases with time as the galaxy is assembled and the BH gains mass (e.g., Small and Blandford 1992; Cavaliere et al. 2000).

Three-dimensional simulations for the cooling and collapse of gas into the first minihalos find that H_2 cooling reduces gas temperatures to a few$\times$100K and produces a quasi-static contraction, with relatively low mass accretion rates of 10^{-3}–10^{-2} $M_\odot$ year^{-1} (e.g. Abel et al. 2002). This external mass-supply rate would still correspond to a super-Eddington growth rate for a BH with a mass of $\lesssim 10^5 M_\odot$. However, there is another limitation: within a Kelvin-Helmholtz time of $\sim 10^5$ years, only a few $\times$ 100 $M_\odot$ of material is accreted to the center (this is shown explicitly in Fig. 4). Much more mass than this is then unlikely to be incorporated onto the central proto-star, before it settles to the main sequence. Radiative feedback from the proto-star (in the form of H_2 dissociation, Lyman α radiation pressure, and ultimately, photoionization heating) on the infalling envelope was found to limit the final mass of the star to $\approx$140 $M_\odot$ (Tan and McKee 2004; McKee and Tan 2008).

On the other hand, in more massive halos *provided that H_2 cooling can be disabled throughout the entire time of the collapse*, the gas temperature is set by atomic H cooling and remains near 10^4K. In a self-gravitating gas, the mass accretion rate is of order $\sim c_s^3/G \propto T^{3/2}/G$ (e.g. Shu 1977). Three-dimensional simulations have confirmed this scaling (e.g. O'Shea and Norman 2007; Shang et al. 2010), and in halos with $T_{\rm vir} \sim 10^4$K, have found mass accretion rates of $\sim 1 M_\odot$ year^{-1} (Shang et al. 2010). As shown explicitly in Fig. 4, with this higher mass accretion rate, the mass that can be accumulated in the nucleus within a Kelvin-Helmholtz time is increased to 10^5 $M_\odot$.

Theoretical models for the accretion on much smaller spatial scales (not resolved in the above simulations) imply that even if $\dot{m} \gg 1$, only a small fraction of the mass supplied to the black hole actually reaches the horizon; most of it is driven away in an outflow (see, e.g., simulations of RIAFs Stone et al. (1999); Stone and Pringle (2001); Hawley and Balbus (2002); Igumenshchev et al. (2003); Proga and Begelman (2003); and analytic models Blandford and Begelman (1999, 2004); Quataert and Gruzinov (2000)). The accretion rate onto a black hole thus probably cannot exceed $\sim\dot{M}_{Edd}$ by a very large factor, even if the mass supply rate from larger radii is large (see Shakura and Sunyaev (1973) for an early discussion of this point).

The above discussion focuses on whether highly super-Eddington accretion is possible. The question of whether the Eddington limit for the luminosity can be exceeded by a modest factor of $\sim$10 is a bit more subtle. Magnetized radiation dominated accretion disks are subject to a "photon bubble" instability that nonlinearly appears to lead to strong density inhomogeneities (see, in particular, Arons (1992); Gammie (1998); Begelman (2001); Blaes and Socrates (2001); Begelman (2002)). Density inhomogeneities allow super-Eddington fluxes from the accretion flow because radiation leaks out of the low density regions while most of the matter is contained in high density regions. Ref. Begelman (2002) estimates that the Eddington limit can potentially be exceeded by a factor of ∼10–100. This would allow much more rapid growth of black holes at high redshifts, circumventing the above arguments that seed black holes at $z \sim 15$ are required. Magneto-hydrodynamic (MHD) simulations of radiation dominated accretion flows have confirmed the rapid growth of unstable short-wavelength modes, with the development of large density variations. Inhomogeneities then allow the radiation to diffuse outward five times more rapidly than in a disk in hydrostatic equilibrium with no magnetic fields (Turner et al. 2005). Explicit models for such slim, porous, accretion disks have been constructed recently (Dotan and Shaviv 2011). In these models, when the external mass accretion rate is 10–20 $\times \dot{M}_{Edd}$, despite the presence of winds, a super-critical fraction, 2.6–3.8 $\dot{M}_{Edd}$, was indeed found to reach the central SMBH.

3.2 *Growth by Rapid Direct Collapse*

An appealing alternative idea is to produce, say, a 10^5 $M_\odot$ SMBH "directly"—i.e. much faster than this would take under Eddington-limited accretion from a stellar seed. This would clearly be helpful to explain the high-redshift SMBHs. In this context, the crucial question is whether gas can accrete at a highly super-Eddington rate onto a black hole, i.e., with $\dot{M} \gg \dot{M}_{Edd}$, where $\dot{M}_{Edd} = 10 L_{Edd}/c^2 \approx 1.7 M_8$ $M_\odot$ year^{-1} is the accretion rate that would produce an Eddington luminosity if accretion onto a black hole of mass $10^8 M_8$ $M_\odot$ proceeded with 10 % radiative efficiency. If so, this could lead to rapid black hole growth at high redshifts. Constraints on BH seeds and their formation redshifts would therefore be much less stringent.

3.2.1 Rapid Collapse of Gas in $T_{\rm vir} \gtrsim 10^4$K Halos

SMBHs may form directly by the collapse of gas clouds at high redshifts, likely via an intermediate stage of a supermassive star or disk. The gas must not only shed angular momentum efficiently and collapse rapidly, but must also then avoid fragmentation. Whether fragmentation of the gas cloud into stars can be avoided is particularly questionable, in view of the large angular momentum barrier that must be overcome to reach small scales in a galactic nucleus (forming an SMBH through a dense stellar cluster is another option, discussed in the next section).

The most promising locations for such rapid "direct collapse" are at the centers of halos with $T_{\rm vir} \sim 10^4$K. In the past several years, many authors have sketched how gas may collapse rapidly, without fragmentation, in these halos. The essential idea is that when contracting gas in a protogalactic nucleus becomes optically thick and radiation pressure supported, it becomes less susceptible to fragmentation and star formation. It is, however, unlikely that radiation pressure becomes important before angular momentum does, implying that the gas forms a viscous accretion disk in the galactic nucleus (fragmentation before the disk forms can also be avoided if the forming fragments collide and "coalesce" before they can separate into discrete dense clumps Kashlinsky and Rees (1983)). On the other hand, if self-gravitating, the resulting disk is strongly gravitationally unstable and becomes prone to fragmentation and star formation (e.g., Shlosman and Begelman 1989; Goodman 2003). Whether this fragmentation can be avoided is unclear. One possibility is to stabilize the disk by keeping its temperature "warm" (i.e. $T \sim 10^4$K, close to the virial temperature). This would fatten the disks (the scale height scales with the ratio of gas and virial temperatures); this scenario may be possible in a virtually metal-free, high redshift halo (Oh and Haiman 2002; Bromm and Loeb 2003; Wise et al. 2008; Regan and Haehnelt 2009).

A suite of recent numerical simulations studied gas collapse in halos with $T_{\rm vir} \sim 10^4$K (Shang et al. 2010). It was found that the gas in such halos, when collapsing in isolation, forms H_2 efficiently, and (unfortunately) cools to temperatures of $T \sim 300$K. Although no fragmentation was seen, the cold gas (well below the virial temperature) is expected to ultimately fragment on smaller scales that have not yet been resolved (Turk et al. 2009). More importantly, even if fragmentation was successfully avoided, there is a problem: the cold gas was found to flow inward at relatively low velocities, near the sound speed of ~2–3 km s^{-1}, with a correspondingly low accretion rate of ~0.01 $M_\odot$ year^{-1}. Shang et al. (2010) speculate that this is explained by a series of weak shocks in the infalling gas, which prevent the gas from accelerating to large Mach numbers (this is similar to the behavior seen in three-dimensional simulations of the so-called "cold mode" of accretion in lower-redshift galaxies Kereš et al. (2005)). Ultimately, the slow infall velocities and cold temperatures produce conditions nearly identical to those in the cores of lower-mass minihalos (mentioned above); extensive ultra-high resolution simulations had concluded that the gas then forms a single ~100 $M_\odot$ star Abel et al. (2002); Bromm et al. (2002); Yoshida et al. (2008) or perhaps fragments even further into several stars Turk et al. (2009); Stacy et al. (2010); Greif et al. (2011a); Prieto et al. (2011), rather than forming a supermassive star or BH.

There have been at least three different ideas on how to avoid H_2-cooling and to keep the gas warm. One is for the gas to "linger" for a sufficiently long time at 10^4K that it collapses to a SMBH, even before H_2 has a chance to reduce the temperature (H_2 is kept dissociated by collisions before the temperature falls below $\sim 4,000$K). For a sufficiently high space—and column-density of neutral hydrogen, the absorption of trapped Lyman α photons can be followed by collisional de-excitation, rather than the resonant scattering of the Lyman α photon, effectively trapping much of the cooling radiation. This could prevent the gas temperature from falling below $\sim$8,000K, and lead to such lingering and to SMBH formation—analogous to opacity-limited fragmentation in colder gas in the context of star formation (Spaans and Silk 2006; Schleicher et al. 2010; Latif et al. 2011b).

Another possibility is that, even in the presence of significant cooling, angular momentum transport by gravitational instabilities, spiral waves, bars, etc., can drive a fraction of the gas to yet smaller scales in the galactic nucleus (e.g. Begelman and Shlosman 2009). Eisenstein and Loeb (1995) argued that this was particularly likely to occur in rare low angular momentum dark matter halos because the disk could viscously evolve before star formation commenced. A similar idea is that a small fraction of the gas, with low specific angular momentum, within the halo may collapse to the center without undergoing fragmentation (Koushiappas et al. 2004; Lodato and Natarajan 2006). It may help that even if most of the gas is initially converted into stars, stellar winds and supernovae will eject a significant amount of this gas back into the nucleus; some of this gas can eventually collapse to smaller scales (Begelman and Rees 1978).

Finally, H_2-cooling may be disabled if the gas is exposed to a sufficiently intense UV flux J, either directly photo-dissociating H_2 (in the Lyman–Werner bands near a photon energy of $\sim$12eV) or photo-dissociating the intermediary H^- (at photon energies $\gtrsim$0.76eV). Requiring the photo-dissociation timescale, $t_{\rm diss} \propto J^{-1}$, to be shorter than the H_2-formation timescale, $t_{\rm form} \propto \rho^{-1}$, generically yields a critical flux that increases linearly with density, $J^{\rm crit} \propto \rho$. In low-mass minihalos, the critical flux is low, $J^{\rm crit} \approx$ 0.01–0.1 (Haiman et al. 1997; Machacek et al. 2001; Mesinger et al. 2006, 2009).[4] Since the gas in halos with $T_{\rm vir} \gtrsim 10^4$K can cool via atomic Lyman α radiation and loose pressure support, it inevitably collapses further. As a result, in these halos, the critical flux is high, $J^{\rm crit} \approx 10^2-10^5$, depending on the assumed spectral shape (Shang et al. (2010); see also Omukai (2001); Bromm and Loeb (2003) who found similar, but somewhat higher values). The existence of this critical flux is illustrated in Fig. 2, using a one-zone model, in which the density evolution is approximated by spherical collapse, and the gas is illuminated by a source with a black-body spectrum (with a temperature of $T = 10^4$K, characteristic of a normal stellar population). Figure 3 shows the radial structure of a $10^8 M_\odot$ halo, at the time of its collapse, when illuminated at various intensities, taken from three-dimensional simulations with the AMR code *enzo*. These profiles clearly show that when the UV flux exceeds a critical value, the core of the halo is prevented from cooling to low temperatures.

[4]Here J denotes the specific intensity just below 13.6eV, in the usual J_{21} units of 10^{-21}erg cm^{-2} $sr^{-1}\, s^{-1}\, Hz^{-1}$.

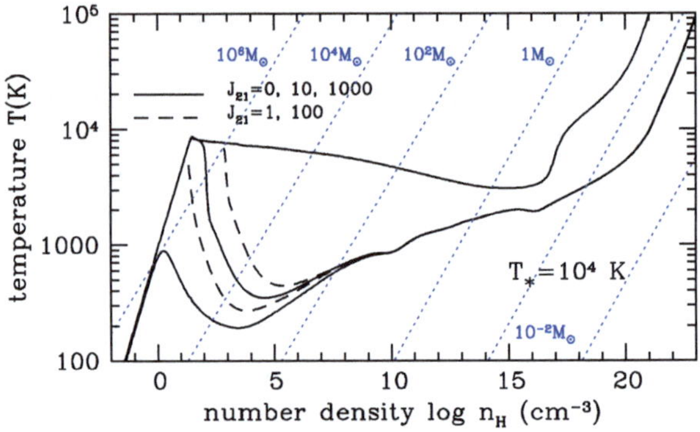

Fig. 2 Temperature evolution of a metal-free cloud, irradiated by a strong UV flux. The models solve for the chemical and thermal evolution, but assume a pre-imposed density evolution, based on the spherical collapse model. Various cases are shown, with UV intensities at the Lyman limit of $J_{21} = 0, 1, 10, 100$ and 10^3, in the usual units of 10^{-21}erg cm^{-2} sr^{-1} s^{-1} Hz^{-1} (solid and dashed curves; see the legend in the panel). Each blue dotted line corresponds to a different constant Jeans mass. The gas is heated adiabatically until a density of $n \approx 10^0$–10^2 cm^{-3}, at which H_2-cooling becomes efficient and cools the gas to a few $\times 100$ K. However, there exists a critical flux, with a value between $J_{21} = 10^2$ and 10^3, above which H_2-cooling is disabled (adapted from Omukai et al. (2008))

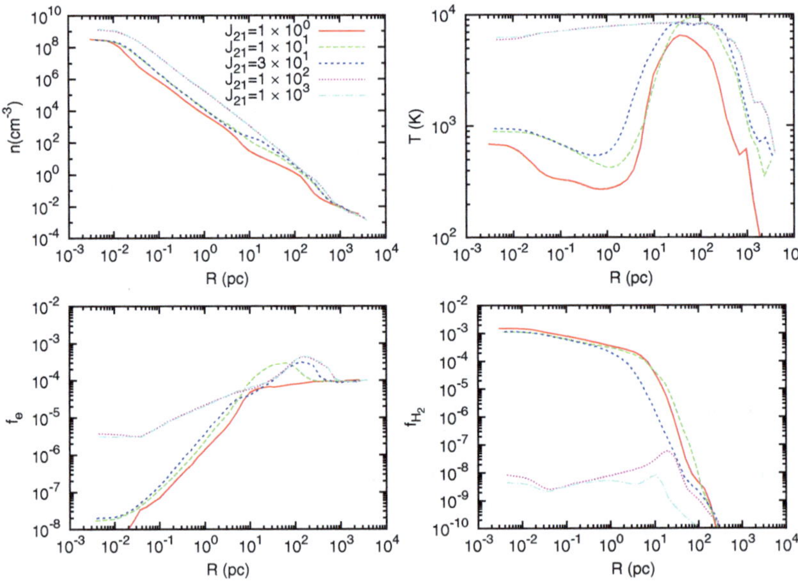

Fig. 3 The results of adaptive mesh refinement (AMR) simulations of a primordial halo, with a total mass of a few $\times 10^7$ $M_\odot$, collapsing at redshift $z \approx 10$–15, exposed to various UV background fluxes. The four panels show snapshots of the spherically averaged profile of the particle number density, gas temperature, e^- fraction and H_2 fraction, at the time of the collapse of the core, for several different values of the UV background intensity J_{21}, as labeled. The existence of a critical flux, here with a value between $J_{21} = 30$ and 10^2, above which H_2-cooling is disabled, is evident (adapted from Shang et al. (2010))

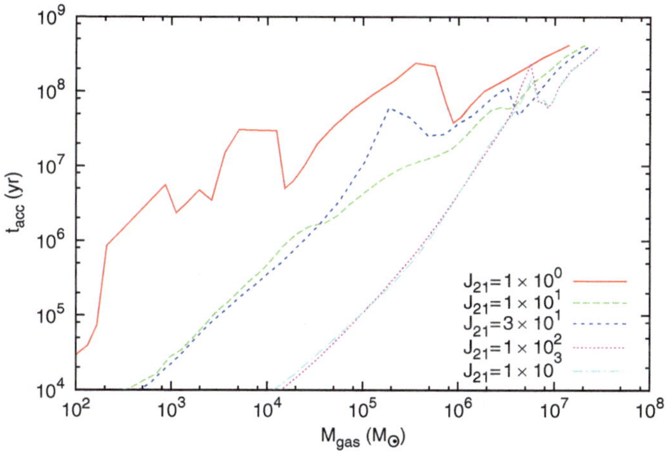

Fig. 4 The local accretion time-scale t_{acc} as a function of the enclosed gas mass M_{gas}, in the same halo depicted in Fig. 3, illuminated with different intensities, as labeled. In the halos exposed to a supercritical flux ($J_{21} = 10^2$ and 10^3), the mass accretion rate, $\dot{M} \approx 1\ M_\odot\ year^{-1}$, is nearly 10^3 times higher than in halos whose gas cools via H_2 ($J_{21} \lesssim 10$). At the center of the brightly illuminated halos, $\approx 10^5 M_\odot$ of gas accumulates within a Kelvin-Helmholtz time of $\approx 10^5 M_\odot$, possibly leading to the formation of a SMBH with a comparable mass (adapted from Shang et al. (2010))

The 3D simulations also provide an estimate of the mass of the central "object" (star or SMBH) that ultimately forms at the core of the halo, based on the measured profile of the mass accretion rate. This is illustrated in Fig. 4. In particular, when the flux exceeds the critical value, and the gas remains warm, the collapse is significantly delayed. However, when the gas ultimately does collapse, it accretes toward the center at the sound speed ($c_s \approx 10$km/s), leading to a mass accretion rate of $\dot{M} \approx 1 M_\odot\, year^{-1}$. This is much higher than in the case of cold ($c_s \sim 1$ km/s) gas in halos with efficient H_2 cooling (as mentioned above, the simulations find $\dot{M} \propto c_s^3$, as expected in self-gravitating gas).

Importantly, the critical flux is high—likely significantly exceeding the expected level of the cosmic UV background at high redshifts. Therefore, only a small subset of all $T_{vir} \gtrsim 10^4$K halos, which have unusually close and bright neighbors, may see a sufficiently high flux. However, given the strong clustering of early halos, there is a sufficient number of these close halo pairs to account for the abundance of the rare $z = 6$ quasars (Dijkstra et al. 2008). A more significant challenge to this idea is that in order to avoid fragmentation, the gas in these halos must also remain essentially free of any metals and dust (Omukai et al. 2008). This requirement could be difficult to reconcile with the presence of a nearby, luminous galaxies.

Finally, an important point to emphasize is that the collapsing gas is optically thick, and the critical flux J_{crit} depends crucially on the details of self-shielding of the Lyman-Werner lines of H_2. Since following radiative transfer in many dozens of lines is computationally expensive, existing works have employed various simplifying approximations. The simplest (and by far most commonly used) approach is to combine a simple power-law fitting formula, $f_{\rm shield} = (N_{\rm H_2}/10^{14}{\rm cm}^{-2})^{-3/4}$, for the H_2 self-shielding factor Draine and Bertoldi (1996) with an estimate for an effective H_2 column density $N_{\rm H_2}$ (most often equated with the product of the local density and Jeans length). These assumptions have recently been scrutinized in Wolcott-Green et al. (2011), which solved radiative line transfer exactly, using post-processing of three-dimensional simulations. This showed, rather promisingly, that when self-shielding is treated more accurately, $J_{\rm crit}$ is reduced by about an order of magnitude. Interestingly, this reduction comes from a product of three very different sources (each of which individually reduces the shielding by a factor of ~two): (1) a numerical inaccuracy of the power-law $f_{\rm shield}$ formula, (2) the inapplicability of this fitting formula at the relevant, relatively high temperatures ($\gtrsim 10^3$K), where excited rotational levels of H_2 are populated, and (3) the Jeans length yielding an overestimate of the effective average (over different sightlines) column density. The order-of-magnitude reduction in J is especially important, since the probability distribution of J, sampled by halos at $z \gtrsim 10$, is very steep near $J \sim 10^4$ (see Dijkstra et al. (2008)). With the original high $J_{\rm crit}$ value, it has been shown Dijkstra et al. (2008) that only one in $\approx 10^6$ halos—only those with an unusually bright and close neighbour—will see a sufficiently high flux. The reduction of the $J_{\rm crit,21}$ value will significantly increase the number of candidates for objects that can avoid H_2-cooling and fragmentation, and makes this scenario much more viable.

3.2.2 The Ultimate Fate of the Gas

Although the detailed evolutionary pathways are still not understood, a possible outcome of the above scenarios is the continued collapse of some gas to smaller scales in the galactic nucleus. As the gas flows in, it becomes optically thick, and the photon diffusion time eventually exceeds the inflow time. Radiation pressure dominates for sufficiently massive objects so that the adiabatic index is $\Gamma \approx 4/3$. Radiation pressure may temporarily balance gravity, forming a supermassive star or disk (SMS; e.g., Hoyle and Fowler 1963; Wagoner 1969; see, e.g., Shapiro and Teukolsky 1983 for a review and additional references to earlier work). The SMS will radiate at the Eddington limit and continue contracting. When the SMS is sufficiently compact ($GM/Rc^2 \approx 10^{-4}M_8^{-1/2}$ for non-rotating stars), general relativistic corrections to the gravitational potential become important, and the star becomes dynamically unstable because its effective polytropic index is $\lesssim 4/3$. For masses $\lesssim 10^5$ $M_\odot$, thermonuclear reactions halt the collapse and generate an explosion (e.g., Fuller et al. 1986), but more massive objects appear to collapse directly to a SMBH (see Shapiro (2004) for a review; and, e.g., Shibata and Shapiro 2002; Saijo et al. 2002 for recent simulations).

If the mass accretion rate is high ($\sim$1 $M_{\odot}$year^{-1}), the outer layers of the SMS do not have time to thermally relax, and a high-pressure core-envelope structure may develop, dubbed a "quasi-star" (Begelman et al. 2006, 2008; Begelman 2010). The envelope initially contains most of the mass, and the central BH embedded in the envelope can grow from this envelope; the key feature of this configuration is that the accretion is limited by the Eddington limit for the entire envelope, rather than just the BH. Interestingly, the mass accretion rate required for this model is comparable to that seen in three-dimensional simulations of $T_{\rm vir} \gtrsim 10^4$K halos, *provided again that the gas in these halos can avoid H_2 cooling* (Shang et al. 2010).

Finally, a possibility for the gas is to ultimately fragment into stars, but not before it reaches very high densities. If the gas is metal-enriched, this scenario may be most likely, and will result in the formation of a dense and compact stellar cluster, which naturally evolves to form a SMBH (Omukai et al. 2008). We turn to this idea in the next section.

3.3 *The Formation of Black Holes in Stellar Clusters*

The negative heat capacity of self-gravitating stellar systems makes them vulnerable to gravitational collapse in which the core of the cluster collapses on a timescale t_{cc} comparable to the two-body relaxation time of the cluster (Binney and Tremaine 1987). If core collapse proceeds unimpeded, the resulting high stellar densities can lead naturally to the runaway collisional growth of a single massive object which may evolve to form a black hole (as in the discussion of SMSs above). This process provides an additional route for the direct formation of SMBHs at high redshifts (or, more likely, intermediate mass seeds).

Early work suggested that the fate of stellar clusters depends sensitively on the number of stars in the cluster. Lee (1987) and Quinlan and Shapiro (1990) found that very dense massive star clusters ($N \gtrsim 10^6$–10^7 stars) were required to have successful core collapse and runaway growth of a single massive object. In less massive clusters, core collapse was halted by binary heating, in which the cluster gains energy at the expense of binaries via three-body interactions (Heggie 1975; Hut et al. 1992). Successful core collapse also requires that t_{cc} is shorter than the timescale for the most massive stars to evolve off the main sequence (Rasio et al. (2004); this requirement implies compact clusters $\lesssim$1 pc in size). Otherwise, mass loss from evolved stars and supernovae prevents the core from collapsing (in much the same way as binary star systems can become unbound by supernovae). In principle, massive stars could evolve into stellar-mass BHs and form a dense cluster of stellar-mass BHs. In the context of high-redshift halo formation, this is naturally expected Madau and Rees (2001), given that the first stars in the first minihalos are believed to be massive. A dense cluster of stellar-mass BHs can, in principle, grow into a more massive IMBH by coalescence due to gravitational radiation; however, this process is effective only in large stellar systems that are found in present-day galactic nuclei (O'Leary et al. 2006).

If the required number of stars was indeed as large as $N \gtrsim 10^6$–10^7, this would be bad news for early SMBH formation: in the cosmological hierarchy, $\gtrsim 10^8$ $M_\odot$ halos—the smallest that could plausibly harbor such star clusters—are very rare ($\gtrsim 2.5\sigma$ fluctuations) at $z \gtrsim 10$. However, recent work has revived earlier ideas that stellar clusters are subject to a "mass segregation instability" that makes even the relatively less massive clusters prone to forming black holes (Spitzer (1969); Vishniac (1978); Begelman and Rees (1978)). Because massive stars in a cluster sink by dynamical friction toward the center (mass segregation), they invariably dominate the dynamics of the cluster core and can undergo core collapse on a timescale much shorter than that of the cluster as a whole (and on a timescale shorter than their main sequence lifetime). Portegies Zwart and McMillan (2002) showed with N-body simulations that the resulting core collapse likely leads to runaway merger and formation of a single black hole, and Gürkan et al. (2004) reached a similar conclusion for much larger $N \sim 10^7$ using Monte Carlo simulations. Including an explicit treatment of stellar collisions, the most recent Monte-Carlo simulations find that the central massive object that forms in core has a mass of $\sim 10^{-3}$ of the whole cluster (Goswami et al. 2012).

In the context of high-redshift BH formation, Omukai et al. (2008) considered the cooling properties of gas in $T_{\rm vir} \gtrsim 10^4$K halos. It is expected that the majority of such halos, when they are assembled, had already undergone some amount of star-formation. In this case, their gas would have at least a trace amount of metals. Omukai et al. (2008) considered the case when such mildly polluted halos are exposed to a large UV flux, which dissociates H_2, and initially prevents cooling. This allows the gas to contract to very high densities, without fragmenting initially. By following the thermal and chemical evolution of such low-metallicity gas, exposed to extremely strong UV radiation, Omukai et al. (2008) found, however, that eventually, gas fragmentation is inevitable above a critical metallicity, whose value is between $Z_{\rm cr} \approx 3 \times 10^{-4} Z_\odot$ (in the absence of dust) and as low as $Z_{\rm cr} \approx 5 \times 10^{-6} Z_\odot$ (with a dust-to-gas mass ratio of about $0.01 Z/Z_\odot$). When the metallicity exceeds these critical values, an ultra-dense cluster (the density at the time of fragmentation is $n \gtrsim 10^{10}$ cm^{-3}) of low-mass stars may form at the halo nucleus (Clark et al. (2008) and Devecchi and Volonteri (2009) argued for similar scenarios, in the central regions of a protogalactic disk). Relatively massive stars in such a cluster can then rapidly coalesce into a single more massive object, which may produce an intermediate-mass BH remnant with a mass up to $M \lesssim 10^3$ $M_\odot$.

The above processes provide a promising channel for the formation of IMBH seeds, which can grow via mergers and/or accretion to form SMBHs. For example, Volonteri et al. (2003) and Islam et al. (2003) have incorporated such early black hole seeds into Monte Carlo simulations of the black hole merger histories. With reasonable prescriptions for the merging and accretion of black holes inside dark halos, these models can account for the observed evolution of the quasar luminosity functions at $z < 5$ and can serve for physically motivated extrapolations to high redshifts to describe the first AGN.

It should be noted that there exist some observational evidence for IMBHs in the local universe. In particular, the presence of IMBHs with masses of order $\sim 10^4$ $M_\odot$ are inferred from stellar kinematics in the globular clusters G1 (Gebhardt et al. (2002); note that the velocity dispersion profile itself does not require a BH Baumgardt et al. (2003) and the evidence comes from higher-order moments of the velocity distribution Gebhardt et al. (2005) instead), ω Cen Noyola et al. (2008) and in M15 (van der Marel et al. (2002), although this object can also be modeled without an IMBH; van der Marel (2004)). Ultra-luminous X-ray sources in nearby galaxies (e.g., Colbert and Mushotzky 1999; Kaaret et al. 2001; Miller and Colbert 2004; Farrell et al. 2009) have also been interpreted as accreting IMBHs. While there are viable non-IMBH interpretations of these sources (e.g., King et al. 2001; Begelman 2002), the X-ray spectrum of one such source during the peak of an outburst implies the presence of a $\gtrsim 2,000$ $M_\odot$ IMBH (assuming that the luminosity is limited to $0.3 \times L_{Edd}$ as in stellar-mass BH X-ray binaries in their hard state; Kaaret et al. (2009)).

3.4 Alternative Models

Since both of the "standard" scenarios discussed above require some optimistic assumptions, it is interesting to consider some more exotic possibilities.

It is commonly believed that the magnetic fields permeating galaxies such as the Milky Way arose by the amplification of a much weaker large-scale seed field. Weak primordial magnetic fields, with strengths of up to $\sim$1nG, can be produced in phase transitions in the early universe, during inflation, or during the electroweak or QCD phase transitions. It has recently been shown that such a primordial magnetic field could produce a variant of the "direct collapse" scenario (Sethi et al. 2010). In particular, if the field is tangled, then ambipolar diffusion will provide an efficient new mechanism to heat the gas as it collapses in protogalactic halos. If the field has a strength above $|B| \gtrsim 3$ (comoving) nG, the collapsing gas is kept warm ($T \sim 10^4$ K) until it reaches the critical density $n_{crit} \approx 10^3 cm^{-3}$ at which the roto-vibrational states of H_2 approach local thermodynamic equilibrium. This is illustrated explicitly by the thermal evolution of fluid elements shown in Fig. 5. H_2-cooling then remains inefficient, and the gas temperature stays near $\sim 10^4$K, even as it continues to collapse to higher densities. The critical magnetic field strength required to permanently suppress H_2-cooling is somewhat higher than upper limit of $\sim$2nG from the cosmic microwave background (CMB). However, it can be realized in the rare $\gtrsim$(2–3)σ regions of the spatially fluctuating B-field; these regions contain a sufficient number of halos to account for the $z \approx 6$ quasar BHs.[5]

[5]Because of the high magnetic Jeans mass, the magnetic pressure has significant dynamical effects, and can prevent gas collapse in halos with masses up to $M \gtrsim$ few $\times 10^{10} M_\odot$. These are $\sim$100

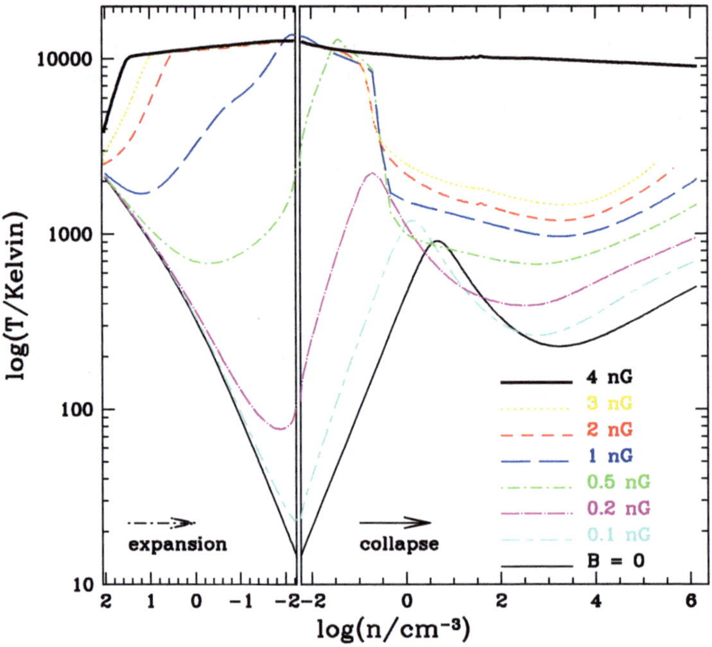

Fig. 5 The temperature evolution of a patch of the intergalactic medium is shown as it initially expands and then turns around and collapses to high density. The different curves correspond to different values of the assumed primordial magnetic field, as labeled. The gas evolves from the left to the right on this figure. The left panel shows the expanding phase, starting from an initial density of $\approx$100 cm^{-3} (corresponding to the mean density at redshift $z \simeq 800$) and ending at the turnaround just below $n = 10^{-2}$ cm^{-3}. The right panel follows the subsequent temperature evolution in the collapsing phase. The figure shows the existence of a critical magnetic field, with a value between $B = 3$ and 4 nG, above which H_2-cooling is disabled, and the gas temperature always remains near 10^4K (adapted from Sethi et al. (2010))

Another "exotic" idea is that the first PopIII stars may be powered by heating by dark matter annihilation, rather than by nuclear fusion (Spolyar et al. 2008). Weakly interacting massive particles (WIMPs), can be such a heat source, as long as they reach sufficiently high density inside the first stars, and if the annihilation products are trapped inside the star. Several authors have studied the impact of this additional heating mechanism on the structure and evolution of such "dark stars" Spolyar et al. (2009); Iocco et al. (2008); Yoon et al. (2008); Taoso et al. (2008); Umeda et al. (2009); Spolyar et al. (2009); Freese et al. (2010); Ripamonti et al. (2010). In particular, these stars can live much longer than "normal" PopIII stars, and because their radiation is soft, they can continue to accrete gas, as long as the dark matter heating persists, and grow to masses of up to $\sim 10^5 M_\odot$. In fact, one of

times more massive than the DM halos in the "usual" direct collapse models discussed in Sect. 3.2 above.

the challenges in these models is to explain why and how the growth of the star stops (Umeda et al. 2009; Freese et al. 2010). An interesting prediction is that these stars are bright, and should be detectable directly by *JWST* (Freese et al. 2010).

4 Observational Considerations

In this section, we first review several recent observations and their implications for the formation of black holes at high redshifts. We then speculate on how future observations may probe the assembly of high-z SMBHs.

4.1 Surveys for High Redshift Quasars

The majority of the $\sim$40 quasars known at $z \sim 6$ to date have been discovered in the SDSS. This is perhaps somewhat surprising, since the SDSS is a relatively shallow survey (with a magnitude limit of $i \sim 22$) capable of detecting only the rarest bright quasars at redshifts as high as $z \sim 6$. Nevertheless, the large solid angle searched for high redshift quasars ($\sim 8,000$ square degrees) has yielded many such objects (Fan 2006). Somewhat deeper surveys covering smaller areas (a few $\sim$100 square degrees), such as the SDSS Deep Stripe Jiang et al. (2009), and the CFHQS Willott et al. (2010) and UKIDSS Lawrence et al. (2007) surveys, have yielded many additional detections. The most important properties (for our purposes) of these sources are that they are probably powered by SMBHs as large as a few $\times\ 10^9\ M_\odot$ and overall, they appear to be indistinguishable from bright quasars at moderate ($z \sim 2$–3) redshifts, with similar spectra and inferred metallicities (e.g. Jiang et al. 2007. In addition, a large reservoir of molecular gas is already present, even in the most distant sources (Walter et al. 2003; Wang et al. 2010, 2011).

Despite the overall similarities, there are some tentative distinctions between these $z \gtrsim 6$ quasars and their $z \sim 2$–3 counterparts. First, there is evidence for increasing Eddington ratios toward high redshift (see below). Second, quasar clustering has been found to strongly increase with redshift (Shen et al. 2007). The observed clustering strength can be used to infer to quasar lifetimes Haiman and Hui (2001); Martini and Weinberg (2001), and implies that the duty cycle of bright quasar activity increases significantly toward high redshifts, to near unity by $z \approx 6$ (Shankar et al. 2010). Finally, two of the $\sim$6 quasars have no detectable emission from hot dust (Jiang et al. 2010). There are no known examples for such apparently hot-dust-free quasars at low redshift; this result therefore suggests that at least these two quasar BHs may have been caught at a young age of the evolution of their host galaxies (i.e., there was insufficient time for a dusty torus to form in the nucleus).

With these exceptions, the high-z SMBHs and their surroundings appear as "fully developed" as their lower redshift counterparts, despite the young age ($\lesssim 10^9$ years) of the universe at $z \gtrsim 6$. These rare quasars are likely harbored by massive ($\sim 10^{12-13}$ M$_\odot$) dark matter halos that form out of 4–5σ peaks of the fluctuating primordial density field. The large halo mass follows directly from the space density of these sources (Haiman and Loeb (2001); another method to confirm the large halo masses is to study the expected Lyman α absorption signatures of cosmological gas infall onto such massive halos, as proposed by Bromm and Loeb (2003)). Indeed, the environment and dynamical history of an individual massive dark matter halo at $z \sim 6$ and $z \sim 3$ can be similar; it is their abundance that evolves strongly with cosmic epoch. This is broadly consistent with the observations: the bright $z \sim 6$ quasars look similar to their $z \sim 3$ counterparts, but their abundance is much reduced (by a factor of $\sim$40).

The fact that these quasars are so rare has important implications. First, they are likely to be the "tip of the iceberg" and accompanied by much more numerous populations of fainter quasars at $z \gtrsim 6$. The slope of the luminosity function is expected to be very steep at $i \sim 22$, and so pushing the magnitude limits further in future surveys should prove rewarding. The most direct constraints on this slope are from combining the CFHQS sample with the more luminous SDSS sample (yielding a total of 40 quasars between redshifts $5.74 < z < 6.42$; Willott et al. (2010)), and from gravitational lensing (Comerford et al. 2002; Wyithe and Loeb 2002a,b; Richards et al. 2004). Combining source counts and lensing yields the strongest limit of $-d \log \Phi / d \log L \lesssim 3$ (Wyithe 2004). Second, the steep slope of the dark halo mass function implies that the masses of the host halos can be "measured" from the abundance quite accurately (see discussion in Sect. 4.2). Conversely, since small changes in the assumed host halo mass results in large changes in the predicted abundance, large uncertainties will remain in other model parameters. In this sense, fainter, but more numerous quasars (or lack thereof) can have more constraining power for models that relate quasars to dark halos.

The most striking feature of the SDSS quasars, however, is the large black hole mass already present at $z \sim 6$. In the rest of this section, we critically assess whether the inferred large black hole masses are robust.

The masses of the black holes powering the SDSS quasars are inferred by assuming that (1) they shine at the Eddington luminosity with a bolometric correction identical to that of lower redshift quasars (this is justified by their similar spectra), and (2) they are neither beamed nor gravitationally lensed (both of these effects would make the quasars appear brighter and allow lower BH masses). These assumptions lead to black hole masses $M_\bullet \approx (2\text{–}6) \times 10^9$ M$_\odot$ for the $z > 6$ quasars known to date. These are reasonable assumptions, which have some empirical justification.

The hypothesis that the quasars are strongly beamed can be ruled out based on their line/continuum ratio. If the quasar's emission was beamed into a solid angle covering a fraction, f, of 4π, it would only excite emission lines within this cone, reducing the apparent line/continuum ratio by a factor f. However, the SDSS quasars have strong lines. Haiman and Cen (2002) found that the line/continuum ratio of the $z = 6.28$ quasar SDSS 1030+0524 is about twice that of the median value in the SDSS sample at $z > 2.25$ (Vanden Berk et al. 2001). This argument, applied to the Mg II line of the $z = 6.41$ quasar SDSS J1148+5251 Willott et al. (2003) yields a similar conclusion.

Another important uncertainty regarding the inferred black hole masses is whether the SDSS quasars may be strongly magnified by gravitational lensing. The optical depth to strong lensing along a random line of sight to $z \sim 6$ is small ($\sim 10^{-3}$; e.g., Kochanek 1998; Barkana and Loeb 2000). Nevertheless, magnification bias can significantly boost the probability of strong lensing. If the intrinsic (unlensed) luminosity function at $z \sim 6$ is steep and/or extends to faint magnitudes, then the probability of strong lensing for the SDSS quasars could be of order unity Comerford et al. (2002); Wyithe and Loeb (2002a,b)). The overwhelming majority (more than 90 %) of strong lensing events would be expected to show up as multiple images with separations at least as large as $0.3''$. It is difficult to produce strong magnification without such multiple images, even in non-standard lensing models (allowing ellipticity and/or external shear Keeton et al. (2005)). However, deep optical observations (e.g. with the *Hubble Space Telescope*) of the highest redshift quasars show no signs of multiple images for any of the $z \gtrsim 6$ sources down to an angle of $0.3''$ (Richards et al. 2004; Willott et al. 2005).[6]

Finally, whether or not the SDSS quasars are shining at the Eddington limit is difficult to decide empirically. Several authors (Willott et al. 2003; Vestergaard 2004; Jiang et al. 2007; Kurk et al. 2007; Vestergaard et al. 2008; Vestergaard and Osmer 2009; Kurk et al. 2009; Willott et al. 2010) have estimated Eddington ratios in samples of high redshift quasars, using observed correlations between the size of the broad line region and the luminosity of the quasar (the correlation is calibrated using reverberation mapping of lower redshift objects; e.g. Kaspi et al. (2000); Vestergaard (2002); Kollmeier et al. (2006)). Values range from ≈ 0.1 to $\gtrsim 1$; in particular, L/L_{Edd} is typically found to increase with redshift, and approaches unity for the $z \gtrsim 6$ quasars.

Inferences about Eddington ratios at high redshifts can also be made by utilizing models of the quasar population as a whole. Such models typically assume the Eddington luminosity at higher redshifts, where fuel is thought to be readily available (Small and Blandford 1992; Haehnelt and Rees 1993). Numerous semi-analytic models for the quasar population (see, e.g., Shankar 2009 and references therein) have found that Eddington ratios of order unity during most of the growth of the black hole mass also yield a total remnant SMBH space density at $z = 0$ that is consistent with observations. Ciotti and Ostriker (2001), and, more recently, Milosavljević et al. (2009) have self-consistently modeled accretion and radiative feedback onto an individual quasar BH, and found that (provided fuel is available) the luminosity is near the Eddington value during the phases when the quasar is on. Despite these arguments, one cannot directly rule out the possibility that the SDSS quasars shine at super-Eddington luminosities (theoretically, this is possible, as discussed

[6]The highest-redshift known lensed quasar is at $z = 4.8$ and was discovered serendipitously in SDSS, initially flagged as a galaxy due to the strong blending of one of the quasar images with a bright galaxy (McGreer et al. 2010).

above). We emphasize that if this were true and the masses were lower than 10^9 $M_\odot$, then the SDSS quasars would have to be luminous for only a short time: maintaining the observed luminosities for $\gtrsim 10^7$ years with a radiative efficiency of $\epsilon \equiv L/\dot{m}c^2 = 0.1$ would bring the black hole masses up to values of 10^9 $M_\odot$ anyway.

4.2 *Local Black Holes as Fossils*

As mentioned above, SMBHs appear ubiquitous in local galaxies, with their masses correlating with the global properties of their host spheroids. Several groups have noted the broad natural implication that the formation of the SMBHs and their host spheroids must be tightly linked (see, e.g., Shankar 2009). Various independent lines of evidence suggest that spheroids are assembled at high redshifts ($z \sim 2$; see Cattaneo and Bernardi (2003) for the age determinations from the Sloan sample and references to older work), which would be consistent with most of the SMBH mass being accreted around this redshift (coinciding with the peak of the activity of luminous quasars). Indeed, starting from the age distribution of local early-type Galaxies, one can reconstruct the cosmic evolution of the quasar luminosity function to within observational errors, using the most naive set of assumptions (namely that the formation of stars and the assembly of the nuclear SMBHs track one another, with the SMBH radiating at a constant $f_{\rm Edd} \sim 0.3$, and that the $M_\bullet - \sigma$ correlation does not evolve with redshift Haiman et al. (2007); Shankar et al. (2009)).

This then has the unwelcome (but unsurprising) implication that the local SMBHs may contain little direct evidence of the formation of their seeds at $z > 6$. Indeed, it seems most plausible that the observed tight correlations, such as between $M_\bullet$ and σ, are established by a feedback process which operates when most of the black hole mass is assembled. However, an upside of this hypothesis is that—with the identification of a specific feedback mechanism—physically motivated extrapolations can be made toward high redshifts. Also, while relative massive local SMBHs ($\sim 10^9$ $M_\odot$) have undergone many mergers, those with the lowest masses ($\sim 10^6$ $M_\odot$) are more likely to have avoided mergers. Therefore, the low-mass end of the $M_\bullet - \sigma$ relation could be a probe of high-z SMBH formation models. To be more explicit: if only a small fraction of high-z halos are seeded with BHs, massive galaxies will have undergone many mergers, and will have a nuclear SMBH at present (i.e. $f_{\rm seed} \approx 1$). On the other hand, many low-mass galaxies may still have no BHs and thus the dwarf-galaxy population can have $f_{\rm seed} \ll 1$ (Menou et al. 2001). Likewise, the direct-collapse models produce SMBHs whose masses are initially well above the $M_\bullet - \sigma$ relation, which can be a diagnostic of such models (Tanaka and Haiman 2009). Such an 'upward curvature in the low-mass end of the relation could indeed be preserved all the way down to $z = 0$ (Volonteri and Natarajan 2009).

More generally, whether the local $M_\bullet - \sigma$ relation holds at higher redshifts, both in normalization and in slope (as discussed by several authors), and also in range (which has received less attention, but see Netzer (2003)), are interesting observational questions. The highest redshift SDSS quasars do appear to satisfy the $M_\bullet - \sigma$ relation of the local SMBHs, at least approximately. If $M_\bullet$ is estimated assuming the Eddington luminosity, and σ from the CO line-width, then, at a given σ, high-z quasars have BH masses a factor of $\sim$4 larger than local galaxies, although this could still be partly caused by selection (with high-z quasars preferentially viewed face-on) Ho (2007) and by the observations probing a smaller (inner) fraction of the DM halo. We also note that if σ is estimated from the circular velocity of the host dark matter halos with the right space density (e.g., Haiman and Loeb 2001), then the SDSS quasars are within the scatter of the $M_\bullet - \sigma$ relations of Gebhardt et al. (2000) and also of Ferrarese (2002). As explained in Sect. 4.1, the halo mass inference is reasonable. The determination of the halo mass and circular velocity from the observed abundance of quasars is also more robust than it may at first appear. This is because, despite the dependence on the poorly known duty cycle, the halo mass function is exponentially steep for the massive $M \sim 10^{13}$ $M_\odot$ halos at $z \sim 6$; therefore, the dependence of the inferred halo mass on the duty cycle (and other uncertainties in the estimated halo abundance) is only logarithmic. The weakest link in the argument is associating the spheroid velocity dispersion with the circular velocity of the dark matter halo. There is evidence Ferrarese (2002) of a correlation between $M_\bullet$ and σ, with the velocity dispersion measured in the dark matter dominated region of SMBH host galaxies; this establishes a direct link to the dark halo and puts the above argument on somewhat firmer ground (although there are still large errors in the inferred correlation, depending on the halo profile one adopts to convert the measured circular velocity to total halo mass).

The (tentative) evidence that high-redshift AGN do not strongly deviate from the local $M_\bullet - \sigma$ relation further supports the idea that the formation of SMBHs and their host galaxies must be tightly coupled by cosmology-independent physical processes (since the SDSS quasars are the rare peaks that have already formed at $z \sim 6$ instead of at $z \sim 2$). Besides the slope and normalization of the $M_\bullet - \sigma$ relation, the *range* (of masses and velocity dispersions) over which observed galaxies satisfy this relation has to match up between low and high redshifts. In particular, the largest black holes observed at high redshifts have inferred masses approaching $M_\bullet \sim 10^{10}$ $M_\odot$. These should also exist at low redshifts, but have not yet been discovered. In the SDSS, the galaxy with the velocity dispersion record has $\sigma = 444$ km s^{-1}Salviander et al. (2008)), whereas a naive application of the local $M_\bullet - \sigma$ relation would predict the presence of $\sigma > 700$ km s^{-1} galaxies (Netzer 2003). This puzzle is alleviated somewhat by the scatter in the relation; it is likely fully resolved by the realization that the $M_\bullet - \sigma$ relation has 'curvature', with BHs in the largest galaxies preferentially more massive than the power-law $M_\bullet - \sigma$ relation would predict (Lauer et al. 2007).

There have been several suggestions in the literature for the nature of the dynamical coupling between the formation of the black hole and its spheroid host. The most promising is radiative or mechanical feedback from the SMBH on the gas supply in the bulge. The essential idea (going back to Silk and Rees (1998)) is that when the black hole in the center of the galaxy grows too large, its outflows and radiation unbind the gas in the bulge or in the disk, quenching further black hole growth via accretion and further star formation. Competition with star formation for the gas supply may also play a role (Di Matteo et al. (2003); Li et al. (2007)). Note that these mechanisms can readily work at any redshift.

There are several alternative possibilities for the origin of the $M_\bullet - \sigma$ relation, which include: (1) filling the dark matter loss cone (Ostriker 2000). In this model, the growth of the SMBH occurs first through the accretion of collisional dark matter particles, and subsequently through the scattering of these particles into orbits that are then perturbed to pass sufficiently close to the black hole's Schwarzschild radius to be captured. This model runs into difficulties with the so-called Sołtan argument; since the SMBHs are fed mostly dark matter rather than gas, there is no associated radiation. (2) Direct capture of stars on high eccentricity orbits by the SMBH (Zhao et al. 2002; Merritt and Poon 2004). This model has a similar problem because black holes more massive than $\gtrsim 10^8$ $M_\odot$ do not tidally disrupt stars, so there is again no radiative output associated with the black hole growth. (3) Stellar captures by the accretion disk feeding the hole (Miralda-Escudé and Kollmeier 2005).

Solving the puzzle of the origin of the $M_\bullet - \sigma$ relation will have important implications for high-z SMBHs: in particular, it will generally determine how the relation evolves with redshift.

4.3 *The Future*

In this section, we briefly summarize the possibility of probing the continuum and line emission from AGN beyond the current redshift horizon of $z \sim 6$. This discussion is necessarily based on models for how the BH population evolves at $z > 6$. Such models can be constructed by assuming that SMBHs populate dark matter halos, e.g., in accordance with the locally measured $M_\bullet - \sigma$ relation (or an extrapolation of the relation to higher redshifts). The relation appears to hold, at least to within a factor of a few, for $z \sim 3$ quasars (this is based on using the Hβ/OIII lines as proxies for black hole mass and σ, respectively; e.g., Shields et al. 2003), and also at $z \sim 6$ (see Sect. 4.1). No doubt the observational constraints will improve as both black hole masses and velocity dispersions are measured in larger samples of distant quasars. Correspondingly, extrapolations to high redshifts will be more reliable as the feedback processes that regulate black hole growth are better understood. Here we summarize predictions from the simplest models.

4.3.1 Broadband Detections

Predictions for the number counts of high redshift AGN have been made using simple semi-analytic models for the near-infrared Haiman and Loeb (1998a) and in the soft X-rays (Haiman and Loeb 1999). In these early models, the quasar black hole was assumed to have a fixed fraction $\sim 10^{-4}$ of the halo mass, shine at the Eddington luminosity, and have a duty cycle of bright activity of $t_q \sim 10^6$ years.

In such models, the surface density of sources is very high in the optical/near-infrared bands, even at $z \sim 10$. For example, in the 1–5μm band, the ~1nJy sensitivity of the *James Webb Space Telescope (JWST)* will allow the detection of an $\sim 10^5\,\mathrm{M}_\odot$ black hole at $z = 10$ (provided that the black hole shines at the Eddington limit with a standard template spectrum Elvis et al. (1994)). Surface densities as high as several sources per square arcminute are predicted at this threshold from $z \gtrsim 5$, with most of these sources at $z \gtrsim 10$ (Haiman and Loeb 1998b). We note, however, that these predictions are very sensitive to the assumed duty cycle of bright activity. For example, for $t_q \sim 10^7$ years, or $M_\bullet \propto M_{\rm halo}^{5/3}$, the $z \sim 10$ counts can be smaller by a factor of 10–100 (depending on what redshift-dependence is assumed for the above scaling relation between black hole and halo mass at high redshift; see Haiman and Loeb (1998a); Haehnelt et al. (1998); Wyithe and Loeb (2003b) for related discussion). It would also be interesting to detect the host galaxies of ultrahigh redshift AGN, which should be feasible with *JWST*'s sensitivity. If the galaxies occupy a fair fraction (~5 %) of the virial radius of their host halos, then a large fraction ($\gtrsim$50 %) of them can potentially be resolved with *JWST*'s planned angular resolution of ~0.06″ (Haiman and Loeb 1998b; Barkana and Loeb 2000). The Large Synoptic Survey Telescope (LSST[7]), with a planned capability of going ~5 magnitudes deeper than SDSS in a ~3 times larger solid angle, will be an ideal instrument for studying high-redshift quasars in the optical/near-infrared.

In the soft X-rays, the 0.5–2 keV flux of 2.5×10^{-17} ergs cm^{-2} s^{-1} reached in a 2 Ms exposure of CDF-North Alexander et al. (2003) corresponds to a larger ($\sim 2 \times 10^7\,\mathrm{M}_\odot$; see Fig. 1 in Haiman and Loeb (1999)) black hole at $z = 10$, but nevertheless, thousands of sources are predicted at $z \gtrsim 5$ per square degree, and tens per square degree at $z > 10$. This would imply that tens of $z > 5$ sources should have been detectable already in the CDFs, whereas only a handful of potential candidates, and no confirmed sources, have been found. In revised models with longer quasar lifetimes and thus a steeper scaling of $M_\bullet$ with $M_{\rm halo}$, these numbers can be sharply decreased Haiman and Loeb (1998b); Haehnelt et al. (1998), which can bring the expected counts into agreement with current non-detections (Wyithe and Loeb 2003b).

The radio sensitivity of the extended Very Large Array and other forthcoming instruments (e.g., Allen Telescope Array and Square Kilometer Array) is also promising for detecting AGN beyond $z \sim 6$. Using the updated scaling of black hole mass with halo mass and redshift Wyithe and Loeb (2003b) and assuming the same radio-loud fraction (~10 %) as at lower redshifts, a simple model predicts that ~ten 10μJy sources per square degree should be detectable at 1–10 GHz (Haiman et al. 2004). The identification of these quasars is a challenge, but should, in principle, be feasible with deep optical/IR observations. To date, only two such radio-selected quasars have been identified (in optical follow-ups of sources in the FIRST and VLA radio catalogs; McGreer et al. (2006); Zeimann et al. (2011)), which falls below the expectations from the simple model by a factor of several.

[7] www.lsst.org

In addition to direct detection of AGN at very high redshifts, it may also be possible to detect lower mass seed black holes at comparable redshifts (or higher). In particular, a plausible model for gamma-ray bursts (GRBs) invokes accretion onto a newly formed $\sim$10 $M_{\odot}$ black hole (the collapsar model; e.g., Woosley 1993). *Swift* has now detected four bursts beyond $z > 6$: GRB090429B at $z = 9.4$ Cucchiara et al. (2011), GRB090423 at $z = 8.2$ Tanvir et al. (2009); Salvaterra et al. (2009), GRB080913 at $z = 6.7$ Greiner et al. (2009), and GRB050904 at $z = 6.3$ Kawai et al. (2006); Haislip et al. (2006), for which the afterglow emission has also been measurable; such afterglows should remain detectable in the infrared out to $z \sim 20$ (Lamb and Reichart 2000; Ciardi and Loeb 2000). Their detection and the characterization of their spectrum and light-curve would open up a new probe of black hole formation and evolution at high redshifts.

In summary, model predictions for the continuum emission of $z > 6$ AGN are very sensitive to how one extrapolates the $M_{\bullet}$–$M_{\rm halo}$ relation to $z \gtrsim 6$. However, this should be viewed as "good news": (1) large numbers of detectable AGN at these redshifts are certainly possible, and (2) their detection will put strong constraints on models for the origin and evolution of the black hole population.

4.3.2 Emission Line Measurements

The strongest recombination lines of H and He from $5 < z < 20$ AGN will fall in the near-infrared bands of *JWST* and could be bright enough to be detectable. Specific predictions have been made for the source counts in the Hα emission line Oh (2001) and for the three strongest HeII lines (Oh et al. 2001; Tumlinson et al. 2001). The key assumption is that most of the ionizing radiation produced by the miniquasars is processed into such recombination lines (rather than escaping into the IGM). Under this assumption, the lines are detectable for a fiducial 10^5 $M_{\odot}$ miniquasar at $z = 10$. The Lyα line is more susceptible to absorption by neutral hydrogen in the IGM near the source but should be detectable for bright sources that are surrounded by a large enough HII region so that Lyα photons shift out of resonance before hitting the neutral IGM (Cen and Haiman 2000). If the Lyman α emission is scattered off expanding shells of material (as expected from galactic winds), this will further shift the photons away from resonance, and make the emission line more detectable (Dijkstra and Wyithe 2010). We also note that in the "trapped" Lyman α model for direct collapse mentioned in Sect. 3.2, the Lyman α emission ultimately emerging from the collapsing halo (before forming the SMBH) appears detectable with *JWST*, as well—as a low-surface brightness diffuse blob (Latif et al. 2011a).

The simultaneous detection of H and He lines would be especially significant. As already argued above, the hardness of the ionizing continuum from the first sources of ultraviolet radiation plays a crucial role in the reionization of the IGM. It would therefore be very interesting to directly measure the ionizing continuum of any $z > 6$ source. While this may be feasible at X-ray energies for exceptionally bright sources, the absorption by neutral gas within the source and in the intervening IGM will render the ionizing continuum of high redshift sources inaccessible to direct observation out to $1\,\mu$m. This is a problem if the ionizing sources are black holes with $M < 10^8\,M_\odot$ at $z \sim 10$ (easily detectable at wavelengths red-ward of redshifted Lyα in the near-infrared by *JWST*, but too faint to see in X-rays). The comparison of Hα and HeII line strengths can be used to infer the ratio of HeII to HI ionizing photons, $Q = \dot{N}_{\rm ion}^{\rm HeII}/\dot{N}_{\rm ion}^{\rm HI}$. A measurement of this ratio would shed light on the nature of the first luminous sources, and, in particular, it could reveal if the source has a soft (stellar) or hard (AGN-like) spectrum. Note that this technique has already been successfully applied to constrain the spectra of sources in several nearby extragalactic HII regions (Garnett et al. 1991). Lyman break galaxies at $z \approx 3$ also appear to have unusually strong (for a normal stellar population) He1640 emission line; however, the lack of X-rays rule out an AGN explanation (the observation could be explained instead by the presence of PopIII stars in these galaxies Jimenez and Haiman (2006) or by an unusual abundant population of Wolf-Rayet stars Brinchmann et al. (2008) that can produce the He1640 line while avoiding an overproduction of metal lines).

Provided the gas in the high redshift AGN is enriched to near-solar levels, several molecular lines may be visible. In fact, CO has already been detected in the hosts of the most distant quasars (Walter et al. 2003; Wang et al. 2010, 2011). The detectability of CO for high redshift sources in general has been considered in simple theoretical models (Silk and Spaans 1997). If AGN activity is accompanied by a star formation rate of $\gtrsim 30\,M_\odot/$ year, the CO lines are detectable at all redshifts $z = 5$–30 by the Sub-Millimeter Array (the redshift independent sensitivity is due to the increasing CMB temperature with redshift), while the Atacama Large Millimeter Array (ALMA) could reveal even fainter CO emission, and other C and O lines in emission, providing spatially resolved images (Schleicher et al. 2010). The detection of these molecular lines will provide valuable information on the stellar content and gas kinematics near the AGN.

4.3.3 Gravitational Waves

The most direct observational constraints on the SMBH assembly at $z > 6$, with especially clear distinctions between the "stellar-mass seed" and "direct-collapse" scenarios, may come from detecting the gravitational waves produced during the SMBH mergers. The *Laser Interferometer Space Antenna* (*LISA*) is expected to be able to detect mergers of SMBHs in the mass range $\sim(10^4$–$10^7)\,M_\odot/(1+z)$ with high S/N out to $z \sim 30$ (Baker et al. 2007). Binary spins and BH masses is expected to be measured with high precision up to $z \sim 10$ Vecchio (2004), especially if spin precession Lang and Hughes (2006) and higher-order harmonics of the waveform McWilliams et al. (2010) are included in the analysis.[8] Many authors have computed

[8]As this chapter was being written, NASA announced a decision to withdraw from the LISA experiment. The European Space Agency is continuing to consider a redesigned version of LISA, with a smaller budget, and a launch date of approximately 2021–2022. Given the very high S/N

the expected *LISA* event rate from high-redshift SMBH merger population models in a range of plausible models. The published estimates (Menou et al. (2001); Wyithe and Loeb (2003a); Sesana et al. (2004, 2005); Islam et al. (2004); Sesana et al. (2007); Lippai et al. (2008); see a review in Arun et al. (2009)), even at lower redshifts, vary by orders of magnitude, from $\sim$1 to as high as $\sim 10^4$ year^{-1}; there is a large range even among models that are explicitly calibrated to fit the evolution of the quasar luminosity function (Lippai et al. 2008).

Closest to the present context of the growth of SMBHs at $z > 6$ are the Monte-Carlo merger tree models in Tanaka and Haiman (2009) (discussed in Sect. 3.1.2 above). These models coupled the merger trees with the orbits of oscillating kicked BHs, to predict detection rates for *LISA*. They have surveyed a wide range of candidate assembly models, including those with rare, massive seeds, or through ultra-early production of numerous Pop-III remnant seeds. As mentioned above, in the latter model, seed BHs need to stop forming below a redshift $z_{\rm cut} \sim 20$, in order to avoid overproducing 10^6 $M_\odot$ BHs.

The simplest SMBH assembly scenarios, which have constant accretion rates, but in which BH seed formation stops abruptly at some redshift, and which meet constraints at both the high-mass and low-mass end of the $z = 6$ SMBH mass function, predict negligibly low *LISA* event rates. The reason for this pessimistic conclusion is as follows: in these models, the BHs that grow into the most massive, highest-redshift quasar-SMBHs accrete at the same (exponential) rate as all the other BHs, typically resulting in a vast overproduction of massive ($m \sim 10^6 M_\odot$) holes. In order to offset this overproduction, seeds must be made very rare, and this diminishes the *LISA* rates. It is difficult to envision a scenario for high ($\gtrsim$10 per year per unit redshift) detection rates unless a vast number of SMBHs in the $10^{5-7} M_\odot$ range lurk in the universe at all redshifts, which the current electromagnetic surveys have missed.

A different class of models, which successfully build the $z \sim 6$ quasar BHs, are those in which the SMBH masses are self-regulated by internal feedback, to always maintain the $M_\bullet - \sigma$ relation. These models can evade this constraint, and produce *LISA* rates as high as 30 year^{-1}. The key difference in these models with higher *LISA* rates is that the SMBH growth is driven by a large number of seed BHs and far lower gas accretion rates than those required in the constant-accretion models. The majority of the *LISA* events occur at $z \approx 6$ and in the low end (10^3–10^4 $M_\odot$) of *LISA*'s mass range for detection.

Also, for these models, the ejected BH mass density can exceed that of the galactic BH population at $z = 6$. Most ejected holes are expected to have low masses (still similar to the original seed mass), but an ejected BH can be as massive as $\sim 10^8 M_\odot$ if large recoil velocities are allowed (e.g. if spins are not always aligned with the orbital angular momentum of the binary).

ratios forecast for the original version of LISA, the redesigned "LISA-lite" mission should still be able to detect low-mass SMBHs out to high redshifts.

Using similar "merger tree" massive black hole formation models, Sesana et al. (2007) analyzed the predicted mass—and redshift—distribution of LISA events. These models have input assumptions similar to the "$M_{\bullet} - \sigma$" models in Tanaka and Haiman (2009), but with varying initial seed masses. These predict a handful of detectable events at $z > 10$. The raw total event rates in the two models are very similar. However, the mass-distribution of events is different (low-mass mergers are missing in the 'heavy seed' model). Another key diagnostic between the 'heavy' and 'stellar-mass' seed models is the mass ratio of the BHs in these detectable events: while the former models predict near-equal mass mergers, in the latter case, one of the merger partners typically have time to grow, resulting in typical mass ratios of $q = 0.1$–0.2.

It is worthwhile to note that essentially all of the work on the gravity wave signal from black hole-black hole in-spiral has assumed efficient (nearly instantaneous) mergers. Stellar-scattering and gas can help drive BHs together on large scales, which can affect detections of the most massive nearby SMBHs by Pulsar Timing Arrays Kocsis and Sesana (2011) and extreme mass ratio inspiral events by *LISA* Yunes et al. (2011). However, SMBH-SMBH coalescences, when they are in *LISA*'s frequency window, are well within the rapidly merging purely gravitational wave-driven regime, at least if the circumbinary gas forms a thin disk (Haiman et al. 2009).

5 Conclusions

In this review, we have summarized theoretical ideas and observational constraints on how massive black holes form at the centers of the earliest protogalaxies, and how such black holes grow via accretion and mergers to give rise to the observed population of black holes at $z \gtrsim 6$ and in the local and moderate redshift universe. As this review shows, this remains a poorly understood but rich and important problem. Perhaps the most direct way of probing the role of mergers in black hole assembly and evolution at $z > 10$ is via their gravity wave signatures, which will hopefully be feasible with the redesigned version of LISA being considered by ESA.

In addition to being of intrinsic interest for understanding the AGN phenomena, sources of gravity waves, etc., there is strong evidence that the formation and evolution of black holes is coupled to the formation and evolution of the host galaxy in which the black hole resides (e.g., the $M_{\bullet} - \sigma$ relation), and thus to the cosmological formation of nonlinear dark matter structures (i.e., the dark halos surrounding these galaxies). We anticipate that this will remain a growth area of research in the coming years, with continued rapid progress on both the observational and theoretical fronts.

Acknowledgements The author wishes to thank the editors of this volume for their patience during the preparation of this article. I would also like to thank my recent collaborators, especially Taka Tanaka, Cien Shang, Mark Dijkstra and Greg Bryan, whose work was especially emphasized here, and Eliot Quataert for permission to draw on material in our earlier joint review. The work described here was supported in part by the NSF, NASA, and by the Polányi Program of the Hungarian National Office for Research and Technology (NKTH).

References

Abel, T., Anninos, P., Zhang, Y., Norman, M.L.: Modeling primordial gas in numerical cosmology. New Astron. **2**, 181–207 (1997)

Abel, T., Bryan, G.L., Norman, M.L.: The formation and fragmentation of primordial molecular clouds. ApJ **540**, 39–44 (2000)

Abel, T., Bryan, G.L., Norman, M.L.: The formation of the first star in the Universe. Science **295**, 93–98 (2002)

Abel, T., Haiman, Z.: The role of H_2 molecules in cosmological structure formation In: Combes, F., Pineau des Forets, G. (eds.) Molecular Hydrogen in Space p. 237. Cambridge Contemporary Astrophysics. Cambridge University Press, Cambridge (2001) (ISBN 0521782244)

Alexander, D.M., Bauer, F.E., Brandt, W.N., Schneider, D.P., Hornschemeier, A. E., Vignali, C., Barger, A.J., Broos, P.S., Cowie, L.L., Garmire, G.P., Townsley, L.K., Bautz, M.W., Chartas, G., Sargent, W.L.W.: The Chandra deep field north survey. XIII. 2 Ms point-source catalogs. AJ **126**, 539–574 (2003)

Alvarez, M.A., Wise, J.H., Abel, T.: Accretion onto the first stellar-mass black holes. ApJL **701**, L133–L137 (2009)

Amaro-Seoane, P., Sesana, A., Hoffman, L., Benacquista, M., Eichhorn, C., Makino, J., Spurzem, R.: Triplets of supermassive black holes: astrophysics, gravitational waves and detection. MNRAS **402**, 2308–2320 (2010)

Armitage, P.J., Natarajan, P.: Accretion during the merger of supermassive black holes. ApJL **567**, L9–L12 (2002)

Arons, J.: Photon bubbles - Overstability in a magnetized atmosphere. ApJ **388**, 561–578 (1992)

Arun, K.G., Babak, S., Berti, E., Cornish, N., Cutler, C., Gair, J., Hughes, S.A., Iyer, B.R., Lang, R.N., Mandel, I., Porter, E.K., Sathyaprakash, B.S., Sinha, S., Sintes, A.M., Trias, M., Van Den Broeck, C., Volonteri, M.: Massive black-hole binary inspirals: results from the LISA parameter estimation taskforce. Classical Quant. Grav. **26**(9), 094027 (2009)

Baker, J.G., Boggs, W.D., Centrella, J., Kelly, B.J., McWilliams, S.T., Miller, M.C., van Meter, J.R.: Modeling kicks from the merger of generic black hole binaries. ApJL **682**, L29–L32 (2008)

Baker, J.G., McWilliams, S.T., van Meter, J.R., Centrella, J., Choi, D.-I., Kelly, B.J., and Koppitz, M.: Binary black hole late inspiral: simulations for gravitational wave observations. Phys. Rev. D **75**(12), 124024 (2007)

Barkana, R., Haiman, Z., Ostriker, J.P.: Constraints on warm dark matter from cosmological reionization. ApJ **558**, 482–496 (2001)

Barkana, R., Loeb, A.: High-redshift galaxies: their predicted size and surface brightness distributions and their gravitational lensing probability. ApJ **531**, 613–623 (2000)

Baumgardt, H., Makino, J., Hut, P., McMillan, S., Portegies Zwart, S.: A dynamical model for the globular cluster G1. ApJL **589**, L25–L28 (2003)

Begelman, M.C.: Black holes in radiation-dominated gas - an analogue of the Bondi accretion problem. MNRAS **184**, 53–67 (1978)

Begelman, M.C.: Super-eddington atmospheres that do not blow away. ApJ **551**, 897–906 (2001)

Begelman, M.C.: Super-eddington fluxes from thin accretion disks? ApJL **568**, L97–L100 (2002)

Begelman, M.C.: Evolution of supermassive stars as a pathway to black hole formation. MNRAS **402**, 673–681 (2010)

Begelman, M.C., Blandford, R.D., Rees, M.J.: Massive black hole binaries in active galactic nuclei. Nature **287**, 307–309 (1980)

Begelman, M.C., Meier, D.L.: Thick accretion disks - Self-similar, supercritical models. ApJ **253**, 873–896 (1982)
Begelman, M.C., Rees, M.J.: The fate of dense stellar systems. MNRAS **185**, 847–860 (1978)
Begelman, M.C., Rossi, E.M., Armitage, P.J.: Quasi-stars: accreting black holes inside massive envelopes. MNRAS **387**, 1649–1659 (2008)
Begelman, M.C., Shlosman, I.: Angular momentum transfer and lack of fragmentation in self-gravitating accretion flows. ApJL **702**, L5–L8 (2009)
Begelman, M.C., Volonteri, M., Rees, M.J.: Formation of supermassive black holes by direct collapse in pre-galactic haloes. MNRAS **370**, 289–298 (2006)
Binney, J.: The physics of dissipational galaxy formation. ApJ **215**, 483–491 (1977)
Binney, J., Tremaine, S.: Galactic Dynamics. Princeton University Press, Princeton (1987)
Blaes, O., Socrates, A.: Local dynamical instabilities in magnetized, radiation pressure-supported accretion disks. ApJ **553**, 987–998 (2001)
Blandford, R.D., Begelman, M.C.: On the fate of gas accreting at a low rate on to a black hole. MNRAS **303**, L1–L5 (1999)
Blandford, R.D., Begelman, M.C.: Two-dimensional adiabatic flows on to a black hole - I. Fluid accretion. MNRAS **349**, 68–86 (2004)
Blecha, L., Loeb, A.: Effects of gravitational-wave recoil on the dynamics and growth of supermassive black holes. MNRAS **390**, 1311–1325 (2008)
Brinchmann, J., Pettini, M., Charlot, S.: New insights into the stellar content and physical conditions of star-forming galaxies at z = 2–3 from spectral modelling. MNRAS **385**, 769–782 (2008)
Bromley, J.M., Somerville, R.S., Fabian, A.C.: High-redshift quasars and the supermassive black hole mass budget: constraints on quasar formation models. MNRAS **350**, 456–472 (2004)
Bromm, V., Coppi, P.S., Larson, R.B.: Forming the first stars in the Universe: The fragmentation of primordial gas. ApJL **527**, L5–L8 (1999)
Bromm, V., Coppi, P.S., Larson, R.B.: The formation of the first stars. I. The primordial star-forming cloud. ApJ **564**, 23–51 (2002)
Bromm, V., Ferrara, A., Heger, A.: First Stars: Formation, Evolution and Feedback Effects, p. 180. Cambridge University Press, Cambridge (2009)
Bromm, V., Loeb, A.: Formation of the first supermassive black holes. ApJ **596**, 34–46 (2003)
Carr, B.J., Bond, J.R., Arnett, W.D.: Cosmological consequences of Population III stars. ApJ **277**, 445–469 (1984)
Cattaneo, A., Bernardi, M.: The quasar epoch and the stellar ages of early-type galaxies. MNRAS **344**, 45–52 (2003)
Cavaliere, A., Giacconi, R., Menci, N.: X-raying the star formation history of the Universe. ApJL **528**, L77–L80 (2000)
Cen, R., Haiman, Z.: Quasar strömgren spheres before cosmological reionization. ApJL **542**, L75–L78 (2000)
Ciardi, B., Loeb, A.: Expected number and flux distribution of gamma-ray burst afterglows with high redshifts. ApJ **540**, 687–696 (2000)
Ciotti, L., Ostriker, J.P.: Cooling flows and quasars. II. Detailed models of feedback-modulated accretion flows. ApJ **551**, 131–152 (2001)
Clark, P.C., Glover, S.C.O., Klessen, R.S.: The first stellar cluster. ApJ **672**, 757–764 (2008)
Colbert, E.J.M., Mushotzky, R.F.: The nature of accreting black holes in nearby galaxy nuclei. ApJ **519**, 89–107 (1999)
Comerford, J.M., Haiman, Z., Schaye, J.: Constraining the redshift z 6 quasar luminosity function using gravitational lensing. ApJ **580**, 63–72 (2002)
Cucchiara, A., Levan, A.J., Fox, D.B., Tanvir, N.R., Ukwatta, T.N., Berger, E., Krühler, T., Küpcü Yoldaş, A., Wu, X.F., Toma, K., Greiner, J., Olivares E., F., Rowlinson, A., Amati, L., Sakamoto, T., Roth, K., Stephens, A., Fritz, A., Fynbo, J.P.U., Hjorth, J., Malesani, D., Jakobsson, P., Wiersema, K., O'Brien, P.T. Soderberg, A.M., Foley, R.J., Fruchter, A.S., Rhoads, J., Rutledge, R.E., Schmidt, B.P., Dopita, M.A., Podsiadlowski, P., Willingale, R., Wolf, C., Kulkarni, S.R., D'Avanzo, P.: A photometric redshift of z 9.4 for GRB 090429B. ApJ **736**, 7 (2011)

Dalal, N., Pen, U.-L., Seljak, U.: Large-scale BAO signatures of the smallest galaxies. JCAP **11**, 7 (2010)
Devecchi, B., Volonteri, M.: Formation of the first nuclear clusters and massive black holes at high redshift. ApJ **694**, 302–313 (2009)
Di Matteo, T., Croft, R.A.C., Springel, V., Hernquist, L.: Black hole growth and activity in a Λ cold dark matter Universe. ApJ **593**, 56–68 (2003)
Dijkstra, M., Haiman, Z., Mesinger, A., Wyithe, J.S.B.: Fluctuations in the high-redshift Lyman-Werner background: close halo pairs as the origin of supermassive black holes. MNRAS **391**, 1961–1972 (2008)
Dijkstra, M., Haiman, Z., Rees, M.J., Weinberg, D.H.: Photoionization feedback in low-mass galaxies at high redshift. ApJ **601**, 666–675 (2004)
Dijkstra, M., Wyithe, J.S.B.: Seeing through the trough: outflows and the detectability of Lyα emission from the first galaxies. MNRAS **408**, 352–361 (2010)
Dotan, C., Shaviv, N.J.: Super-eddington slim accretion discs with winds. MNRAS **413**, 1623–1632 (2011)
Draine, B.T., Bertoldi, F.: Structure of stationary photodissociation fronts. ApJ **468**, 269 (1996)
Efstathiou, G.: Suppressing the formation of dwarf galaxies via photoionization. MNRAS **256**, 43P–47P (1992)
Efstathiou, G., Rees, M.J.: High-redshift quasars in the cold dark matter cosmogony. MNRAS **230**, 5P (1988)
Eisenstein, D.J., Loeb, A.: Origin of quasar progenitors from the collapse of low-spin cosmological perturbations. ApJ **443**, 11–17 (1995)
Elvis, M., Wilkes, B.J., McDowell, J.C., Green, R.F., Bechtold, J., Willner, S.P., Oey, M.S., Polomski, E., Cutri, R.: Atlas of quasar energy distributions. ApJS **95**, 1–68 (1994)
Escala, A., Larson, R.B., Coppi, P.S., Mardones, D.: The role of gas in the merging of massive black holes in galactic nuclei. II. Black hole merging in a nuclear gas disk. ApJ **630**, 152–166 (2005)
Fan, X.: Evolution of high-redshift quasars. New Astron. Rev. **50**, 665–671 (2006)
Fan, X., Strauss, M.A., Schneider, D.P., Becker, R.H., White, R.L., Haiman, Z., Gregg, M., Pentericci, L., Grebel, E.K., Narayanan, V.K., Loh, Y.-S., Richards, G.T., Gunn, J.E., Lupton, R.H., Knapp, G.R., Ivezić, Ž., Brandt, W.N., Collinge, M., Hao, L., Harbeck, D., Prada, F., Schaye, J., Strateva, I., Zakamska, N., Anderson, S., Brinkmann, J., Bahcall, N.A., Lamb, D.Q., Okamura, S., Szalay, A., York, D.G.: A survey of $z > 5.7$ quasars in the sloan digital sky survey. II. Discovery of three additional quasars at $z > 6$. AJ **125**, 1649–1659 (2003)
Farrell, S.A., Webb, N.A., Barret, D., Godet, O., Rodrigues, J.M.: An intermediate-mass black hole of over 500 solar masses in the galaxy ESO243-49. Nature **460**, 73–75 (2009)
Ferrarese, L.: Beyond the bulge: a fundamental relation between supermassive black holes and dark matter halos. ApJ **578**, 90–97 (2002)
Freese, K., Ilie, C., Spolyar, D., Valluri, M., Bodenheimer, P.: Supermassive dark stars: detectable in JWST. ApJ **716**, 1397–1407 (2010)
Fuller, G.M., Woosley, S.E., Weaver, T.A.: The evolution of radiation-dominated stars. I - Nonrotating supermassive stars. ApJ **307**, 675–686 (1986)
Galli, D., Palla, F.: The chemistry of the early Universe. A&A **335**, 403–420 (1998)
Gammie, C.F.: Photon bubbles in accretion discs. MNRAS **297**, 929–935 (1998)
Garnett, D.R., Kennicutt, R.C. Jr., Chu, Y.-H., Skillman, E.D.: He II emission in extragalactic H II regions. ApJ **373**, 458–464 (1991)
Gebhardt, K., Bender, R., Bower, G., Dressler, A., Faber, S.M., Filippenko, A.V., Green, R., Grillmair, C., Ho, L.C., Kormendy, J., Lauer, T.R., Magorrian, J., Pinkney, J., Richstone, D., Tremaine, S.: A relationship between nuclear black hole mass and galaxy velocity dispersion. ApJL **539**, L13–L16 (2000)
Gebhardt, K., Rich, R.M., Ho, L.C.: A 20,000 $M_{\odot}$ black hole in the stellar cluster G1. ApJL **578**, L41–L45 (2002)

Gebhardt, K., Rich, R.M., Ho, L.C.: An intermediate-mass black hole in the globular cluster G1: Improved significance from New Keck and Hubble Space Telescope observations. ApJ **634**, 1093–1102 (2005)

Gnedin, N.Y., Ostriker, J.P.: Reionization of the Universe and the early production of metals. ApJ **486**, 581 (1997)

Goodman, J.: Self-gravity and quasi-stellar object discs. MNRAS **339**, 937–948 (2003)

Goswami, S., Umbreit, S., Bierbaum, M., Rasio, F.A.: Formation of massive black holes in dense star clusters. II. IMF and primordial mass segregation. ApJ **752**, 43 (2012)

Gould, A., Rix, H.-W.: Binary black hole mergers from planet-like migrations. ApJL **532**, L29–L32 (2000)

Greif, T., Springel, V., White, S., Glover, S., Clark, P., Smith, R., Klessen, R., Bromm, V.: Simulations on a moving mesh: the clustered formation of population III protostars. ApJ **737**, 75 (2011a)

Greif, T.H., White, S.D.M., Klessen, R.S., Springel, V: The delay of population III star formation by supersonic streaming velocities. ApJ **736**, 147 (2011b)

Greiner, J., Krühler, T., Fynbo, J.P.U., Rossi, A., Schwarz, R., Klose, S., Savaglio, S., Tanvir, N.R., McBreen, S., Totani, T., Zhang, B.B., Wu, X.F., Watson, D., Barthelmy, S.D., Beardmore, A.P., Ferrero, P., Gehrels, N., Kann, D.A., Kawai, N., Yoldaş, A.K., Mészáros, P., Milvang-Jensen, B., Oates, S.R., Pierini, D., Schady, P., Toma, K., Vreeswijk, P.M., Yoldaş, A., Zhang, B., Afonso, P., Aoki, K., Burrows, D.N., Clemens, C., Filgas, R., Haiman, Z., Hartmann, D.H., Hasinger, G., Hjorth, J., Jehin, E., Levan, A.J., Liang, E.W., Malesani, D., Pyo, T.-S., Schulze, S., Szokoly, G., Terada, K., Wiersema, K.: GRB 080913 at Redshift 6.7. ApJ **693**, 1610–1620 (2009)

Gualandris, A., Merritt, D.: Ejection of supermassive black holes from galaxy cores. ApJ **678**, 780–797 (2008)

Guedes, J., Madau, P., Kuhlen, M., Diemand, J., Zemp, M.: Simulations of recoiling massive black holes in the via lactea halo. ApJ **702**, 890–900 (2009)

Guedes, J., Madau, P., Mayer, L., Callegari, S.: Recoiling massive black holes in gas-rich galaxy mergers. ApJ **729**, 125 (2011)

Gürkan, M.A., Freitag, M., Rasio, F.A.: Formation of massive black holes in dense star clusters. I. Mass segregation and core collapse. ApJ **604**, 632–652 (2004)

Haehnelt, M.G., Natarajan, P., Rees, M.J.: High-redshift galaxies, their active nuclei and central black holes. MNRAS **300**, 817–827 (1998)

Haehnelt, M.G., Rees, M.J.: The formation of nuclei in newly formed galaxies and the evolution of the quasar population. MNRAS **263**, 168–178 (1993)

Haiman, Z.: Constraints from gravitational recoil on the growth of supermassive black holes at high redshift. ApJ **613**, 36–40 (2004)

Haiman, Z.: Galaxy formation: caught in the act? Nature **430**, 979–980 (2004)

Haiman, Z., Abel, T., Rees, M.J.: The radiative feedback of the first cosmological objects. ApJ **534**, 11–24 (2000)

Haiman, Z., Bryan, G.L.: Was star formation suppressed in high-redshift minihalos? ApJ **650**, 7–11 (2006)

Haiman, Z., Cen, R.: A constraint on the gravitational lensing magnification and age of the redshift z=6.28 quasar SDSS 1030+0524. ApJ **578**, 702–707 (2002)

Haiman, Z., Ciotti, L., Ostriker, J.P.: Reasoning from fossils: learning from the local black hole population about the evolution of quasars. ApJ **606**, 763–773 (2004)

Haiman, Z., Holder, G.P.: The reionization history at high redshifts. I. Physical models and new constraints from cosmic microwave background polarization. ApJ **595**, 1–12 (2003)

Haiman, Z., Hui, L.: Constraining the lifetime of quasars from their spatial clustering. ApJ **547**, 27–38 (2001)

Haiman, Z., Jimenez, R., Bernardi, M.: Reconstructing the cosmic evolution of quasars from the age distribution of local early-type galaxies. ApJ **658**, 721–730 (2007)

Haiman, Z., Kocsis, B., Menou, K.: The population of viscosity- and gravitational wave-driven supermassive black hole binaries among luminous active galactic nuclei. ApJ **700**, 1952–1969 (2009)

Haiman, Z., Loeb, A.: Detection of the first star clusters with NGST. In: Smith, E.P., Koratkar, A. (eds.) Science with the NGST, vol. 133 of Astronomical Society of the Pacific Conference Series, p. 251 (1998)

Haiman, Z., Loeb, A.: Observational signatures of the first quasars. ApJ **503**, 505 (1998)

Haiman, Z., Loeb, A.: Determining the redshift of reionization from the spectra of high-redshift sources. ApJ **519**, 479–485 (1999)

Haiman, Z., Loeb, A.: What is the highest plausible redshift of luminous quasars? ApJ **552**, 459–463 (2001)

Haiman, Z., Quataert, E.: The formation and evolution of the first massive black holes. In: Barger, A.J. (ed.), Supermassive Black Holes in the Distant Universe, vol. 308 of Astrophysics and Space Science Library, p. 147 (2004)

Haiman, Z., Quataert, E., Bower, G.C.: Modeling the counts of faint radio-loud quasars: constraints on the supermassive black hole population and predictions for high redshift. ApJ **612**, 698–705 (2004)

Haiman, Z., Rees, M.J., Loeb, A.: Destruction of molecular hydrogen during cosmological reionization. ApJ **476**, 458 (1997)

Haiman, Z., Thoul, A.A., Loeb, A.: Cosmological formation of low-mass objects. ApJ **464**, 523 (1996)

Haislip, J.B., Nysewander, M.C., Reichart, D.E., Levan, A., Tanvir, N., Cenko, S.B., Fox, D.B., Price, P.A., Castro-Tirado, A.J. Gorosabel, J., Evans, C.R., Figueredo, E., MacLeod, C.L., Kirschbrown, J.R., Jelinek, M., Guziy, S., Postigo, A.D.U., Cypriano, E.S., Lacluyze, A., Graham, J., Priddey, R., Chapman, R., Rhoads, J., Fruchter, A.S., Lamb, D.Q., Kouveliotou, C., Wijers, R.A.M.J., Bayliss, M.B., Schmidt, B.P., Soderberg, A.M., Kulkarni, S.R., Harrison, F.A., Moon, D.S., Gal-Yam, A., Kasliwal, M.M., Hudec, R., Vitek, S., Kubanek, P., Crain, J.A., Foster, A.C., Clemens, J.C., Bartelme, J.W., Canterna, R., Hartmann, D.H., Henden, A.A., Klose, S., Park, H.-S., Williams, G.G., Rol, E., O'Brien, P., Bersier, D., Prada, F., Pizarro, S., Maturana, D., Ugarte, P., Alvarez, A., Fernandez, A.J.M., Jarvis, M.J., Moles, M., Alfaro, E., Ivarsen, K.M., Kumar, N.D., Mack, C.E., Zdarowicz, C.M., Gehrels, N., Barthelmy, S., Burrows, D.N.: A photometric redshift of z=6.39 $\pm$ 0.12 for GRB 050904. Nature **440**, 181–183 (2006)

Hawley, J.F., Balbus, S.A.: The dynamical structure of nonradiative black hole accretion flows. ApJ **573**, 738–748 (2002)

Heger, A., Fryer, C.L., Woosley, S.E., Langer, N., Hartmann, D.H.: How massive single stars end their life. ApJ **591**, 288–300 (2003)

Heggie, D.C.: Binary evolution in stellar dynamics. MNRAS **173**, 729–787 (1975)

Hirasawa, T.: Formation of protogalaxies and molecular processes in hydrogen gas. Progr. Theor. Phys. **42**, 523–543 (1969)

Ho, L.C.: The CO tully-fisher relation and implications for the host galaxies of high-redshift quasars. ApJ **669**, 821–829 (2007)

Hoffman, L., Loeb, A.: Dynamics of triple black hole systems in hierarchically merging massive galaxies. MNRAS **377**, 957–976 (2007)

Hoyle, F., Fowler, W.A.: On the nature of strong radio sources. MNRAS **125**, 169 (1963)

Hughes, S.A., Blandford, R.D.: Black hole mass and spin coevolution by mergers. ApJL **585**, L101–L104 (2003)

Hut, P., McMillan, S., Goodman, J., Mateo, M., Phinney, E.S., Pryor, C., Richer, H.B., Verbunt, F., Weinberg, M.: Binaries in globular clusters. PASP **104**, 981–1034 (1992)

Igumenshchev, I.V., Narayan, R., Abramowicz, M.A.: Three-dimensional magnetohydrodynamic simulations of radiatively inefficient accretion flows. ApJ **592**, 1042–1059 (2003)

Iocco, F., Bressan, A., Ripamonti, E., Schneider, R., Ferrara, A., Marigo, P.: Dark matter annihilation effects on the first stars. MNRAS **390**, 1655–1669 (2008)

Islam, R.R., Taylor, J.E., Silk, J.: Massive black hole remnants of the first stars in galactic haloes. MNRAS **340**, 647–656 (2003)

Islam, R.R., Taylor, J.E., Silk, J.: Massive black hole remnants of the first stars - III. Observational signatures from the past. MNRAS **354**, 629–640 (2004)

Jenkins, A., Frenk, C.S., White, S.D.M., Colberg, J.M., Cole, S., Evrard, A.E., Couchman, H.M.P., Yoshida, N.: The mass function of dark matter haloes. MNRAS **321**, 372–384 (2001)

Jiang, L., Fan, X., Bian, F., Annis, J., Chiu, K., Jester, S., Lin, H., Lupton, R.H., Richards, G.T., Strauss, M.A., Malanushenko, V., Malanushenko, E., Schneider, D.P.: A survey of $z \approx 6$ quasars in the sloan digital sky survey deep stripe. II. Discovery of six quasars at $z_{AB} \gtrsim 21$. AJ **138**, 305–311 (2009)

Jiang, L., Fan, X., Brandt, W.N., Carilli, C.L., Egami, E., Hines, D.C., Kurk, J.D., Richards, G.T., Shen, Y., Strauss, M.A., Vestergaard, M., Walter, F.: Dust-free quasars in the early Universe. Nature **464**, 380–383 (2010)

Jiang, L., Fan, X., Vestergaard, M., Kurk, J.D., Walter, F., Kelly, B.C., Strauss, M.A.: Gemini near-infrared spectroscopy of luminous $z \approx 6$ quasars: chemical abundances, black hole masses, and Mg II absorption. AJ **134**, 1150 (2007)

Jimenez, R., Haiman, Z.: Significant primordial star formation at redshifts z 3 4. Nature **440**, 501–504 (2006)

Kaaret, P., Feng, H., Gorski, M.: A major x-ray outburst from an ultraluminous x-ray source in M82. ApJ **692**, 653–658 (2009)

Kaaret, P., Prestwich, A.H., Zezas, A., Murray, S.S., Kim, D.-W., Kilgard, R.E., Schlegel, E.M., Ward, M.J.: Chandra high-resolution camera observations of the luminous x-ray source in the starburst galaxy M82. MNRAS **321**, L29–L32 (2001)

Kang, H., Shapiro, P.R.: Radiative shocks and hydrogen molecules in pregalactic gas - The effects of postshock radiation. ApJ **386**, 432–451 (1992)

Kang, H., Shapiro, P.R., Fall, S.M., Rees, M.J.: Radiative shocks inside protogalaxies and the origin of globular clusters. ApJ **363**, 488–498 (1990)

Kashlinsky, A., Rees, M.J.: Formation of population III stars and pregalactic evolution. MNRAS **205**, 955–971 (1983)

Kaspi, S., Smith, P.S., Netzer, H., Maoz, D., Jannuzi, B.T., Giveon, U.: Reverberation measurements for 17 quasars and the size-mass-luminosity relations in active galactic nuclei. ApJ **533**, 631–649 (2000)

Kawai, N., Kosugi, G., Aoki, K., Yamada, T., Totani, T., Ohta, K., Iye, M., Hattori, T., Aoki, W., Furusawa, H., Hurley, K., Kawabata, K.S., Kobayashi, N., Komiyama, Y., Mizumoto, Y., Nomoto, K., Noumaru, J., Ogasawara, R., Sato, R., Sekiguchi, K., Shirasaki, Y., Suzuki, M., Takata, T., Tamagawa, T., Terada, H., Watanabe, J., Yatsu, Y., Yoshida, A.: An optical spectrum of the afterglow of a γ-ray burst at a redshift of $z = 6.295$. Nature **440**, 184–186 (2006)

Keeton, C.R., Kuhlen, M., Haiman, Z.: Gravitational lensing magnification without multiple imaging. ApJ **621**, 559–573 (2005)

Kereš, D., Katz, N., Weinberg, D.H., Davé, R.: How do galaxies get their gas? MNRAS **363**, 2–28 (2005)

King, A.R., Davies, M.B., Ward, M.J., Fabbiano, G., Elvis, M.: Ultraluminous x-ray sources in external galaxies. ApJL **552**, L109–L112 (2001)

Kitayama, T., Ikeuchi, S.: Formation of subgalactic clouds under ultraviolet background radiation. ApJ **529**, 615–634 (2000)

Kochanek, C.S.: Gravitational lenses and NGST. In: Smith, E.P., Koratkar, A. (eds.) Science With The NGST, vol. 133 of Astronomical Society of the Pacific Conference Series, p. 96 (1998)

Kocsis, B., Sesana, A.: Gas-driven massive black hole binaries: signatures in the nHz gravitational wave background. MNRAS **411**, 1467–1479 (2011)

Kollmeier, J.A., Onken, C.A., Kochanek, C.S., Gould, A., Weinberg, D.H., Dietrich, M., Cool, R., Dey, A., Eisenstein, D.J., Jannuzi, B.T., Le Floc'h, E., Stern, D.: Black hole masses and eddington ratios at $0.3 < z < 4$. ApJ **648**, 128–139 (2006)

Komatsu, E., Smith, K.M., Dunkley, J., Bennett, C.L., Gold, B., Hinshaw, G., Jarosik, N., Larson, D., Nolta, M.R., Page, L., Spergel, D.N., Halpern, M., Hill, R.S., Kogut, A., Limon, M., Meyer, S.S., Odegard, N., Tucker, G.S., Weiland, J.L., Wollack, E., Wright, E.L.: Seven-year Wilkinson Microwave Anisotropy Probe (WMAP) observations: cosmological interpretation. ApJS **192**, 18 (2011)

Koushiappas, S.M., Bullock, J.S., Dekel, A.: Massive black hole seeds from low angular momentum material. MNRAS **354**, 292–304 (2004)

Kurk, J.D., Walter, F., Fan, X., Jiang, L., Jester, S., Rix, H.-W., Riechers, D.A.: Near-infrared spectroscopy of SDSS J0303 - 0019: A low-luminosity, high-eddington-ratio quasar at z 6. ApJ **702**, 833–837 (2009)

Kurk, J.D., Walter, F., Fan, X., Jiang, L., Riechers, D.A., Rix, H.-W., Pentericci, L., Strauss, M.A., Carilli, C., Wagner, S.: Black hole masses and enrichment of z 6 SDSS quasars. ApJ **669**, 32–44 (2007)

Lamb, D.Q., Reichart, D.E.: Gamma-ray bursts as a probe of the very high redshift Universe. ApJ **536**, 1–18 (2000)

Lang, R.N., Hughes, S.A.: Measuring coalescing massive binary black holes with gravitational waves: the impact of spin-induced precession. Phys. Rev. D **74**(12), 122001 (2006)

Latif, M.A., Schleicher, D.R.G., Spaans, M., Zaroubi, S.: Lyman α emission from the first galaxies: signatures of accretion and infall in the presence of line trapping. MNRAS **413**, L33–L37 (2011)

Latif, M.A., Zaroubi, S., Spaans, M.: The impact of Lyman α trapping on the formation of primordial objects. MNRAS **411**, 1659–1670 (2011)

Lauer, T.R., Faber, S.M., Richstone, D., Gebhardt, K., Tremaine, S., Postman, M., Dressler, A., Aller, M.C., Filippenko, A.V., Green, R., Ho, L.C., Kormendy, J., Magorrian, J., Pinkney, J.: The masses of nuclear black holes in luminous elliptical galaxies and implications for the space density of the most massive black holes. ApJ **662**, 808–834 (2007)

Lawrence, A., Warren, S.J., Almaini, O., Edge, A.C., Hambly, N.C., Jameson, R.F., Lucas, P., Casali, M., Adamson, A., Dye, S., Emerson, J.P., Foucaud, S., Hewett, P., Hirst, P., Hodgkin, S.T., Irwin, M.J., Lodieu, N., McMahon, R.G., Simpson, C., Smail, I., Mortlock, D., Folger, M.: The UKIRT Infrared Deep Sky Survey (UKIDSS). MNRAS **379**, 1599–1617 (2007)

Lee, H.M.: Dynamical effects of successive mergers on the evolution of spherical stellar systems. ApJ **319**, 801–818 (1987)

Lepp, S., Shull, J.M.: Molecules in the early universe. ApJ **280**, 465–469 (1984)

Li, Y., Haiman, Z., M.-M. Mac Low. Correlations between central massive objects and their host galaxies: from bulgeless spirals to ellipticals. ApJ **663**, 61–70 (2007)

Li, Y., Hernquist, L., Robertson, B., Cox, T.J., Hopkins, P.F., Springel, V., Gao, L., Di Matteo, T., Zentner, A.R., Jenkins, A., Yoshida, N.: Formation of z 6 quasars from hierarchical galaxy mergers. ApJ **665**, 187–208 (2007)

Lippai, Z., Frei, Z., Haiman, Z.: Prompt shocks in the gas disk around a recoiling supermassive black hole binary. ApJL **676**, L5–L8 (2008)

Lodato, G., Natarajan, P.: Supermassive black hole formation during the assembly of pre-galactic discs. MNRAS **371**, 1813–1823 (2006)

Lukić, Z., Heitmann, K., Habib, S., Bashinsky, S., Ricker, P.M.: The halo mass function: high-redshift evolution and universality. ApJ **671**, 1160–1181 (2007)

Machacek, M.E., Bryan, G.L., Abel, T.: Simulations of pregalactic structure formation with radiative feedback. ApJ **548**, 509–521 (2001)

Madau, P., Ferrara, A., Rees, M.J.: Early metal enrichment of the intergalactic medium by pregalactic outflows. ApJ **555**, 92–105 (2001)

Madau, P., Quataert, E.: The effect of gravitational-wave recoil on the demography of massive black holes. ApJL **606**, L17–L20 (2004)

Madau, P., Rees, M.J.: Massive black holes as Population III remnants. ApJL **551**, L27–L30 (2001)

Mahmood Khan, F., Just, A., Merritt, D.: Efficient merger of binary supermassive black holes in merging galaxies. ApJ **732**, 89 (2011)

Maio, U., Koopmans, L.V.E., Ciardi, B.: The impact of primordial supersonic flows on early structure formation, reionization and the lowest-mass dwarf galaxies. MNRAS **412**, L40–L44 (2011)

Martini, P.: QSO lifetimes. In: Ho, L.C. (ed.) Coevolution of Black Holes and Galaxies. Carnegie Observatories Centennial Symposia, p.169. Cambridge University Press, as part of the Carnegie Observatories Astrophysics Series (2004) (ISBN: 0521824494)

Martini, P., Weinberg, D.H.: Quasar clustering and the lifetime of quasars. ApJ **547**, 12–26 (2001)

Matsuda, T., Sat = o, H., Takeda, H.: Cooling of pre-galactic gas clouds by hydrogen molecule. Progr. Theor. Phys. **42**, 219–233 (1969)
McGreer, I.D., Becker, R.H., Helfand, D.J., White, R.L.: Discovery of a z = 6.1 radio-loud quasar in the NOAO deep wide field survey. ApJ **652**, 157–162 (2006)
McGreer, I.D., Hall, P.B., Fan, X., Bian, F., Inada, N., Oguri, M., Strauss, M.A., Schneider, D.P., Farnsworth, K.: SDSS J094604.90+183541.8: A gravitationally lensed quasar at z = 4.8. AJ **140**, 370–378 (2010)
McKee, C.F., Tan, J.C.: The formation of the first stars. II. Radiative feedback processes and implications for the initial mass function. ApJ **681**, 771–797 (2008)
McWilliams, S.T., Thorpe, J.I., Baker, J.G., Kelly, B.J.: Impact of mergers on LISA parameter estimation for nonspinning black hole binaries. Phys. Rev. D **81**(6), 064014 (2010)
Menou, K., Haiman, Z., Narayanan, V.K.: The merger history of supermassive black holes in galaxies. ApJ **558**, 535–542 (2001)
Merritt, D., M. Milosavljević. Massive black hole binary evolution. Living Rev. Relat. **8**, 8 (2005)
Merritt, D., Poon, M.Y.: Chaotic loss cones and black hole fueling. ApJ **606**, 788–798 (2004)
Mesinger, A., Bryan, G.L., Haiman, Z.: Ultraviolet radiative feedback on high-redshift protogalaxies. ApJ **648**, 835–851 (2006)
Mesinger, A., Bryan, G.L., Haiman, Z.: Relic HII regions and radiative feedback at high redshifts. MNRAS **399**, 1650–1662 (2009)
Miller, M.C., Colbert, E.J.M.: Intermediate-mass black holes. Int. J. Modern Phys. D **13**, 1–64 (2004)
Milosavljević, M., Bromm, V., Couch, S.M., Oh, S.P.: Accretion onto "Seed" black holes in the first galaxies. ApJ **698**, 766–780 (2009)
Milosavljević, M., Couch, S.M., Bromm, V.: Accretion onto intermediate-mass black holes in dense protogalactic clouds. ApJL **696**, L146–L149 (2009)
Miralda-Escudé, J., Kollmeier, J.A.: Star captures by quasar accretion disks: a possible explanation of the M-σ relation. ApJ **619**, 30–40 (2005)
Naoz, S., Noter, S., Barkana, R.: The first stars in the Universe. MNRAS **373**, L98–L102 (2006)
Navarro, J.F., Frenk, C.S., White, S.D.M.: A universal density profile from hierarchical clustering. ApJ **490**, 493 (1997)
Navarro, J.F., Steinmetz, M.: The effects of a photoionizing ultraviolet background on the formation of disk galaxies. ApJ **478**, 13 (1997)
Netzer, H.: The largest black holes and the most luminous galaxies. ApJL **583**, L5–L8 (2003)
Noyola, E., Gebhardt, K., Bergmann, M.: Gemini and hubble space telescope evidence for an intermediate-mass black hole in ω centauri. ApJ **676**, 1008–1015 (2008)
Oh, S.P.: Reionization by Hard Photons. I. X-Rays from the First Star Clusters. ApJ **553**, 499–512 (2001)
Oh, S.P., Haiman, Z.: Second-generation objects in the Universe: radiative cooling and collapse of halos with virial temperatures above 10^4 K. ApJ **569**, 558–572 (2002)
Oh, S.P., Haiman, Z., Rees, M.J.: HE II recombination lines from the first luminous objects. ApJ **553**, 73–77 (2001)
O'Leary, R.M., Rasio, F.A., Fregeau, J.M., Ivanova, N., O'Shaughnessy, R.: Binary mergers and growth of black holes in dense star clusters. ApJ **637**, 937–951 (2006)
Omukai, K.: Primordial star formation under far-ultraviolet radiation. ApJ **546**, 635–651 (2001)
Omukai, K., Schneider, R., Haiman, Z.: Can supermassive black holes form in metal-enriched high-redshift protogalaxies? ApJ **686**, 801–814 (2008)
O'Shea, B.W., Norman, M.L.: Population III star formation in a ΛCDM Universe. I. The effect of formation redshift and environment on protostellar accretion rate. ApJ **654**, 66–92 (2007)
Ostriker, J.P.: Collisional dark matter and the origin of massive black holes. Phys. Rev. Lett. **84**, 5258–5260 (2000)
Ostriker, J.P., Gnedin, N.Y.: Reheating of the Universe and Population III. ApJL **472**, L63+ (1996)
Palla, F., Salpeter, E.E., Stahler, S.W.: Primordial star formation - The role of molecular hydrogen. ApJ **271**, 632–641 (1983)

Pelupessy, F.I., Di Matteo, T., Ciardi, B.: How rapidly do supermassive black hole "Seeds" grow at early times? ApJ **665**, 107–119 (2007)
Portegies Zwart, S.F., McMillan, S.L.W.: The runaway growth of intermediate-mass black holes in dense star clusters. ApJ **576**, 899–907 (2002)
Press, W.H., Schechter, P.: Formation of galaxies and clusters of galaxies by self-similar gravitational condensation. ApJ **187**, 425–438 (1974)
Preto, M., Berentzen, I., Berczik, P., Spurzem, R.: Fast coalescence of massive black hole binaries from mergers of galactic nuclei: implications for low-frequency gravitational-wave astrophysics. ApJL **732**, L26+ (2011)
Prieto, J., Padoan, P., Jimenez, R., Infante, L.: Population III stars from turbulent fragmentation at redshift 11. ApJL **731**, L38+ (2011)
Proga, D., Begelman, M.C.: Accretion of low angular momentum material onto black holes: two-dimensional magnetohydrodynamic case. ApJ **592**, 767–781 (2003)
Quataert, E., Gruzinov, A.: Constraining the accretion tate onto Sagittarius A* using linear polarization. ApJ **545**, 842–846 (2000)
Quinlan, G.D., Shapiro, S.L.: The dynamical evolution of dense star clusters in galactic nuclei. ApJ **356**, 483–500 (1990)
Rasio, F.A., Freitag, M., Gürkan, M.A.: Formation of massive black holes in dense star clusters. In: Ho, L.C. (ed.) 'Coevolution of Black Holes and Galaxies', from the Carnegie Observatories Centennial Symposia, p. 138. Cambridge Press as part of the Carnegie Observatories Astrophysics Series (2004)
Reed, D.S., Bower, R., Frenk, C.S., Jenkins, A., Theuns, T.: The halo mass function from the dark ages through the present day. MNRAS **374**, 2–15 (2007)
Rees, M.J.: Black hole models for active galactic nuclei. ARA&A **22**, 471–506 (1984)
Rees, M.J., Ostriker, J.P.: Cooling, dynamics and fragmentation of massive gas clouds - Clues to the masses and radii of galaxies and clusters. MNRAS **179**, 541–559 (1977)
Regan, J.A., Haehnelt, M.G.: The formation of compact massive self-gravitating discs in metal-free haloes with virial temperatures of 13000–30000K. MNRAS **393**, 858–871 (2009)
Richards, G.T., Strauss, M.A., Pindor, B., Haiman, Z., Fan, X., Eisenstein, D., Schneider, D.P., Bahcall, N.A., Brinkmann, J., Brunner, R.: A snapshot survey for gravitational lenses among $z \gtrsim = 4.0$ quasars. I. The $z \gtrsim 5.7$ sample. AJ **127**, 1305–1312 (2004)
Ripamonti, E., Iocco, F., Ferrara, A., Schneider, R., Bressan, A., Marigo, P.: First star formation with dark matter annihilation. MNRAS, page 883 (2010)
Saijo, M., Baumgarte, T.W., Shapiro, S.L., Shibata, M.: Collapse of a rotating supermassive star to a supermassive black hole: post-newtonian simulations. ApJ **569**, 349–361 (2002)
Salvaterra, R., Della Valle, M., Campana, S., Chincarini, G., Covino, S., D'Avanzo, P., Fernández-Soto, A., Guidorzi, C., Mannucci, F., Margutti, R., Thöne, C.C., Antonelli, L.A., Barthelmy, S.D., de Pasquale, M., D'Elia, V., Fiore, F., Fugazza, D., Hunt, L.K., Maiorano, E., Marinoni, S., Marshall, F.E., Molinari, E., Nousek, J., Pian, E., Racusin, J.L., Stella, L., Amati, L., Andreuzzi, G., Cusumano, G., Fenimore, E.E., Ferrero, P., Giommi, P., Guetta, D., Holland, S.T., Hurley, K., Israel, G.L., Mao, J., Markwardt, C.B., Masetti, N., Pagani, C., Palazzi, E., Palmer, D.M., Piranomonte, S., Tagliaferri, G., Testa, V.: GRB090423 at a redshift of z 8.1. Nature **461**, 1258–1260 (2009)
Salviander, S., Shields, G.A., Gebhardt, K., Bernardi, M., Hyde, J.B.: In search of the largest velocity dispersion galaxies. ApJ **687**, 828–834 (2008)
Saslaw, W.C., Valtonen, M.J., Aarseth, S.J.: The gravitational slingshot and the structure of extragalactic radio sources. ApJ **190**, 253–270 (1974)
Schleicher, D.R.G., Spaans, M., Glover, S.C.O.: Black hole formation in primordial galaxies: chemical and radiative conditions. ApJL **712**, L69–L72 (2010)
Schleicher, D.R.G., Spaans, M., Klessen, R.S.: Probing high-redshift quasars with ALMA. I. Expected observables and potential number of sources. A&A **513**, A7+ (2010)
Sesana, A., Haardt, F., Madau, P., Volonteri, M.: Low-frequency gravitational radiation from coalescing massive black hole binaries in hierarchical cosmologies. ApJ **611**, 623–632 (2004)

Sesana, A., Haardt, F., Madau, P., Volonteri, M.: The gravitational wave signal from massive black hole binaries and its contribution to the LISA data stream. ApJ **623**, 23–30 (2005)
Sesana, A., Volonteri, M., Haardt, F.: The imprint of massive black hole formation models on the LISA data stream. MNRAS **377**, 1711–1716 (2007)
Sethi, S.K., Haiman, Z., Pandey, K.: Supermassive black hole formation at high redshifts through a primordial magnetic field. ApJ **721**, 615 (2010)
Shakura, N.I., Sunyaev, R.A.: Black holes in binary systems. Observational appearance. A&A **24**, 337–355 (1973)
Shang, C., Bryan, G.L., Haiman, Z.: Supermassive black hole formation by direct collapse: keeping protogalactic gas H_2 free in dark matter haloes with virial temperatures $T_{vir} \gtrsim 10^4$ K. MNRAS **402**, 1249–1262 (2010)
Shankar, F.: The demography of supermassive black holes: Growing monsters at the heart of galaxies. New Astron. Rev. **53**, 57–77 (2009)
Shankar, F., Bernardi, M., Haiman, Z.: The evolution of the $M_{BH} - \sigma$ relation inferred from the age distribution of local early-type galaxies and active galactic nuclei evolution. ApJ **694**, 867–878 (2009)
Shankar, F., Crocce, M., Miralda-Escudé, J., Fosalba, P., Weinberg, D.H.: On the radiative efficiencies, eddington ratios, and duty cycles of luminous high-redshift quasars. ApJ **718**, 231–250 (2010)
Shankar, F., Weinberg, D.H., Miralda-Escudé, J.: Self-consistent models of the AGN and black hole populations: duty cycles, accretion rates, and the mean radiative efficiency. ApJ **690**, 20–41 (2009)
Shankar, F., Weinberg, D.H., Shen, Y.: Constraints on black hole duty cycles and the black hole-halo relation from SDSS quasar clustering. MNRAS **406**, 1959–1966 (2010)
Shapiro, P.R., Giroux, M.L., Babul, A.: Reionization in a cold dark matter universe: The feedback of galaxy formation on the intergalactic medium. ApJ **427**, 25–50 (1994)
Shapiro, P.R., Kang, H.: Hydrogen molecules and the radiative cooling of pregalactic shocks. ApJ **318**, 32–65 (1987)
Shapiro, S.L.: Formation of supermassive black holes: simulations in general relativity. In: Ho, L.C. (ed.) 'Coevolution of Black Holes and Galaxies', from the Carnegie Observatories Centennial Symposia, p. 103. Cambridge Press as part of the Carnegie Observatories Astrophysics Series (2004)
Shapiro, S.L.: Spin, accretion, and the cosmological growth of supermassive black holes. ApJ **620**, 59–68 (2005)
Shapiro, S.L., Teukolsky, S.A.: Black holes, white dwarfs, and neutron stars. The physics of compact objects. Book: Research supported by the National Science Foundation, p. 663. Wiley-Interscience, New York (1983)
Shaver, P.A., Wall, J.V., Kellermann, K.I., Jackson, C.A., Hawkins, M.R.S.: Decrease in the space density of quasars at high redshift. Nature **384**, 439–441 (1996)
Shen, Y., Strauss, M.A., Oguri, M., Hennawi, J.F., Fan, X., Richards, G.T., Hall, P.B., Gunn, J.E., Schneider, D.P., Szalay, A.S., Thakar, A.R., Vanden Berk, D.E., Anderson, S.F., Bahcall, N.A., Connolly, A.J., Knapp, G.R.: Clustering of high-redshift (z >= 2.9) quasars from the sloan digital sky survey. AJ **133**, 2222–2241 (2007)
Sheth, R.K., Mo, H.J., Tormen, G.: Ellipsoidal collapse and an improved model for the number and spatial distribution of dark matter haloes. MNRAS **323**, 1–12 (2001)
Shibata, M., Shapiro, S.L.: Collapse of a rotating supermassive star to a supermassive black hole: fully relativistic simulations. ApJL **572**, L39–L43 (2002)
Shields, G.A., Gebhardt, K., Salviander, S., Wills, B.J., Xie, B., Brotherton, M.S., Yuan, J., Dietrich, M.: The black hole-bulge relationship in quasars. ApJ **583**, 124–133 (2003)
Shlosman, I., Begelman, M.C.: Evolution of self-gravitating accretion disks in active galactic nuclei. ApJ **341**, 685–691 (1989)
Shu, F.H.: Self-similar collapse of isothermal spheres and star formation. ApJ **214**, 488–497 (1977)
Sijacki, D., Springel, V., Haehnelt, M.G.: Growing the first bright quasars in cosmological simulations of structure formation. MNRAS **400**, 100–122 (2009)

Silk, J., Rees, M.J.: Quasars and galaxy formation. A&A **331**, L1–L4 (1998)
Silk, J., Spaans, M.: Molecular lines as diagnostics of high-redshift objects. ApJL **488**, L79+ (1997)
Silverman, J.D., Green, P.J., Barkhouse, W.A., Kim, D.-W., Kim, M., Wilkes, B.J., Cameron, R.A., Hasinger, G., Jannuzi, B.T., Smith, M.G., Smith, P.S., Tananbaum, H.: The luminosity function of x-ray-selected active galactic nuclei: evolution of supermassive black holes at high redshift. ApJ **679**, 118–139 (2008)
Small, T.A., Blandford, R.D.: Quasar evolution and the growth of black holes. MNRAS **259**, 725–737 (1992)
Spaans, M., Silk, J.: Pregalactic black hole formation with an atomic hydrogen equation of state. ApJ **652**, 902–906 (2006)
Spitzer, L. Jr.: Equipartition and the formation of compact nuclei in spherical stellar systems. ApJL **158**, L139+ (1969)
Spolyar, D., Bodenheimer, P., Freese, K., Gondolo, P.: Dark stars: a new look at the first stars in the Universe. ApJ **705**, 1031–1042 (2009)
Spolyar, D., Freese, K., Gondolo, P.: Dark matter and the first stars: A new phase of stellar evolution. Phys. Rev. Lett. **100**(5), 051101 (2008)
Stacy, A., Bromm, V., Loeb, A.: Effect of streaming motion of baryons relative to dark matter on the formation of the first stars. ApJL **730**, L1+ (2011)
Stacy, A., Bromm, V., Loeb, A.: Rotation speed of the first stars. MNRAS **413**, 543–553 (2011)
Stacy, A., Greif, T.H., Bromm, V.: The first stars: formation of binaries and small multiple systems. MNRAS **403**, 45–60 (2010)
Stone, J.M., Pringle, J.E.: Magnetohydrodynamical non-radiative accretion flows in two dimensions. MNRAS **322**, 461–472 (2001)
Stone, J.M., Pringle, J.E., Begelman, M.C.: Hydrodynamical non-radiative accretion flows in two dimensions. MNRAS **310**, 1002–1016 (1999)
Tan, J.C., McKee, C.F.: The formation of the first stars. I. Mass infall rates, accretion disk structure, and protostellar evolution. ApJ **603**, 383–400 (2004)
Tanaka, T., Haiman, Z.: The assembly of supermassive black holes at high redshifts. ApJ **696**, 1798–1822 (2009)
Tanvir, N.R., Fox, D.B., Levan, A.J., Berger, E., Wiersema, K., Fynbo, J.P.U., Cucchiara, A., Krühler, T., Gehrels, N., Bloom, J.S., Greiner, J., Evans, P.A., Rol, E., Olivares, F., Hjorth, J., Jakobsson, P., Farihi, J., Willingale, R., Starling, R.L.C., Cenko, S.B., Perley, D., Maund, J.R., Duke, J., Wijers, R.A.M.J., Adamson, A.J., Allan, A., Bremer, M.N., Burrows, D.N., Castro-Tirado, A.J., Cavanagh, B., de Ugarte Postigo, A., Dopita, M.A., Fatkhullin, T.A., Fruchter, A.S., Foley, R.J., Gorosabel, J., Kennea, J., Kerr, T., Klose, S., Krimm, H.A., Komarova, V.N., Kulkarni, S.R., Moskvitin, A.S., Mundell, C.G., Naylor, T., Page, K., Penprase, B.E., Perri, M., Podsiadlowski, P., Roth, K., Rutledge, R.E., Sakamoto, T., Schady, P., Schmidt, B.P., Soderberg, A.M., Sollerman, J., Stephens, A.W., Stratta, G., Ukwatta, T.N., Watson, D., Westra, E., Wold, T., Wolf, C.: A γ-ray burst at a redshift of z 8.2. Nature **461**, 1254–1257 (2009)
Taoso, M., Bertone, G., Meynet, G., Ekström, S.: Dark matter annihilations in Population III stars. Phys. Rev. D **78**(12), 123510 (2008)
Tegmark, M., Silk, J., Rees, M.J., Blanchard, A., Abel, T., Palla, F.: How small were the first cosmological objects? ApJ **474**, 1 (1997)
Thoul, A.A., Weinberg, D.H.: Hydrodynamic simulations of galaxy formation. II. Photoionization and the formation of low-mass galaxies. ApJ **465**, 608 (1996)
Tseliakhovich, D., Barkana, R., Hirata, C.: Suppression and spatial variation of early galaxies and minihalos. ArXiv e-prints (2010)
Tseliakhovich, D., Hirata, C.: Relative velocity of dark matter and baryonic fluids and the formation of the first structures. Phys. Rev. D **82**(8), 083520 (2010)
Tumlinson, J., Giroux, M.L., Shull, J.M.: Probing the first stars with hydrogen and helium recombination emission. ApJL **550**, L1–L5 (2001)

Turk, M.J., Abel, T., O'Shea, B.: The formation of Population III binaries from cosmological initial conditions. Science **325**, 601 (2009)

Turner, E.L.: Quasars and galaxy formation. I - The Z greater than 4 objects. AJ **101**, 5–17 (1991)

Turner, N.J., Blaes, O.M., Socrates, A., Begelman, M.C., Davis, S.W.: The effects of photon bubble instability in radiation-dominated accretion disks. ApJ **624**, 267–288 (2005)

Umeda, H., Yoshida, N., Nomoto, K., Tsuruta, S., Sasaki, M., Ohkubo, T.: Early black hole formation by accretion of gas and dark matter. Jo Cosmol. Astro Part. Phys. **8**, 24 (2009)

van deer Marel, R. P.: Intermediate-mass black holes in the universe: a review of formation theories and observational constraints. In: Ho, L.C. (ed.) 'Coevolution of Black Holes and Galaxies', from the Carnegie Observatories Centennial Symposia, p. 37. Cambridge Press as part of the Carnegie Observatories Astrophysics Series (2004)

van der Marel, R.P., Gerssen, J., Guhathakurta, P., Peterson, R.C., Gebhardt, K.: Hubble space telescope evidence for an intermediate-mass black hole in the globular cluster M15. I. STIS spectroscopy and WFPC2 photometry. AJ **124**, 3255–3269 (2002)

Vanden Berk, D.E., Richards, G.T., Bauer, A., Strauss, M.A., Schneider, D.P., Heckman, T.M., York, D.G., Hall, P.B., Fan, X., Knapp, G.R., Anderson, S.F., Annis, J., Bahcall, N.A., Bernardi, M., Briggs, J.W., Brinkmann, J., Brunner, R., Burles, S., Carey, L., Castander, F.J., Connolly, A.J., Crocker, J.H., Csabai, I., Doi, M., Finkbeiner, D., Friedman, S., Frieman, J.A., Fukugita, M., Gunn, J.E., Hennessy, G.S., Ivezić, Ž., Kent, S., Kunszt, P.Z., Lamb, D.Q., Leger, R.F., Long, D.C., Loveday, J., Lupton, R.H., Meiksin, A., Merelli, A., Munn, J.A., Newberg, H.J., Newcomb, M., Nichol, R.C., Owen, R., Pier, J.R., Pope, A., Rockosi, C.M., Schlegel, D.J., Siegmund, W.A., Smee, S., Snir, Y., Stoughton, C., Stubbs, C., SubbaRao, M., Szalay, A.S., Szokoly, G.P., Tremonti, C., Uomoto, A., Waddell, P., Yanny, B., Zheng, W.: Composite quasar spectra from the sloan digital sky survey. AJ **122**, 549–564 (2001)

Vecchio, A.: LISA observations of rapidly spinning massive black hole binary systems. Phys. Rev. D **70**(4), 042001 (2004)

Vestergaard, M.: Determining central black hole masses in distant active galaxies. ApJ **571**, 733–752 (2002)

Vestergaard, M.: Early growth and efficient accretion of massive black holes at high redshift. ApJ **601**, 676–691 (2004)

Vestergaard, M., Fan, X., Tremonti, C.A., Osmer, P.S., Richards, G.T.: Mass functions of the active black holes in distant quasars from the sloan digital sky survey data release 3. ApJL **674**, L1–L4 (2008)

Vestergaard, M., Osmer, P.S.: Mass functions of the active black holes in distant quasars from the large bright quasar survey, the bright quasar survey, and the color-selected sample of the SDSS fall equatorial stripe. ApJ **699**, 800–816 (2009)

Vishniac, E.T.: A necessary condition for equilibrium in stellar systems with a continuous mass spectrum. ApJ **223**, 986–990 (1978)

Volonteri, M.: Formation of supermassive black holes. Astron. Astrophys. Rev. **18**, 279–315 (2010)

Volonteri, M., Haardt, F., Madau, P.: The assembly and merging history of supermassive black holes in hierarchical models of galaxy formation. ApJ **582**, 559–573 (2003)

Volonteri, M., Lodato, G., Natarajan, P.: The evolution of massive black hole seeds. MNRAS **383**, 1079–1088 (2008)

Volonteri, M., Natarajan, P.: Journey to the $M_{\rm BH} - \sigma$ relation: the fate of low-mass black holes in the Universe. MNRAS **400**, 1911–1918 (2009)

Volonteri, M., Rees, M.J.: Rapid growth of high-redshift black holes. ApJ **633**, 624–629 (2005)

Volonteri, M., Rees, M.J.: Quasars at z=6: The survival of the fittest. ApJ **650**, 669–678 (2006)

Wagoner, R.V.: Physics of massive objects. ARA&A **7**, 553 (1969)

Walter, F., Bertoldi, F., Carilli, C., Cox, P., Lo, K.Y., Neri, R., Fan, X., Omont, A., Strauss, M.A., Menten, K.M.: Molecular gas in the host galaxy of a quasar at redshift z = 6.42. Nature **424**, 406–408 (2003)

Wang, R., Carilli, C.L., Neri, R., Riechers, D.A., Wagg, J., Walter, F., Bertoldi, F., Menten, K.M., Omont, A., Cox, P., Fan, X.: Molecular gas in $z \approx 6$ quasar host galaxies. ApJ **714**, 699–712 (2010)

Wang, R., Wagg, J., Carilli, C.L., Walter, F., Riechers, D.A., Willott, C., Bertoldi, F., Omont, A., Beelen, A., Cox, P., Strauss, M.A., Bergeron, J., Forveille, T., Menten, K.M., Fan, X.: CO (2-1) line emission in redshift 6 quasar host galaxies. ApJ **739**, 34 (2011)

Wasserburg, G.J., Qian, Y.-Z.: A model of metallicity evolution in the early Universe. ApJL **538**, L99–L102 (2000)

White, S.D.M., Rees, M.J.: Core condensation in heavy halos - A two-stage theory for galaxy formation and clustering. MNRAS **183**, 341–358 (1978)

Willott, C.J., Albert, L., Arzoumanian, D., Bergeron, J., Crampton, D., Delorme, P., Hutchings, J.B., Omont, A., Reylé, C., Schade, D.: Eddington-limited accretion and the black hole mass function at redshift 6. AJ **140**, 546–560 (2010)

Willott, C.J., Delorme, P., Reylé, C., Albert, L., Bergeron, J., Crampton, D., Delfosse, X., Forveille, T., Hutchings, J.B., McLure, R.J., Omont, A., Schade, D.: The canada-france high-z quasar survey: nine new quasars and the luminosity function at redshift 6. AJ **139**, 906–918 (2010)

Willott, C.J., McLure, R.J., Jarvis, M.J.: A $3\times10^9 M_\odot$ black hole in the quasar SDSS J1148+5251 at $z=6.41$. ApJL **587**, L15–L18 (2003)

Willott, C.J., Percival, W.J., McLure, R.J., Crampton, D., Hutchings, J.B., Jarvis, M.J., Sawicki, M., Simard, L.: Imaging of SDSS $z > 6$ quasar fields: gravitational lensing, companion galaxies, and the host dark matter halos. ApJ **626**, 657–665 (2005)

Wise, J.H., Abel, T.: Resolving the formation of protogalaxies. III. Feedback from the first stars. ApJ **685**, 40–56 (2008)

Wise, J.H., Turk, M.J., Abel, T.: Resolving the formation of protogalaxies. II. Central gravitational collapse. ApJ **682**, 745–757 (2008)

Wolcott-Green, J., Bryan, G.L., Haiman, Z.: Photodissociation of $\mathbf{H_2}$ in protogalaxies: self-shielding in 3D simulations. MNRAS **418**, 838 (2011)

Woosley, S.E.: Gamma-ray bursts from stellar mass accretion disks around black holes. ApJ **405**, 273–277 (1993)

Wyithe, J.S.B.: The shallow slope of the z 6 quasar luminosity function: limits from the lack of multiple-image gravitational lenses. MNRAS **351**, 1266–1276 (2004)

Wyithe, J.S.B., Loeb, A.: Gravitational lensing of the sloan digital sky survey high-redshift quasars. ApJ **577**, 57–68 (2002)

Wyithe, J.S.B., Loeb, A.: Magnification of light from many distant quasars by gravitational lenses. Nature **417**, 923–925 (2002)

Wyithe, J.S.B., Loeb, A.: Low-frequency gravitational waves from massive black hole binaries: predictions for LISA and pulsar timing arrays. ApJ **590**, 691–706 (2003)

Wyithe, J.S.B., Loeb, A.: Self-regulated growth of supermassive black holes in galaxies as the origin of the optical and x-ray luminosity functions of quasars. ApJ **595**, 614–623 (2003)

Yoo, J., Miralda-Escudé, J.: Formation of the black holes in the highest redshift quasars. ApJL **614**, L25–L28 (2004)

Yoon, S.-C., Iocco, F., Akiyama, S.: Evolution of the first stars with dark matter burning. ApJL **688**, L1–L4 (2008)

Yoshida, N., Abel, T., Hernquist, L., Sugiyama, N.: Simulations of early structure formation: primordial gas clouds. ApJ **592**, 645–663 (2003)

Yoshida, N., Omukai, K., Hernquist, L.: Protostar formation in the early Universe. Science **321**, 669 (2008)

Yoshida, N., Sokasian, A., Hernquist, L., Springel, V.: Early structure formation and reionization in a warm dark matter cosmology. ApJL **591**, L1–L4 (2003)

Yu, Q.: Evolution of massive binary black holes. MNRAS **331**, 935–958 (2002)

Yu, Q., Tremaine, S.: Observational constraints on growth of massive black holes. MNRAS **335**, 965–976 (2002)

Yunes, N., Kocsis, B., Loeb, A., Haiman, Z.: Imprint of accretion disk-induced migration on gravitational waves from extreme mass ratio inspirals. Phys Rev Lett **107**, 1103 (2011)

Zeimann, G.R., White, R.L., Becker, R.H., Hodge, J.A., Stanford, S.A., Richards, G.T.: Discovery of a radio-selected z 6 quasar. ApJ **736**, 57 (2011)
Zhang, J., Fakhouri, O., Ma, C.-P.: How to grow a healthy merger tree. MNRAS **389**, 1521–1538 (2008)
Zhao, H., Haehnelt, M.G., Rees, M.J.: Feeding black holes at galactic centres by capture from isothermal cusps. New Astron. **7**, 385–394 (2002)

Part III
Tools & Techniques

Evolutionary Synthesis Models as a Tool and Guide Towards the First Galaxies

Daniel Schaerer

Abstract We summarize the principles and fundamental ingredients of evolutionary synthesis models, which are stellar evolution, stellar atmospheres, the IMF, star-formation histories, nebular emission, and also attenuation from the ISM and IGM. The chapter focusses in particular on issues of importance for predictions of metal-poor and Population III dominated galaxies.

We review recent predictions for the main physical properties and related observables of star-forming galaxies based on up-to-date inputs. The predicted metallicity dependence of these quantities and their physical causes are discussed. The predicted observables include in particular the restframe UV–to-optical domain with continuum emission from stars and the ionized ISM, as well as emission lines from H, He, and metals.

Based on these predictions we summarize the main observational signatures (emission line strengths, colors etc.), which can be used to distinguish "normal" stellar populations from very metal-poor objects or even Pop III.

Evolutionary synthesis models provide an important and fundamental tool for studies of galaxy formation and evolution, from the nearby Universe back to first galaxies. They are used in many applications to interpret existing observations, to predict and guide future missions/instruments, and to allow direct comparisons between state-of-the-art galaxy simulations and observations.

D. Schaerer (✉)
Observatoire de Genève, Université de Genève, 51 Ch. des Maillettes, 1290 Versoix, Switzerland

CNRS, IRAP, 14 Avenue E. Belin, 31400 Toulouse, France
e-mail: daniel.schaerer@unige.ch

T. Wiklind et al. (eds.), *The First Galaxies*, Astrophysics and Space Science Library 396, DOI 10.1007/978-3-642-32362-1_7,

1 Introduction

Evolutionary synthesis models, first pioneered by Tinsley (1968, 1980), are a simple, but fundamental tool to predict the emission of integrated stellar populations, such as those of distant galaxies. Their basic immediate objective is to predict the total spectrum or the spectral energy distribution (hereafter SED) emitted e.g. by a galaxy or by another ensemble of stars.

Synthesis models are generally used to interpret observations of integrated stellar populations, i.e. to infer their physical properties—such as the total stellar mass, star formation rate, age, attenuation etc.—from comparisons between models and observations. Often synthesis models are also used to predict/guide future observations, since from our knowledge of star formation, stellar evolution, and atmospheres, one is able to predict a large number of observables for a very broad range of parameters (ages, star formation histories, stellar initial mass functions, metallicities, redshifts etc.). This is particularly true in the present context related to the first galaxies, which are at the limit or beyond the reach of present-day facilities, and where predictions are needed to plan future missions and devise observational strategies to search for these distant, "exotic" objects.

Focussing on emission from stars and the ISM in the ultraviolet, optical, and near-IR domain (taken in their rest frame)—the spectral range where usually stars dominate the integrated emission of galaxies—we will describe and summarize the main ingredients (inputs) of synthesis models, as well as the basic assumptions and related uncertainties (Sect. 2). In Sect. 3 we show how and why physical properties of stellar populations and corresponding observables vary with metallicity, and which main differences are expected between "first galaxies", metal-free (PopIII), metal-poor, and present-day stellar populations. Methods used to distinguish/select observationally PopIII-dominated and similar objects from "normal" galaxies are reviewed in Sect. 4. Brief conclusions are given in Sect. 5.

2 Synthesis Models: Basic Ingredients and Assumptions

Popular, widely used synthesis models include the models of Bruzual and Charlot (Bruzual and Charlot 2003), the *Starburst99* models specialized for young stellar populations/starbursts (Leitherer et al. 1999), the PÉGASE models of Fioc and Rocca-Volmerange (1999), and also recent models including a special treatment of TP-AGB stars (Maraston 2005; Maraston et al. 2006). Recent reviews on synthesis models, including some basics as well as topics for current and future improvements, have e.g. been presented by Bruzual (2003, 2011), Maraston (2011a), and Leitherer and Ekström (2011). Papers presenting tests and confrontation of synthesis model predictions with basic observations of young stellar populations include e.g. Cid Fernandes et al. (2001), Bruzual (2003), and Cid Fernandes and González Delgado (2010).

Schematically, the following needs to be known (or assumed) to be able to predict the spectrum of integrated stellar populations:

1. **Stellar evolution:** A description of the evolution (in time) of stars in the HR-diagram as a function of their initial mass, metallicity (chemical composition), and other parameters which may govern their evolution (e.g. initial rotation rate, magnetic field).
2. **Stellar atmospheres:** A description of the emergent spectrum (over the spectral range of interest) of individual stars at all phases of their evolution.
3. **The stellar initial mass function (IMF),** which determines the relative distribution of stars of different masses at the time of formation.
4. **The star formation history (SFH)** of the galaxy, describing the history of the amount of stars (commonly expressed in mass formed per unit time), i.e. the star formation rate (SFR) as a function of time.
5. **Nebular emission,** i.e. the emission from HII regions nearby massive star-forming regions, which—in general—cannot be separated from the stellar emission.
6. **Attenuation** within the intervening interstellar medium (ISM) of the galaxy. For a simple prescription this implies that we need to know the attenuation law (i.e. it's dependence on wavelength), and the amount of attenuation at a given reference wavelength.
7. **Intergalactic medium opacity:** Finally, since the photons emitted by a distant galaxy also travel through the intergalactic medium (IGM), its transmission properties must be known/specificed.

Items 1–3 describe the properties of stars and ensembles thereof. So-called simple stellar populations (SSPs), corresponding to an ensemble of stars formed at the same time, represent the basic units. Predictions for SSPs are widely distributed in the literature. In practice SSPs may represent stellar populations of stellar clusters where the age spread between stars is small. For any arbitrary, given star formation history (4), the predicted spectrum can be derived from SSPs by convolution. Nebular emission (5) is important to properly describe star-forming galaxies, where the contribution from young massive stars is significant. The remaining items (6, 7) describe the way the emitted spectrum is altered both at the galaxy scale and through the IGM on its way to the observer.

Evolutionary synthesis models traditionally describe the emission from stars (plus emission from surrounding HII regions in some extensions). By construction such models are thus usually tailored to the (rest-frame) UV–optical–near-IR part of the electromagnetic spectrum, where emission from stars (+nebulae) dominate. This is the domain on which the present text is focussed. Extensions of these models to other wavelength domains, e.g. to X-rays or to the IR–radio, have also been constructed by Mas-Hesse and Kunth (1991), and Cerviño et al. (2002). Since focussed on primeval galaxies and on their observability, we will here emphasize in particular very low metallicities (necessarily more important in the early Universe) and relatively massive stars, which dominate the rest-frame UV emission in strongly star-forming galaxies.

Let us now briefly discuss each of these "ingredients" of evolutionary synthesis models with a special emphasis on first/distant galaxies.

2.1 Stellar Evolution

Stellar evolution models have seen a "boom" in the 1990s, when new radiative opacities were published, triggering extensive evolutionary track calculations for a wide range of stellar masses and metallicities. The most widely used of these models are the Geneva and Padova tracks (see Schaller et al. 1992; Meynet et al. 1994; Bertelli et al. 1994), which have been extensively used since then for evolutionary synthesis models and other applications. An illustration of such tracks for zero metallicity, and a comparison with 1/50 $Z_{\odot}$ and solar metallicity ($Z_{\odot}$) is shown in Fig. 1. One of the main, if not the main property distinguishing PopIII stars from others is clearly apparent from this figure: the fact that massive ($M \gtrsim 5$ $M_{\odot}$) stars are much hotter than their counterparts at non-zero metallicity. For the most massive stars their effective temperature can reach up to $\sim 10^5$ K on the zero-age main sequence (ZAMS). These differences has several important observational consequences, which are discussed below. A detailed discussion of the peculiarities of interior models of PopIII stars and the literature before 2001 is given in Marigo et al. (2001).

Since the stellar evolution models computed in the 1990s including numerous physical processes among which in particular mass loss, large efforts have been undertaken to describe other processes affecting the interior evolution of stars, such as stellar rotation,s magnetic fields, and the various transport mechanisms related to it. For reviews on these issues see e.g. Maeder and Meynet (2000), Maeder and Meynet (2012). The impact of other, maybe more "exotic" phenomena, like possible variations of the fine-structure constant on stellar properties at very low metallicity have e.g. been explored by Ekström et al. (2010).

Several arguments, both observational and theoretical, indicate that the importance of rotation becomes stronger for low metallicities (see e.g. Maeder et al. 1999; Chiappini et al. 2006). However, the predicted stellar tracks depend strongly on the initial rotational velocity (a free parameter) and its history (which is predicted from the model following also the evolution of angular momentum), which are very uncertain, and difficult to constrain at (very) low metallicity. In consequence, the impact of stellar rotation on predictions for integrated stellar populations remains currently poorly known. Preliminary explorations of the impact of stellar rotation, mostly at solar metallicity, have been presented by Vázquez et al. (2007), Leitherer and Ekström (2011), and Levesque et al. (2012).

Recent PopIII tracks including the effects of stellar rotation and magnetic fields are presented in Ekström et al. (2008) and Yoon et al. (2012). For the initial rotational velocities chosen in the former paper, rotation has a small impact on the evolution in the HR-diagram and hence on predictions from synthesis models. However, the surface abundances of these stars are significantly modified, with several implications on nucleosynthesis and chemical evolution. Effects with

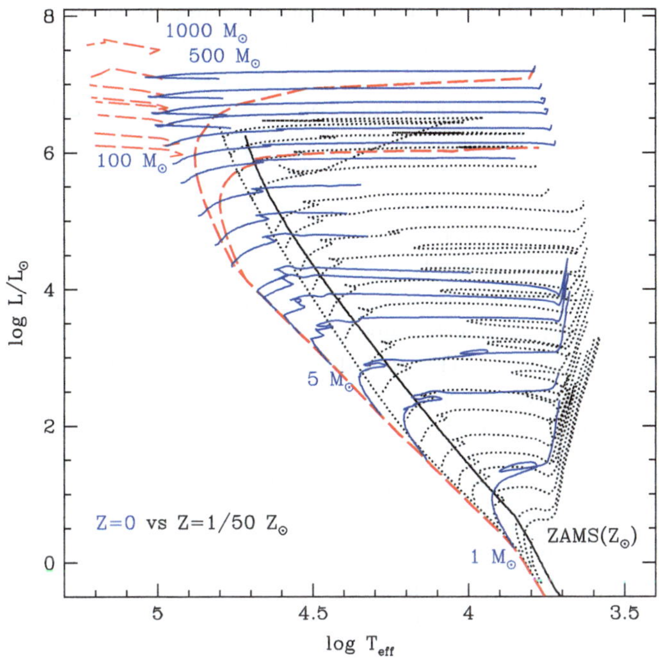

Fig. 1 HR-diagram for metal free ($Z = 0$, *solid and long-dashed lines*) and low metallicity ($Z = 1/50 Z_{\odot}$, *dotted*) stars. Isochrones of 2 and 4 Myr for $Z = 0$ tracks without mass loss are also plotted (*long-dashed*). The *short dashed high mass tracks* evolving blueward of the ZAMS are computed assuming strong mass loss. The position of the ZAMS at solar metallicity ($Z_{\odot}$) from Schaller et al. (1992) is shown by the *solid line*. Other tracks and isochrones are also shown (see legend in Schaerer 2002). Note the important shift of the ZAMS to high $T_{\rm eff}$ from low metallicity to $Z = 0$, as well as the rapid redward evolution of the massive stars. From Schaerer (2002)

potential impact on evolutionary synthesis models can be found for very high rotation rates (cf. Yoon et al. 2012), when stars are very strongly mixed, following a nearly homogeneous evolution, which implies much hotter temperatures and a blue-ward evolution. In this case properties such as their UV and ionizing flux are significantly altered, leaving imprints on the spectra predicted from stellar populations containing such stars. However, the distribution of rotational velocities remains poorly know, especially at very low metallicity, and hence the proportion of stars with properties significantly altered by rotation are not known. Schaerer (2003) has explored in a simple way the possible impact of very hot, homogeneous PopIII stars on the hardness of the ionizing spectrum of stellar populations. It is quite evident that we currently do not have a clear view of the impact stellar rotation and magnetic fields may have on evolutionary synthesis models, quite independently of metallicity. Progress in these areas is ongoing.

Other issues of general importance for synthesis models include e.g. the importance of thermally-pulsating asymptotic giant branch (TP-AGB) stars, whose contribution to the integrated light is being debated (see Maraston 2005, 2011b, and references therein), and the recurrent question of the importance of binary stars (cf. Vanbeveren et al. 1998; Schaerer and Vacca 1998; Belkus et al. 2003; Dionne and Robert 2006; Eldridge 2011) For our present objective, first galaxies, uncertainties related to TP-AGB stars are not a major concern, since we are dealing with young stellar populations and focussing on the UV to optical (rest-frame) domain, where these stars can safely be neglected. The evolution of close massive binary stars can alter the predicted spectra of young stellar populations, as illustrated e.g. recently by Eldridge (2011). However, the physics of these stars is even more complex than that of single stars, and depends on a number of additional, poorly constrained parameters. In any case, it is important that the predictions from stellar evolution models be compared to and tested against direct stellar observations in our Galaxy and in the nearby Universe. This includes e.g. observations of individual stars in clusters, their basic stellar parameters and surface abundances, color-magnitude diagrams, integrated colors of clusters, statistics of stars of different types and trends with metallicity etc. Such comparisons have e.g. been carried out with the (non-rotating) Padova and Geneva models (cf. Maeder and Meynet 1994; Bruzual 2003; Bruzual and Charlot 2003; Massey 2003; Meynet and Maeder 2005), and are being carried out for the latest tracks including stellar rotation. They serve thus also to "calibrate" uncertainties or unknowns in the stellar models, and hence to place synthesis model predictions on the best ground.

2.2 *Stellar Atmospheres*

It is well known that strong departures from local thermodynamic equilibrium (LTE) occur in the atmospheres of hot stars. To properly predict the emergent spectra of massive stars it is therefore essential to use non-LTE model atmospheres. This statement holds in particular also for low- and zero metallicities, as departures from LTE significantly alter the level populations of H and He, which—together with electron scattering—are the main opacity sources in stars of such composition. While generally stellar winds also affect the predicted spectra of massive star (e.g. Gabler et al. 1989), the mass outflow from very low metallicity stars is very low, and its effect can hence be safely neglected to compute the observable properties of PopIII and similar massive stars (cf. Schaerer 2002; Kudritzki 2002; Krtička and Kubát 2006). Using non-LTE plane parallel model atmospheres is therefore sufficient for this purpose. For less extreme metallicities, appropriate model atmospheres describing main sequence stars, Wolf-Rayet stars with strong winds, and also cooler stars must be combined (cf. Conti et al. 2008) to achieve the most reliable synthesis model predictions (e.g. Leitherer et al. 1999; Smith et al. 2002).

The predicted ionizing fluxes of H, He, and He^+ from PopIII stars as a function of effective temperature is shown in Fig. 2. For comparison, the same values predicted using simple black-body spectra are also shown. While for $T_{\rm eff} \gtrsim 40$ kK black bodies provide a good approximation to the total number of H ionising photons (i.e. at energies >13.6 eV), this is not the case for the shape of the spectra at higher

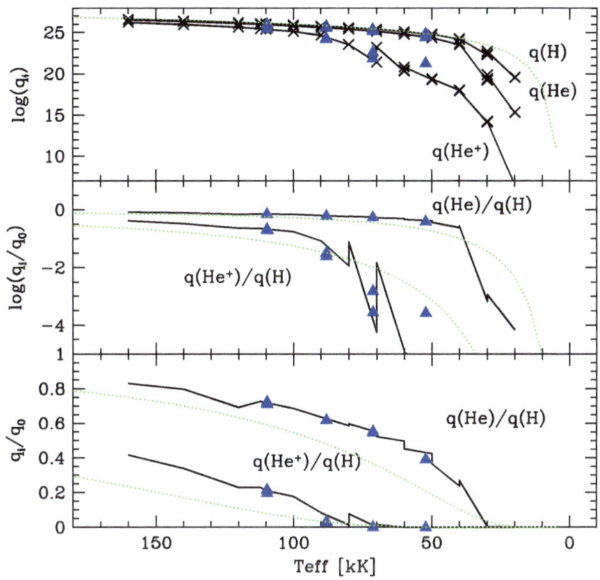

Fig. 2 Ionising photon flux per unit surface (*top panel*) and hardness of the ionising spectrum (*middle and bottom panel*) as a function of effective temperature predicted from non-LTE plane parallel *TLUSTY* Pop III models of various $T_{\rm eff}$ and log g (*solid black lines*). *Triangles* show calculations from the spherically expanding *CMFGEN* models for Pop III stars. *Green dashed lines* show predictions from blackbody spectra. Similar to Fig. 2 from Schaerer (2002)

energies, for the number of He and He^{+} ionizing photons, and for other features such as the Lyman break (cf. Schaerer 2002). Also, using black-body spectra is a bad approximation to compute observable properties such as UV fluxes and emission line strengths (equivalent widths), since black-bodies do not reproduce the main bound-free edges, which significantly shape true stellar spectra. See Fig. 3 for an illustration. In short, appropriate state-of-the-art stellar atmosphere models should be used for reliable predictions both of individual stars and integrated stellar populations.

2.3 *IMF*

As for other ingredients summarized above, the stellar initial mass function (IMF) is a separate topic, to which many papers and conferences are devoted (see e.g. Gilmore and Howell 1998; Bastian et al. 2010; Treyer et al. 2011)

In the nearby Universe the IMF is well described by a log-normal function for masses below $\sim$1 $M_{\odot}$, and by a power-law above that (Chabrier 2003), as already found e.g. in the pioneering study of Salpeter (1955). Other authors approximate the IMF by piecewise power laws (Kroupa 2001). In our Galaxy and in other nearby systems it is found that the IMF is only weakly dependent on environment (Chabrier 2003; Bastian et al. 2010). For obvious reasons it is not possible to constrain the IMF in the same manner (e.g. by direct star counts) outside the local Group. Hence our empirical knowledge on the IMF and possible variations with metallicity, stellar density, UV radiation field and other factor which may influence it, is very limited.

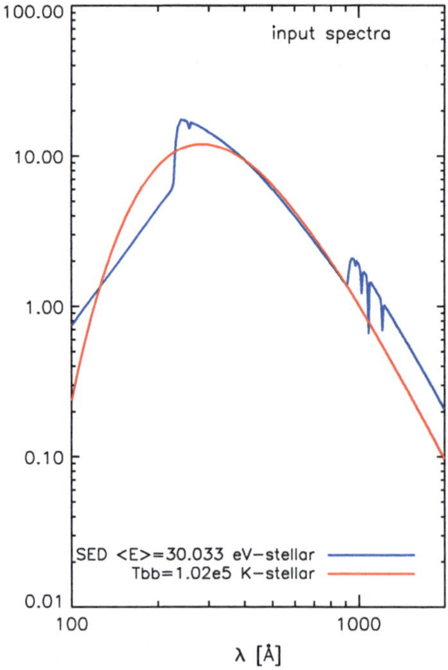

Fig. 3 Comparison of input stellar spectra for photoionization models showing a realistic SED (*blue line*; Pop III, Salpeter IMF 1–100 $M_\odot$ at zero age) and a black body spectrum (*red line*) with the same average photon energy ($\overline{E} = 30.033$ eV) in the Lyman continuum. The SED model is normalized at 1,240 Å and rescaled to match the same ionizing flux $Q(H)$. Despite having the same Lyman continuum flux and average mean energy, the black body differs significantly, e.g. in the He^+ continuum ($\lambda < 228$ Å), and in the observable UV. Figure from Raiter et al. (2010b)

For many/most observables measurable for high-redshift, star-forming galaxies only the upper part of the IMF is relevant, since massive stars largely dominate the luminosity in integrated stellar populations. As a reasonable rule of thumb one may therefore simply rescale the results derived for one given IMF to those expected for another IMF differing in the domain of low stellar masses (i.e. typically at $M \ll 5$ $M_\odot$). This is e.g. used to "correct" results from synthesis models computed for a Salpeter IMF extrapolated down to 0.1 $M_\odot$ to those expected for a more realistic IMF below $\lesssim$1 $M_\odot$. However, the applicability of this "rule" depends on the age and star-formation history, and on the observable used.

Several studies, following different arguments have suggested deviations from the "local" IMF in distant galaxies with deviations depending on environment, galaxy type, gas density, metallicity and other factors. For an overview over this vast topic the reader is referred to a recent conference proceedings on the IMF (Treyer et al. 2011). For metal-poor environments and Population III stars, simulations have long suggested a preferential mass scale much higher than in the local Universe, with "typical" masses of the order of ~10–100 $M_\odot$. See the chapters of Glover and Johnson in this book, and also recent conference proceedings on the first stars (e.g. Hunt et al. 2008; Whalen et al. 2010). This difference, basically due to significantly reduced cooling of metal-poor or free gas, is predicted to occur below a certain critical metallicity $Z_{crit} \leq 10^{-5\pm1}$ $Z_\odot$ (Schneider et al. 2002, 2003). If correct, this implies a different, more massive IMF, at least below this metallicity threshold. Accordingly various parametrisations of the IMF have been used in this domain (see e.g. Larson 1998; Tumlinson 2006; Raiter et al. 2010b). However, as already mentioned above, our knowledge of the IMF, especially in such extreme conditions remains very limited, and synthesis models can simply assume different cases and examine their implications.

Synthesis models also generally make the assumption of a continuous, well-populated IMF, which is correct for stellar systems with a large enough number of stars. However, in "small" stellar populations the analytic statistical description of the IMF will break down, and sampling of the IMF with a small/finite number of stars may lead to significant differences, due to the progressive absence of massive stars. Examples of such "stochastic" IMF effects on colors, ionizing fluxes and others have been illustrated by various authors (see Cerviño et al. 2000; Cerviño and Luridiana 2006; Bruzual 2002, and references therein). Typically stochastic effects are found to be significant for cluster masses below $\lesssim 10^5$ $M_\odot$ or for low star formation rates SFR $\lesssim 1$ $M_\odot$ year^{-1}, although this limit again depends on the observable of interest (Cerviño et al. 2000; Fumagalli et al. 2011). For small entities, such as the first galaxies, stochasticity may thus be important for some cases.

The question of a stochastically sampled IMF is also related to the concept of the integrated galactic initial mass function (IGIMF) introduced by Kroupa and collaborators (see e.g. Pflamm-Altenburg et al. 2007; Weidner and Kroupa 2006). Various applications of synthesis models implementing a stochastic IMF can e.g. be found in Pindao et al. (2002), Fumagalli et al. (2011), and Eldridge (2011). For example, a stochastic IMF leads to a significant scatter in the relative Hα and UV output for populations with a low SFR (i.e. forming a small number of stars), as illustrated in Fig. 4. This is due to the fact that these two emissions originate from stars of somewhat different mass regimes, Hα being due to more massive stars than the UV continuum. Since stochastic effects are of increasing importance for low mass / low SFR galaxies, this issue should be relevant for studies of the first galaxies. The implications remain largely to be worked out.

2.4 Star Formation History

Star formation histories are a key ingredient for evolutionary synthesis models. As already mentioned above, simple stellar populations (SSPs) represent the basic units, and for any arbitrary star formation history (SFH), the integrated spectrum can be derived from SSPs by convolution. Historically, simple parametrisations of the SFH have been used, the most common one being a family of exponentially decreasing SFHs with SFR$(t) \propto \exp(-t/\tau)$, where $\tau > 0$ is a characteristic timescale. It is well known that such star formation histories are able of reproducing the observed spectro-photometric properties of present-day galaxies of all Hubble types (Larson and Tinsley 1978; Tinsley 1968; Kennicutt 1998).

Constant SFR, corresponding to $\tau = \infty$, is a limiting case often used to derive calibrations for the SFR from various observables (e.g. UV continuum, H recombination lines, bolometric luminosity etc.). See e.g. Kennicutt (1998), and Schaerer (2000).

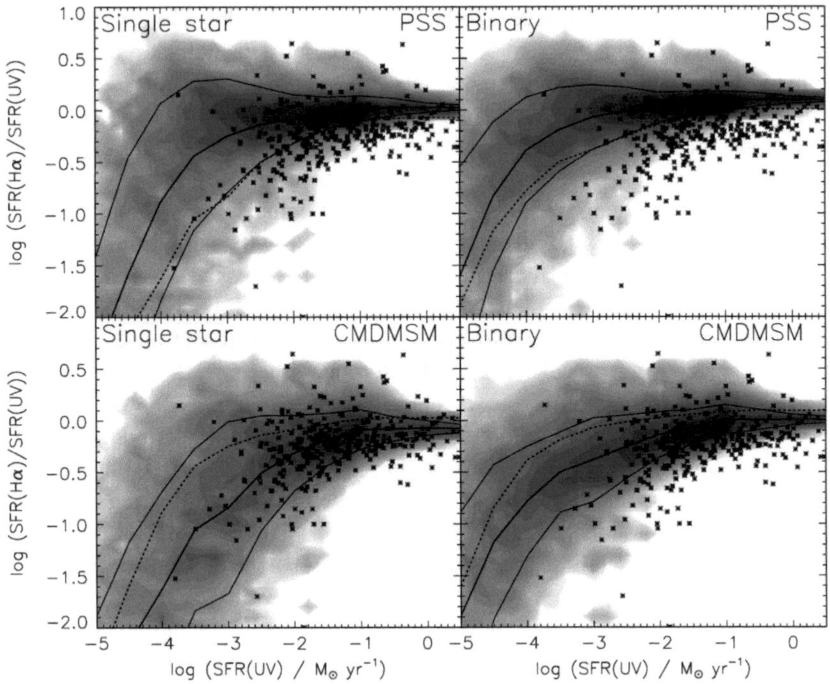

Fig. 4 The ratio of SFR measured by Hα and UV fluxes versus the SFR from UV flux. The *asterisks* are the observations of Lee et al. (2009a), while the *shaded region* show the density of our individual realisations of synthetic galaxies. The *thick solid lines* indicate the mean ratios for the synthetic galaxies and their 1σ limits. The *dashed lines* show the mean ratios for the other IMF filling method with the same stellar population. The *upper and lower panels* distinguish a stochastic sampling of the IMF and a sampling including a description of the cluster mass function. *Left/right panels* distinguish synthesis model including single/(single+binary) stars. From Eldridge (2011)

Depending on the application, other, more complex SFHs have been explored. Many numerical simulations of galaxy formation and evolution (semi-analytical models, hydrodynamic models and others) are now "coupled" with evolutionary synthesis models to predict observable properties consistent with their (complex) star formation histories.

For distant ($z > 2$) galaxies SF histories and corresponding timescales are currently vividly debated. For example, simulations suggest rapid growth of galaxies with increasing star formation rates during the first Gyr of the Universe, i.e. at $z > 6$ (Finlator et al. 2007, 2011). From the apparent tightness of the mass-SFR relation of galaxies at high redshift some authors argue for rising star formation histories for galaxies down to $z \sim$ 2–3 (Maraston et al. 2010; Papovich et al. 2011). At somewhat lower redshift, other authors suggest e.g. "delayed" star formation histories SFR(t) $\propto t \exp(-t/\tau)$ showing both phases of increasing (for $t < \tau$) and decreasing SFR ($t > \tau$), and suggest relatively long timescales of several 100 Myr (Wuyts et al. 2011). From the clustering of $z \sim$ 4–5 Lyman break galaxies

(LBGs), Lee et al. (2009b) argue for relatively short duty cycles (<0.4 Gyr). Even shorter timescales and declining SFHs are favored from SED analysis of Lyman break galaxies at $z \sim 3$–6 with spectral templates accounting for nebular emission (Schaerer and de Barros 2011, de Barros et al. 2012, in preparation). The question of star formation histories of distant galaxies is closely related to the debate about the main mode of star formation, i.e. schematically about the relative importance of (cold) accretion driven star formation and mergers (see e.g. Khochfar and Silk 2011, and references therein) In any case the question of the "typical" star formation history of distant galaxies is not yet settled, and a large diversity of histories—more complicated than simple, parametrised functions—must occur in nature. Furthermore it is clear that predictions from evolutionary synthesis models strongly depend on this important quantity.

2.5 Nebular Emission

Stars (and stellar populations) described above provide a dominant, but not the sole source of emission in the (rest-frame) UV, optical, and near-IR light of galaxies. Emission from ionized regions (the so-called HII regions) of the surrounding ISM is another important contribution, which needs to be taken into account to describe/predict the spectra or SEDs of galaxies in this part of the electromagnetic spectrum. Hence this should, in general, be treated (added) to the prediction from "standard" evolutionary synthesis models describing stellar emission only.

Most galaxy types across the Hubble sequence show at least some signs of emission lines in the optical. Their strength increases towards late types and irregular galaxies, basically due to an increasing ratio of present over average past star formation (cf. Kennicutt 1998). Nearly by definition, nebular emission is associated with star-forming galaxies, since—as long as stars more massive than 5–10 $M_\odot$ are formed—star formation always implies the emission of UV photons with energies >13.6 eV capable of ionizing H and other elements in the ISM, causing a plethora of recombination lines. In addition several processes in HII regions, including recombination and 2-photon emission from the 2 ^{2}S level of hydrogen, produce a continuum emission longward of Lyα whose emissivity increases with wavelength (Osterbrock and Ferland 2006). Hence, nebular emission implies a priori both line and continuum emission. An illustration of the three emission components, stars, nebular lines, and nebular continuum, observed in metal-poor, nearby galaxies is shown in Fig. 5.

To first order, the strength/luminosity of both lines and nebular continuum emission depends on the flux/luminosity of ionizing photons in the Lyman continuum denoted here by Q(H), as can be derived from well known nebular physics, assuming e.g. so-called case A or B and typical nebular densities and temperatures (Osterbrock and Ferland 2006). In this approximation it is straightforward to predict the emission from the major H and He recombination lines and continuum emission for a given source spectrum, such as calculated by evolutionary synthesis

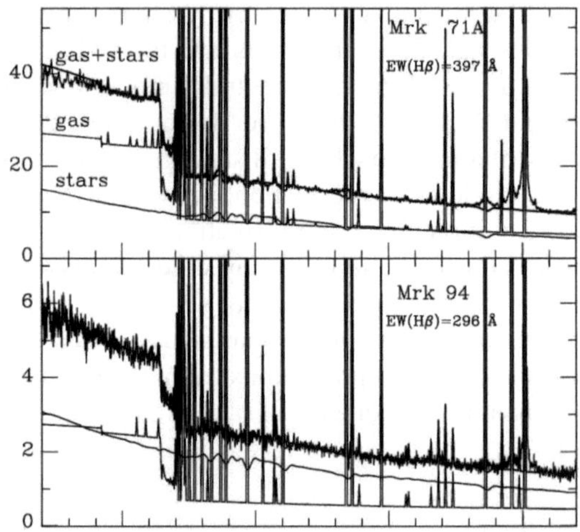

Fig. 5 Observed spectra of two metal-poor HII galaxies from the sample of Guseva et al. (2006) focussing on the region between the UV and optical domain. Superposed are the stellar and gaseous (nebular continuum) contribution. The flux is given in F_λ units between 3,200 and 5,200 Å with spacing of 100 Å between tickmarks. The Balmer jump (not a "break" here), due to nebular free-bound emission, is clearly seen around 3,650 Å; no stellar source is known to show such emission. Figure adapted from Guseva et al. (2006)

models. For other emission lines from HII regions, mostly forbidden metal-lines (in the optical domain), full photoionization models need to be computed, or other prescriptions be used.

To second order nebular emission depends on the conditions in the nebula/ISM, which are primarily described by the electron temperature and density (T_e, n_e), and most importantly on the ionization parameter U. The latter depends on the radiation field, geometry, and density, and in general the nebular conditions also depend on metallicity. It is well known that HII regions and galaxies at low metallicity show higher excitation and stronger optical emission lines, as illustrated e.g. in nebular diagnostic diagrams (Baldwin et al. 1981; Veilleux and Osterbrock 1987), an example of which is shown in Fig. 6.

Since both the excitation of the ionized gas and the ionizing flux increase with decreasing metallicity, nebular emission must on average be stronger in unevolved/early galaxies than in present-day galaxies. How this affects various observables will be discussed in detail below.

Indeed, for Population III stars and for ensembles thereof, Schaerer (2002) has demonstrated that nebular continuum emission dominates the spectrum at wavelength longward of Lyα, as illustrated in Fig. 7. At higher metallicities the effect can also be significant, as discussed below. In some metal-poor, nearby galaxies, nebular continuum emission is readily observable in the UV-optical domain (see Fig. 5), and signs of nebular emission are frequently observed in spectra and photometry of starburst galaxies, e.g. in diffuse, outer regions, or in the near-IR

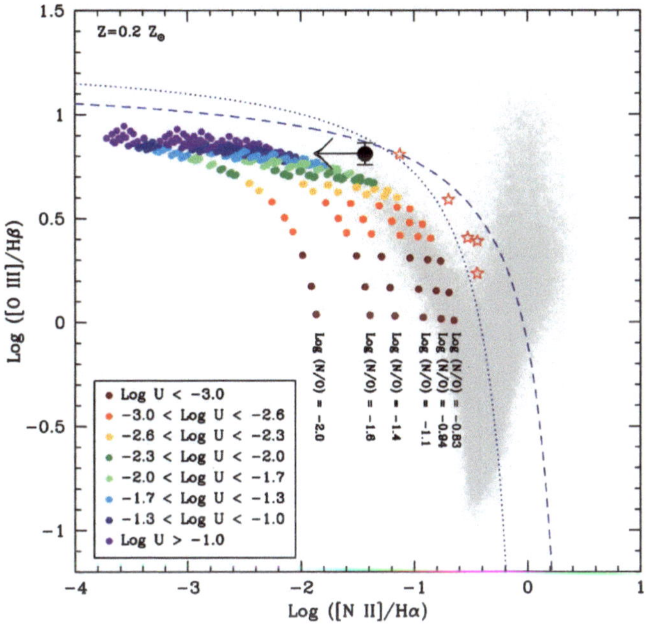

Fig. 6 The [NII] /Hα versus [OIII] /Hβ diagnostic diagram. *Small grey points* show ~96,000 objects from the Sloan Digital Sky Survey, the *large black circles* shows the peculiar, low-metallicty $z \sim 2$ galaxy BX418 discovered by Erb et al. (2010). *Colored points* are predictions from photoionization models, coded by ionization parameter as labeled at *lower left*. From Erb et al. (2010)

domain in young star-forming galaxies such as HII galaxies, blue compact dwarfs, and similar objects (cf. Izotov et al. 1997; Papaderos et al. 2003; Izotov et al. 2001; Vanzi et al. 2002; Guseva et al. 2004). At high redshift a peculiar, lensed galaxy (the so-called Lynx arc at $z = 3.357$) has been found, whose spectrum appears to be dominated by nebular continuum emission and lines (Fosbury et al. 2003). A similar explanation has been suggested for a $z = 5.5$ galaxy (Raiter et al. 2010a), although alternative explanations based on more conventional galaxy SEDs can also be found (Vanzella et al. 2010). In any case, the objects and conditions mentioned here represent probably just somewhat extreme examples illustrating the potential effects of nebular emission, and their detailed importance needs to analyzed on a case-by-case basis. However, there is growing evidence that nebular emission plays a role for the interpretation of observations of many/most Lyman break and Lyα emitter galaxies at $z \gtrsim 3$ (cf. Schaerer and de Barros 2011, and discussion below).

Several evolutionary synthesis models, such as PÉGASE, GALEV and others have long included nebular emission (continuum emission and also lines in some cases, and relying on different assumptions) (cf. Fioc and Rocca-Volmerange 1999; Leitherer et al. 1999; Charlot and Longhetti 2001; Anders and Fritze-v. Alvensleben 2003; Zackrisson et al. 2008). However, other widely used codes (e.g. Bruzual and Charlot 2003) neglect this component. To interpret emission line observations of HII

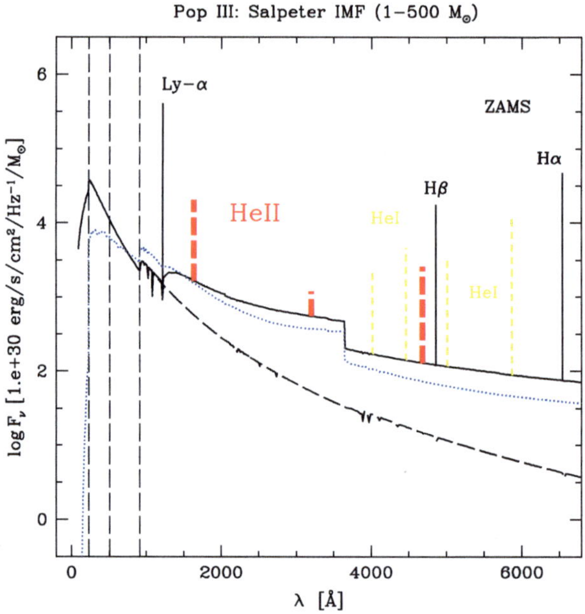

Fig. 7 Spectral energy distribution of a young (zero age) PopIII starburst with an IMF extending to 500 $M_{\odot}$. The SED shown in *full lines* includes H and He recombination lines, nebular, and stellar continuum. The pure stellar continuum (neglecting nebular emission) is shown by the *dashed line*. For comparison the SED of the $Z = 1/50 Z_{\odot}$ population is shown by the *dotted line*. The *vertical dashed lines* indicate the ionisation potentials of H, He^0, and He^+. Note the presence of the unique He II features (shown as *thick dashed lines*) and the importance of nebular continuous emission. From Schaerer (2002)

regions and galaxies, the stellar spectra predicted by evolutionary synthesis codes have often been used as input for photoionisation models (Garcia-Vargas et al. 1995; Stasińska and Leitherer 1996; Stasińska et al. 2001; Kewley et al. 2001). Recently, nebular emission has again received considerable attention, in particular for the interpretation of photometric observations of distant Lyman break galaxies (LBGs) and Lyα emitters (LAE) as discussed below (Schaerer and de Barros 2009, 2010; Ono et al. 2010; Schaerer and de Barros 2011; Acquaviva et al. 2011).

Now that all emission sources have been "assembled", we need to briefly discuss absorption/attenuation professes occurring along the lines-of-sight to the observer.

2.6 Attenuation Law

To compare predictions of evolutionary synthesis model with observations, the effect of interstellar reddening (extinction and/or attenuation) obviously needs to be taken into account. Simple prescriptions are generally used, describing e.g. the mean observed attenuation law of star-forming galaxies (e.g. the "Calzetti law" Calzetti 2001) or various laws describing extinction in the Galaxy or the SMC. At very high redshift (Gallerani et al. 2010) have recently argued for an attenuation law differing somewhat from the Calzetti law, probably due to different dust composition from supernovae dominating in the early Universe.

In the nearby Universe, stellar emission is generally found to be less attenuated than emission lines (Calzetti 2001), which is attributed to geometrical effects. If this also applies to distant star-forming galaxies remains, however, controversial and may vary from case to case (e.g. Hainline et al. 2009; Yoshikawa et al. 2010). Other complication arise e.g. in the presence of multiple stellar populations suffering from different attenuations. More complex prescriptions and additional assumptions are needed to describe such cases (see e.g. Charlot and Fall 2000; Granato et al. 2000).

In concrete applications of synthesis models to fitting (restframe) UV-optical photometric observations of star-forming galaxies attenuation/extinction is often degenerate with the age of the stellar population, since both lead to a redder SED on average. This well-known age-reddening degeneracy is illustrated e.g. in studies of simple stellar populations (see Cid Fernandes and González Delgado 2010), distant star-forming galaxies (Pelló et al. 1999), distant red galaxies (Pozzetti and Mannucci 2000), and in many other works. Obviously, the use of other, more sensitive indicators of age and/or attenuation, e.g. spectral lines, the Balmer break etc. (cf. Cid Fernandes et al. 2001), can reduce such degeneracies. A more detailed discussion of this and other issues related to the fitting of observed SEDs and spectra is beyond the scope of this chapter.

2.7 IGM

Finally, the collective effect of the Lyman forest on predicted spectra of distant galaxies must also to taken into account. Indeed for redshifts above $z \gtrsim 3$–4 the IGM significantly reduces the flux shortward of Lyα in the restframe of the galaxy causing the Lyman (continuum) break at 912 Å to shift rapidly to Lyα ($\sim$1,216 Å). Generally the average IGM attenuation is described by a simple expression derived from statistical analysis of the Lyα forest, e.g. following (Madau 1995) or recent updates of this work (Faucher-Giguère et al. 2008).

A related, important question in several contexts concerns the transmission of Lyα photons emitted by high redshift galaxies. Indeed, as strong Lyα emission is one of the expected signatures of primordial and very metal-poor stellar populations (cf. below), knowing its escape fraction out of galaxies and the subsequent transmission through the surrounding IGM are of fundamental importance. By the same token, observations of Lyα emission (e.g. the Lyα luminosity function of galaxies) and its evolution with redshift are used to constrain the IGM transmission and hence the reionization history of the Universe. For more information on these related topics we refer the reader to a vast literature including e.g. Haiman (2002), Fan et al. (2006), Schaerer (2007), Hayes et al. (2011), Dayal et al. (2011), Lidman et al. (2012) and references therein.

3 From Present-Day Metallicities Back to the First Galaxies

The *Starburst99* evolutionary synthesis models (Leitherer et al. 1999) and the models of Schaerer (2002, 2003) allow one to examine in detail the dependence of the expected observational properties of star-forming galaxies with metallicities from current (solar) metallicities to metal-poor and metal-free cases. Furthermore Schaerer (2002, 2003) and Raiter et al. (2010a) also present calculations for a variety of different IMFs, such that IMF or combined IMF/metallicity changes can be examined. We here summarize the behavior of some of the main observables based on these models.

3.1 UV-Optical: Stellar and Nebular Continuum Emission

Figure 8 shows variations of the UV spectrum, including the ionizing (<912 Å) and non-ionizing part ($\lambda > 912$ Å) of a young stellar population with metallicity. Three main features are immediately clear: First, due to the increase of the average stellar temperature, the stellar spectrum becomes harder/bluer with decreasing metallicity Z. Second, the contribution of nebular continuum emission increases concomitantly. At young ages, the continuum longward of Lyα(>1,216 Å) will be dominated by nebular emission. Finally, H and He emission lines become also stronger with decreasing Z (cf. below). The models in Fig. 8 assume so-called case B recombination, i.e. in particular that all ionizing photons are absorbed within the surrounding HII region. This case obviously maximizes the nebular emission, whereas in a more general case where a fraction f_{esc} of Lyman continuum photons escape, nebular emission will be decreased; see e.g. Inoue (2011) and Fig. 9.

The color of the UV continuum and its variations can be quantified e.g. by the exponent β of a power law adjusted to a specific spectral region. This so-called "β-slope" determined from models and measured by spectroscopy or photometry is often used in the literature, in particular to estimate the amount of dust attenuation in high-z star-forming galaxies (see e.g. Meurer et al. 1999; Bouwens et al. 2009, and see Chapter 2.2).

Figure 10 shows the dependence of the predicted β-slope (here measured between 1,300 and 1,800 Å) of a young (zero age) population on metallicity and on the IMF (coded by different colors). Figure 11 shows the same quantity as function of age for instantaneous bursts and constant SFR. From Fig. 10 we see how the *stellar spectrum* steepens (becomes bluer) with decreasing Z. However, once nebular emission is taken into account (here with $f_{esc} = 0$) the UV spectrum becomes significantly flatter and the trend with metallicity even inverses! In short, the UV slope cannot be a good metallicity indicator. Figure 11 shows the rapid evolution of β on timescales of ~50–100 Myr. For constant SFR, β reaches the typical asymptotic value of $\beta \sim -2$ to -2.4, the precise value depending on

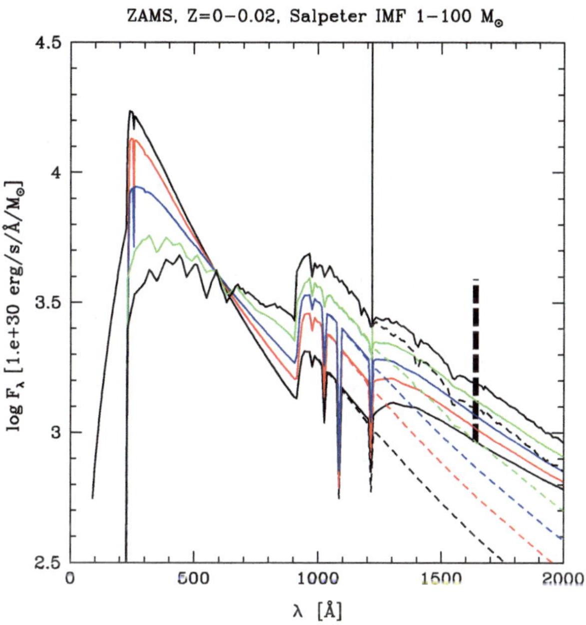

Fig. 8 Predicted SEDs including Lyα and He II λ1640 emission lines for zero age main sequence (ZAMS) models at different metallicities. The metallicities $Z = 0$. (Pop III), 10^{-7}, 10^{-5}, 0.0004, and 0.02 (solar) are from top to bottom in the EUV ($\lambda < 912$ Å), and reversed at longer wavelengths. The *dashed lines* are the pure stellar emission, the *solid lines* show the total (stellar + nebular) emission. A Salpeter IMF from 1–100 $M_{\odot}$ is assumed here for all metallicities. From Schaerer (2003)

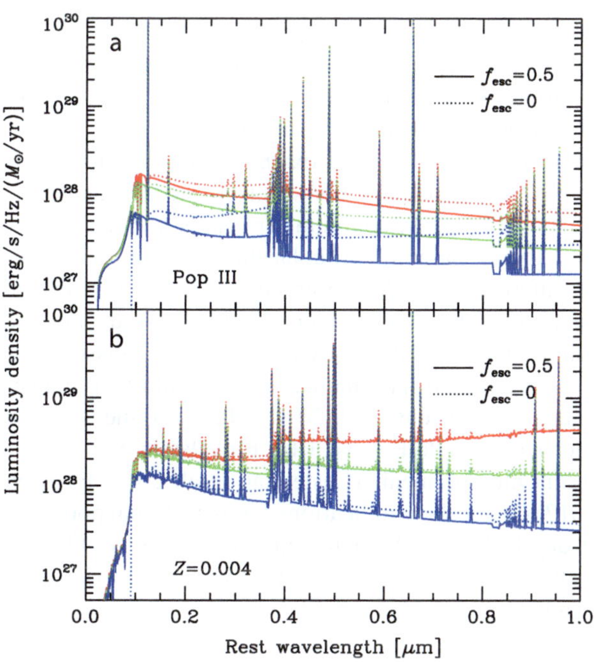

Fig. 9 Predicted model spectra of a PopIII galaxy (*top panel*) and a moderate sub-solar metallicity ($Z = 1/50 Z_{\odot}$, *bottom panel*), at ages of 10, 100, and 500 Myr (from bottom to top in each panel), evolving with a constant SFR. *Solid and dotted lines* show cases computed for 50 % and no escape of Lyman continuum photons. Figure from Inoue (2011)

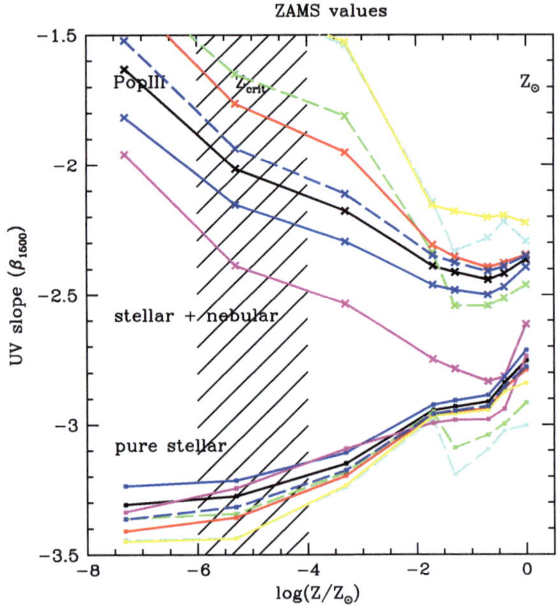

Fig. 10 Predicted UV slope β_{1500} for different IMFs (*color codes*) plotted as a function metallicity. The values are shown for very young (ZAMS) populations, which correspond to the bluest possible slopes (i.e. minimal β values). The *upper set of lines* shows the UV slopes of the total spectrum (stellar + nebular continuum), the *lower lines* using the pure stellar spectrum. *Black lines* show the predictions for a Salpeter IMF slope. All color codes are given in Raiter et al. (2010b), from where the figure is taken

the wavelength base used to define β. Shifts of ~0.1–0.3 in β can be typical, as illustrated e.g. by Raiter et al. (2010b). For PopIII we see that the nebular contribution vanishes after $\gtrsim$ 20 Myr in a burst; between ages of ~10–100 Myr an integrated population can reach a very blue spectrum with $\beta \sim -2.5$–3. Otherwise, and for more extended periods of star formation it is difficult, if not impossible, to obtain a UV spectrum steeper than $\beta \sim -2.5$, except if a significant fraction of Lyman continuum photons are "leaking" (i.e. $f_{esc} > 0$). Overall, the two figures show in particular the following: First, in principle metallicity cannot be inferred from the observed UV slope (cf. Schaerer and Pelló 2005). Second, very steep/blue UV spectra ($\beta \ll -2.5$) are not predicted for very metal-poor, primordial, populations, except if nebular emission is "suppressed".

Some authors have indicated the possibility of unusually blue UV slopes in $z \sim 7$ galaxies (Bouwens et al. 2010). However, the significance of these findings is low (Schaerer and de Barros 2010; Finkelstein et al. 2010), and independent and more recent measurements (Castellano et al. 2011; Bouwens et al. 2011; Dunlop et al. 2012) do not show indications for exceptional populations or conditions (PopIII and strong leakage of Lyman continuum photons). Other observational aspects related to the UV slope of distant star-forming galaxies are discussed in the Chapter of Dunlop.

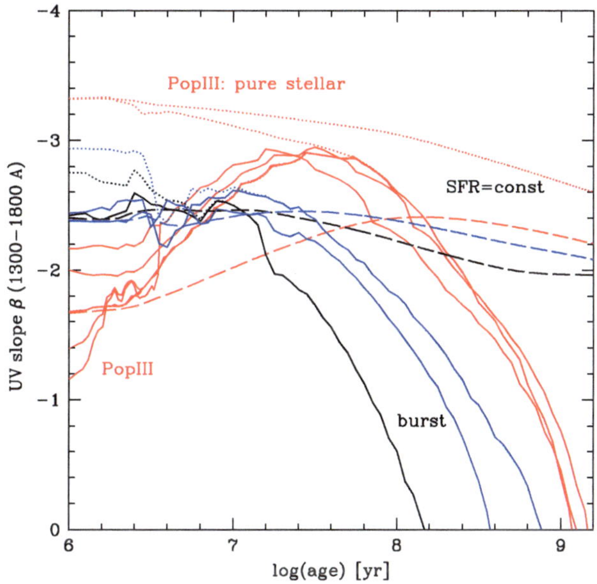

Fig. 11 Temporal evolution of the UV slope β_{1500} derived from synthesis models of different metallicities and for instantaneous bursts (*solid lines*) and constant SF (*long dashed lines*). *Black lines* show solar metallicity models, *red lines* show metallicities between $Z = 10^{-5}$ and zero (PopIII) and *blue lines* show intermediate cases of $Z = 0.004$ and 0.0004. The *dotted lines* show β if nebular continuous emission is neglected, that is, assuming pure stellar emission. Note especially the strong degeneracies of β in age and metallicity for bursts, the insensitivity of β on Z for constant SF, and the rather red slope for young very metal-poor bursts. From Schaerer and Pelló (2005)

3.2 *Ionizing Photon Production*

To quantity the contribution of galaxies to cosmic reionization, and to predict their observable spectra it is important to know the amount of energy or photons emitted in the Lyman continuum, i.e. at $\lambda < 912$ Å or at energies $E > 13.6$ eV. This quantity is straightforwardly predicted by evolutionary synthesis models, and is generally expressed as a photon flux (e.g. in units of photons per second) normalized for example per unit stellar mass, per unit SFR, or so. Another, observationally relevant way to express the Lyman continuum flux is by normalizing it to the UV output of the same stellar population, since the latter is a direct observable. This prediction is shown, for a constant star formation rate over relatively long timescales, as a function of metallicity and for different IMFs in Fig. 12, taken from Raiter et al. (2010b).

As expected Q_H/L_{1500} increases with decreasing metallicity, since the ionizing flux depends very strongly on the effective stellar temperature and hence increases more rapidly than the UV luminosity. The IMF dependence also behaves as expected, with the IMFs favouring the most massive stars showing also the highest the Q_H/L_{1500} ratios, since Q_H increases more rapidly with stellar mass than the

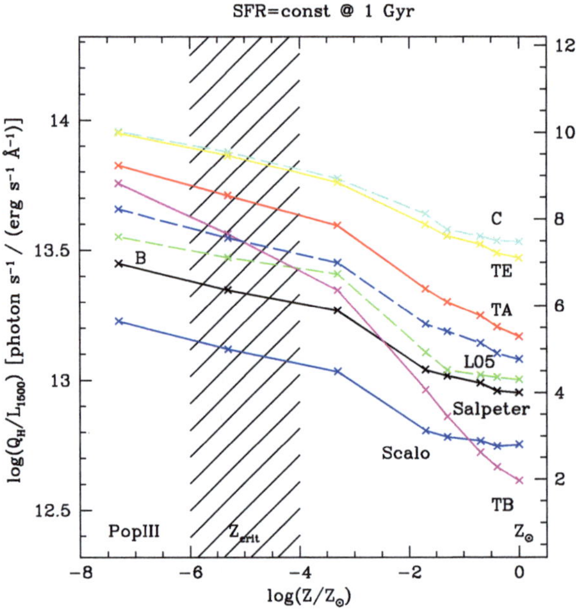

Fig. 12 Predicted Lyman continuum flux as function of metallicity. The relative output of hydrogen ionizing photons to UV continuum light, measured at 1,500 Å restframe, Q_H/L_{1500}, is shown as a function of metallicity for constant star formation over 1 Gyr. Q_H/L_{1500} is given in L_λ units on the left side of each panel, and in L_ν units on the right. Results for different IMFs are shown using the color codes and labels summarised in Table 1 of Raiter et al. (2010b). The *shaded area* indicates the critical metallicity range where the IMF is expected to change from a "normal" Salpeter-like regime to a more massive IMF (see text). From Raiter et al. (2010b)

UV luminosity. For constant SFR and for a fixed IMF, the increase of the relative ionizing power from solar metallicity to Pop III is typically a factor 2–3. When considering an IMF change from Salpeter to a massive IMF (i.e. all cases except Salpeter and Scalo) the increase of Q_H/L_{1500} is larger, approximately 0.6–1 dex between solar and zero metallicity. For a very young (zero age) population the Lyman continuum production increases typically by $\lesssim$40 %.

Obviously, the ionizing photon output also depends on age and the assumed star formation history, not illustrated here (see e.g. Schaerer 2002, 2003; Raiter et al. 2010b, where other normalisations are also used). For example, for zero age stellar populations the ionizing photon output per UV flux, Q_H/L_{1500}, is higher than shown here, typically by up to a factor 2–4 depending on the IMF (see Fig. 1 of Raiter et al. 2010b). For very "massive" IMFs (i.e. IMFs favoring very massive stars) the predictions shown in Fig. 12 for constant star formation over long timescales (up to 1 Gyr) are of course very similar to those for much shorter, probably more realistic timescales for the first galaxies, since in any case the lifetimes of the bulk of the stars is much shorter than 1 Gyr. For general ages and star formation histories these quantities can be derived from the available data files of the models.

3.3 Lyα Emission

Given predictions for the Lyman continuum flux from a population, the flux in the different H and He recombination lines and nebular continuum emission from the surrounding HII region can easily be computed, as described above (Sect. 2.5), using simple case A or B recombination theory or computing detailed photoionization models. The equivalent widths of these lines—a very useful measure of their strength—can also be computed, given the stellar and nebular continuum.

Since intrinsically the strongest line, and conveniently located in the rest frame UV domain, the Lyα line is well known to be central to many studies of distant/primeval galaxies, in particular since the early work of Partridge and Peebles (1967) and the discovery of large populations of high redshift galaxies (e.g. Hu et al. 1998; Ouchi et al. 2008, and many other papers on such observations). Before the era of synthesis models and based on relatively crude assumptions, Partridge and Peebles (1967) estimated that up to $\sim$10 % of the bolometric luminosity of primeval galaxies could be emitted in the Lyα line. For constant star formation the most recent update, using our current knowledge of stellar evolution, atmospheres and the Salpeter IMF, place this number at $\sim$3 % for solar metallicity, as shown in Fig. 13. For the same IMF this fraction is expected to be significantly higher at low metallicity, especially below $Z \lesssim 1/50\ Z_{\odot}$, where departures from case B (due to collisional effects because of the high electron temperature in the nebula) significantly increase the ionization and hence also recombination rates, as recently showed by Raiter et al. (2010b). As illustrated in Fig. 13, the Lyα line can therefore carry up to $\sim$20–40 % (depending on the IMF) of the bolometric luminosity in primordial gas. As already discussed above for the ionizing photon flux, the precise fraction of the Lyα luminosity obviously also depends on the age and star formation history. In any case, for the most extreme cases shown here (with IMFs dominated by massive stars), the predictions in Fig. 13 do not vary much with age, since the bulk of stars in such populations have anyway short lifetimes.

The maximum strength of Lyα, measured by its equivalent width $W_{\mathrm{Ly}\alpha}$, is recognized as an interesting diagnostic of young, metal-poor/metal-free stellar populations in the first galaxies. Indeed, very high equivalent widths—well beyond the maximum $W_{\mathrm{Ly}\alpha}$ of $\sim$200–250 Å predicted for solar metallicities—are expected for such populations. As such, or together with other unique spectral features of He II discussed below, this line is of great interest to search for "unusual" stellar populations (see e.g. Malhotra and Rhoads 2002). While various initial predictions of the Lyα equivalent widths have yielded somewhat different results (see e.g. Tumlinson et al. 2001; Bromm et al. 2001), it is now clear that non-LTE atmosphere models, the treatment of nebular emission, and departures from case B, are essential ingredients to properly predict the strength and equivalent widths of Lyα emission from very metal-poor populations (see Schaerer 2002, 2003; Raiter et al. 2010b).

With these ingredients, the maximum $W_{\mathrm{Ly}\alpha}$ predicted as a function of metallicity and for different IMFs reaches $\sim$2,000 Å in the rest-frame (for a Salpeter IMF), or higher for IMFs favoring more massive stars, as illustrated in Fig. 14. In

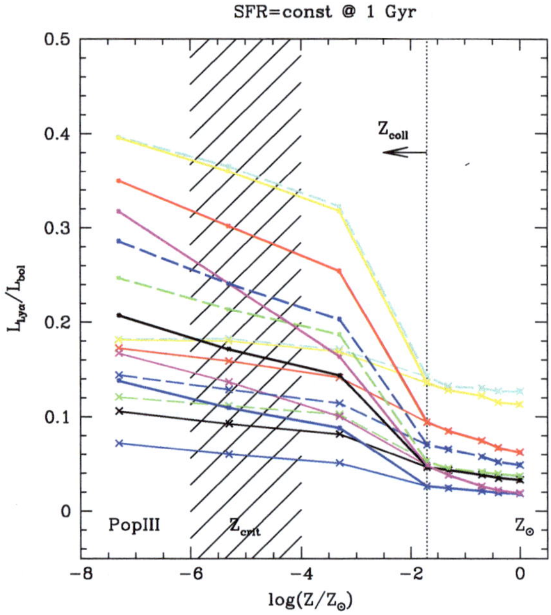

Fig. 13 Fraction of the Lyα luminosity to the total bolometric luminosity, $L(\mathrm{Ly}\alpha)/L_{\mathrm{bol}}$ for SFR = const as a function of metallicity and IMF. Results for different IMFs are shown using the same colour codes as in same color codes as in Figs. 10, 12. *Thin lines* show the results using standard case B recombination; *thick lines* the recent results accounting for departures from case B at very low metallicity. Note the strong increase of the predicted $L(\mathrm{Ly}\alpha)/L_{\mathrm{bol}}$ values from solar to very low metallicity. From Raiter et al. (2010b)

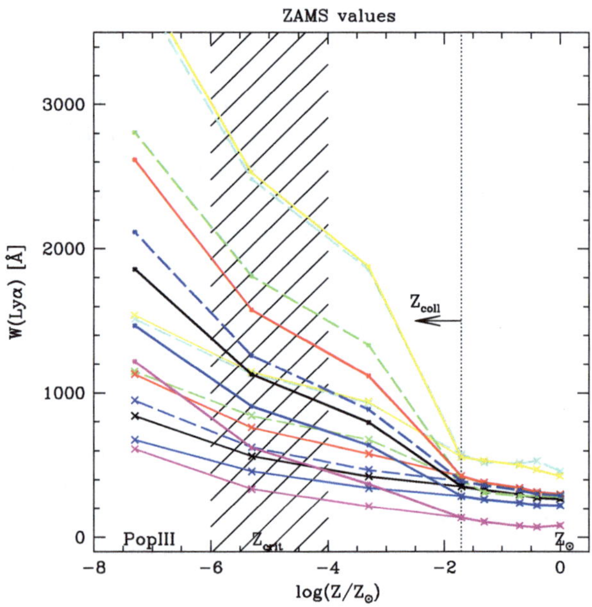

Fig. 14 Predicted Lyα equivalent width as a function of metallicity for very young ($\lesssim$ 1–2 Myr). *Thin lines* show standard case B predictions, *thick lines* the predicted $W(\mathrm{Ly}\alpha)$ accounting to first order for departure from case B, leading to an increase by up to a factor ∼1.5–2.5 at low metallicities ($Z \lesssim Z_{\mathrm{coll}}$). Same color codes as in Figs. 10, 12, 13. From Raiter et al. (2010b)

principle, observations of such large equivalent widths of Lyα, if attributable to photoionization from stars (as opposed to non-thermal sources), should be strong sign-posts of extreme conditions (metallicity and/or IMFs) expected in the first galaxies. Possible complications for the application of such a diagnostic include the effect of the IGM and dust/radiation transfer inside galaxies, which can significantly reduce the Lyα flux (and more so than the nearby UV continuum), and hence reduce $W_{\mathrm{Ly}\alpha}$. Furthermore the Lyα line emission varies rapidly with time (on timescales of $\sim 10^7$ Myr). Neufeld (1991) has suggested that radiation transfer in a clumpy ISM could increase $W_{\mathrm{Ly}\alpha}$, which, if applicable, could complicate the interpretation of high $W_{\mathrm{Ly}\alpha}$ objects. In any case it is clear that galaxies with unusually high Lyα equivalent widths ($\gtrsim$200–250 Å) are interesting candidates worth examine further in searches for very metal-poor and PopIII galaxies.

3.4 Hardness of the Ionizing Spectrum and Helium Line Emission from the First Galaxies

As already mentioned above, the first stellar populations are expected to contain unusually hot massive stars whose ionizing spectra will be harder than that of "normal" present-day massive stars. As a consequence these stars will emit in particular more ionizing photons above 54 eV ($\lambda < 228$ Å), the energy required to fully ionize Helium in the ISM surrounding these stars. This will give rise to recombination lines of He II observable in the UV and optical domain (e.g. He II λ1640, He II λ4,686), which should therefore be a fairly unique signature of these energetic, metal-poor stars.

The use of He II lines to identify PopIII stars/galaxies was discussed by Tumlinson and Shull (2000), Tumlinson et al. (2001), Bromm et al. (2001), Oh et al. (2001), and Schaerer (2002, 2003), who present predictions for the strength of these lines. Schaerer (2003) and more recently Raiter et al. (2010b) have discussed the transition from zero metallicity to present-day conditions. The predicted hardness of the ionizing flux of starbursts at different metallicities and the behavior of the He II λ1,640 line as an example, are shown in Figs. 15 and 16. The main result is indeed a strong increase of the hardness of the ionizing flux, typically by 2 orders of magnitudes from metallicities $Z \sim 1/50\ Z_{\odot}$ to PopIII (zero metallicity). The (maximum) equivalent width of He II λ1640 (again taking the "dilution" by both the stellar and nebular continuum into account) is generally very low. Values above $\gtrsim$5 Å (rest frame) are only expected for young, very metal-poor stellar populations. One should, however, remember that these predictions are based on simple assumptions (case B, ionization bounded regions), which may not always apply.

Johnson et al. (2009) have computed the emission in H and He II recombination lines from hydrodynamic simulations taking into account the (time-dependent) leakage of Lyman continuum and He$^+$ ionizing photons. They find a smaller leakage for the higher energy photons, as expected, since for stellar sources these less abundant photons are absorbed closer to the source than H ionizing photons. They therefore conclude that the He II equivalent width should indeed be a fairly robust indicator for PopIII.

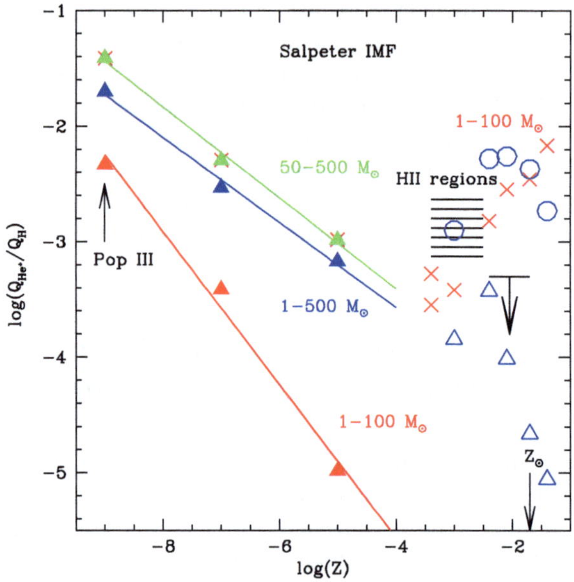

Fig. 15 Hardness $Q(\text{He}^+)/Q(\text{H})$ of the He^+ ionising flux for constant star formation as a function of metallicity (in mass fraction) for three different power-law IMFs. At metallicities above $Z \geq 4.10^{-4}$ the predictions from our models (*crosses*), as well as those of Leitherer et al. (1999, *open circles*), and Smith et al. (2002, *open triangles*) are plotted. The *shaded area* and the upper limit (at higher Z) indicates the range of the empirical hardness estimated from H II region observations (see discussion in Schaerer 2003). From Schaerer (2003)

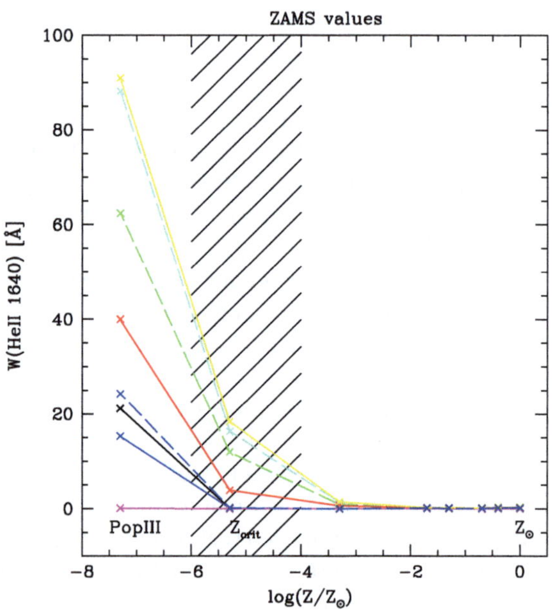

Fig. 16 Predicted He II λ1640 equivalent width as a function of metallicity for very young ($\leq$1–2 Myr) bursts. Same color codes as in Figs. 10, 12, 13. Note that photoionization models predict generally fainter He II λ1,640 emission, hence lower equivalent widths, except for high ISM densities. From Raiter et al. (2010b)

As for Lyα, the prediction of the intrinsic emission in Helium lines may also be more complicated than expected from simple "photon-counting" assumed for case B (and implemented in most synthesis models). The basic reason for this is that both H and He in the ISM compete for ionizing photons, which leads to a lower He^+ ionization rate in regions of low ionization parameter. This effect, already discussed

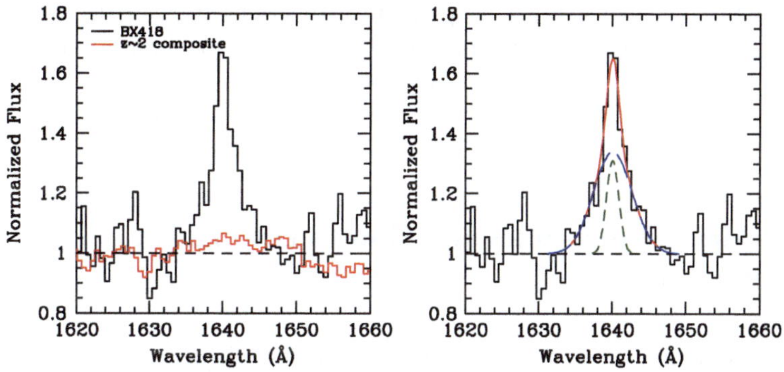

Fig. 17 *Left:* He II λ1,640 emission in the peculiar $z \sim 2$ galaxy BX418 (*black*), compared to the broad and far weaker emission in the composite spectrum of 966 $z \sim 2$ galaxies (*red*). *Right:* The line is well fit by a superposition of two Gaussian components (*solid red line*). The broad component (*long-dashed dark blue line*) has FWHM $\sim$1,000 km s^{-1}, while the narrow component (*short-dashed green line*) is unresolved. The broad emission is attributed to W-R stellar winds, and the narrow component to nebular He II emission. The line has a total equivalent width of 2.7 Å, considerably smaller than expectations for very metal-poor stellar populations. Figure from Erb et al. (2010)

by Stasińska and Tylenda (1986) for planetary nebulae, implies that the He II line flux may be lower than predicted by evolutionary synthesis models, as shown by Raiter et al. (2010b). Detailed photoionization models are necessary to properly account for this effect.

Finally—as always—it is also useful and important to examine what is known empirically about He II emission. As well known, some stars (mostly the so-called Wolf-Rayet stars, evolved massive stars) show emission in He II lines, and these lines are visible in the integrated spectra of some galaxies (sometimes eluded to as Wolf-Rayet galaxies, cf. Conti 1991; Schaerer et al. 1999; Brinchmann et al. 2008). The He II λ1640 line is also seen in some spectra (individual or stacked) of high redshift galaxies (e.g. Shapley et al. 2003; Erb et al. 2010), as shown in Fig. 17. The observed equivalent widths remain, however, small compared to the values expected at very low metallicities. Due to the strong winds in the atmospheres of Wolf-Rayet stars these lines are broad, extending typically up to several thousand km/s (Conti et al. 2008). In spectra of sufficient signal-to-noise and spectral resolution it should therefore be possible to separate any stellar emission from nebular He II emission (cf. Fig. 17). Nebular He II emission is observed in some low-metallicity HII regions and starburst galaxies in the nearby/low-z Universe (see e.g. Schaerer et al. 1999, for a compilation and references therein). The hardness of the ionizing flux inferred from the relative He II λ4686/Hβ intensities is shown by the shaded region in Fig. 15; it is lower by $\sim$1–1.5 orders of magnitude than what is expected for PopIII dominated objects! However, the origin of nebular He II emission in nearby objects remains difficult to understand in many cases and several sources/mechanisms may contribute to it (see e.g. Schaerer 1996; Izotov et al. 2001; Thuan et al. 2004; Kehrig et al. 2011; Shirazi and Brinchmann 2012).

Despite these open questions, the prediction of very hard spectra from primordial stars and stellar populations remains quite solid, with the main uncertainty probably being the IMF of these stars. Signatures from He II emission should therefore be a crucial tool for observational searches of the first stellar generations.

4 The Main Observables and How to Distinguish Population III?

From the predictions from evolutionary synthesis models discussed in the previous Section it is already apparent how PopIII or very metal-poor populations can be distinguished from those of "normal" metallicities, mostly using H and He recombination lines. Other diagnostics can be derived from integrated colors, at least to some extent. Obviously, direct measurements of metallicity are a third way to tackle this question observationally. We here briefly summarize/describe these various diagnostics.

4.1 Hydrogen and Helium Lines

The predicted behavior of recombination lines of H and He^+ with metallicity has already been discussed above. In short, unusually strong Lyα emission (say EW(Lyα) $\gg$ 250 Å) and/or strong He II emission are among the best indicators expected for very metal-poor stellar populations, due to their unusually high temperatures plus possibly a more "massive" IMF (i.e. favoring more massive stars).

4.2 Metal Lines

The relative line intensities of emission lines from all elements can be predicted from photoionization models, such as *Cloudy* (Ferland et al. 1998). Figure 18 shows the predicted spectra and strengths of several of the strongest lines from photoionization models as a function of metallicity. The left panel shows the overall SED including H and He lines for nebulae computed with blackbody spectra from calculations of Panagia (2005); the right panel the relative intensity of selected metal lines in the (rest frame) UV and optical domain computed using realistic SEDs from synthesis models (Inoue 2011). As is well known, the intensity of the forbidden oxygen lines peaks at low metallicities ($Z/Z_\odot \sim$ 1/50), and it is found to decrease monotonously to lower metallicity. Measuring e.g. a line ratio of [O III] λ5007/Hβ <0.1 would indicate metallicities below $<10^{-3}$ $Z_\odot$ (or alternatively $Z/Z_\odot \gtrsim 4$), which could be feasible with NIRSpec on the James Webb Space Telescope (JWST),

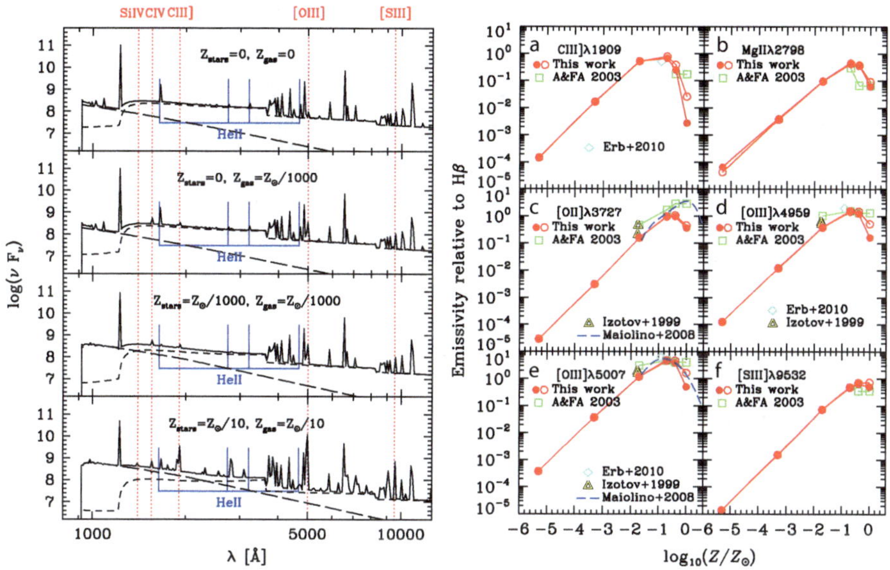

Fig. 18 *Left:* Predicted spectra (lines plus nebular and stellar continuum, *solid*) from photoionization models at different metallicities. He II and few other lines are identified. From Panagia (2005). *Right:* Predicted emissivities relative to Hβ of selected strong metal lines as a function of metallicity from photoionization models (*filled circles*). Other empirical data are shown by *open symbols and the dashed line*. From Inoue (2011)

according to Panagia (2005) and Inoue (2011). For more information on the feasibility of such observations see Chapter 3.3.

4.3 Peculiar Colors

Several recent papers have examined/proposed various criteria which could be used to distinguish PopIII from other stellar populations at high redshift, based on photometry/colors (see Inoue 2011; Zackrisson et al. 2011a,b).

As already discussed above (Sect. 3.1), the UV slope is difficult to use as a metallicity indicator due to degeneracies between age/star-formation history and metallicity, and due to the contribution of nebular continuum emission which considerably "dampens" the metallicity dependence of the UV slope on metallicity. In cases of a large escape fraction of the Lyman continuum flux the latter difficulty is of course not present. Predicted magnitudes and colors for different populations and variable escape fractions are presented in Inoue (2011), and Zackrisson et al. (2011b).

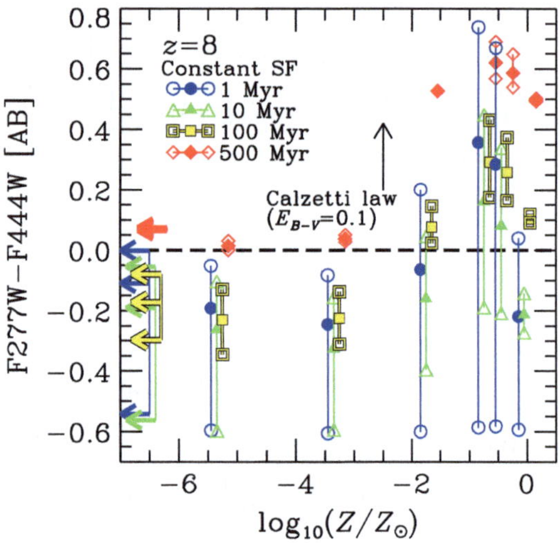

Fig. 19 Predicted JWST/NIRCAM F277W–F444W colors of $z = 8$ galaxies as a function of metallicity. Different symbols/colors illustrate variations of the star formation history. *Vertical lines* connect models with varying Lyman continuum escape fraction. The *horizontal dashed line* is a proposed criterium to separate very metal-poor cases from higher metallicity cases. From Inoue (2011)

Zackrisson et al. (2011b) have proposed that PopIII galaxies at $z \sim 8$ with low escape fractions can be distinguished from metal-rich objects based on two colors between 4.4 and 7.7 μm, which could be measured with the NIRCam and MIRI instrument on the JWST. The basis for such a distinction is the presence of strong H lines (Hα) and the absence of [O III] $\lambda\lambda$4959,5007 in these filters at these redshifts.

Other criteria to select very metal-poor galaxies exploit the expectation of very blue colors between the UV and optical domain, which can to some extent be quite independent of nebular emission, i.e. of the escape fraction (Inoue 2011). Indeed, in case of strong nebular emission, the UV continuum should be strong with a Balmer jump in emission, and emission lines weak in the optical (due to low metallicity). If in contrast the escape fraction was large, one recovers again the intrinsically blue stellar spectrum. In both cases one therefore expects a fairly blue UV–optical color. At $z \sim 8$, for example, the color could be measured between 2.7 and 4.4 μm with the JWST; the corresponding predictions are shown in Fig. 19. They show how such a blue color could be exploited to select very metal-poor ($Z \lesssim 10^{-2}$ $Z_\odot$) galaxies, although in some conditions an overlap also exists with higher metallicities.

PopIII or very metal-poor galaxies with strong Lyα emission(cf. Sect. 3.3) could also be detected due to the excess caused by this line. For example at $z \sim 8$–9 this can cause an unusually blue $J - H$ color, as shown by Pello and Schaerer (2003) and discussed in detail by Zackrisson et al. (2011a). In any case, it is clear that nebular emission (both lines and continua) can significantly contribute to broad-band fluxes (Schaerer and de Barros 2009, 2010, 2011), and Schaerer et al. (2011) have recently shown that SED modeling techniques can recover the strength of Lyα emission from current broad-band photometric surveys of $z \sim 3$–6 galaxies. This demonstrates that various photometric criteria and SED fitting methods should also be able to select peculiar objects, such as very metal-poor galaxies and PopIII dominated objects.

5 Conclusion

As should be clear from the onset, evolutionary synthesis models are an important, fundamental tool to interpret a wide variety of extra-galactic observations, from the nearby Universe to the most distant, first galaxies. They are commonly used to render complex state-of-the-art simulations of the Universe "visible", i.e. to translate physical properties of simulated galaxies into observables. Finally, they are also key for many "prospective" studies, such as for the preparation of new missions, to guide observers searching for Population III objects etc.

Conceptually simple, synthesis models basically gather what is known about star-formation (the IMF and star-formation history), stellar evolution and atmospheres, and some additional emission processes, to predict the temporal evolution of the spectrum of an integrated stellar population. For this reason, synthesis models can basically only be as good/reliable as their input physics is. Therefore regular updates are necessary, and it is important to keep problems and limitations of these "ingredients" and assumptions in mind.

For the spectral modeling of first galaxies, very metal-poor galaxies, and Population III objects, one of the major unknowns is certainly the stellar initial mass function. The evolution of massive stars especially at low metallicities, observationally inaccessible in the nearby Universe, remains also uncertain and efforts are ongoing to properly account for the effects of rotation, magnetic fields and related processes. The star formation histories of young, distant galaxies are probably complex and fairly stochastic at least on short time scales. Some of these questions have been addressed above, at least schematically (Sect. 2).

In any case, combining the best of our knowledge evolutionary synthesis models represent a very useful and important tool for studies of galaxy formation and evolution, from the nearby Universe back to first galaxies.

Acknowledgements I have benefitted from many interesting and stimulating discussions and collaborations with colleagues during the past years. I wish to thank them here collectively for this. I acknowledge support from the Swiss National Science Foundation.

References

Acquaviva, V., Gawiser, E., Guaita, L.: ApJ **737**, 47 (2011)
Anders, P., Fritze-v.Alvensleben, U.: A&A **401**, 1063 (2003)
Baldwin, J.A., Phillips, M.M., Terlevich, R.: PASP **93**, 5 (1981)
Bastian, N., Covey, K.R., Meyer, M.R.: ARA&A **48**, 339 (2010)
Belkus, H., Van Bever, J., Vanbeveren, D., van Rensbergen, W.: A&A **400**, 429 (2003)
Bertelli, G., Bressan, A., Chiosi, C., Fagotto, F., Nasi, E.: A&AS **106**, 275 (1994)
Bouwens, R.J., Illingworth, G.D., Franx, M., et al.: ApJ **705**, 936 (2009)
Bouwens, R.J., Illingworth, G.D., Oesch, P.A., et al.: ArXiv e-prints (2011)
Bouwens, R.J., Illingworth, G.D., Oesch, P.A., et al.: ApJL **708**, L69 (2010)
Brinchmann, J., Kunth, D., Durret, F.: A&A **485**, 657 (2008)

Bromm, V., Kudritzki, R.P., Loeb, A.: ApJ **552**, 464 (2001)
Bruzual, G.: IAU Symposium. In: Geisler, D.P., Grebel, E.K., Minniti, D., (eds.) Extragalactic Star Clusters, vol. 207, p. 616 (2002)
Bruzual, G., Charlot, S.: MNRAS **344**, 1000 (2003)
Bruzual, A.G.: In: Pérez-Fournon, I., Balcells, M., Moreno-Insertis, F., Sánchez, F. (eds.) in Galaxies at High Redshift, pp. 185–221 (2003)
Bruzual, A.G.: in Revista Mexicana de Astronomia y Astrofisica Conference Series, vol. 40, Revista Mexicana de Astronomia y Astrofisica Conference Series, 36–41 (2011)
Calzetti, D.: PASP **113**, 1449 (2001)
Castellano, M., Fontana, A., Grazian, A., et al.: ArXiv e-prints (2011)
Cerviño, M., Luridiana, V.: A&A **451**, 475 (2006)
Cerviño, M., Luridiana, V., Castander, F.J.: A&A **360**, L5 (2000)
Cerviño, M., Mas-Hesse, J.M., Kunth, D.: A&A **392**, 19 (2002)
Chabrier, G.: PASP **115**, 763 (2003)
Charlot, S., Fall, S.M.: ApJ **539**, 718 (2000)
Charlot, S., Longhetti, M.: MNRAS **323**, 887 (2001)
Chiappini, C., Hirschi, R., Meynet, G., et al.: A&A **449**, L27 (2006)
Cid Fernandes, R., González Delgado, R.M.: MNRAS **403**, 780 (2010)
Cid Fernandes, R., Sodré, L., Schmitt, H.R., Leão, J.R.S.: MNRAS **325**, 60 (2001)
Conti, P.S.: ApJ **377**, 115 (1991)
Conti, P.S., Crowther, P.A., Leitherer, C.: From Luminous Hot Stars to Starburst Galaxies. Cambridge University Press, Cambridge (2008)
Dayal, P., Maselli, A., Ferrara, A.: MNRAS **410**, 830 (2011)
Dionne, D., Robert, C.: ApJ **641**, 252 (2006)
Dunlop, J.S., McLure, R.J., Robertson, B.E., et al.: MNRAS **420**, 901 (2012)
Ekström, S., Coc, A., Descouvemont, P., et al.: , A&A **514**, A62 (2010)
Ekström, S., Meynet, G., Chiappini, C., Hirschi, R., Maeder, A.: A&A **489**, 685 (2008)
Eldridge, J.J.: ArXiv e-prints (2011)
Erb, D.K., Pettini, M., Shapley, A.E., et al.: ApJ **719**, 1168 (2010)
Fan, X., Carilli, C.L., Keating, B.: ARA&A **44**, 415 (2006)
Faucher-Giguère, C.-A., Prochaska, J.X., Lidz, A., Hernquist, L., Zaldarriaga, M.: ApJ **681**, 831 (2008)
Ferland, G.J., Korista, K.T., Verner, D.A., et al.: PASP **110**, 761 (1998)
Finkelstein, S.L., Papovich, C., Giavalisco, M., et al.: ApJ **719**, 1250 (2010)
Finlator, K., Davé, R., Oppenheimer, B.D.: MNRAS **376**, 1861 (2007)
Finlator, K., Oppenheimer, B.D., Davé, R.: MNRAS **410**, 1703 (2011)
Fioc, M., Rocca-Volmerange, B.: A&A **351**, 869 (1999)
Fosbury, R.A.E., Villar-Martín, M., Humphrey, A., et al.: ApJ **596**, 797 (2003)
Fumagalli, M., da Silva, R.L., Krumholz, M.R.: ApJL **741**, L26 (2011)
Gabler, R., Gabler, A., Kudritzki, R.P., Puls, J., Pauldrach, A.: A&A **226**, 162 (1989)
Gallerani, S., Maiolino, R., Juarez, Y., et al.: A&A **523**, A85 (2010)
Garcia-Vargas, M.L., Bressan, A., Diaz, A.I.: A&AS **112**, 13 (1995)
Gilmore, G., Howell, D., (eds.) Astronomical society of the pacific conference series, vol. 142. The Stellar Initial Mass Function (38th Herstmonceux Conference), 1998
Granato, G.L., Lacey, C.G., Silva, L., et al.: ApJ **542**, 710 (2000)
Guseva, N.G., Izotov, Y.I., Thuan, T.X.: ApJ **644**, 890 (2006)
Guseva, N.G., Papaderos, P., Izotov, Y.I., Noeske, K.G., Fricke, K.J.: A&A **421**, 519 (2004)
Haiman, Z.: ApJL **576**, L1 (2002)
Hainline, K.N., Shapley, A.E., Kornei, K.A., et al.: ApJ **701**, 52 (2009)
Hayes, M., Schaerer, D., Östlin, G., et al.: ApJ **730**, 8 (2011)
Hu, E.M., Cowie, L.L., McMahon, R.G.: ApJL **502**, L99+ (1998)
Hunt, L.K., Madden, S., Schneider, R., (eds.), IAU symposium, vol. 255, Low-Metallicity Star Formation: From the First Stars to Dwarf Galaxies (2008)
Inoue, A.K.: MNRAS **415**, 2920 (2011)

Izotov, Y.I., Chaffee, F.H., Schaerer, D.: A&A **378**, L45 (2001)
Izotov, Y.I., Lipovetsky, V.A., Chaffee, F.H., et al.: ApJ **476**, 698 (1997)
Johnson, J.L., Greif, T.H., Bromm, V., Klessen, R.S., Ippolito, J.: MNRAS **399**, 37 (2009)
Kehrig, C., Oey, M.S., Crowther, P.A., et al.: A&A **526**, A128 (2011)
Kennicutt, R.C., Jr.: ARA&A **36**, 189 (1998)
Kewley, L.J., Dopita, M.A., Sutherland, R.S., Heisler, C.A., Trevena, J.: ApJ **556**, 121 (2001)
Khochfar, S., Silk, J.: MNRAS **410**, L42 (2011)
Kroupa, P.: MNRAS **322**, 231 (2001)
Krtička, J., Kubát, J.: A&A **446**, 1039 (2006)
Kudritzki, R.P.: ApJ **577**, 389 (2002)
Larson, R.B.: MNRAS **301**, 569 (1998)
Larson, R.B., Tinsley, B.M.: ApJ **219**, 46 (1978)
Lee, J.C., Gil de Paz, A., Tremonti, C., et al.: ApJ **706**, 599 (2009a)
Lee, K.-S., Giavalisco, M., Conroy, C., et al.: ApJ **695**, 368 (2009b)
Leitherer, C., Ekström, S.: ArXiv e-prints (2011)
Leitherer, C., Schaerer, D., Goldader, J.D., et al.: ApJS **123**, 3 (1999)
Lidman, C., Hayes, M., Jones, D.H., et al.: MNRAS **2199** (2012)
Levesque, E.M., Leitherer, C., Ekström, S., Meynet, G., Schaerer, D.: ApJ **751**, 67 (2012)
Madau, P.: ApJ **441**, 18 (1995)
Maeder, A., Grebel, E.K., Mermilliod, J.-C.: A&A **346**, 459 (1999)
Maeder, A., Meynet, G.: A&A **287**, 803 (1994)
Maeder, A., Meynet, G.: ARA&A **38**, 143 (2000)
Maeder, A., Meynet, G.: Rev. Mod. Phys. **84**, 25 (2012)
Malhotra, S., Rhoads, J.E.: ApJL **565**, L71 (2002)
Maraston, C.: MNRAS **362**, 799 (2005)
Maraston, C.: in Proceedings of the International Astronomical Union, IAU Symposium *Tracing the Ancestry of Galaxies (on the land of our ancestors)*. In : Carnigan, C., Combes, F., Freeman, K.C., (eds.), vol. 277, p. 158 (2011a)
Maraston, C.: in Astronomical Society of the Pacific Conference Series, Why Galaxies Care about AGB Stars II: Shining Examples and Common Inhabitants. In: Kerschbaum, F., Lebzelter, T., Wing, R.F., (eds.) vol. 445, p. 391 (2011b)
Maraston, C., Daddi, E., Renzini, A., et al.: ApJ **652**, 85 (2006)
Maraston, C., Pforr, J., Renzini, A., et al.: MNRAS **407**, 830 (2010)
Marigo, P., Girardi, L., Chiosi, C., Wood, P.R.: A&A **371**, 152 (2001)
Mas-Hesse, J.M., Kunth, D.: A&AS **88**, 399 (1991)
Massey, P.: ARA&A **41**, 15 (2003)
Meurer, G.R., Heckman, T.M., Calzetti, D.: ApJ **521**, 64 (1999)
Meynet, G., Maeder, A.: A&A **429**, 581 (2005)
Meynet, G., Maeder, A., Schaller, G., Schaerer, D., Charbonnel, C.: A&AS **103**, 97 (1994)
Neufeld, D.A.: ApJL **370**, L85 (1991)
Oh, S.P., Haiman, Z., Rees, M.J.: ApJ **553**, 73 (2001)
Ono, Y., Ouchi, M., Shimasaku, K., et al.: ApJ **724**, 1524 (2010)
Osterbrock, D.E., Ferland, G.J.: Astrophysics of Gaseous Nebulae and Active Galactic Nuclei (2006)
Ouchi, M., Shimasaku, K., Akiyama, M., et al.: ApJS **176**, 301 (2008)
Panagia, N.: in Astrophysics and Space Science Library, The Initial Mass Function 50 Years Later. In: Corbelli, E., Palla, F., Zinnecker, H. (eds.), vol. 327, 479 (2005)
Papaderos, P., Izotov, Y.I., Noeske, K.G., et al.: Ap&SS **284**, 619 (2003)
Papovich, C., Finkelstein, S.L., Ferguson, H.C., Lotz, J.M., Giavalisco, M.: MNRAS **412**, 1123 (2011)
Partridge, R.B., Peebles, P.J.E.: ApJ **147**, 868 (1967)
Pelló, R., Kneib, J.P., Le Borgne, J.F., et al.: A&A **346**, 359 (1999)
Pello, R., Schaerer, D.: in Revista Mexicana de Astronomia y Astrofisica Conference Series, In: Rodriguez Espinoza, J.M., Garzon Lopez, F., Melo Martin, V., (eds.) Revista Mexicana de Astronomia y Astrofisica Conference Series, vol. 16, pp. 225–228 (2003)

Pflamm-Altenburg, J., Weidner, C., Kroupa, P.: ApJ **671**, 1550 (2007)
Pindao, M., Schaerer, D., González Delgado, R.M., Stasińska, G.: A&A **394**, 443 (2002)
Pozzetti, L., Mannucci, F.: MNRAS **317**, L17 (2000)
Raiter, A., Fosbury, R.A.E., Teimoorinia, H.: A&A **510**, A109+ (2010a)
Raiter, A., Schaerer, D., Fosbury, R.A.E.: A&A **523**, A64 (2010b)
Salpeter, E.E.: ApJ **121**, 161 (1955)
Schaerer, D.: ApJL **467**, L17 (1996)
Schaerer, D.: In: Hammer, F., Thuan, T.X., Cayatte, V., Guiderdoni, B., Thanh Van, J.T.: (eds.) Building Galaxies; From the Primordial Universe to the Present, 389 (2000)
Schaerer, D.: A&A **382**, 28 (2002)
Schaerer, D.: A&A **397**, 527 (2003)
Schaerer, D.: ArXiv e-prints **706** (2007)
Schaerer, D., Contini, T., Pindao, M.: A&AS **136**, 35 (1999)
Schaerer, D., de Barros, S.: A&A **502**, 423 (2009)
Schaerer, D., de Barros, S.: A&A **515**, A73+ (2010)
Schaerer, D., de Barros, S.: ArXiv e-prints (2011)
Schaerer, D., de Barros, S., Stark, D.P.: A&A **536**, A72 (2011)
Schaerer, D., Pelló, R.: MNRAS **362**, 1054 (2005)
Schaerer, D., Vacca, W.D.: ApJ **497**, 618 (1998)
Schaller, G., Schaerer, D., Meynet, G., Maeder, A.: A&AS **96**, 269 (1992)
Schneider, R., Ferrara, A., Natarajan, P., Omukai, K.: ApJ **571**, 30 (2002)
Schneider, R., Ferrara, A., Salvaterra, R., Omukai, K., Bromm, V.: Nature **422**, 869 (2003)
Shapley, A.E., Steidel, C.C., Pettini, M., Adelberger, K.L.: ApJ **588**, 65 (2003)
Shirazi, M., Brinchmann, J.: ArXiv e-prints (2012)
Smith, L.J., Norris, R.P.F., Crowther, P.A.: MNRAS **337**, 1309 (2002)
Stasińska, G., Leitherer, C.: ApJS **107**, 661 (1996)
Stasińska, G., Schaerer, D., Leitherer, C.: A&A **370**, 1 (2001)
Stasińska, G., Tylenda, R.: A&A **155**, 137 (1986)
Thuan, T.X., Bauer, F.E., Papaderos, P., Izotov, Y.I.: ApJ **606**, 213 (2004)
Tinsley, B.M.: ApJ **151**, 547 (1968)
Tinsley, B.M.: Fundamentals Cosmic Phys. **5**, 287 (1980)
Treyer, M., Wyder, T., Neill, J., Seibert, M., Lee, J.: (eds.), Astronomical Society of the Pacific Conference Series, vol. 440, UP2010: Have Observations Revealed a Variable Upper End of the Initial Mass Function? (2011)
Tumlinson, J.: ApJ **641**, 1 (2006)
Tumlinson, J., Giroux, M.L., Shull, J.M.: ApJL **550**, L1 (2001)
Tumlinson, J., Shull, J.M.: ApJL **528**, L65 (2000)
Vanbeveren, D., De Donder, E., van Bever, J., van Rensbergen, W., De Loore, C.: New Astron. **3**, 443 (1998)
Vanzella, E., Grazian, A., Hayes, M., et al.: A&A **513**, A20 (2010)
Vanzi, L., Hunt, L.K., Thuan, T.X.: A&A **390**, 481 (2002)
Vázquez, G.A., Leitherer, C., Schaerer, D., Meynet, G., Maeder, A.: ApJ **663**, 995 (2007)
Veilleux, S., Osterbrock, D.E.: ApJS **63**, 295 (1987)
Weidner, C., Kroupa, P.: MNRAS **365**, 1333 (2006)
Whalen, D.J., Bromm, V., Yoshida, N., (eds.) American Institute of Physics Conference Series, vol. 1294, THE First Stars and Galaxies: Challenges for the Next Decade (2010)
Wuyts, S., Förster Schreiber, N.M., Lutz, D., et al.: ApJ **738**, 106 (2011)
Yoon, S.-C., Dierks, A., Langer, N.: ArXiv e-prints (2012)
Yoshikawa, T., Akiyama, M., Kajisawa, M., et al.: ApJ **718**, 112 (2010)
Zackrisson, E., Bergvall, N., Leitet, E.: ApJL **676**, L9 (2008)
Zackrisson, E., Inoue, A.K., Rydberg, C.-E., Duval, F.: MNRAS **418**, L104 (2011a)
Zackrisson, E., Rydberg, C.-E., Schaerer, D., Östlin, G., Tuli, M.: ApJ **740**, 13 (2011b)

Exploring the Universe with Metal-Poor Stars

Anna Frebel

Abstract The early chemical evolution of the Galaxy and the Universe is vital to our understanding of a host of astrophysical phenomena. Since the most metal-poor Galactic stars (with metallicities down to [Fe/H] ~ -5.5) are relics from the high-redshift Universe, they probe the chemical and dynamical conditions of the Milky Way and the origin and evolution of the elements through nucleosynthesis. They also provide constraints on the nature of the first stars, their associated supernovae and initial mass function, and early star and galaxy formation. The Milky Way's dwarf satellites contain a large fraction ($\sim$30 %) of the known most metal-poor stars that have chemical abundances that closely resemble those of equivalent halo stars. This suggests that chemical evolution may be universal, at least at early times, and that it is driven by massive, energetic SNe. Some of these surviving, ultra-faint systems may show the signature of just one such Pop III star; they may even be surviving first galaxies. Early analogs of the surviving dwarfs may thus have played an important role in the assembly of the old Galactic halo whose formation can now be studied with stellar chemistry. Following the cosmic evolution of small halos in simulations of structure formation enables tracing the cosmological origin of the most metal-poor stars in the halo and dwarf galaxies. Together with future observations and additional modeling, many of these issues, including the reionization history of the Milky Way, may be constrained this way. The chapter concludes with an outlook about upcoming observational challenges and ways forward to use metal-poor stars to constrain theoretical studies.

A. Frebel (✉)
Massachusetts Institute of Technology, 77 Massachusetts Avenue, Bldg 37-664C, 02139 Cambridge MA, USA
e-mail: afrebel@mit.edu

T. Wiklind et al. (eds.), *The First Galaxies*, Astrophysics and Space Science Library 396, DOI 10.1007/978-3-642-32362-1_8,

1 Introduction

As Carl Sagan once remarked, *If you wish to make an apple pie from scratch, you must first create the Universe*. An apple contains at least 16 different elements, and the human body is even more complex, having at least trace amounts of nearly 30 elements,[1] all owing to a 14-billion year long manufacturing process called cosmic chemical evolution. Thus, the basis of chemically complex and challenging undertakings such as cooking and baking, not to mention the nature of life, will ultimately be gained through an understanding of the formation of the elements that comprise organic material. It is thus important to examine how the constituents of an apple, and by extension the stuff of life and the visible Universe were created: baryonic matter in the form of elements heavier than primordial hydrogen and helium.

This chapter aims at describing that the chemical abundances observed in the most metal-poor stars can be employed to unravel a variety of details about the young Universe, such as early star formation environments, supernovae (SNe) nucleosynthesis, and the formation process(es) of the Galactic halo. To illustrate the meaning of low-metallicity, Fig. 1 shows the progression from metal-rich to the most metal-poor stars; spectra around the strongest optical Fe line at 3,860 Å are shown of the Sun and three metal-poor main-sequence turn-off stars. The number of atomic absorption lines detectable in the spectra decreases with increasing metal-deficiency. In HE 1327−2326, the star with the currently low Fe abundance, only the intrinsically strongest metal lines remain observable. As can be seen in the figure, these are extremely weak. If a main-sequence star with even lower Fe value (or somewhat hotter temperature) was discovered, no Fe lines would be measurable anymore. In the case of a metal-deficient giant, the lines would be somewhat stronger due to its cooler temperature and thus allow for the discovery of an object with $[\mathrm{Fe/H}] \lesssim -6$.

Because these most metal-poor stars represent easily accessible *local equivalents* of the high-redshift universe, and as such, provide a unique tool to address a wide range of near and far-field cosmological topics. In short, metal-poor stars enable scientific progress in three areas that bridge our understanding of the current state of the Galactic halo and its old stellar population with that of the evolution of local dwarf galaxies, to the formation of large galaxies like the Milky Way more generally, as well as the beginning of star and galaxy formation in the early universe.

1. *Stellar Archaeology:* Constrains the astrophysical sites and conditions of nucleosynthesis and the major physical processes that drove early star formation. Abundance measurements of many elements throughout the periodic table of metal-poor halo stars enable the detailed documentation of the earliest chemical enrichment events.

[1] http://chemistry.about.com/cs/howthingswork/f/blbodyelements.htm

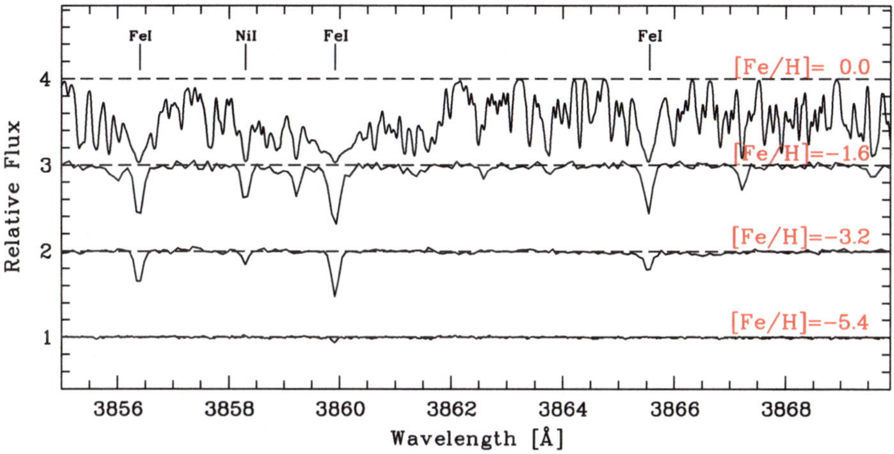

Fig. 1 Spectral comparison of stars in the main-sequence turn-off region with different metallicities. Several absorption lines are marked. The variations in line strength reflect the different metallicities. From *top to bottom*: Sun with [Fe/H] = 0.0, G66-30 [Fe/H] = −1.6 (Norris et al. 1997) G64-12 [Fe/H] = −3.2 (Aoki et al. 2006), and HE1327-2326 [Fe/H] = −5.4 (Frebel et al. 2005a). Figure taken from Frebel (2010)

2. *Dwarf Archaeology:* Provides constraints on galaxy formation on small scales, and metal mixing and feedback processes. By comparing abundances of metal-poor stars in ultra-faint dwarf galaxies to those of equivalent halo stars, the universality of the (beginning of) chemical evolution can be tested, what the relation is between the dwarfs and the "building blocks" of the Galactic halo, and whether they are the survivors of the first galaxies.
3. *Near-Field Cosmology:* Determines the role of metal-poor stars as tracers of the accretion history of the Milky Way halo. The coupling of low-metallicity stellar abundances with results from cosmological simulations enables the study of the formation mechanism(s) of large galaxies like the Milky Way with its old halo and satellites.

For stellar archaeology, large numbers of Galactic metal-poor halo stars, mostly found in objective-prism surveys in both hemispheres such as the HK survey of Beers and collaborators, the Hamburg/ESO survey (Christlieb and collaborators), and more recently SDSS, are needed to gain detailed insights into the history and evolution of our Galaxy (e.g., Beers and Christlieb 2005; Frebel and Norris 2011). For dwarf archaeology, observations of any stars in dwarf satellite galaxies orbiting the Milky Way are required, although these are more difficult to obtain than those of the halo stars. Finally, near-field cosmology encourages the systematic use of metal-poor stars for studying galaxy formation and cosmological aspects. In a universe dominated by cold dark matter (CDM), like the one we live in, galaxy formation proceeds hierarchically through the accretion of smaller objects onto the main halo. Simulations show that successive growth is reflected in the abundance of dark matter substructures in the halos of large galaxies like the Milky Way, and it is believed that the luminous satellites of our Galaxy are the visible counterparts to at least some of these substructures. Thus, the collective body of metal-poor stars now found in the halo as well as the dwarf galaxies enable addressing a number of important, outstanding questions that show how closely connected.the three topics are.

- What is the nature of Pop III stars? Are the yields of the first SNe different from today's? Can the signatures of theorized pair-instability SNe be found in metal-poor stars?
- What drove early star formation? How/where did the first low-mass stars and the first galaxies form?
- What are the main nucleosynthesis processes and sites that are responsible for forming the elements from the Big Bang until today?
- How did chemical evolution proceed? How do stellar chemistry and halo kinematics correlate? How can the abundances be used to learn about the halo formation process?
- Was the old halo built from accreted satellites? Can accreted dwarf galaxies be identified in the halo? Did the first stars form in dwarf galaxies?

Each section of this chapter discusses a significant aspect in which metal-poor stars offer unique insight into the young universe. Section 3 sets the overall stage by introducing the first stars, the halo metallicity distribution function and considerations regarding early low-mass star formation. Section 4 describes the role of metal-poor stars in the Galactic as tracers of the earliest enrichment events and chemical evolution. This concept is extended to dwarf galaxies in Sect. 5. Understanding the formation history of the Milky Way with the help of metal-poor stars is outlined in Sect. 6. Conclusions and an outlook at given in Sect. 7.

2 Further Reading and Definitions

This chapter is partially based on an article by Frebel (2010)[2] that, among other topics, introduces metal-poor stars as probes for theoretical works related to the early universe. Here, this discussion is extended to showcase the versatility and potential of the oldest stars for dwarf archaeology and near-field cosmology. The main aim is to outline the broad picture of studying galaxy formation processes with stellar chemistry. For a more in-depth discussion of stellar abundances, abundance derivations, details on nucleosynthesis and chemical evolution, kinematics, stellar age determinations and cosmogony, the reader is referred to Frebel and Norris (2011) and references therein.

---

[2]The 2009 Biermann Award Lecture, originally published in Astronomische Nachrichten by VCH/Wiley, 2010, 331, 474.

Since there exist a large range of metal-poor stars in terms of their metallicities and chemical signatures, Beers and Christlieb (2005) suggested a classification scheme. Extensive use will be made of their term "extremely metal-poor stars", referring to stars with [Fe/H] < -3.0. This nomenclature shows that the main metallicity indicator used to determine any stellar metallicity is the iron abundance, [Fe/H], which is defined as $[A/B] = \log_{10}(N_A/N_B)_\star - \log_{10}(N_A/N_B)_\odot$ for the number N of atoms of elements A and B, and $\odot$ refers to the Sun. For example, [Fe/H] $= -3.0$ is 1/1,000 of solar Fe abundance. With few exceptions, [Fe/H] traces the overall metallicity of the objects fairly well.

3 Exploring The Early Universe with Metal-Poor Stars

3.1 The First Stars

According to cosmological simulations that are based on the Λ cold dark matter model of hierarchical structure growth in the Universe, the first stars formed in small minihalos some few hundred million years after the Big Bang. Due to the lack of cooling agents in the primordial gas, significant fragmentation was largely suppressed so that these first objects were very massive (of the order to $\sim$100 $M_\odot$; e.g., Bromm and Yoshida (2011) and references therein). This is in contrast to low-mass stars dominating today's mass function. These objects are referred to as Population III (Pop III) as they formed from metal-free gas. Recent modeling of first star formation suggests that these early behemoth were rapidly rotating (Stacy et al. 2010) and new observations have provided evidence in support of this claim (Chiappini et al. 2011). Moreover, significant fragmentation of the star forming cloud may occur that could lead to multiple first stars in a given minihalo (Clark et al. 2011).

The stars soon exploded as SNe to either collapse into black holes (progenitor masses of $25 < M_\odot < 140$ and $M_\odot > 260$) or to die as energetic pair-instability SNe (PISN; $140 < M_\odot < 260$; Heger and Woosley 2002). During their deaths, these objects provided vast amounts of ionizing radiation (and some of the first metals in the case of the PISNe) that changed the conditions of the surrounding material for subsequent star formation even in neighboring minihalos. Hence, the second generation of stars might have been less massive ($M_\star \sim 10\,M_\odot$). Partially ionized gas supports the formation of the H_2, and then the HD molecule, which in turn facilitates more effective cooling than what is possible in neutral gas. Also, any metals or dust grains left behind from PISNe would have similar cooling effects. This may then have led to the first more regular metal-producing SNe, although not all higher mass SNe must necessarily end in black hole formation. Umeda and Nomoto (2003) suggested that some 25 $M_\odot$ stars undergo only a partial fallback, so that only some of the newly created metals get ejected into the surrounding gas.

By that time, most likely enough metals were present to ensure sufficient gas fragmentation to allow for low-mass ($<1\,M_\odot$) star formation. Stars that formed from any metal-enriched material are referred to as Population II (Pop II) stars. More metal-rich stars like the Sun that formed in a much more metal-rich Universe are called Population I. Studying the "chemical fingerprints" of the oldest, most metal-poor stars (extreme Pop II) reveals information about the first nucleosynthesis events in the Universe; indeed, several metal-poor star abundance patterns have been fitted with calculated Pop III SN yields (see Sect. 4.3). Moreover, evidence for the existence of PISNe could potentially be obtained if their characteristic signature (a pronounced effect in the abundance signature of elements with odd or even atomic number) were found in metal-poor stars. This has, however, not yet occurred.

3.2 The Metallicity Distribution Function of the Galactic Halo

The metallicity distribution function (MDF) represents the integrated chemical evolution of a system which began with the first stars and was continued by many stellar generations at various astrophysical sites and over different timescales.

To establish the MDF of a given system, a large, complete sample of stars with good metallicity estimates is required. Over the past two decades, the quest to find the most metal-poor stars to study the chemical evolution of the Galaxy led to a significant number of stars with metallicities down to [Fe/H] ~ -4.0 (see Beers and Christlieb 2005 for a more detailed review). Those stars were initially selected as candidates from a large survey, such as the HK survey (Beers et al. 1992) and the Hamburg/ESO survey (Wisotzki et al. 1996). A large survey is required to provide numerous low-resolution spectra to search for weak-lined stellar candidates indicating metal deficiency. Those spectra have to cover the strong Ca II K line at 3,933 Å because the strength of this line indicates the metallicity of the star, and can be measured even in low-quality spectra. This is shown in Fig. 2. If this line is sufficiently weak as a function of the star's estimated effective temperature, an object is selected as a candidate metal-poor star. For all candidates, medium-resolution spectra ($R \sim 2,000$) are required to more accurately determine the Ca II K line strength for a more robust estimate for the Fe abundance. This line is still the best indicator for the overall metallicity [Fe/H] of a metal-poor star in such spectra. In the Sloan Digital Sky Survey and LAMOST survey, the survey spectra themselves are already of medium-resolution, allowing for a quicker and more direct search for metal-poor stars. Photometric surveys like Skymapper with extensive filter sets designed for stellar work (Keller et al. 2007) will also yield large numbers of high-quality candidates.

To confirm the metallicity, and to measure elemental abundances from their respective absorption lines besides that of iron, high-resolution ($R > 20,000$) optical spectroscopy is required (see bottom panel of Fig. 2). Only then the various elements become accessible for studying the chemical evolution of the Galaxy. Those elements include carbon, magnesium, calcium, titanium, nickel, strontium,

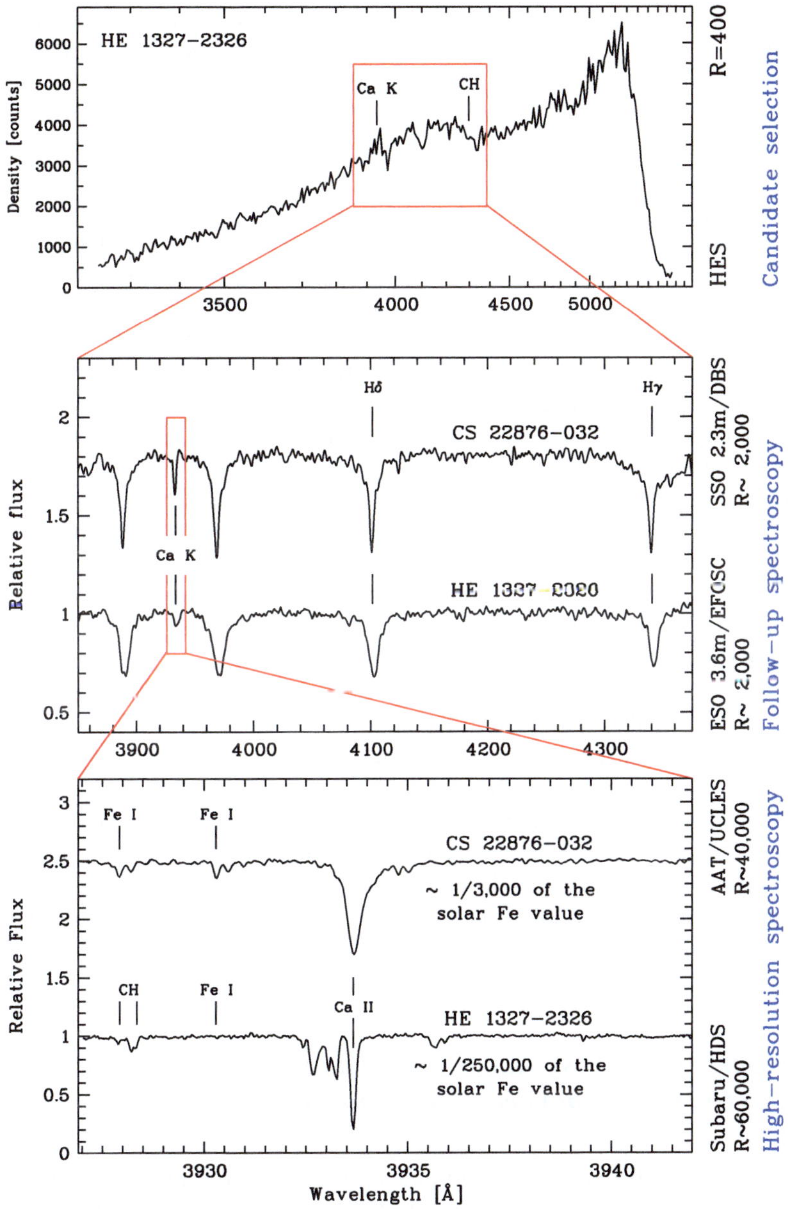

Fig. 2 The three observational steps to find metal-poor stars illustrated by means of HE 1327−2326. *Top panel*: HES objective-prism spectrum. *Middle panel*: Medium-resolution spectrum of HE 1327−2326 in comparison with CS 22876−032 ([Fe/H] = −3.7; Norris et al. 2000 and references therein). From this data we measured [Fe/H] = −4.3 for HE 1327−2326 because interstellar Ca blended with the Ca II K line. *Bottom panel*: High-resolution spectra of both objects. Only with the high-resolution data was it possible to determine the true iron abundance, [Fe/H] = −5.4, for HE 1327−2326. Figure taken from Frebel et al. (2005b)

and barium, and trace different enrichment mechanisms, events and timescales. Abundance ratios [X/Fe] as a function of [Fe/H] can then be derived for the lighter elements ($Z < 30$) and neutron-capture elements ($Z > 38$). The resulting abundance trends will be further described in Sect. 4.2. The final number of elements thereby depends on the type of metal-poor star, the wavelength coverage of the data, and the data quality itself.

Schörk et al. (2009) and Li et al. (2010) presented MDFs for halo stars that are corrected for various selection effects and other biases. The number of known metal-poor stars declines significantly with decreasing metallicity (below [Fe/H] < -2.0) as illustrated in Fig. 3. Only very few stars are known ($\lesssim$30) with metallicities below [Fe/H] < -3.5, but it is these objects that enable the most insight into the early universe and the beginning of chemical evolution.

The bias-corrected MDF shows how rare metal-poor stars really are, but also, that past targeted ("biased") searches for metal-poor stars have been extremely successful at identifying these rare objects (e.g., Frebel et al. 2006; Christlieb et al. 2008). The most important achievements in terms of the most iron-deficient stars was the push to a significantly lower stellar metallicity [Fe/H] almost a decade ago: From a longstanding [Fe/H] $= -4.0$ (CD $-38°$ 245; Bessell and Norris 1984) to [Fe/H] $= -5.2$ (HE 0107$-$5240; Christlieb et al. 2002)[3]), and down to [Fe/H] $= -5.4$ more recently (HE 1327$-$2326; Frebel et al. 2005a). Overall, only three stars are known with iron abundances of [Fe/H] < -4.0. The third star, HE 0557$-$4840 Norris et al. 2007 with [Fe/H] < -4.8, bridges the gap between [Fe/H] $= -4.0$ and the two hyper Fe-poor objects.

Beyond these individual stars that form the very tail of the MDF, a number of studies involving samples of $\sim$10–30 stars in the metallicity range of -4.0[Fe/H] < -2.5 have been carried out over the years. These works have delivered many important details about chemical enrichment and nucleosynthesis, and greatly improved our understanding of the early universe (e.g., McWilliam et al. 1995, Ryan et al. 1996, Cayrel et al. 2004, Barklem et al. 2005, Aoki et al. 2008, Cohen et al. 2008, Lai et al. 2008, Hollek et al. 2012).

3.3 Early Low-Mass Star Formation and the Connection to Carbon-Enhanced Metal-Poor Stars

Early Pop II stars began to form from the enriched material left behind by the first stars. The actual formation process(es) of these initial low-mass ($M \leq 0.8\ M_\odot$) Pop II stars (i.e. the most metal-poor stars) that live longer than a Hubble time, are, however, not well understood so far. Ideas for the required cooling processes necessary to induce sufficient fragmentation of the near-primordial gas include

[3]Applying the same non-LTE correction to the Fe I abundances of HE 0107$-$5240 and HE 1327$-$2326 leads to a final abundance of [Fe/H] $= -5.2$ for HE 0107$-$5240.

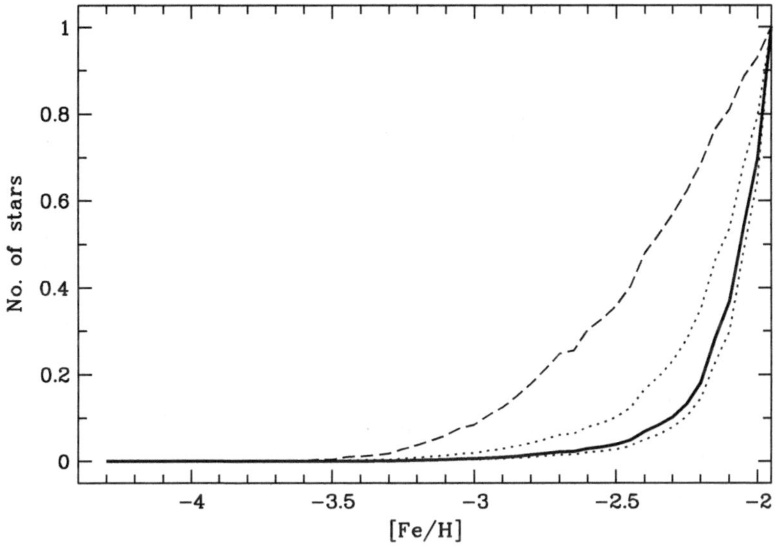

Fig. 3 Cumulative metallicity distribution function of the Galactic halo based on metal-poor giants selected form the Hamburg/ESO survey (Schörk et al. 2009). The solid line represents the bias-corrected MDF, whereas the dotted lines show the level of uncertainty in the MDF based on the various correction functions. The dashed line shows the uncorrected, "as observed" MDF—The difference to the solid line shows how successful searches for the most metal-poor stars have been

cooling through metal enrichment ("critical metallicity") or dust, cooling based on enhanced molecule formation due to ionization of the gas, as well as more complex effects such as turbulence and magnetic fields (Bromm et al. 2009).

Fine-structure line cooling through neutral carbon and singly-ionized oxygen has been suggested as a main cooling agent facilitating low-mass star formation (Bromm and Loeb 2003). These elements were likely produced in vast quantities in Pop III objects (e.g. Meynet et al. 2006; Chiappini et al. 2011). Gas fragmentation is then induced once a critical metallicity of the interstellar medium (ISM) is achieved. The existence of such a critical metallicity can be probed with large numbers of carbon and oxygen-*poor* metal-poor stars. Frebel et al. (2007b) developed an "observer-friendly" description of the critical metallicity that incorporates the observed C and/or O stellar abundances; $D_{\text{trans}} = \log(10^{[\text{C/H}]} + 0.3 \times 10^{[\text{O/H}]}) \geq -3.5$. Any low-mass stars still observable today then has to have C and/or O abundances above the threshold of $D_{\text{trans}} = -3.5$ (see Fig. 1 in Frebel et al. 2007b). At metallicities of [Fe/H] $\gtrsim -3.5$, most stars have C and/or O abundances that are above the threshold since they follow the solar C and O abundances simply scaled down to their respective Fe values. Naturally, this metallicity range is not suitable for directly probing the first low-mass stars. Below [Fe/H] ~ -3.5, however, the observed C and/or O levels must be higher than the Fe-scaled solar abundances to be above the critical metallicity. Indeed, none of the known lowest-metallicity stars has a D_{trans}

below the critical value, consistent with this cooling theory. Some stars, however, have values very close to $D_{\text{trans}} = -3.5$. HE 0557−4840, at [Fe/H] $= -4.75$ (Norris et al. 2007), falls just onto the critical limit (M. Bessell 2009, priv. comm.). A star in the ultra-faint dwarf galaxy Boötes I has $D_{\text{trans}} = -3.2$ (at [Fe/H] $= -3.7$; and assuming that its oxygen abundance is twice that of carbon). Another interesting case is the most metal-poor star in the classical dwarf galaxy Sculptor, which has an upper limit of carbon of [C/H] < -3.6 at [Fe/H] $= -3.8$ (Frebel et al. 2010b). Despite some still required up-correction of carbon to account for atmospheric carbon-depletion of this cool giant, the star could potentially posses a sub-critical D_{trans} value.

Overall, more such "borderline" examples are crucial to test for the existence of a critical metallicity. If fine-structure line cooling were the dominant process for low-mass star formation, two important consequences would follow: (1) Future stars to be found with [Fe/H] $\lesssim -4.0$ are predicted to have these substantial C and/or O overabundances with respect to Fe. (2) The so-far unexplained large fraction of metal-poor objects that have large overabundances of carbon with respect to iron ([C/Fe] > 1.0) may reflect an important physical cause. About 20 % of metal-poor stars with Fe/H $\lesssim -2.5$ exhibit this behavior (e.g. Beers and Christlieb 2005). Moreover, at the lowest metallicities, this fraction is even higher. All three stars with [Fe/H] < -4.0 are extremely C-rich, well in line with the prediction of the line cooling theory.

This may, however, not the only way for forming low-mass stars. Cooling through dust grains might also have been responsible for the transition from Pop III to Pop II star formation. Dust created in high-density regions during the first SNe explosions or mass loss during the evolution of Pop III stars may induce fragmentation processes (e.g., Schneider et al. 2006) that lead to the formation of subsolar-mass stars. The critical metallicity in this scenario is a few orders of magnitude below that of C and O line cooling. If some metal-poor stars are found to be significantly below $D_{\text{trans}} = -3.5$, their existence may still be consistent with the critical value set by dust cooling. Irrespective of the differences in cooling channels, such criteria will need to be incorporated in large-scale simulations to take environmental influences, such as the available gas mass, into account.

4 Stellar Archaeology

4.1 Validating Stellar Archaeology

The concept of stellar archaeology is based on long-lived low-mass metal-poor main-sequence and giant stars whose chemical abundances are thought to reflect the composition of the gas cloud during their formation period. A vital assumption is that the stellar surface compositions have not been significantly altered by any internal mixing processes given that these stars are fairly unevolved despite their old age. But are there other means by which the surface composition could be modified? Accretion of interstellar matter while a star orbits in the Galaxy for ~10 Gyr has long been suggested as a mechanism to affect the observed abundance patterns. Iben (1983) calculated a basic "pollution limit" of [Fe/H] $= -5.7$ based on Bondi-Hoyle accretion. He predicted that no stars with Fe abundances below this value could be identified as such since they would have accreted too much enriched material.

Assuming that stars with such low-metallicities exist (for example low-mass Pop III stars if the IMF was Salpeter-like, and not top-heavy), significant amounts of interstellar accretion could masquerade the primordial abundances of those putative low-mass Pop III stars. Analogously, stars with very low abundances, say [Fe/H] < -5.0, could principally be affected also. To assess the potential accretion level, Frebel et al. (2009) carried out a kinematic analysis of a sample of metal-poor stars to assess their potential accretion histories over the past 10 Gyr in a Milky Way-like potential. The amount of accreted Fe was calculated based on the total accreted material over 10 Gyr. The overall chemical evolution with time was taken into account assuming the ISM to have scaled solar abundances. The stellar abundances were found to be little affected by accretion given their generally fast space velocities. The calculated, "accreted abundances" were often lower than the observed measurements by several orders of magnitude. Johnson and Khochfar (2011), on the other hand, investigated direct accretion onto primordial low-mass stars. If these stars had a weak solar-like wind it would prevent the accretion of any material, at least in the early universe. This would also be true for any low-metallicity stars, although it was not considered by Frebel et al. (2009).

Generally, these studies show that accretion does not significantly alter the observed abundance patterns, even in an extreme case in which a star moves once through a very large, dense cloud. The concept of stellar archaeology can thus be deemed viable. Nevertheless, since there is a large accretion dependency on the space velocity it becomes obvious that kinematic information is vital for the identification of the lowest-metallicity stars in the Milky Way and the interpretation of their abundances. A way forward would be an extensive assessment of potential gas accretion histories for stars throughout the hierarchical assembly of a large galaxy. However, the uncertainties regarding the existence and extent of stellar winds may prevent strong conclusions.

4.2 Chemical Evolution of the Galaxy

Generally, there are several main groups of elements observed in metal-poor stars, with each group having a common, main production mechanism; (1) α-elements (e.g. Mg, Ca, Ti) are produced through α-capture during various burning stages of late stellar evolution, before and during SN explosions. These yields appear very robust with respect to parameters such as mass and explosion energy; (2) Fe-peak elements ($23 < Z < 30$) are synthesized in a host of different nucleosynthesis processes before and during SN explosions such as radioactive decay of heavier nuclei or direct synthesis in explosive burning stages, neutron-capture onto lower-mass Fe-peak elements during helium and later burning stages and α-rich freeze-out processes. Their yields also depend on the explosion energy; (3) Light and heavy neutron-capture elements ($Z > 38$) are either produced in the slow (s-) process occurring in thermally pulsing AGB stars (and then transferred to binary companions or deposited into the ISM through stellar winds) or in the rapid (r-) process most likely occurring in core-collapse SN explosions. For more details on SN nucleosynthesis see e.g., Woosley and Weaver (1995).

The α-element abundances in metal-poor halo stars with [Fe/H] < -1.5 are enhanced by $\sim$0.4 dex with respect to Fe as seen in Fig. 4. This reflects a typical core-collapse SN signature because at later times (in chemical space at about [Fe/H] ~ -1.5) the onset of SN Ia provides a significant contribution to the overall Galactic Fe inventory. As a consequence, the [α/Fe] ratio decreases down to the solar value at [Fe/H] $\sim$ 0.0. The general uniformity of light element abundance trends down to [Fe/H] ~ -4.0 led to the conclusion that the ISM must have been already well-mixed at very early times (Cayrel et al. 2004). Otherwise it would be hard to understand why so many of the most metal-poor stars have almost identical abundance patterns. However, despite the well-defined abundance trends, some stars, particularly those in the lowest metallicity regime show significant deviations. Among those are some stars with unusually high or low α-element abundances.

Among the Fe-peak elements, many have subsolar abundance trends at low metallicity (e.g. [Cr,Mn/Fe]) which become solar-like as the metallicity increases. This is shown in Fig. 5. It is not clear whether these large underabundances are of cosmic origin or have to be attributed to modeling effects such as that of non-LTE (Sobeck et al. 2007; Bergemann and Gehren 2008). Trends of other elements are somewhat overabundant at low metallicity (Co) or relatively unchanged throughout (Sc, Ni). All elements with $Z < 30$ hereby have relatively tight abundance trends.

On the contrary, the abundances of the neutron-capture elements in metal-poor stars are "all over the place". Sr has an extremely large scatter ($\sim$3 dex). This indicates that different nucleosynthetic processes must have contributed to its Galactic inventory, or that neutron-capture yields are very environmentally-sensitive. Ba has even more scatter at [Fe/H] ~ -3.0 (see Fig. 6). Some of this spread may be explained through the existence of massive rotating low-metallicity stars that produce large amounts of s-process like Sr and Ba in the early universe (Chiappini et al. 2011). Other heavier neutron-capture elements, such as r-process element Eu, have somewhat less scatter. At the lowest metallicities, core-collapse SNe must have dominated the chemical evolution (below [Fe/H] $= -3.0$). Hence, the r-process is likely responsible for the neutron-capture elements at this early time. Metal enrichment through the s-process began to significantly contribute at somewhat later times, driven by the evolutionary timescales of stars with $\sim$2–8 $M_{\odot}$ to become AGB stars. According to Simmerer et al. (2004), the s-process was in full operation at [Fe/H] ~ -2.6.

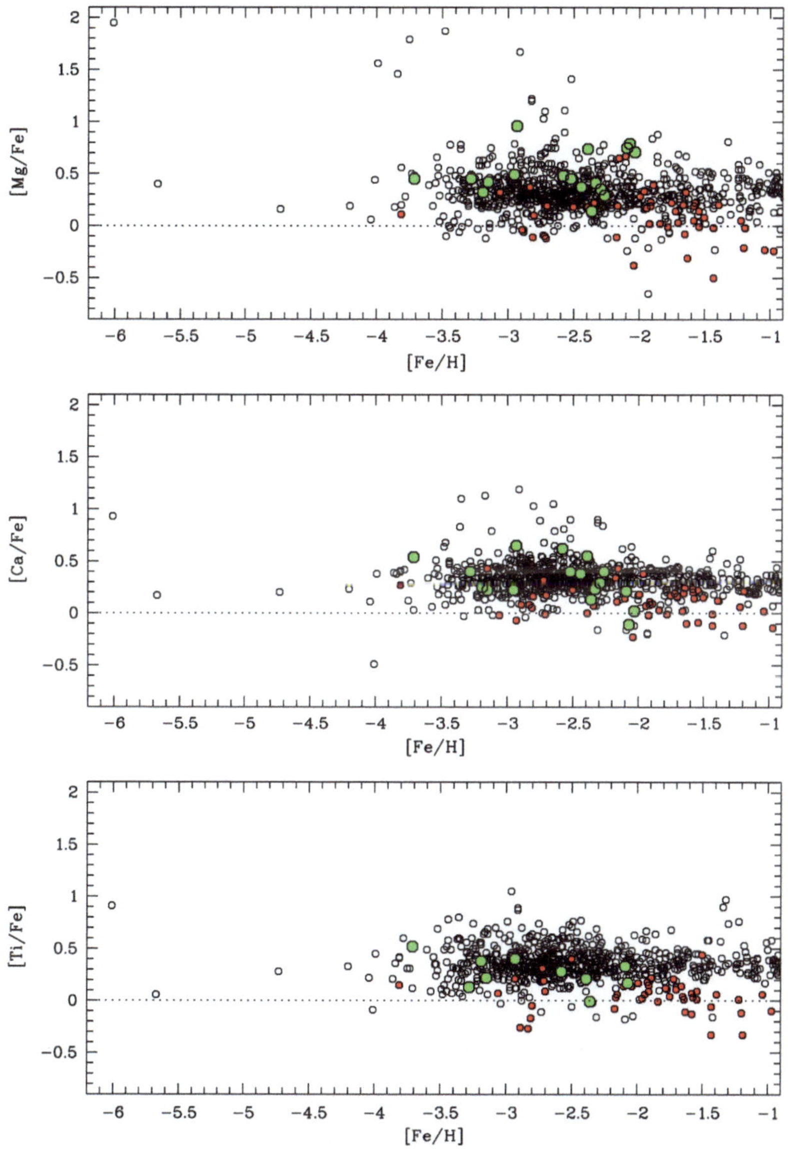

Fig. 4 Light element abundance trends of Mg, Ca, and Ti. *Black open circles* represent halo stars, *red filled circles* are stars in the classical dwarf galaxies, and *green filled circles* show stars in the ultra-faint dwarf galaxies. The scatter in the data likely reflects systematic differences between literature studies. Assuming this, systematic uncertainties in abundance analyses may be around ~0.3 dex. Figure taken from Frebel 2010

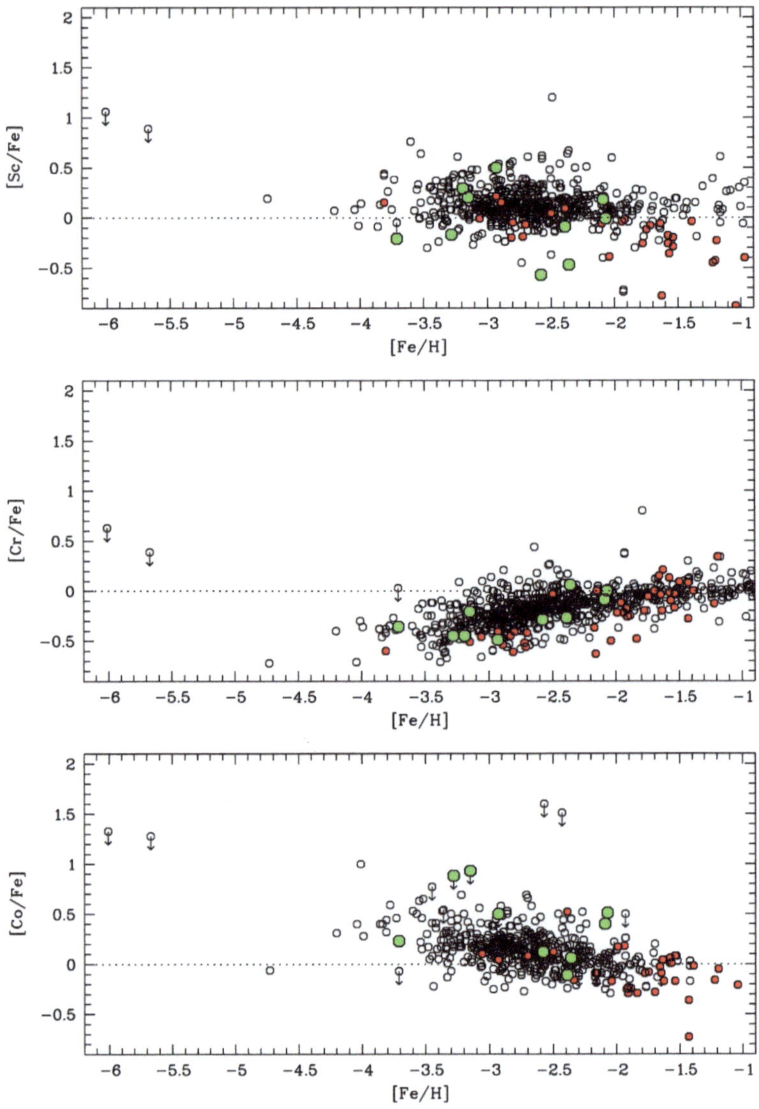

Fig. 5 Same as Fig. 4, but for the Fe-peak elements Sc, Cr, and Co. Figure taken from Frebel 2010

4.3 *Tracing Individual SN Explosions with the Most Iron-Poor Stars*

The faint ($V = 15.2$) red giant HE 0107−5240 has [Fe/H] = −5.2 (Christlieb et al. 2002). The brighter ($V = 13.5$) subgiant HE 1327−2326 has an even lower iron abundance of [Fe/H] = −5.4 (Frebel et al. 2005a; Aoki et al. 2006). The latter

Fig. 6 Same as Fig. 4, but for the neutron-capture elements Sr, Ba, and Eu. For these elements, the scatter is much beyond systematic differences between individual studies and thus indicates a cosmic origin. Figure taken from Frebel 2010

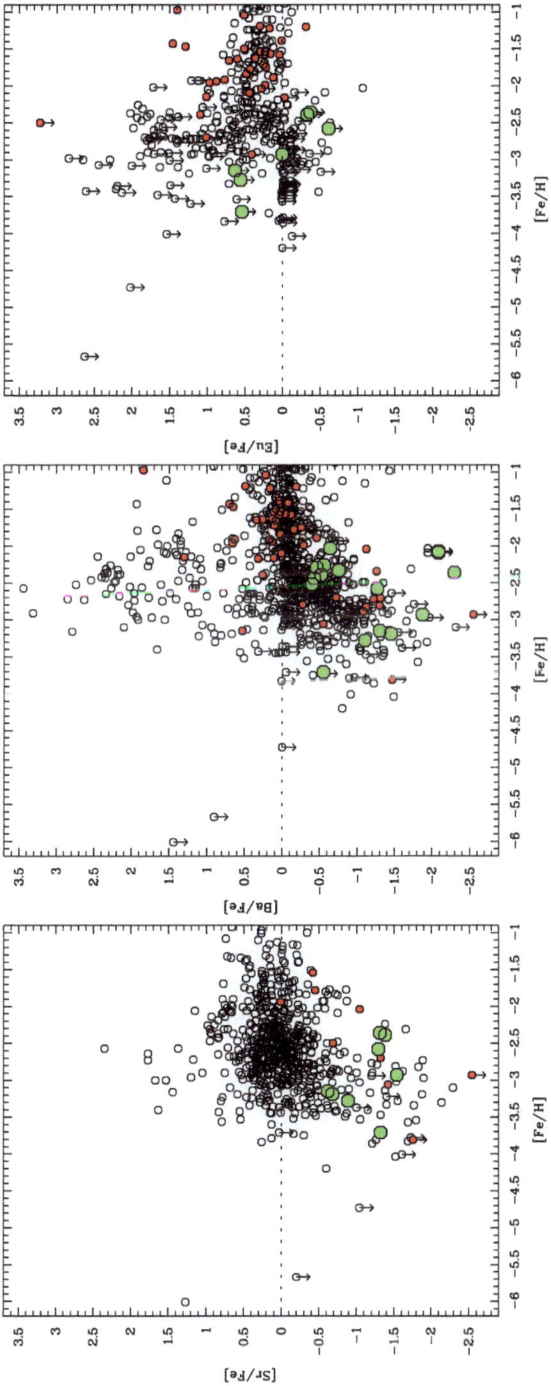

value corresponds to $\sim 1/250,000$ of the solar iron abundance[4] A third star with [Fe/H] $= -4.75$ (Norris et al. 2007) was found in 2006. The metallicity of the giant HE 0557−4840 is in between the two [Fe/H] < -5.0 stars and the next most metal-poor stars are [Fe/H] ~ -4.0. Hence, it sits right in the previously claimed "metallicity gap" (between [Fe/H] ~ -4.0 and [Fe/H] ~ -5.0; e.g. Shigeyama et al. 2003) showing that the scarcity of stars below [Fe/H] $= -4.0$ has no physical cause but is merely an observational incompleteness. All three objects were found in the Hamburg/ESO survey making it the so far most successful database for metal-poor stars.

The most striking features in both [Fe/H] < -5.0 stars are the extremely large overabundances of CNO elements. HE 0557−4840 partly shares this signature by also having a fairly large [C/Fe] ratio. Other elemental ratios [X/Fe] are somewhat enhanced in HE 1327−2327 with respect to the stars with $-4.0 <$ [Fe/H] < -2.5, but less so for the two giants. Despite expectations, Li could not be detected in the relatively unevolved subgiant HE 1327−2326.[5] The upper limit is $\log\epsilon(\mathrm{Li}) < 1.6$, where $\log\epsilon(\mathrm{A}) = \log_{10}(N_{\mathrm{A}}/N_{\mathrm{H}}) + 12$. This is surprising, given that the primordial Li abundance is often inferred from similarly unevolved metal-poor stars (Ryan et al. 1999). Furthermore, the upper limit found from HE 1327−2326, however, strongly contradicts the WMAP value ($\log\epsilon(\mathrm{Li}) = 2.6$) from the baryon-to-photon ratio (Spergel et al. 2007). This may indicates that the star formed from extremely Li-poor material. No neutron-capture element is detected in HE 0107−5240 or HE 0557−4840, whereas, unexpectedly, Sr is observed in HE 1327−2326. Massive rotating stars may be responsible for this neutron-capture element (Chiappini et al. 2011).

How can all those signatures be understood in terms of early chemical enrichment? HE 0107−5240 and HE 1327−2326 immediately became benchmark objects to constrain various theoretical studies of the early Universe, such as the formation of the first stars (e.g., Yoshida et al. 2006), the chemical evolution of the early ISM (e.g., Karlsson and Gustafsson 2005) or calculations of Pop III SN yields. Their highly individual abundance patterns have been successfully reproduced by several different SNe scenarios. This makes HE 0107−5240 and HE 1327−2326 early, extreme Pop II stars that possibly display the "fingerprint" of just one Pop III SN. Umeda and Nomoto (2003) first matched the yields of a faint $25\,\mathrm{M}_{\odot}$ SN that underwent a mixing and fallback process to the observed abundances of HE 0107−5240. To achieve a simultaneous enrichment of a lot of C and only little Fe, large parts of the Fe-rich SN ejecta have to fall back onto the newly created black hole. Using yields from a SN with similar explosion energy and mass cut, Iwamoto et al. (2005) then reproduced the abundance pattern of HE 1327−2326 also.

---

[4]Interestingly, the entire mass of iron in HE 1327−2326 is actually 100 times less than that in the Earth's core. At the same time the star is of course of the order of a million times more massive than the Earth.

[5]The other stars are giants. Thus, the surface Li is already destroyed due to the thick convection zone transporting Li to deeper, hotter layers where it burns.

Trying to fit the observed stellar abundances, Heger and Woosley (2010) are employing an entire grid of Pop III SN yields to search for the best match to the data. A similar progenitor mass range as the (Umeda and Nomoto 2003) 25 $M_{\odot}$ was found to be the best match to have provided the elemental abundances to the ISM from which these Pop II stars formed.

Limongi et al. (2003) were able to reproduce the abundances of HE 0107−5240 through pollution of the birth cloud by at least two SNe. Suda et al. (2004) proposed that the abundances of HE 0107−5240 would originate from a mass transfer of CNO elements from a postulated companion, and from accretion of heavy elements from the ISM. However, neither HE 0107−5240 nor HE 1327−2326 show radial velocity variations that would indicate binarity. Meynet et al. (2006) explored the influence of stellar rotation on elemental yields of 60 $M_{\odot}$ near-zero-metallicity SNe. The stellar mass loss rate of rotating massive Pop III stars qualitatively reproduces the CNO abundances observed in HE 1327−2326 and other carbon-rich metal-poor stars.

More generally, the observed abundances of the most metal-poor stars with typical halo signatures in the range have successfully been reproduced with Pop III SN yields. Tominaga et al. (2007) model the averaged abundance pattern of four non-carbon-enriched stars with $-4.2 < [Fe/H] < -3.5$ with the elemental yields of a massive, energetic ($\sim$30–50 $M_{\odot}$) Pop III hypernova. The abundances can also be fitted with integrated yields of Pop III SNe (Heger and Woosley 2010). Special types of SNe or unusual nucleosynthesis yields can then be considered for stars with chemically peculiar abundance patterns. It is, however, often difficult to completely explain the entire abundance pattern of a given star. Additional metal-poor stars as well as a better understanding of nucleosynthesis and the explosion mechanism and the impact of the initial conditions on SNe yields are required to arrive at a more solid picture of the details of early SNe nucleosynthesis.

5 Dwarf Archaeology

Simulations of the hierarchical assembly of galaxies within the cold dark matter (CDM) framework (Diemand et al. 2007; Springel 2005) show that the Milky Way halo was successively built up from small dark matter substructures, often referred to as galactic building blocks, as long ago suggested by Searle and Zinn (1978). Figure 7 shows the substructure around the six simulated, high-resolution "Aquarius" halos at $z = 0$ (Springel et al. 2008; Lunnan et al. 2012). The satellites around the main halos are clearly visible—these are smaller halos that survived the violent accretion process until today.[6] They can be regarded as the counterparts to today's satellite population of the Milky Way.

[6]In this simulation, prescriptions for the population of dark halos with luminous matter have already been applied. See Lunnan et al. (2012) for further details and references therein.

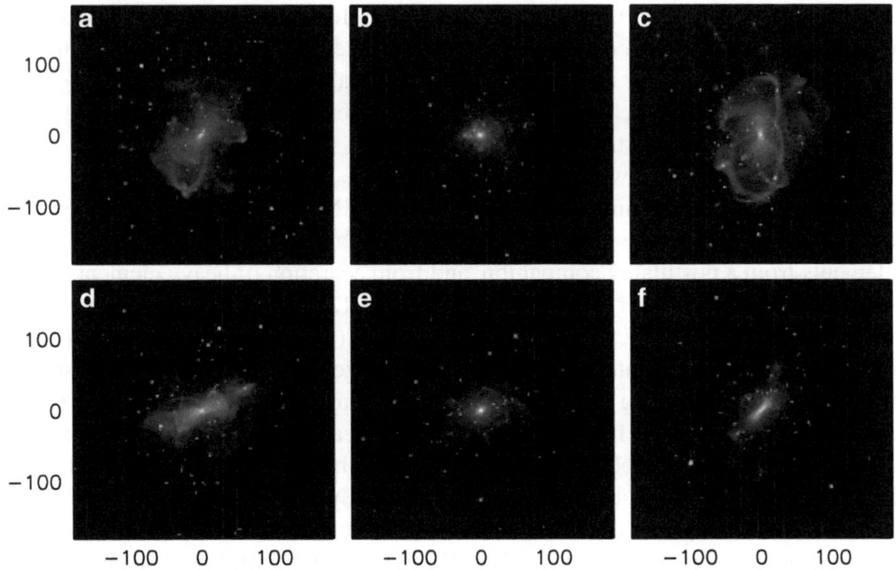

Fig. 7 Spatial distribution of the main halo and the corresponding substructure of the six Aquarius halos (A–F). Figure taken from Lunnan et al. (2012)

However, these simulations generally only include dark matter, and it remains unclear to what extent small dark halos contain luminous matter in the form of stars and gas. This question is particularly important with respect to the so-called "missing-satellite" problem which reflects the mismatch of the number of observed dwarf galaxies surrounding the Milky with the predicted number of substructures for a Milky Way-like halo. Studying the onset of star formation and associated chemical evolution in dwarf galaxies thus provides some of the currently missing information for our understanding of how the observed properties of small satellites relate to the (dark matter) substructures that build up larger galaxies. Thus, the study of the entire stellar population of a dwarf galaxy for the purpose of inferring details about the nature and origin of the first galaxies and early galaxy assembly is termed "dwarf archaeology".

5.1 Chemical Evolution of Dwarf Galaxies

The connection between the surviving dwarfs and those that dissolved to form the halo can partially be addressed by examining in detail the stellar chemical compositions of present-day dwarf galaxies. Establishing detailed chemical histories of these systems can provide constraints on their dominant chemical enrichment events and timescales. From such information, conclusions about the formation process of the Milky Way can be derived. Specifically, detailed knowledge of the most metal-poor stars in a given system allow insight into its earliest phases of star formation before the effects of internal chemical evolution were imprinted in stars born later with higher metallicity. Despite the fact that dwarf galaxies are regarded as "simple" systems, many of them show extended star formation histories with similar, albeit slower, chemical evolution to the Milky Way (for further details on this topic the reader is referred to the recent reviews by Tolstoy et al. 2009 and Koch 2009). Altogether, they are old, gas-less systems, and a tight correlation between the average metallicity and systemic luminosity.

Kirby et al. (2008) extended the metallicity–luminosity relationship to the ultra-faint dwarfs ($L < 10^5\,L_\odot$; discovered in the Sloan Digital Sky Survey; SDSS). They also showed that the lowest luminosity dwarfs have very low average metallicities (down to $<$[Fe/H]$>$ ~ -2.6), with several systems having values lower than those of the most metal-poor globular clusters. Many individual stars even have [Fe/H] < -3, while interestingly, no stars with [Fe/H] > -1.0 are found. The combined MDF of all the ultra-faint dwarfs extends down to [Fe/H] ~ -3.8, and the shape appears similar to that of the Milky Way halo (for the low-metallicity tail, that is), although perhaps with a slightly larger relative fraction towards the most metal-poor stars. Thus, it is not surprising that the ultra-faint dwarf galaxies contain a relatively many of extremely metal-poor stars, with large internal [Fe/H] abundance spreads of up to 3 dex. These spreads were first found in lower spectral resolution data (e.g., Simon and Geha 2007; Norris et al. 2008) and later confirmed with high-resolution follow-up of individual stars. At the same time, it became apparent that $\sim$30 % of the known stars with [Fe/H] < -3.5 are now found in dwarf galaxies. Segue 1, the most metal-poor system, contains $\sim$15 % alone (Frebel and Norris (2011, in prep.)).

Such high-resolution spectra are needed for a full chemical abundance analysis. Given the faint magnitude of stars in dwarf galaxies, these data are difficult to acquire, but good progress has been made over the past few years. In fact, most stars with $V < 19.2$ available in the ultra-faint dwarfs have now been observed this way. More ultra-faint dwarfs are expected to be found soon in surveys such as Pan-Starrs, Skymapper and LSST, thus providing new targets that can again be observed with high-resolution on current 6–10 m telescopes.

Generally, these kinds of data enable measurements of chemical abundances and upper limits of 10–15 elements for a given star. For example, the three brightest stars in each of Ursa Major II (UMa II) and Coma Berenices (ComBer) and two stars in Hercules are the first stars in the ultra-faint dwarf galaxies to have been observed with high-resolution spectroscopy. Two of them (in UMa II) are also the first known extremely metal-poor stars which are not members of the halo field population. Details on the observations and analysis techniques are given in Frebel et al. (2010a) and Koch et al. (2008). For the UMa II and ComBer stars, chemical abundances and upper limits of up to 26 elements were determined for each star. The abundance results demonstrate that the evolution of the elements in the ultra-faint dwarfs is very similar to that of the Milky Way, and likely dominated by massive stars. The α-elements in these two ultra-faint dwarf stars are overabundant, showing the halo-typical core-collapse SNe signature (see Fig. 4).

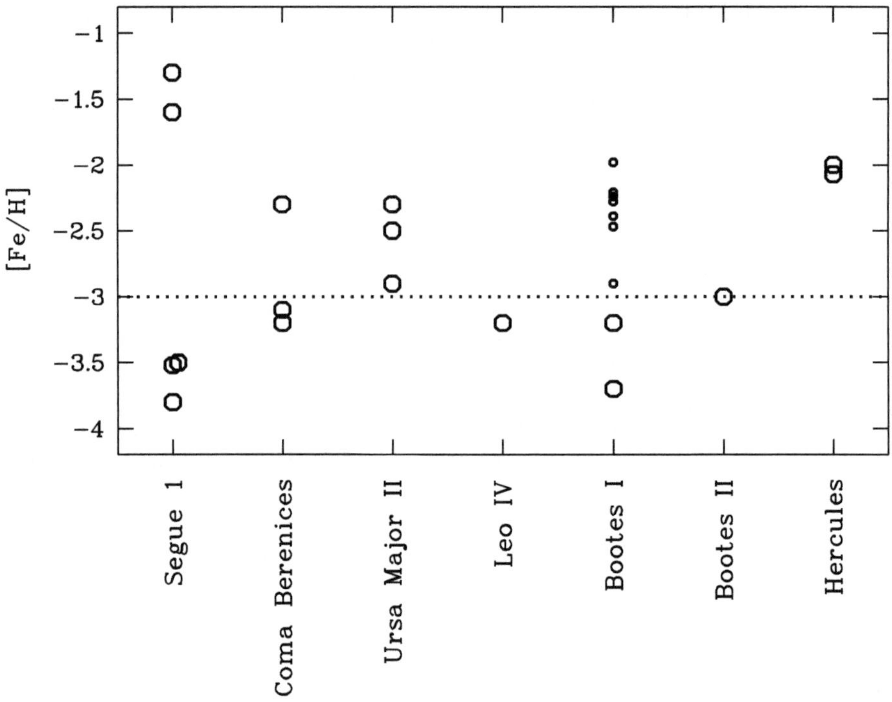

Fig. 8 Current status of available high-resolution spectroscopy of stars in the ultra-faint dwarf galaxies. Many stars have [Fe/H] < -3.0 *Large circles* indicate extensive abundances studies of many elements, whereas *smaller circles* refer to studies of only a few elements. Data from Koch et al. (2008), Feltzing et al. (2009), Frebel et al. (2010a), Simon et al. (2010), Norris et al. (2010b), Norris et al. (2010a) and Frebel and Norris (2011, in prep.)

The neutron-capture abundances are extremely low in the ultra-faint dwarf stars (see green (large shaded) circles in Fig. 6). The observed Sr and Ba values are up to two orders of magnitude below the abundances found in typical MW halo stars with similar Fe content. However, a very large spread (up to 3 dex) in these elements is found among halo stars themselves. The large depletions in the dwarf galaxy stars are thus not inconsistent with the halo data since similarly low values are found in numerous objects. Interestingly though, the low neutron-capture abundances may represent a typical signature of stars in dwarf galaxies. Similarly low values are also found in Hercules (Koch et al. 2008) and Draco (Fulbright et al. 2004) despite their sometimes relatively high Fe values of [Fe/H] ~ -2.0 (in Hercules).

By applying improved search techniques also to the more luminous dwarf galaxies, the first extremely metal-poor star in a classical dwarfs Sculptor was recently discovered (in a sample of 380 stars; Kirby et al. 2009). The metallicity of [Fe/H] ~ -3.8 was confirmed from a high-resolution spectrum taken with Magellan/MIKE (Frebel et al. 2010b). Only nine stars in the halo have even lower Fe abundances than this object. Shortly afterwards, Tafelmeyer et al. (2010) presented

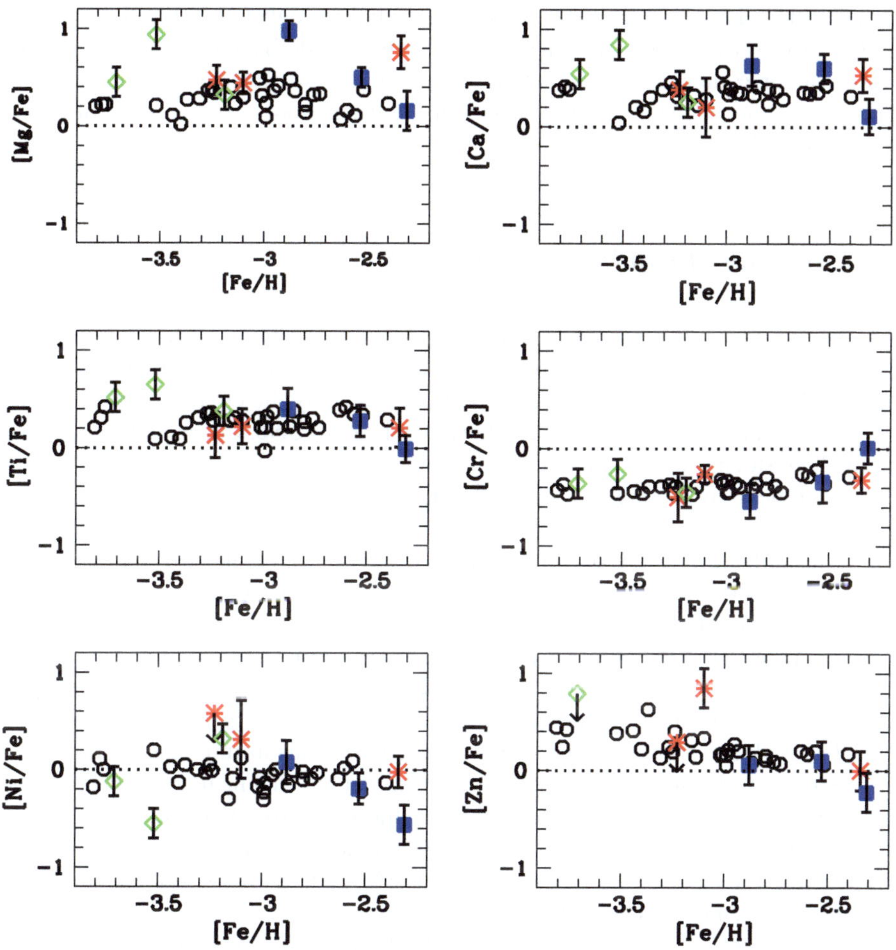

Fig. 9 Chemical abundance comparison of Galactic halo stars (*open circles*, from Cayrel et al. 2004) and stars in the ultra-faint dwarfs (*asterisks*: Ursa Major II, *filled squares*: Coma Berenices, *diamonds*: Segue 1, Bootes I and Leo IV). The abundances generally agree very well, suggesting a similar enrichment history of the gas from which all these stars formed. Based on Fig. 20 from Frebel and Norris (2011)

additional, similarly metal-poor stars in Sculptor, and also Fornax. These discoveries show that such low-metallicity stars are indeed present in the more luminous dwarfs (see also Starkenburg et al. 2010), contrary to earlier claims.

The new stars underpin that metallicity spreads of $\sim$3 dex or more are present in many classical dwarfs. The chemical abundances obtained from the high-resolution spectrum reveal a similar picture to what has been found in the ultra-faint dwarf stars. For example, the Sculptor star at [Fe/H] ~ -3.8, also has a remarkably similar chemical make-up compared to that of the Milky Way halo at the lowest metallicities. This is in contrast to what is found at higher metallicities in these brighter dwarfs. As can be seen in Fig. 4 (red (small shaded) circles), those stars have lower stellar α-abundance ratios than comparable halo stars, indicating different enrichment mechanisms and timescales in these gas-poor systems (e.g., Shetrone et al. 2003; Geisler et al. 2005).

5.2 Individual Pop III Enrichments?

The large number of extremely metal-poor stars in the faintest dwarf galaxies is very surprising given the very limited total stellar material in these systems with $L_{tot} \lesssim 10^5 L_\odot$. This represents an opportunity to study the environment of a low-metallicity galaxy which should have hosted only a few SNe. Consequently, signs of individual, stochastic enrichment events may be preserved in the metal-poor stars.

Enrichment by a single $13\,\mathrm{M}_\odot$ has been suggested for the Hercules dwarf ($L \sim 10^5 L_\odot$), based on the chemical abundances of two stars with [Fe/H] ~ -2.0 and unusually low levels of Ba (Koch et al. 2008). With its low luminosity of ($L_V = 14,000 L_\odot$) and low metallicity, Simon et al. (2010) determined that Leo IV contains only 0.042 $\mathrm{M}_\odot$ of Fe (assuming a stellar mass-to-light ratio of 1 $\mathrm{M}_\odot/\mathrm{L}_\odot$). Canonical SN nucleosynthesis yield calculations predict a maximum Fe yield of $\sim 0.1 M_\odot$ (e.g. Heger and Woosley 2010). The total Fe amount in Leo IV, and by extension that of all the other observed elements, could thus stem from a single star that exploded soon after the system formed. The star forming gas cloud at that time must have been only $\sim 40,000 M_\odot$. Given that the chemical abundances of the brightest star that is accessible to high-resolution spectroscopic studies, resemble those of metal-poor halo stars, it was suggested that indeed a single Pop III star was responsible for Leo IV's enrichment. Various SN nucleosynthesis models invoking different progenitor masses and explosion energies (e.g., Tominaga et al. (2007); Heger and Woosley 2010) have shown that the typical halo abundance signature can be explained this way.

Alternatively, given the shallow potential of these kinds of systems, several SNe could have contributed to the enrichment, but part of the enriched gas was soon lost (e.g., through stellar winds or SN explosions). Future observations of stars in ultra-faint dwarfs will provide more detailed insight. Nevertheless, the currently available data already suggest that these systems played an important role for our understanding of the first stars and chemical enrichment events that shaped the nature of these relatively small, early systems.

5.3 Are the Faintest Dwarf Satellites Surviving First Galaxies?

Given that the ultra-faint dwarf galaxies are the most metal-poor objects we know of today, they are ideal probes of the physical, chemical, and dynamical processes at work in the early universe. Since these systems have much simpler star formation histories than the halo of the Milky Way, their stellar populations should preserve the fossil record of the first supernova (SN) explosions in their long-lived, low-mass stars (see also Sect. 5.2). Hence, the metal-poor stars in the ultra-faint dwarf galaxies should be used to empirically constrain the formation process of the first galaxies, and early galaxy assembly more generally.

Guided by recent hydrodynamical cosmological "ab-initio"simulations (Greif et al. 2008, 2010) of first galaxy formation, Frebel and Norris (2011) developed a set of stellar abundance signatures that characterizes the nucleosynthetic history of such an early system if it was observed in the present-day universe. In particular, the simulations suggest that a first galaxy can be regarded a chemical "one-shot" event, where only one (long-lived) stellar generation forms after the first, Population III, SN explosions. The system would thus be dominated by an [α/Fe] enrichment due to enrichment by massive stars as seen in the halo at low metallicity. With no stars present displaying an erstwhile enrichment by AGB stars of SN Ia, the α-enhancement would also be present in stars with higher metallicities ([Fe/H] > -1.5). These criteria thus constrain the strength of negative feedback effects inside the first galaxies.

A comparison of the abundances of about a dozen stars in the ultra-faint dwarfs with this one-shot criterion indicates that some of these faintest satellites could be surviving first galaxies. Several systems (Ursa Major II, and also Coma Berenices, Bootes I, Leo IV, Segue 1) largely fulfill the requirements (most notably the high α/Fe ratios), indicating that their high-redshift predecessors did experience strong feedback effects that shut off star formation soon after the formation of the system. More observations are needed to firm up these initial conclusions, and also additional simulations of early metal mixing, turbulence and the extend of feedback in early, assembling galaxies. This will provide clues to the connection of the first galaxies, the surviving, metal-poor dwarf galaxies, and the building blocks of the Milky Way.

6 Near-Field Cosmology

A long-standing problem, originally noted by e.g. Moore et al. (1999) and Klypin et al. (1999), is that the observed number of Milky Way satellites appears to be significantly lower than the number of dark matter substructures expected based on the CDM theory (the so-called "missing satellite problem"). This calls into question the validity of the CDM picture on the scales of individual galaxies. Many ideas have been proposed to reconcile the theory with the observations, including the possibility that the dark matter may instead be dynamically "warm", rather than cold (achieved by making the individual dark matter particles less massive). Another class of models proposes instead that effects related to feedback from baryonic processes or heating by cosmic radiation fields may inhibit star formation in dark matter halos of sufficiently low mass, rendering them invisible. Either way,

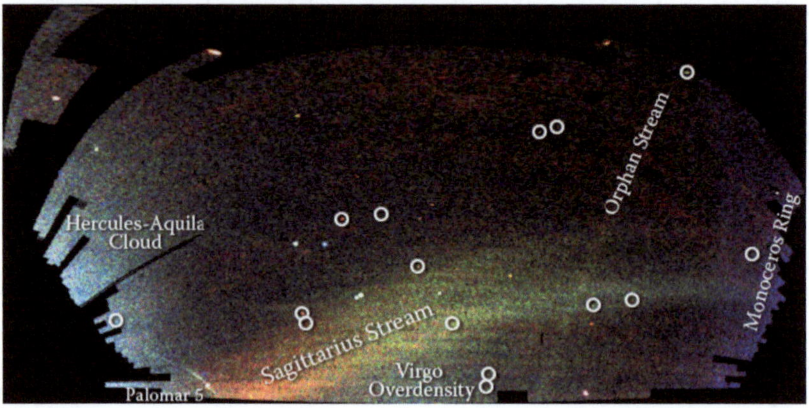

Fig. 10 Field of Streams: A map of stars in the outer regions of the Milky Way Galaxy, shown in a Mercator-like projection. The color indicates the distance of the stars, while the intensity indicates the density of stars on the sky. Structures visible in this map include streams of stars torn from the Sagittarius dwarf galaxy, a smaller "orphan" stream crossing the Sagittarius stream, the "Monoceros Ring" that encircles the Milky Way disk, trails of stars being stripped from the globular cluster Palomar 5, and excesses of stars found towards the constellations Virgo and Hercules. Circles enclose new Milky Way companions discovered by the SDSS; two of these are faint globular star clusters, while the others are faint dwarf galaxies. Image credit: V. Belokurov and the Sloan Digital Sky Survey

a definitive solution to the missing satellite problem will inform our theories of the nature of the dark matter and the assembly of galaxies.

Progress has been made through detailed studies of different types of dwarf galaxies (ultra-faint, classical dwarf spheroidal, more massive dwarfs such as the Magellanic Clouds) which orbit the Milky Way. Extensive photometric and spectroscopic data of these satellites, paired with the discovery of stellar streams in the Galactic halo arising from past and even ongoing mergers of the massive host with smaller galaxies, have revealed much about the complex nature of the life and times of dwarf galaxies and their role in shaping their parent galaxy. Figure 10 shows a number of streams that are currently present on the stellar halo of the Milky Way due to past accretion events.

Nevertheless it is surprising that the Milky Way overall has so few satellite galaxies when its sister galaxy Andromeda appears to have a significantly larger population signaling a potentially very different assembly history. By learning about the Galaxy and its assembly, as well as pushing for large-scale simulations to address cosmic variance and where the Milky Way really fits within the zoo of galaxies, we will soon be able to quantify the host of observations that suggest the unique nature of our Galaxy.

6.1 Did the Stellar Halo Form From Accreted Dwarf Galaxies?

The overarching questions of near-field cosmology concern the nature of the "building blocks" of large galaxies, and to what extent dwarf galaxies play a role in the assembly of old stellar halos. This comes at a time when in particular the population of Milky Way ultra-faint dwarf galaxies have been shown to be extremely metal-deficient systems that host ~30 % of the known most metal-poor stars. Moreover, they extend the metallicity–luminosity relationship of the classical dwarfs down to $L_{\rm tot} \sim 10^3\, L_{\odot}$ (see Kirby et al. (2008) for more details). Future observations will reveal how far this relationship can be extended.

If the old, metal-poor halo was indeed assembled from dwarf galaxies, the metallicities of stars in dwarf galaxies must reach values as low as (or lower) what is currently found in the Galactic halo population. Moreover, the abundance ratio of those low-metallicity stars must be roughly equal to those of equivalent stars in the halo.

Earlier works missed finding extremely metal-poor stars in the classical dwarfs, which posed a significantly problem to the idea of an accreted halo through systems similar to the surviving dwarfs. However, it has now been shown that this claim stems merely from biases in earlier search techniques (Cohen and Huang 2009; Kirby et al. 2008; Starkenburg et al. 2010; Kirby et al. 2011). With improved methods for identifying the lowest-metallicity objects, extremely metal-poor stars have already been identified in several dwarf galaxies (Kirby et al. 2008; Geha et al. 2009; Frebel et al. 2010b; Tafelmeyer et al. 2010). The higher-metallicity stars in the classical dwarfs all have abundance ratios different from comparable halo stars. Most strikingly, the α-element abundances are not enhanced, which must be due to the onset on SN Ia enrichment at a time when the overall chemical evolution of these small systems was less evolved than the halo. It also reflects that any of these stars are the product of the galaxy's own chemical evolution that has nothing in common in that of the Milky Way. However, there is increasing evidence that a transition to more halo-typical abundance ratios may take place around a metallicity of [Fe/H] $= -3.0$ (Cohen and Huang 2009; Aoki et al. 2009) in these dwarfs. This means that if the halo formed from accreted satellites, any systems like the classical dwarfs must have been accreted at a relatively early time when the chemical evolution was not much progressed. Given that the dwarfs we observe today were not accreted, they kept forming stars and continued chemical evolution until they lost all their gas.

Due to their simple nature, the ultra-faint systems are expected to retain signatures of the earliest stages of chemical enrichment in their stellar populations. If the halo was assembled from early analogs of the surviving dwarfs, most of the metal-poor halo stars should have been deposited there during late-time accretion events (Frebel et al. 2010a; Simon et al. 2010; Font et al. 2006). Indeed the chemical abundances of individual stars in the faintest galaxies suggest a close connection to equivalent, extremely metal-poor halo stars in the Galaxy.

The recent high-resolution studies (Feltzing et al. 2009; Frebel et al. 2010a; Norris et al. 2010a,b; Simon et al. 2010) provide evidence that the abundance patterns of light elements ($Z < 30$) in these dwarfs are remarkably similar to the Milky Way halo, even for stars with [Fe/H] > -3.0. this is illustrated in Fig. 9 which shows a very detailed comparison between halo and ultra-faint dwarf galaxy stars. The similarity between the data sets is clearly seen. However, given the limited number of stars it is still unclear up to what metallicity the halo-typical abundances are found in these systems. There are indications that the chemical evolution in the ultra-faint dwarfs may have been very inhomogeneous, and also, that a number of stars show strong carbon-enhancement (Norris et al. 2010a). Taken altogether, these features are found among the lowest metallicity halo stars, making a plausible case for an accretion-built halo, at least for stars with [Fe/H] < -2.5.

These observational results about the halo assembly are broadly consistent with the predictions of CDM cosmological models (e.g. Robertson et al. 2005; Johnston et al. 2008). While the majority of the mass that formed the inner part of the stellar halo (at [Fe/H] ~ -1.2 to -1.6) was provided at early times by much larger systems such as the Magellanic Clouds, the outer halo assembled at later times. In fact, as shown in Fig. 9, it is still assembling today, with ongoing accretion event leading to a variety of streams and substructure. The similarity of the chemical abundances, suggesting the same chemical enrichment history in these stars prior to their formation, make it plausible to assume that the ultra-faint dwarf galaxies contributed (some) individual metal-poor stars to the Galaxy. These stars are now found primarily in the outer Galactic halo.

However, these systems may not have been sufficiently numerous to account for the entire metal-poor end of the Fe metallicity distribution of the Milky Way halo. Since the classical dSphs have more stellar mass and have been shown to also contain at least some of the most metal-poor stars (e.g., Kirby et al. 2009; Frebel et al. 2010b), they could have been a major source of the lowest-metallicity halo stars. Additional observations of more extremely metal-poor stars in the various dwarf galaxies are highly desirable in the quest to determine individual MDFs for each of these galaxies, and how those would compare with each other and that of the Milky Way.

6.2 *Towards Constraining the Reionization History of the Milky Way*

Cosmological simulations of the growth of structure have shown that galaxy assembly proceeds hierarchically. The results of these studies have indicated a mismatch between the predicted number of low-mass, dark matter substructures and the actual observed number of satellite galaxies around the Milky Way; the "missing satellite problem". The underlying cause of this discrepancy can be investigated with cosmological simulations. By examining the impact of various physical processes on the evolution of faint galaxies, the nature of the surviving luminous satellites and their stellar content can be understood in more detail. An important question is this regard is which effects critically determine the number of small satellites hosting luminous matter throughout the build-up of a large galaxy, and whether there would be enough of them to form a metal-poor stellar halo of their host, similar to what is found for the Milky Way.

The epoch of reionization in the early universe may have had an important influence on the star forming ability of small halos. Assuming that reionization of the universe was patchy and extended in redshift (Zahn et al. 2007; McQuinn et al. 2007), the number of smallest, resolved satellites ($10^6\,M_\odot$ halos) that survive until $z = 0$ can be reduced relative to models in which reionization is considered to be homogeneous and instantaneous (Lunnan et al. 2012). This is about a 10 % effect. Figure 11 shows the observed luminosity function of dwarf galaxies and several models for different reionization histories for the six Aquarius halos. Not considering reionization at all affects the luminous content of faint system and leads indeed to a missing satellite problem, i.e. an overproduction of the smallest halos in CDM simulations. This is the case for all six halos. The general halo-to-halo scatter (factor of 2–3; see also Fig. 7), however, suggests that the missing satellite problem is in part due to cosmic variance. Moreover, taking the effects of reionization into account has a pronounced impact of just the faintest halos. While larger halos (equivalents to today's brighter dwarf galaxies) are relatively insensitive to a non-instantaneous reionization prescription, the number of surviving small galaxies changes significantly with the reionization model for different mean reionization redshift and Thompson optical depths still in agreement with the WMAP value (Komatsu et al. 2011).

Given these initial results, it appears promising to use the brighter end of the luminosity function to constrain the halo-to-halo scatter, while the faint end will enable to discriminate reionization histories. However, a solid understanding of how "normal" the Milky Way is and where it falls within these halo-to-halo variations and will be required. Future simulations may be able to quantify the nature of the Galaxy with respect to the majority of other large galaxies. Knowing the substructure abundance of galaxy halos is critical for interpreting observations of the satellite populations of all large galaxies, including the Milky Way and Andromeda. Moreover, estimates would become possible of the stellar contribution to the halo as a function of halo mass and the associated merger and accretion history.

Altogether, the opportunity to constrain the reionization history of the Milky Way through a careful analyses of the faintest satellite population will help understanding the nature of these systems themselves. Knowing the fraction of surviving small halos with truly old stars that formed before reionization is of critical importance so it can be established that today's metal-poor stars indeed trace the earliest times and enrichment events. Using the cosmological simulations together with prescriptions for luminous matter, feedback processes and chemical evolution (e.g., Cooper et al. 2010) will shed more light on the cosmological origin of the most metal-poor stars. Stellar and dwarf archaeology meet near-field cosmology for exactly these kinds of questions, and only the combination of high-quality observations and powerful cosmological simulations will enable the progress that is required to understand the early star and galaxy formation and the evolution of our own Milky Way.

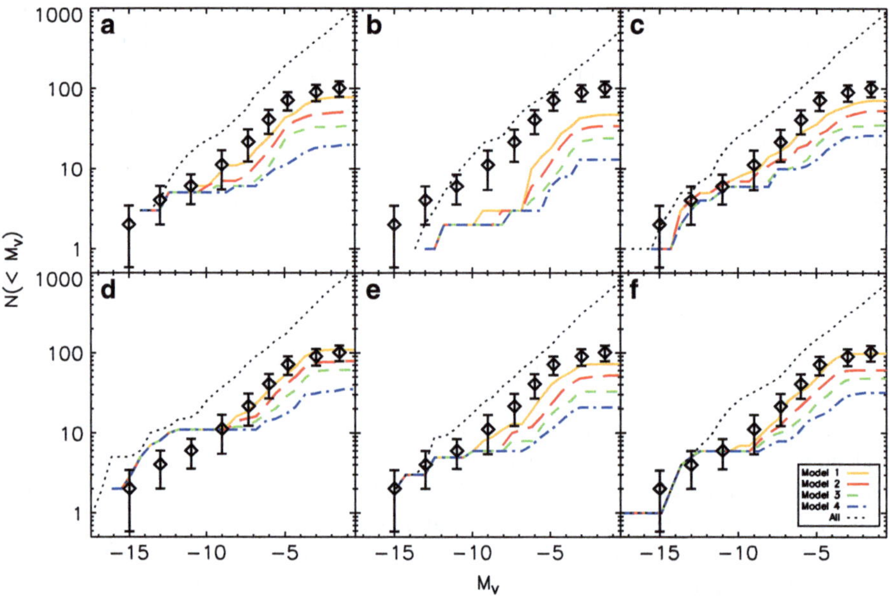

Fig. 11 Observed luminosity function of dwarf galaxies (*black diamonds*) in comparison with models employing four different physically motivated reionization histories for the six Aquarius halos. Note the scatter in the numbers of simulated satellites among the halos (factor of 2–3) as well as similar variations of the faint end caused by the different reionization histories. The *dotted lines* depict satellite population unaffected by reionization. From Lunnan et al. (2012)

7 Outlook

Old metal-poor stars have long been used to learn about the conditions of the early Universe. This includes the origin and evolution of the chemical elements, the relevant nucleosynthesis processes and sites and the overall chemical and dynamical history of the Galaxy. By extension, metal-poor stars provide constraints on the nature of the first stars and their initial mass function, the chemical yields of first/early SNe, as well as early star and galaxy formation processes including the formation of the galactic halo.

Supported by these large-scale survey efforts, the field of near-field cosmology has been catapulted into an "excited state" because it constrains many astrophysical problems related to galaxy formation. It is thus very timely to extensively pursue research into the oldest stars and stellar populations in the dwarf galaxies. They not only enable us to study the history of galaxies with different masses and luminosities, but also facilitate an in-depth study of what role small(er) galaxies may have played in the build-up of the stellar halo.

7.1 Observational Challenges

Finding more of the most metal-poor stars (e.g., stars with [Fe/H] < -5.0) is of great importance for addressing the topics described in this chapter in more detail. However, as outlined these stars are extremely rare (Schörk et al. 2009) and difficult to find. The most promising way forward is to survey larger volumes further out in the Galactic halo. But how feasible is it to identify stars with even lower metallicities? Frebel et al. (2009) calculated the minimum observable Fe and Mg abundances in the Galaxy by combining the critical metallicity of $[\mathrm{C/H}]_{min} = -3.5$ (the criterion for the formation of the first low-mass stars by Bromm and Loeb (2003) with the maximum carbon-to-iron ratio found in any metal-poor star. The resulting minimum Fe value is $[\mathrm{Fe/H}]_{min} = -7.3$. Analogously, the minimum Mg value is $[\mathrm{Mg/H}]_{min} = -5.5$. If $[\mathrm{C/H}]_{min}$ was lower, e.g., $[\mathrm{C/H}]_{min} = -4.5$, as suggested by recent dust cooling computations, the minimum observable Fe and Mg abundances would accordingly be lower. Spectrum synthesis calculations suggest these low abundance levels are indeed measurable from each of the strongest Fe and Mg lines in suitably cool metal-deficient giants.

Future surveys such as Skymapper (Keller et al. 2007) and LAMOST will provide an abundance of new metal-poor candidates as well as new, faint dwarf galaxies. By accessing such stars in the outer Galactic halo and dwarf galaxies we will be able to gain a more complete census of the chemical and dynamical history of our own Galaxy. Since the lowest metallicity stars are expected to be in the outer halo (e.g., Carollo et al. 2012), their corresponding kinematic properties may prevent them from accreting too much enriched material from the ISM during their lives so that their surface composition would not be altered (i.e., increased; Frebel et al. 2009). Hence, selecting for the most metal-poor candidates will increasingly rely on our ability to combine chemical abundances with kinematic information. Future missions such as GAIA will provide accurate proper motions for many object that currently have no kinematic information available, including for most of the currently known metal-poor giants.

However, many, if not most, of these future metal-poor candidates will be too faint to be followed up with the high-resolution spectroscopy necessary for detailed abundance analyses. This is already an issue for many current candidates leaving the outer halo a so far largely unexplored territory: The limit for high-resolution work is $V \sim 19$ mag, and one night's observing with 6–10 m telescopes is required for the minimum useful signal-to-noise (S/N) ratio of such a spectrum. With the light-collecting power of the next generation of optical telescopes, such as the Giant Magellan Telescope, the thirty Meter Telescope or the European ELT, and if they are equipped with high-resolution spectrographs, it would be possible to not only reach out into the outer halo in search of the most metal-poor stars, but also provide spectra with very high-S/N ratio of somewhat brighter stars. For example, the so-called r-process enhanced stars which provide crucial empirical constraints on the nature of this nucleosynthesis process require exquisite data quality e.g. for uranium and lead measurements that are currently only possible for the very brightest stars (e.g., Frebel et al. 2007b).

It appears that the hunt for the metal-deficient stars in dwarf galaxies may have just begun since these dwarfs host nearly a third of the known low-metallicity stars. The detailed abundance patterns of the stars in UMa II, ComBer, Leo IV, etc. are strikingly similar to that of the Milky Way stellar halo, thus renewing the support for dwarf galaxies as the building blocks of the halo. Future discoveries of additional faint dwarf galaxies will enable the identification of many more metal-poor stars in new, low-luminosity systems. But also the brighter dSphs have to be revisited for their metal-poor content (Kirby et al. 2009). More stars at the lowest metallicities are clearly desired to better quantify the emerging chemical signatures and to solidify our understanding of the early Galaxy assembly process. Together with advances in the theoretical understanding of early star and galaxy formation and SNe yields, a more complete picture of the evolution of the Milky Way Galaxy and other systems may soon be obtained. Only in this way can the hierarchical merging paradigm for the formation of the Milky Way be put on firm observational ground.

7.2 *Constraining the Theoretical Framework*

An important next step towards a full understanding of the infant universe is to combine the phenomenological approach of collecting new observational data with theoretical knowledge about the formation of the first generations of stars and galaxies. This way, tools can be developed that allow for cosmologically motivated interpretations of the abundance patterns of metal-poor stars, in both the halo and dwarf galaxies. The new field of dwarf archaeology promises a more complete understanding of early enrichment events and the processes that led to galaxy formation at the end of the cosmic dark ages. At the same time, this will have profound implications for the search of the major enrichment mechanisms in the early universe and the physical origin of these stars.

Understanding the Milky Way as a whole is therefore of crucial importance. Most importantly, new results have raised the question of how well the Milky Way represents a typical, large spiral galaxy. Our Galaxy is often used as a reference, especially when comparing its general properties with those derived from simulations of the formation of large galaxies. But recent works, both observationally and theoretically, have shown that our Galaxy has at least several unusual features. The existence of the long discovered Magellanic Cloud satellites have recently garnered significant attention (Boylan-Kolchin et al. 2010; Liu et al. 2011) in this respect. Observational analyses using Sloan Digital Sky Survey data of many other large spiral galaxies confirmed that galaxies like the Milky Way are very unlikely to have two companions as bright as the Magellanic Clouds. Indeed, less than 5 % of galaxies host two such bright companions, and more than 80 percent host no such satellites at all. Previously, Boylan-Kolchin et al. (2010) have examined this issue using the Millennium simulation of Springel et al. (2005) and finding that the Milky Way is unusual in hosting the Magellanic Clouds.

Furthermore, the evolutionary differences between the Milky Way and Andromeda galaxy need to be to established. The Milky Way did not undergo major mergers with other galaxies since nearly 11 billion years, whereas Andromeda underwent many mergers in a recent past (a few billion years). Only then, can it be understood how the different formation histories influence the extent of surviving substructure and differences in properties such as the stellar mass, disk radius, and metal-deficient halo between the two sister galaxies. Assessing the degree of "normality" of the Milky Way will be vital for understanding of whether or not the Magellanic Clouds were only recently accreted by the Milky Way, as proposed by Besla et al. (2007) based on new proper motion measurements.

A major step forward would be a detailed understanding of the missing satellite problem. This is of great interest in cosmology, both observationally and theoretically, but also many associated fields e.g, regarding galaxy formation or dwarf galaxy studies. Is the overproduction of halo substructure at $z = 0$ in DM simulations really based on physical processes associated with the gas that lights up dark halos, or is it merely an artifact of past simulations? It has been shown that cosmic variance may play an important role, and without properly quantify this effect, perhaps no conclusions can be drawn at this point in time. Nevertheless, the fact that the Milky Way may not be an ordinary galaxy could partly explain the missing satellite problem as all simulation results are always compared to the observations associated with the Galaxy.

Progress can be made by carrying out simulations that aim at incorporating near-field cosmological constraints to address, e.g., the underlying physical causes of the missing satellite problems beyond cosmic variance. To learn about the effects that influence the number of small subhalos over the course of the universe, studies are now being carried out that aim at quantifying the impact of physically motivated, patchy, reionization histories on the faintest halos (e.g., Lunnan et al. 2012). While these studies may partly resolve the missing satellite problem, the overall halo-to-halo differences in the populations is of a similar level, preventing strong, global conclusions. The way forward it to quantify the level and extend of variations of the substructure around Milky Way galaxies. Only then can a number of specific details about the origin and evolution of subhalos with different masses equivalent to those of a variety of observed dwarf galaxies, such as massive Magellanic Cloud-sized objects, classical dwarf Spheroidal galaxies, and even fainter systems, be understood.

In summary, the details of the many physical processes that govern the evolution of a luminous halo at $z = 0$ need to be known to conclusively address the formation of large galaxies within the hierarchical assembly paradigm. They will need to be coupled to simulation results, quantified, and compared to the observations. Hopefully, in the not too distant future, there will be unparalleled opportunities to study the assembly of galaxy halos in detail, connected to the results of the latest observations of dwarf galaxies, halo stars and stellar streams found in the Milky Way. For example, [Fe/H] variations in dwarf galaxies and well-established abundance trends are important to constrain the chemical evolution in dwarf galaxies. These quantities are crucial for the development of prescriptions for the chemical enrichment throughout the hierarchical merging process. Moreover, they constrain, e.g., metal mixing, turbulence and feedback effects in early star forming halos, as simulated in hydrodynamical "ab-initio" simulations (e.g., Greif et al. 2010).

All these works will soon become possible in great detail as large-scale parallel supercomputers will enable ever more realistic simulations of structure formation as part of the early universe, and on large scales down to $z = 0$. On a similar timescale increasingly detailed observations of stars in the halo and ultra-faint dwarfs will become available (e.g., Skymapper). However, observations alone will likely not be able to uncover the underlying physical processes to conclusively confirm the details of the assembly history of the Milky Way stellar halo and the cosmological origin of the ancient $\sim$13 Gyr old stars which must have formed long before the Milky Way was fully assembled.

Acknowledgements A.F. gratefully acknowledges a Clay Fellowship which is administered by the Smithsonian Astrophysical Observatory.

References

Aoki, W., Arimoto, N., Sadakane, K., Tolstoy, E., Battaglia, G., Jablonka, P., Shetrone, M., Letarte, B., Irwin, M., Hill, V., Francois, P., Venn, K., Primas, F., Helmi, A., Kaufer, A., Tafelmeyer, M., Szeifert, T., Babusiaux, C.: A&A **502**, 569 (2009)
Aoki, W., Beers, T.C., Sivarani, T., Marsteller, B., Lee, Y.S., Honda, S., Norris, J.E., Ryan, S.G., Carollo, D.: ApJ **678**, 1351 (2008)
Aoki, W., Frebel, A., Christlieb, N., Norris, J.E., Beers, T.C., Minezaki, T., Barklem, P.S., Honda, S., Takada-Hidai, M., Asplund, M., Ryan, S.G., Tsangarides, S., Eriksson, K., Steinhauer, A., Deliyannis, C.P., Nomoto, K., Fujimoto, M.Y., Ando, H., Yoshii, Y., Kajino, T.: ApJ **639**, 897 (2006)
Barklem, P.S., Christlieb, N., Beers, T.C., Hill, V., Bessell, M.S., Holmberg, J., Marsteller, B., Rossi, S., Zickgraf, F.-J., Reimers, D.: A&A **439**, 129 (2005)
Beers, T.C., Christlieb, N.: ARA&A **43**, 531 (2005)
Beers, T.C., Preston, G.W., Shectman, S.A.: AJ **103**, 1987 (1992)
Bergemann, M., Gehren, T.: A&A **492**, 823 (2008)
Besla, G., Kallivayalil, N., Hernquist, L., Robertson, B., Cox, T.J., van der Marel, R.P., Alcock, C.: ApJ **668**, 949 (2007)
Bessell, M.S., Norris, J.: ApJ **285**, 622 (1984)
Boylan-Kolchin, M., Besla, G., Hernquist, L.: ArXiv:1010.4797 (2010)
Bromm, V., Loeb, A.: Nature **425**, 812 (2003)
Bromm, V., Yoshida, N.: ArXiv:1102.4638 (2011)
Bromm, V., Yoshida, N., Hernquist, L., McKee, C.F.: Nature **459**, 49 (2009)
Carollo, D., Beers, T.C., Bovy, J., Sivarani, T., Norris, J.E., Freeman, K.C., Aoki, W., Lee, Y.S., Kennedy, C.R.: ApJ **744**, 195 (2012)
Cayrel, R., Depagne, E., Spite, M., Hill, V., Spite, F., François, P., Plez, B., Beers, T., Primas, F., Andersen, J., Barbuy, B., Bonifacio, P., Molaro, P., Nordström, B.: A&A **416**, 1117 (2004)
Chiappini, C., Frischknecht, U., Meynet, G., Hirschi, R., Barbuy, B., Pignatari, M., Decressin, T., Maeder, A.: Nature **472**, 454 (2011)
Christlieb, N., Bessell, M.S., Beers, T.C., Gustafsson, B., Korn, A., Barklem, P.S., Karlsson, T., Mizuno-Wiedner, M., Rossi, S.: Nature **419**, 904 (2002)
Christlieb, N., Schörk, T., Frebel, A., Beers, T.C., Wisotzki, L., Reimers, D.: A&A **484**, 721 (2008)

Clark, P.C., Glover, S.C.O., Smith, R.J., Greif, T.H., Klessen, R.S., Bromm, V.: Science **331**, 1040 (2011)
Cohen, J.G., Christlieb, N., McWilliam, A., Shectman, S., Thompson, I., Melendez, J., Wisotzki, L., Reimers, D.: ApJ **672**, 320 (2008)
Cohen, J.G. Huang, W.: ApJ **701**, 1053 (2009)
Cooper, A.P., Cole, S., Frenk, C.S., White, S.D.M., Helly, J., Benson, A.J., De Lucia, G., Helmi, A., Jenkins, A., Navarro, J.F., Springel, V., Wang, J.: MNRAS **406**, 744 (2010)
Diemand, J., Kuhlen, M., Madau, P.: ApJ **667**, 859 (2007)
Feltzing, S., Eriksson, K., Kleyna, J., Wilkinson, M.I.: A&A **508**, L1 (2009)
Font, A.S., Johnston, K.V., Bullock, J.S., Robertson, B.E.: ApJ **646**, 886 (2006)
Frebel, A.: Astronomische Nachrichten **331**, 474 (2010)
Frebel, A., Aoki, W., Christlieb, N., Ando, H., Asplund, M., Barklem, P.S., Beers, T.C., Eriksson, K., Fechner, C., Fujimoto, M.Y., Honda, S., Kajino, T., Minezaki, T., Nomoto, K., Norris, J.E., Ryan, S.G., Takada-Hidai, M., Tsangarides, S., Yoshii, Y.: Nature **434**, 871 (2005a)
Frebel, A., Aoki, W., Christlieb, N., Ando, H., Asplund, M., Barklem, P.S., Beers, T.C., Eriksson, K., Fechner, C., Fujimoto, M.Y., Honda, S., Kajino, T., Minezaki, T., Nomoto, K., Norris, J.E., Ryan, S.G., Takada-Hidai, M., Tsangarides, S., Yoshii, Y.: in IAU Symposium **228**, 207 (2005b)
Frebel, A., Christlieb, N., Norris, J.E., Beers, T.C., Bessell, M.S., Rhee, J., Fechner, C., Marsteller, B., Rossi, S., Thom, C., Wisotzki, L., Reimers, D.: ApJ **652**, 1585 (2006)
Frebel, A., Christlieb, N., Norris, J.E., Thom, C., Beers, T.C., Rhee, J.: ApJ **660**, L117 (2007a)
Frebel, A., Johnson, J.L., Bromm, V.: MNRAS **380**, L40 (2007b)
Frebel, A., Johnson, J.L., Bromm, V.: MNRAS, 392, L50 (2009)
Frebel, A., Kirby, E.N., Simon, J.D.: Nature **464**, 72 (2010a)
Frebel, A., Norris, J.E.: ArXiv e-prints 1102.1748 (2011)
Frebel, A., Simon, J.D., Geha, M., Willman, B.: ApJ **708**, 560 (2010b)
Fulbright, J.P., Rich, R.M., Castro, S.: ApJ **612**, 447 (2004)
Geha, M., Willman, B., Simon, J.D., Strigari, L.E., Kirby, E.N., Law, D.R., Strader, J.: ApJ **692**, 1464 (2009)
Geisler, D., Smith, V.V., Wallerstein, G., Gonzalez, G., Charbonnel, C.: AJ **129**, 1428 (2005)
Greif, T.H., Glover, S.C.O., Bromm, V., Klessen, R.S.: ApJ **716**, 510 (2010)
Greif, T.H., Johnson, J.L., Klessen, R.S., Bromm, V.: MNRAS **387**, 1021 (2008)
Heger, A., Woosley, S.E.: ApJ **567**, 532 (2002)
Heger, A., Woosley, S.E.: ApJ **724**, 341 (2010)
Hollek, J.K., Frebel, A., Roederer, I.U., Sneden, C., Shetrone, M., Beers, T., Kang, S.-J., Thom, C., Lambert, D.: ApJ **742**, 54 (2012)
Iben, I.: Memor. Soc. Astronom. Ital. **54**, 321 (1983)
Iwamoto, N., Umeda, H., Tominaga, N., Nomoto, K., Maeda, K.: Science **309**, 451 (2005)
Johnson, J.L., Khochfar, S.: MNRAS **413**, 1184 (2011)
Johnston, K.V., Bullock, J.S., Sharma, S., Font, A., Robertson, B.E., Leitner, S.N.: ApJ **689**, 936 (2008)
Karlsson, T., Gustafsson, B.: A&A **436**, 879 (2005)
Keller, S.C., Schmidt, B.P., Bessell, M.S., Conroy, P.G., Francis, P., Granlund, A., Kowald, E., Oates, A.P., Martin-Jones, T., Preston, T., Tisserand, P., Vaccarella, A., Waterson, M.F.: Publications of the Astronomical Society of Australia **24**, 1 (2007)
Kirby, E.N., Guhathakurta, P., Bolte, M., Sneden, C., Geha, M.C.: ApJ **705**, 328 (2009)
Kirby, E.N., Guhathakurta, P., Sneden, C.: ApJ **682**, 1217 (2008)
Kirby, E.N., Lanfranchi, G.A., Simon, J.D., Cohen, J.G., Guhathakurta, P.: ApJ **727**, 78 (2011)
Kirby, E.N., Simon, J.D., Geha, M., Guhathakurta, P., Frebel, A.: ApJ **685**, L43 (2008b)
Klypin, A., Kravtsov, A.V., Valenzuela, O., Prada, F.: ApJ **522**, 82 (1999)
Koch, A.: Astronomische Nachrichten **330**, 675 (2009)
Koch, A., McWilliam, A., Grebel, E.K., Zucker, D.B., Belokurov, V.: ApJ **688**, L13 (2008)
Komatsu, E., Smith, K.M., Dunkley, J., Bennett, C.L., Gold, B., Hinshaw, G., Jarosik, N., Larson, D., Nolta, M.R., Page, L., Spergel, D.N., Halpern, M., Hill, R.S., Kogut, A., Limon, M., Meyer, S.S., Odegard, N., Tucker, G.S., Weiland, J.L., Wollack, E., Wright, E.L.: ApJS **192**, 18 (2011)

Lai, D.K., Bolte, M., Johnson, J.A., Lucatello, S., Heger, A., Woosley, S.E.: ApJ **681**, 1524 (2008)
Li, H.N., Christlieb, N., Schörk, T., Norris, J.E., Bessell, M.S., Yong, D., Beers, T.C., Lee, Y.S., Frebel, A., Zhao, G.: A&A **521**, A10 (2010)
Limongi, M., Chieffi, A., Bonifacio, P.: ApJ **594**, L123 (2003)
Liu, L., Gerke, B.F., Wechsler, R.H., Behroozi, P.S., Busha, M.T.: ApJ **733**, 62 (2011)
Lunnan, R., Vogelsberger, M., Frebel, A., Hernquist, L., Lidz, A., Boylan-Kolchin, M.: ApJ **746**, 109 (2012)
McQuinn, M., Lidz, A., Zahn, O., Dutta, S., Hernquist, L., Zaldarriaga, M.: MNRAS **377**, 1043 (2007)
McWilliam, A., Preston, G.W., Sneden, C., Searle, L.: AJ **109**, 2757 (1995)
Meynet, G., Ekström, S., Maeder, A.: A&A **447**, 623 (2006)
Moore, B., Ghigna, S., Governato, F., Lake, G., Quinn, T., Stadel, J., Tozzi, P.: ApJ **524**, L19 (1999)
Norris, J.E., Beers, T.C., Ryan, S.G.: ApJ **536**, 456 (2000)
Norris, J.E., Christlieb, N., Korn, A.J., Eriksson, K., Bessell, M.S., Beers, T.C., Wisotzki, L., Reimers, D.: ApJ **670**, 774 (2007)
Norris, J.E., Gilmore, G., Wyse, R.F.G., Wilkinson, M.I., Belokurov, V., Evans, N.W., Zucker, D.B.: ApJ **689**, L113 (2008)
Norris, J.E., Gilmore, G., Wyse, R.F.G., Yong, D., Frebel, A.: ApJ **722**, L104 (2010a)
Norris, J.E., Ryan, S.G., Beers, T.C., Deliyannis, C.P.: ApJ **485**, 370 (1997)
Norris, J.E., Yong, D., Gilmore, G., Wyse, R.F.G.: ApJ **711**, 350 (2010b)
Robertson, B., Bullock, J.S., Font, A.S., Johnston, K.V., Hernquist, L.: ApJ **632**, 872 (2005)
Ryan, S.G., Norris, J.E., Beers, T.C.: ApJ **471**, 254 (1996)
Ryan, S.G., Norris, J.E., Beers, T.C.: ApJ **523**, 654 (1999)
Schneider, R., Omukai, K., Inoue, A.K., Ferrara, A.: MNRAS **369**, 1437 (2006)
Schörk, T., Christlieb, N., Cohen, J.G., Beers, T.C., Shectman, S., Thompson, I., McWilliam, A., Bessell, M.S., Norris, J.E., Meléndez, J., Ramírez, S., Haynes, D., Cass, P., Hartley, M., Russell, K., Watson, F., Zickgraf, F., Behnke, B., Fechner, C., Fuhrmeister, B., Barklem, P.S., Edvardsson, B., Frebel, A., Wisotzki, L., Reimers, D.: A&A **507**, 817 (2009)
Searle, L., Zinn, R.: ApJ **225**, 357 (1978)
Shetrone, M., Venn, K.A., Tolstoy, E., Primas, F., Hill, V., Kaufer, A.: AJ **125**, 684 (2003)
Shigeyama, T., Tsujimoto, T., Yoshii, Y.: ApJ **586**, L57 (2003)
Simmerer, J., Sneden, C., Cowan, J.J., Collier, J., Woolf, V.M., Lawler, J.E.: ApJ **617**, 1091 (2004)
Simon, J.D., Frebel, A., McWilliam, A., Kirby, E.N., Thompson, I.B.: ApJ **716**, 446 (2010)
Simon, J.D., Geha, M.: ApJ **670**, 313 (2007)
Sobeck, J.S., Lawler, J.E., Sneden, C.: ApJ **667**, 1267 (2007)
Spergel, D.N., Bean, R., Doré, O., Nolta, M.R., Bennett, C.L., Dunkley, J., Hinshaw, G., Jarosik, N., Komatsu, E., Page, L., Peiris, H.V., Verde, L., Halpern, M., Hill, R.S., Kogut, A., Limon, M., Meyer, S.S., Odegard, N., Tucker, G.S., Weiland, J.L., Wollack, E., Wright, E.L.: ApJS **170**, 377 (2007)
Springel, V.: MNRAS **364**, 1105 (2005)
Springel, V., Wang, J., Vogelsberger, M., Ludlow, A., Jenkins, A., Helmi, A., Navarro, J.F., Frenk, C.S., White, S.D.M.: MNRAS **391**, 1685 (2008)
Springel, V., White, S.D.M., Jenkins, A., Frenk, C.S., Yoshida, N., Gao, L., Navarro, J., Thacker, R., Croton, D., Helly, J., Peacock, J.A., Cole, S., Thomas, P., Couchman, H., Evrard, A., Colberg, J., Pearce, F.: Nature **435**, 629 (2005)
Stacy, A., Greif, T.H., Bromm, V.: MNRAS **403**, 45 (2010)
Starkenburg, E., Hill, V., Tolstoy, E., González Hernández, J.I., Irwin, M., Helmi, A., Battaglia, G., Jablonka, P., Tafelmeyer, M., Shetrone, M., Venn, K., de Boer, T.: A&A **513**, A34 (2010)
Suda, T., Aikawa, M., Machida, M.N., Fujimoto, M.Y., Iben, I.J.: ApJ **611**, 476 (2004)
Tafelmeyer, M., Jablonka, P., Hill, V., Shetrone, M., Tolstoy, E., Irwin, M.J., Battaglia, G., Helmi, A., Starkenburg, E., Venn, K.A., Abel, T., Francois, P., Kaufer, A., North, P., Primas, F., Szeifert, T.: A&A **524**, 58 (2010)
Tafelmeyer, M., Jablonka, P., Hill, V., Shetrone, M., Tolstoy, E., Irwin, M.J., Battaglia, G., Helmi, A., Starkenburg, E., Venn, K.A., Abel, T., Francois, P., Kaufer, A., North, P., Primas, F., Szeifert, T.: A&A **527**, C1+ (2011)

Tolstoy, E., Hill, V., Tosi, M.: ARA&A **47**, 371 (2009)
Tominaga, N., Umeda, H., Nomoto, K.: ApJ **660**, 516 (2007)
Umeda, H., Nomoto, K.: Nature **422**, 871 (2003)
Wisotzki, L., Köhler, T., Groote, D., Reimers, D.: A&AS **115**, 227 (1996)
Woosley, S.E., Weaver, T.A.: ApJS **101**, 181 (1995)
Yoshida, N., Omukai, K., Hernquist, L., Abel, T.: ApJ **652**, 6 (2006)
Zahn, O., Lidz, A., McQuinn, M., Dutta, S., Hernquist, L., Zaldarriaga, M., Furlanetto, S.R.: ApJ **654** 12 (2007)

The Potential Impact of Future Observatories

Massimo Stiavelli

Abstract I will pose a number of open questions on galaxy formation and discuss how facilities currently under development such as the James Webb Space Telescope, the large 30m+ ground based telescope, and the Atacama Large Millimeter Array will be able to address them. I will argue that thanks to these facilities, progress in the next 20 years will be as dramatic as progress in the past 20 years.

1 Introduction

The last decade has witnessed extraordinary improvements in our understanding of the formation and evolution of galaxies thanks to the simultaneous availability and synergy between space based observatories such has the Hubble, Spitzer, and Chandra space telescopes and the large ground based telescopes. The next two decades promise to be equally transformational thanks to a new generations of instruments such as the James Webb Space Telescope (hereafter JWST), the Atacama Large Millimeter Array (ALMA), and the 30m+ class telescopes on the ground (indicated here generically as Extremely Large Telescopes or ELTs). These new facilities, building on top of the corpus of knowledge accumulated in the last two decades enables us to address some of the fundamental issues about galaxies, about their evolution, and their early environments.

Our standard cosmological framework is based on the hot Big Bang theory which implies that Hydrogen in the Universe recombines at a redshift of about 1,300. The Gunn–Peterson test (Gunn and Peterson 1965) on quasars at redshift 6 or below shows that there is a negligible column density of neutral Hydrogen between us and those quasars. Thus, the Universe must have reionized by redshift $\sim$6 and we are in the position to investigate what sources caused reionization. It is commonly believed

M. Stiavelli (✉)
Space Telescope Science Institute, 3700 San Martin Dr., MD 21218 Baltimore, USA
e-mail: mstiavel@stsci.edu

T. Wiklind et al. (eds.), *The First Galaxies*, Astrophysics and Space Science Library 396,
DOI 10.1007/978-3-642-32362-1_9,

that the early galaxies probably caused reionization. However, we still don't know what is the contribution of Active Galactic Nuclei (hereafter AGNs) nor what the first galaxies and what are their basic properties.

Galaxies in the local Universe often host massive black holes in their centers. Moreover, scaling laws have been found, connecting the black hole mass to basic properties of the stellar spheroids. We do not when and how the first supermassive black holes formed, nor what factors drive the co-evolution of galaxies and AGNs.

It has long been known that the giant elliptical galaxies in the local Universe have, at least in their cores, very old and metal rich stellar populations, displaying also an excess of alpha elements. Such objects would be hard to form by merging of present day spiral galaxies that have younger and less metal rich stellar populations characterized by solar ratios of alpha elements over iron. However, our underlying theoretical understanding of galaxy formation is based on the idea of hierarchical merging and we expect that at some point during the history of ellipticals these objects were indeed formed by merging. Still the assembly of the mass of a galaxy doesn't have to be correlated with the formation of its stars. Thus, a valid question to ask is when and how did stars in L⋆ galaxies form. Environment and the resulting galaxy interactions and mergers have probably played a major role in shaping the galaxies we see today in the local Universe. Thus, it will be important to test hierarchical formation of galaxies and the interplay between galaxies and IGM.

In the following sections, I will describe how some of the planned facilities will contribute to addressing the questions discussed above.

2 What Sources Caused Reionization?

The Hubble Ultra Deep Field (UDF) and the Great Observatories Origins Deep Survey (GOODS) have enabled the identification of more than 500 candidate galaxies at $z = 6$ with a luminosity function extending almost three magnitudes below L⋆ (Bouwens et al. 2006). UDF followup with WFC3 have produced the first samples of candidates at $z = 7$ (Oesch et al. 2010; McLure et al. 2010; Finklestein et al. 2010), and 8 (Bouwens et al. 2010; McLure et al. 2010; Finklestein et al. 2010) and even and handful of objects at $z = 9$–10 (Bouwens et al. 2011) (see also Fig. 1). Spectroscopic identification is missing for the majority of these objects but the sample appears not to suffer from major contamination for $AB < 27$ (Malhotra et al. 2005) and it is likely that the contamination by interlopers or brown dwarfs at fainter magnitudes will be even lower. The ionizing output of these objects needs to be estimated assuming a Spectral Energy Distribution (SED) and an escape fraction. Their contribution to reionization can be estimated from the measured fluxes by assuming a value for the escape fraction, the SED, and the metallicity (Stiavelli et al. 2004). In Table 1 we give the minimum escape fraction needed for reionization assuming a low metallicity or a solar metallicity for galaxies at $z = 6,7$, or 8 taken individually or for all objects at 6–8. The table shows that reionization could be achieved by assuming an escape fraction of 20 %, a clumping factor

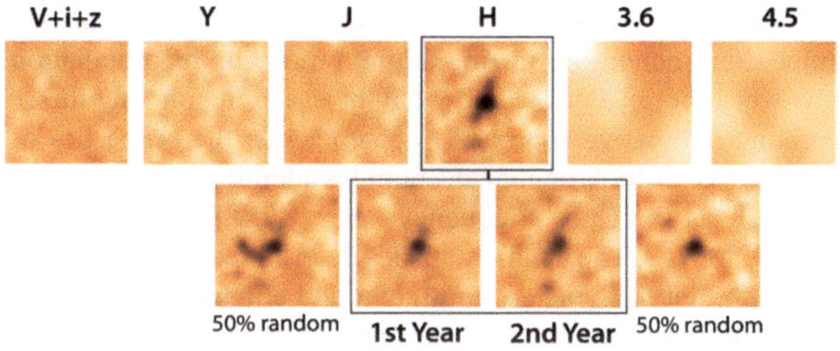

Fig. 1 The *upper panels* shows Hubble Space Telescope images in different bands for a z = 10 candidate object (Bouwens et al. 2011). The object is visible only in the H = F160W band images obtained with the IR Channel of the HST Wide Field Camera 3 instrument. The object is not detected in shallower Spitzer's IRAC images (marked 3.6 and 4.5) making it more likely that it is a truely high redshift galaxy rather than a low redshift interloper. The *lower panels* show different decompositions of the data showing that the objects is visible in subsets of the whole data set and for this reason it is less likely to be an artifact. It took about 75 h with WFC3 to achieve this detection. The James Webb Space Telescope will achieve a more significant detection at S/N $\simeq$ 10 in about 3 h

Table 1 Surface brightness at 1,400Å restframe and required escape fraction f of ionizing radiation needed to achieve reionization. The value of f were computed using the methods highlighted by (Stiavelli et al. 2004)

Redshift	Observed μ_{AB}	Needed f for $C = 5$, low-Z	Needed f for $C = 5$, solar
6	25.4	> 0.19	N/A
7	26.1	> 0.37	N/A
8	26.6	> 0.58	N/A
6+7+8	24.7	> 0.14	> 0.86

of 5 (Pawlik et al. 2004) and that these galaxies are metal poor (say, one hundreth solar)—but not necessarily metal-free. However, this conclusion relies on assuming a metallicity that is plausible but not measured and gives a value for the Compton optical depth that is lower than that derived by WMAP (Komatsu et al. 2011). Moreover, the conclusion would be upset if the escape fraction was significantly lower and more similar to what is measured at low redshift. Thus, most people assume that the contribution of faint dwarf galaxies is essential for reionization (Yan and Windhorst 2004; Trenti et al. 2010; Bouwens et al. 2012).

JWST and the ELTs will be able to clarify the nature of these objects by obtaining deeper exposures and also by obtaining spectra for the brightest galaxies at these redshifts. The best candidates for spectroscopic follow-up—at least initially—may be the ones coming from broader area searches (Trenti et al. 2011; Yan et al. 2011). The metallicity could be measured from line ratios and—when coupled to the observed continuum—would enable one to estimate more accurately the ionizing output of a galaxy. Modeling of the Balmer lines and Lyman α would allow one to estimate the fraction of ionizing output producing recombination lines. The difference between these values would yield the escaping ionizing flux.

An interesting alternative approach to characterize the properties of the population of ionizers is to look at ionized bubbles identified by high redshift 21cm experiments such as the LOw Frequency ARrray (LOFAR) and the Murchison Widefield Array (MWA). The bubble statistics determined with these missions could be correlated with that of Lyman α emitters found either from the ground or using JWST and the Tunable Filter imager (Lidz et al. 2009) probing the sizes of ionized bubbles during reionization.

It is likely that by identifying faint sources down to AB 31 and studying the spectral properties of galaxies down to AB 28-29, JWST and the ELTs will identify the sources responsible for completing reionization completing our basic understanding of this process.

3 What Are the First Galaxies?

The definition of "first galaxy" is ambiguous. First of all, the redshift at which an object will form first in a given volume is going to be on average lower than the first object forming in a larger volume. One could interpret "first galaxy" as the first galaxy forming without being significantly affected by other objects but this is probably not a practical definition if first stars really form in minihalos as they will preceed the formation of most galaxies. Thus, it may be more practical to adopt a working definition of "first galaxies" as the first population of objects with mass exceeding 10^8 solar masses forming in the high redshift Universe. Generally speaking we could identify this population using one or a combination of three methods based, respectively, on very low metal content, dramatic evolution of the luminosity function, and absence of an older stellar population. Clearly only the first method is reasonably secure as one could have fast evolution of the luminosity function and absence of an older population also for later objects. However, even the metallicity argument is not absolutely full proof as one could imagine cases where a second generation of galaxies may have lower metallicity than the first (Stiavelli and Trenti 2010).

JWST has been designed to study the luminosity function to the same relative depth as that measured in the UDF at $z = 6$ (i.e. 3 magnitudes below $M\star$) up to $z = 20$. With this sensitivity JWST will be able to distinguish between luminosity and number evolution and enable a detailed comparison of observations and models. As an example, based on dark halo statistics, a JWST deep field reaching down to AB 31 would identify several tens of galaxies at $z > 10$ (Trenti and Stiavelli 2008).

The sensitivity of NIRSpec (3×10^{-19} erg s^{-1} cm^{-2} in 10^5 s) is such that JWST will be able to measure metallicity of galaxies, using the OIII] 1665 line down to 10^{-3} solar for galaxies 1.5 magnitudes below $M\star$. This metallicity appears to be close to the metallicity of gas in first generation galaxies (Wise and Abel 2008)

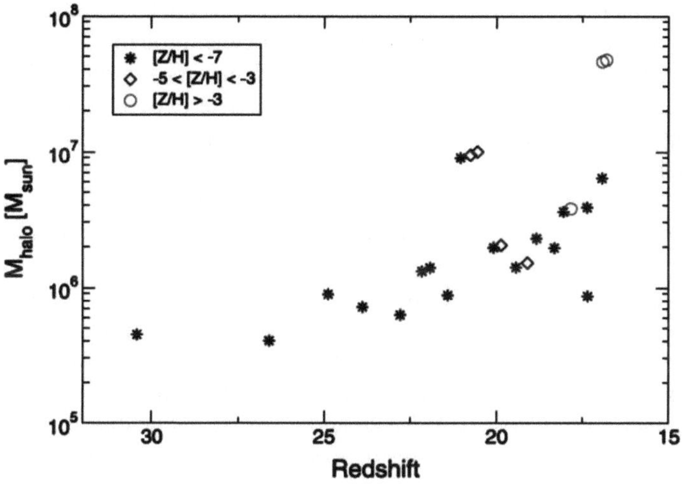

Fig. 2 Star formation redshifts and corresponding halo mass for a proto-galaxy simulation (Wise and Abel 2008). Up until redshift 20 most stars have metallicity close to primordial but already at redshift 15 some stars have metallicity around 10^{-3} solar

(see also Fig. 2) and would enable us to establish whether the galaxies we observe are evolved systems. Finally, MIRI could be used to identify older stellar population on the basis of the SED of the objects.

The detection of the first galaxies with ALMA is not expected because of their low luminosity and very low metallicity which makes it less likely to form dust and molecules. However, when opening new areas in parameter space as ALMA is doing one often has surprises. A detection by ALMA would be exciting and would likely highlight a missing ingredient in our understanding of the formation of these objects. The surprise would be even more welcome in this case since a detection of, e.g., CO, would come with a precise redshift measurement which might otherwise elude us.

4 When and How Did the First Supermassive Black Holes Form?

We know that supermassive black holes are essential components of massive galaxies at low redshift. How did black holes form and in which environments is still largely an open question. It is intriguing that X-ray selected AGNs—similarly to red galaxies—show indications of downsizing, with the most luminous AGNs that peak at higher redshift than the low luminosity ones (Hasinger et al. 2005; Marconi et al. 2004). Whether or not this is a clue to formation it is still early to assess but it is certainly worrysome that no major X-ray mission is currently under development to replace Chandra and XMM.

Presently, a number of different formation scenarios are possible and discrimination between them will require efforts by the next generation of instruments and the availability of statistically meaningful samples of quasars at high redshift enabling us to probe the earliest epochs of quasar growth. The existence of massive black holes already at redshift 6 puts pressure on most formation scenarios as these objects would require constant growth at the Eddington rate from the earliest redshifts if they started out from a stellar mass black hole as in the classical model for black hole growth (Hills 1975). Two scenarios have been proposed to address this issue but they both have weaknesses (Haiman 2009).

In one scenario the black hole growth is kick started by the massive remnants left by Population III stars (Madau and Rees 2001). Some Population III stars may leave black hole remnants with mass of 100 $M_\odot$. Ionizing radiation from a Population III star is predicted to be sufficient to expel all gas from the small host halos even without a supernova explosion so we should expect a delay after the end of life of the star before the black hole can begin accreting. The luminosity of the accreting black hole will remain low, comparable to that of the progenitor star, because both essentially radiate at the Eddington luminosity. The gain of two orders of magnitude in mass shortens the process by a few Salpeter times making the process more plausible. These black holes would become detectable with JWST (or perhaps with future X-ray missions) only after significant mass growth. However, whether or not these black holes can grow sufficiently is being debated because of gravitational recoil following the merger of two halos and leading to possible expulsion of the black holes from their host halo due to gravitational wave kick (Volonteri et al. 2010; Volonteri 2010). The main issue is then whether the remnants of the earliest Population III stars can grow sufficiently before the first major merger as to be safe against ejection. This may well be possible given their small numbers and isolated environments (Trenti and Stiavelli 2008) (see also Fig. 3).

The second scenario is based on the idea that a 10^8 $M_\odot$ halo could collapse monolithically to form a supermassive black hole of 10^3–10^4 $M_\odot$ (Begelman et al. 2006). These objects are also likely not to be detectable directly with JWST (AB = 33.5–35) until they have grown a factor 10 or so. The main problem of this scenario is that the gas in such halo would need to have primordial composition to avoid fragmentation. This favors certain scenario for Lyman–Werner feedback quenching Population III formation in mini-halos as well as an isolated location to avoid contamination by winds in neighboring halos (Trenti et al. 2009; Johnson et al. 2008). Both these factors tend to favor a preferential location in voids.

In principle, the two scenarios might be tested by studying the faint end luminosity function of quasars and mini-quasars at very high redshift but this is going to be observationally challenging. Telling apart these scenarios would provide a good probe of our models of Population III formation by constraining the formation of Population III in mini-halos, the Lyman–Werner background onset and the survivability of halos with pristine chemical composition.

Further progress in this area will require improved theoretical understanding and observation of the luminosity function of mini-AGNs and its variation with redshift. Unfortunately, the JWST will be unable to detect the direct collapse black holes

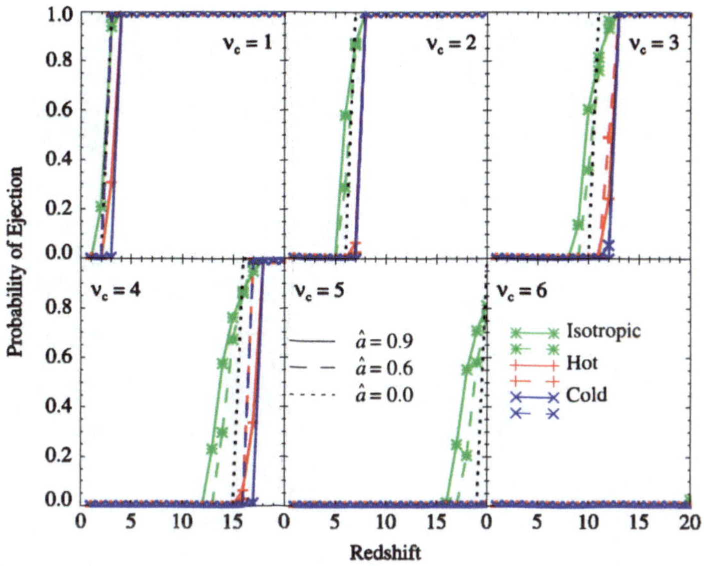

Fig. 3 Probability of ejection as a function of redshift for different values of black hole specific angular momentum and spin orientation distribution (Volonteri et al. 2010). Different panels represent different peak overdensity ratios ν_c. As expected the rarest peaks have a lower probability of ejection

and thus it will make discrimination between models indirect and model dependent. Clustering properties of the first AGNs that become detectable might lead to an indirect discrimination between the various scenarios. Preliminary evidence at redshift 6 is suggesting of some degree of clustering (Kim et al. 2009; Utsumi et al. 2010) but this will need to be better established and extended to higher redshift before strong statements on formation scenarios can be made.

Some of the early accreting massive black holes could be targets of ALMA. Indeed, QSOs at $z \simeq 6$ have already been detected in molecular lines with present day facilities (Bertoldi et al. 2003). The X-ray flux produced by an accreting black hole can trigger the formation of H_2 and CO. For an Eddington rate accretion, black holes down to $\sim 10^6\ M_\odot$ should be visible to redshift 10–20 (Spaans and Meijerink 2008).

5 When and How Did Stars in L⋆ Galaxies Form?

A major finding of the Great Observatories Origins Survey (GOODS) (Giavalisco 2005; Dickinson 2008) and followup studies has been the existence of a red and a blue sequence of galaxies and their implications on the properties of the local galaxies and on the establishment of the Hubble sequence. This is a field that will

Fig. 4 The *upper panels* show a portion of the Hubble Ultra Deep Field observed in the near-infrared with the WFC3/IR instrument (*left*) and a zoom-in on a particular galaxy (*right*). The *lower panels* are a simulation of the same area as observed with the NIRCam instrument on the James Webb Space Telescope. The simulation was done deconvolving the HST images by the HST PSF and convolving the resulting image with the JWST PSF. From this simulation it is clear the advantage in angular resolution that JWST will have over HST and the capability to study spatially resolved galaxies at high-redshift

see major progress thanks to the new facilities under development. Objects that are presently selected through photometric techniques and have stellar masses estimated through infrared luminosities will have their spectroscopic redshifts, dynamical masses, and metallicities measured by the JWST NIRSpec instrument as well as by the ELTs. Measuring gas metallicities for a large sample of objects as a function of redshift will be an important step. However, gas can be accreted or ejected so that the metallicity of the interstellar medium in a given object may not be a monotonic function of time. In contrast, the metals are locked in stars when they form and the stellar metallicity is much more likely to be a monotonic function and therefore a more significant tracer of evolution. For the first time, we will be able to track in detail what stellar populations do and where they form, especially for those objects bright enough to enable us to measure the metallicity of the stellar population through absorption lines.

ALMA will become an important tool of the trade, not only for studying dust and molecules but also as a redshift machine. In addition to Luminous Infrared Galaxies, ALMA will be also able to study typical star forming $L_\star$ galaxies at redshift $z \sim 3$.

Pushing to sub-L★ objects this type of observations may be pushing the instrument to its limits (Greve and Sommer-Larsen 2008).

Stellar population studies presently carried out by Spitzer will be feasible with JWST for much fainter galaxies (see Fig. 4). A simple rule of thumb is that anything detectable in imaging by Spitzer will be suitable for JWST MIRI follow-up spectroscopy. Clearly, fainter objects presently inaccessible by Spitzer will be within reach of JWST MIRI imaging.

6 What Are the Factors that Drive the Co-evolution of Galaxies and AGNs?

AGNs are considered one of the premier tools for quenching star formation in galaxies at late times, moving objects from the blue to the red sequence. While we don't know if they are the only factor at play in this transition, the existence of mass correlations between black holes and spheroids is suggestive that such an interplay must be in action at least to some extent. The basic idea of mutually constrained growth of galaxies and black holes goes back several years (Silk and Rees 1998) but it is not yet well understood. Moreover, we really don't know whether parallel evolution of black holes and galaxy spheroids is true for every galaxy or whether the tight connection between galaxy and black hole mass is achieved only through statistical processes (Heckman 2009). There is some evidence that black holes grow mostly by mass accretion after redshift ~3 (Marconi et al. 2004). If that's the case most of this growth needs to take place in gas rich galaxies and the gas availability makes it, in principle, possible to have parallel growth of the black hole and the stellar component. The analysis of the black hole mass vs velocity dispersion relation to $z \sim 0.37$ shows some indications of evolution toward having more massive black holes at higher redshift (Treu et al. 2004). On the same trend, analysis of QSO hosts and radiogalaxies at redshift ~2 finds also an enhanced black hole mass (by a factor of 4) for a given bulge mass (Peng et al. 2006; McLure et al. 2006). Even at redshift 6 there is evidence for black holes more massive than the global correlation with host galaxy mass would predict, this time by a factor ~15 (Wang et al. 2010). The evolution of the black hole to spheroid mass ratio increasing with increasing redshift would be interesting and a major step forward if established solidly. However, there have been also results finding evolution with the opposite sign and an argument based on integral constraints on the comoving mass density of black holes suggests that there is little room for fast evolution of black holes (Hopkins et al. 2006).

Clearly, it would be desireable to study the mutual evolution with the most reliable indicators, this may imply measuring the galaxy velocty dispersion for high redshift objects. This should be easily enabled by JWST and the ELTs and I would expect significant evolution in this field.

7 Can We Test Hierarchical Formation of Galaxies?

Our basic theoretical understanding of structure formation is based on the assumption that some form of non-baryonic, cold dark matter dominate the mass density, Moreover, our models for the first stars and galaxies depend critically on the dark matter particle mass. The problem can be separated into two different issues, one is the initial power spectrum of dark matter fluctuations and the other the hierarchical clustering picture.

A way to study the fluctuation spectrum on the small scales more relevant for galaxies may be the power spectrum of 21-cm anisotropies (Loeb and Zaldarriaga 2004). Another approach to constrain the power spectrum and to place limit on the mass of the (warm) dark matter particle is based on the flux power spectrum of the high-redshift Lyman-α forest (Viel et al. 2008).

In recent years several groups have tackled also the issue of hierarchical clustering. The study of clustering of Lyman break galaxies has provided some evidence that the basic model of biased galaxy formation is indeed correct (Giavalisco and Dickinson 2001).

The correlation function and the halo occupation number are important parameters for understanding galaxy formation. The occupation number in particular gives indications about substructure in halos and how fast galaxies sharing the same halo are able to merge. Several groups have attempted to study these parameter with a variety of methods, ranging from Lyman break galaxy samples from GOODS and UDF (Lee et al. 2006; Overzier et al. 2006), to SDSS samples of red galaxies (Reid and Spergel 2009), to Hershel-selected samples of galaxies (Cooray et al. 2010), to detailed studies of small samples of objects (Lowenthal et al. 2009).

LSST should provide statistically significant samples enabling detailed derivation of these parameters. LSST will also enable detailed derivation of dark matter "maps" from weak lensing which will then be compared to the distribution of visible galaxies extending the pioneering work based on the COSMOS survey (Leauthaud et al. 2012) (Fig. 5).

8 What Is the Interplay Between Galaxies and IGM?

The study of galaxy formation would not be complete if one stopped at considering only dark halos and visible galaxies because intergalactic gas plays a fundamental role.

Detailed gas distributions from the ELTs and ALMA coupled with weak-lensing derived dark matter maps and galaxy distributions from LSST will complete the picture providing detailed test cases to be compared with numerical simulations and theoretical models. One of the issues in this field is that of gas accretion from the IGM onto galaxies. This is often exemplified in terms of cold or hot accretion, with the former being very hard to observe as it is expected to take place along dense cold filaments and to have therefore very small solid angle.

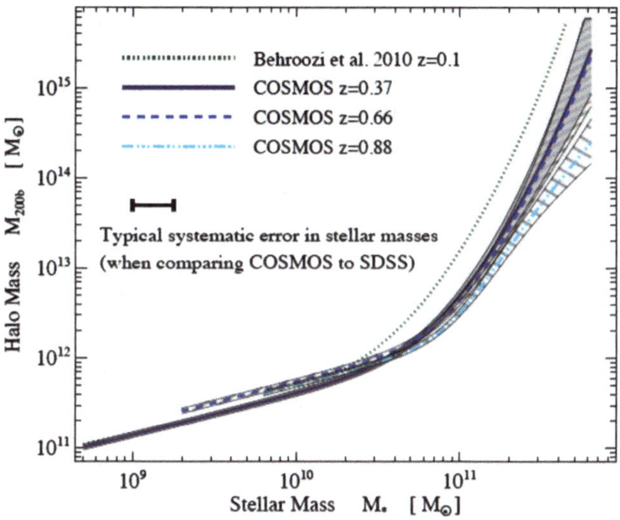

Fig. 5 The evolution with redshift of the correlation between halo mass and stellar mass as measured by the COSMOS survey (Leauthaud et al. 2012). COSMOS has provided the first dark matter maps obtained by weak lensing

The first exploratory studies of the interactions of galaxies and IGM have already started by they face difficulties due to the faintness of many of the targets and small number statistics on the lines of sight which make it hard to find cold filaments if they exist. A recent study based on 89 galaxies at $z \sim 2-3$ compares a simple model for the gas distribution to the data and concludes that there is little evidence for cold accretion while large scale outflows appear in better agreement with observations (Steidel et al. 2010). Of course, it is in principle possible that both cold accretion and warm outflows coexist, with the latter simply much easier to observe. Only by increasing the lines of sight density using fainter background objects or, possibly, ALMA, we will be able to clarify this issue.

9 Summary

In the last twenty years our knowledge of galaxy formation has dramatically improved. Reading papers published in 1981 one can see that the issue of which type of dark matter if any was present and consequently whether structure formation was hierarchical or following the pancake model was still open. We have gone a long way since then. Answered many questions and posed many new ones. It is likely that theoretical progress and new observations with the facilities under development will prove the next twenty years to be equally revolutionary.

The James Webb Space Telescope and the ELTs will allow us to obtain data for galaxies at redshift 3–5 that were only possibly in the local Universe until recently. We will be able to measure resolved velocity fields for high-redshift galaxies and measure their dynamical mass. We will measure gas and stellar metallicities as a function of position within galaxies, an essential ingredient for tracking how galaxies merge and understanding where star formed for the main Hubble types we recognize at low redshift. We will also understand better how black holes and their activity shape their host galaxies.

ALMA will be revolutionary for understanding the role of dust, finally with angular resolution matched to that available in the visible and near-IR and sensitivity sufficient to push investigations to more normal galaxies. This will allow us to address one of the open issues of today, namely whether or not dust enshrouded star formation is an important component of star formation at $z \geq 5$.

Finally LSST will provide us with an unprecedent sky coverage and sufficient statistical muscle to understand many issues of evolution in properties as well as to map the dark matter distribution.

References

Begelman, M.C., Volonteri, M., Rees, M.J.: Formation of supermassive black holes by direct collapse in pre-galactic halos. Mon. Not. R. Astron. Soc. **370**, 289–298 (2006)

Bertoldi, F., Cox, P., Neri, R., et al.: High-excitation CO in a quasar host galaxy at z = 6.42. Astron. Astrophys. **409**, L47–L50 (2003)

Bouwens, R.J., Illingworth, G.D., Blakeslee, J.P., Franx, M.: Galaxies at $z \sim 6$: The UV Luminosity function and Luminosity density from 506 HUDF, HUDF parallel ACS field, and GOODS i-Dropouts. Astrophys. J. **653**, 53–85 (2006)

Bouwens, R.J., Illingworth, G.D., Oesch, P.A., et al.: Discovery of $z \sim 8$ Galaxies in the Hubble ultra deep field from Ultra-Deep WFC3/IR observations. Astrophys. J. **709**, L133–L137 (2010)

Bouwens, R.J., Illingworth, G.D., Labbe, I., et al.: A candidate redshift $z \sim 10$ galaxy and rapid changes in that populatin at an age of 500 Myr. Nature **469**, 504–507 (2011)

Bouwens, R.J., Illingworth, G.D., Oesch, P.A., Trenti, M., Labbe', I., Franx, M., Stiavelli, M., Carollo, C.M., van Dokkum, P., Magee, D.: Lower-Luminosity Galaxies could reionize the Universe: very steep faint-end slopes to the UV Luminosity functions at z ≥ 5-8 from the HUDF09 WFC3/IR observations. Astrophys. J. **752**, L5 (2012)

Cooray, A., Amblard, A., Wang, L. et al.: HerMES: Halo occupation number and bias properties of dusty galaxies from angular clustering measurements. Astron. Astrophys. **518**, L22–L26 (2010)

Dickinson, M.: Galaxy Mass Assembly. In: "Panoramic Views of Galaxy Formation and Evolution". ASP Conference Series **399**, 161–168 (2008)

Finklestein, S.L., Papovich, C., Giavalisco, M., Reddy, N.A., Ferguson, J.C., Koekemoer, A., Dickinson, M.: On the Stellar populations and evolution of Star forming Galaxies at $6.3 < z \leq 8.6$. Astrophys. J. **719**, 1250–1273 (2010)

Giavalisco, M.: Lessons from the GOODS. New Astron. Rev. **49**, 440–446 (2005)

Giavalisco, M., Dickinson, M.: Clustering Segregation with Ultraviolet Luminosity in Lyman break Galaxies at $z \sim 3$ and its implications. Astrophys. J. **550**, 177–194 (2001)

Greve, R.R., Sommer-Larsen, J.: CO line emission fro Lyman break galaxies. Astron. Astrophys. **480**, 335–338 (2008)

Gunn, J.E., Peterson, B.A.: On the density of neutral hydrogen in intergalactic space. Astrophys. J. **142**, 1633–1641 (1965)

Haiman, Z.: Observing the first stars and Black Holes. In: Thronson, H.A., Stiavelli, M., Tielens, A. (eds.) "Astrophysics in the Next Decade", pp. 385–418. Springer, New York (2009)

Hasinger, G., Miyaji, T., Schmidt, M.: Luminosity-dependent evolution of soft X-ray selected AGN. Astron. Astrophys. **441**, 417–434 (2005)

Heckman, T.M.: The co-evolution of Galaxies and Black Holes: current status and future prospects. In: Thronson, H.A., Stiavelli, M., Tielens, A. (eds.) "Astrophysics in the Next Decade", pp. 335–356. Springer, New York (2009)

Hills, J.G.: Possible power source of Seyfert galaxies and QSOs. Nature **254**, 295–298, (1975)

Johnson, J., Greif, T., Bromm, V.: Occurrence of metal-free galaxies in the early Universe. Mon. Not. R. astron. Soc. **388**, 26–38 (2008)

Hopkins, P.F., Somerville, R.S., Hernquist, L., Cox, T.J., Robertson, B., Li, X.: The relation between quasar and merging galaxy luminosity functions and the merger-driven star formation history of the Universe. Astrophys. J. **652**, 864–888 (2006)

Kim, S., Stiavelli, M., Trenti, M., et al.: The environments of high-redshift quasi-stellar objects. Astrophys. J. **695**, 809–817 (2009)

Komatsu, E., Smith, K.M., Dunkley, J., et al.: Seven-year Wilkinson Microwave Anisotropy Probe (WMAP) observations: cosmological interpretation. Astrophys. J. Suppl. **192**, 18 (2011)

Leauthaud, A., Tinker, J., Bundy, K., et al.: New constraints on the evolution of the stellar-to-dark matter connection: a combined analysis of galaxy-galaxy lensing, clustering, and stellar mass functions from $z = 0.2$ to $z = 1$. Astrophys. J. **744**, 159 (2012)

Lee, K.-S., Giavalisco, M., Gnedin, O.Y., Somerville, R.S., Ferguson, H.C., Dickinson, M., Ouchi, M.: The large-scale and small-scale clustering of Lyman Break Galaxies at $3.5 \leq z \leq 5.5$ from the GOODS survey. Astrophys. J. **642**, 63–80 (2006)

Lidz, A., Zahn, O., Furlanetto, S.R., McQuinn, M., Hernquist, L., Zaldarriaga, M.: Probing reionization with the 21cm galaxy cross-power spectrum. Astrophys. J. **690**, 252–266 (2009).

Loeb, A., Zaldarriaga, M.: Measuring the small-scale power spectrum of cosmic density fluctuations through 21cm tomography prior to the epoch of structure formation. Phys. Rev. Lett. **92**, 1301–1304 (2004)

Lowenthal, J.D., Koo, D.C., Simard, L., van Kampen, E.: Dynamics of Lyman Break Galaxies and their host Halos. Astrophys. J. **703**, 192–221 (2009)

Madau, P., Rees, M.J.: Massive black holes as Population III remnants. Astrophys. J. **551**, L27–L30 (2001)

Malhotra, S., Rhoads, J.E., Pirzkal, N., et al.: An overdensity of Galaxies at $z = 5.9 \pm 0.2$ in the Hubble ultra deep field confirmed using the ACS Grism. Astrophys. J. **626**, 666–679 (2005)

Marconi, A., Risaliti, G., Gilli, R., Hunt, L.K., Maiolino, R., Salvati, M.: Local supermassive black holes, relics of active galactic nuclei, and the X-ray background. Mon. Not. Roy. Astron. Soc. **351**, 169–185 (2004)

McLure, R.J., Jarvis, M.J., Targett, T.A., Dunlop, J.S., Best, P.N.: On the evolution of the black hole: spheroid mass ratio. New Astron. Rev. **50**, 782–785 (2006)

McLure, R.J., Dunlop, J.S., Cirasuolo, M., Koekemoer, A.M., Sabbi, E., Stark, D.P., Targett, T.A., Ellis, R.S.: Galaxies at $z \sim 6 - 9$ from the WFC3/IR imaging of the Hubble ultra deep field. Mon. Not. R. Astron. Soc. **403**, 960–983 (2010)

Oesch, P.A., Bouwens, R.J., Illingworth, G.D., et al.: $z \sim 7$ Galaxies in the HUDF: first epoch WFC3/IR results. Astrophys. J. **709**, L16–L20 (2010)

Overzier, R.A., Bouwens, R.J., Illingworth, G.D., Franx, M.: Clustering of i_{775} dropouts galaxies at $z \sim 6$ in GOODS and the UDF. Astrophys. J. **648**, L5–L8 (2006)

Pawlik, A.H., Schaye, J., van Scherpenzeel, E.: Keeping the Universe ionized: photoheating and the clumping factor of the high-redshift intergalactic medium. Mon. Not. R. Astron. Soc. **394**, 1812–1824 (2004)

Peng, C.Y., Impey, C.D., Rix, H.-W., et al.: Probing the coevolution of supermassive black holes and galaxies using gravitationally lensed quasar hosts. Astrophys. J. **649**, 616–634 (2006)

Reid, B.A., Spergel, D.N.: Constraining the Luminous Red Galaxy Halo occupation distribution using Counts-In-Cylinders. Astrophys. J. **698**, 143–154 (2009)

Silk, J., Rees, M.J.: Quasars and galaxy formation. Astron. Astrophys. **331**, L1–L4 (1998)

Spaans, M., Meijerink, R.: On the detection of high-redshift black holes with ALMA through CO and H_2 emission. Astrophys. J. **678**, L5–L8 (2008)
Steidel, C.C., Erb, D.K., Shapley, A.E., Pettini, M., Reddy, N., Bogosavljevic, M., Rudie, G.C., Rakic, O.: The structure and kinematics of the circumgalactic medium from far-ultraviolet spectra of $z \simeq 2 - 3$ galaxies. Astrophys. J. **717**, 289–322 (2010)
Stiavelli, M., Fall, S.M., Panagia, N.: Observable properties of cosmological reionization sources. Astrophys. J. **600**, 508–519 (2004)
Stiavelli, M., Trenti, M.: The clustering properties of the first galaxies. Astrophys. J. **716**, L190–L194 (2010)
Trenti, M., Stiavelli, M.: Cosmic variance and its effect on the Luminosity function determination in deep High-z surveys. Astrophys. J. **676**, 767–780 (2008)
Trenti, M., Stiavelli, M., Shull, M.J.: Metal-free gas supply at the edge of reionization: late-epoch Population III star formation. Astrophys. J. **700**, 1672–1679 (2009)
Trenti, M., Stiavelli, M., Bouwens, R.J.: The Galaxy Luminosity function during the reionization epoch. Astrophys. J. **714**, L202–L207 (2010)
Trenti, M., Bradley, L.D., Stiavelli, M., et al.: The brightest of reionizing galaxies survey: design and preliminary results. Astrophys. J. **727**, L39–L44 (2011)
Treu, T., Malkan, M.A., Blandford, R.D.: The relation between black hole mass and velocity dispersion at $z \sim 0.37$. Astrophys. J. **615**, L97–L100 (2004)
Utsumi, Y., Goto, T., Kashikawa, N., Miyazaki, S., Komiyama, Y., Furusawa, H., Overzier, R.: A large number of $z > 6$ galaxies around a QSO at $z = 6.43$: evidence for a protocluster? Astrophys. J. **721**, 1680–1688 (2010)
Viel, M., Becker, G., Bolton, J.S., Haehnelt, M.G., Rauch, M., Sargent, W.L.W.: How cold is cold dark matter? Small-scales constraints from the flux power spectrum of the high-redshift Lyman-α forest. Phys. Rev. Lett. **100**, 1304–1307 (2008)
Volonteri, M., Gültekin, K., Dotti, M.: Gravitational recoil: effects on massive black hole occupation fraction over cosmic time. Mon. Not. R. Astron. Soc. **404**, 2143–2150 (2010)
Volonteri, M.: Formation of supermassive black holes. Astron. Astrophys. Rev. **18**, 279–315 (2010)
Wang, R., Carilli, C.L., Neri, R., et al.: Molecular gas in $z \sim 6$ quasar host galaxies. Astrophys. J. **714**, 699–712 (2010)
Wise, J.H., Abel, T.: Resolving the formation of protogalaxies. III. Feedback from the first stars. Astrophys. J. **685**, 40–56 (2008)
Yan, H., Windhorst, R.A.: Candidates of $z \simeq 5.5$–7 galaxies in the Hubble Space Telescope ultra deep field. Astrophys. J. **612**, L93–L96 (2004)
Yan, H., Yan, L., Zamojski, M.A., et al.: Probing very bright end of Galaxy Luminosity function at $z \geq 7$ using Hubble Space Telescope pure parallel observations. Astrophys. J. **728**, L22–L25 (2011)

Index

T. Wiklind et al. (eds.), *The First Galaxies*, Astrophysics and Space Science Library 396,
DOI 10.1007/978-3-642-32362-1, © Springer-Verlag Berlin Heidelberg 2013

MIX
Papier aus verantwortungsvollen Quellen
Paper from responsible sources
FSC® C105338

If you have any concerns about our products,
you can contact us on
ProductSafety@springernature.com

In case Publisher is established outside the EU,
the EU authorized representative is:
Springer Nature Customer Service Center GmbH
Europaplatz 3, 69115 Heidelberg, Germany

Printed by Libri Plureos GmbH
in Hamburg, Germany